Genetics, Genomics and Breeding of Crop Plants

Genetics, Genomics and Breeding of Crop Plants

Editor: Harold Salazar

www.callistoreference.com

Callisto Reference,
118-35 Queens Blvd., Suite 400,
Forest Hills, NY 11375, USA

Visit us on the World Wide Web at:
www.callistoreference.com

ISBN: 978-1-64116-767-3 (Hardback)

Cataloging-in-Publication Data

Genetics, genomics and breeding of crop plants / edited by Harold Salazar.
    p. cm.
Includes bibliographical references and index.
ISBN 978-1-64116-767-3
1. Genetics. 2. Genomics. 3. Plant breeding. 4. Crops. I. Salazar, Harold.
QH430 .G46 2023
576.5--dc23

Table of Contents

Preface

A crop refers to a type of plant that can be developed and harvested majorly for subsistence or income. Crops have been categorized into six different varieties depending on the applications which include fiber crops, industrial crops, food crops, ornamental crops, feed crops and oil crops. The selection of crop genetics in plants is commonly utilized to find out the optimum genes for increasing crop varieties. It is mainly focused on forecasting their performance and suitable circumstances. This is a useful technique for plant breeding for selecting crop plants according to the requirements. Crop breeding is mainly concerned with the creation, selection and fixation of superior phenotypes to develop better lines or cultivars. This book aims to shed light on the genetics, genomics and breeding of crop plants. It elucidates new techniques and their applications in a multidisciplinary manner. This book, with its detailed analyses and data, will prove immensely beneficial to professionals and students involved in this area at various levels.

Significant researches are present in this book. Intensive efforts have been employed by authors to make this book an outstanding discourse. This book contains the enlightening chapters which have been written on the basis of significant researches done by the experts.

Finally, I would also like to thank all the members involved in this book for being a team and meeting all the deadlines for the submission of their respective works. I would also like to thank my friends and family for being supportive in my efforts.

Editor

Conventional and Molecular Techniques from Simple Breeding to Speed Breeding in Crop Plants

Sunny Ahmar [1,†], Rafaqat Ali Gill [2,†], Ki-Hong Jung [3,*] [iD], Aroosha Faheem [4], Muhammad Uzair Qasim [1], Mustansar Mubeen [5] and Weijun Zhou [6,*]

[1] National Key Laboratory of Crop Genetic Improvement, College of Plant Science and Technology, Huazhong Agricultural University, Wuhan 430070, Hubei, China; sunny.ahmar@yahoo.com (S.A.); uzairqasim1149@yahoo.com (M.U.Q.)
[2] Oil Crops Research Institute, Chinese Academy of Agriculture Sciences, Wuhan 430070, China; drragill@caas.cn
[3] Graduate School of Biotechnology & Crop Biotech Institute, Kyung Hee University, Yongin 17104, Korea
[4] State Key Laboratory of Agricultural Microbiology and State Key Laboratory of Microbial Biosensor, College of Life Sciences Huazhong Agriculture University, Wuhan 430070, China; arushafaheem@hotmail.com
[5] State Key Laboratory of Agricultural Microbiology and Provincial Key Laboratory of Plant Pathology of Hubei Province, College of Plant Science and Technology, Huazhong Agricultural University, Wuhan 430070, China; mustansar01@yahoo.com
[6] Institute of Crop Science and Zhejiang Key Laboratory of Crop Germplasm, Zhejiang University, Hangzhou 310058, China
*   Correspondence: khjung2010@khu.ac.kr (K.-H.J.); wjzhou@zju.edu.cn (W.Z.)
†   These authors contributed equally to this work.

Abstract: In most crop breeding programs, the rate of yield increment is insufficient to cope with the increased food demand caused by a rapidly expanding global population. In plant breeding, the development of improved crop varieties is limited by the very long crop duration. Given the many phases of crossing, selection, and testing involved in the production of new plant varieties, it can take one or two decades to create a new cultivar. One possible way of alleviating food scarcity problems and increasing food security is to develop improved plant varieties rapidly. Traditional farming methods practiced since quite some time have decreased the genetic variability of crops. To improve agronomic traits associated with yield, quality, and resistance to biotic and abiotic stresses in crop plants, several conventional and molecular approaches have been used, including genetic selection, mutagenic breeding, somaclonal variations, whole-genome sequence-based approaches, physical maps, and functional genomic tools. However, recent advances in genome editing technology using programmable nucleases, clustered regularly interspaced short palindromic repeats (CRISPR), and CRISPR-associated (Cas) proteins have opened the door to a new plant breeding era. Therefore, to increase the efficiency of crop breeding, plant breeders and researchers around the world are using novel strategies such as speed breeding, genome editing tools, and high-throughput phenotyping. In this review, we summarize recent findings on several aspects of crop breeding to describe the evolution of plant breeding practices, from traditional to modern speed breeding combined with genome editing tools, which aim to produce crop generations with desired traits annually.

Keywords: food security; food scarcity; conventional breeding; CRISPR/Cas9; CRISPR/Cpf1; high-throughput phenotyping; speed breeding

1. Introduction

Since the early 1900s, plant breeding has played a fundamental role in ensuring food security and safety and has had a profound impact on food production all over the world [1,2]. In recent years, however, problems related to food quality and quantity globally have arisen as a consequence of the excessive food requirement for the rapidly increasing human population. Furthermore, radical changes in weather conditions caused by global climate change are causing heat and drought stress; consequently, farmers around the world are facing significant yield losses [3]. Global epidemics, such as the Irish potato blight of the 1840s and the Southern corn leaf blight in the United States in the 1970s, were disastrous events leading to the deaths of millions of people due to food shortage [4,5]. In recent years, the ratio of food production to consumption has decreased considerably, while both urbanization rates and demographic growth have increased globally. In this era of fast development and rapid growth, people prefer to consume processed foods, where nutritional quality is compromised. The world is expected to reach 10 billion by 2050, but no satisfactory strategies are in place to feed this massive population [6,7]. Developed countries have increased their agricultural productivity, partially meeting their food requirements, but this has resulted in increased stress on food manufacturing departments [8].

Plant breeding can be used to develop plants with desired traits [9]. Artificial plant selection has been used by humans for the past 10,000 years, selecting and breeding plants with higher nutritional values [10] (Figure 1). Traditional agricultural methods aimed to improve the nutritional status of different food plants. Recent scientific developments provide a wide range of possibilities and innovations in plant breeding [11]. To satisfy the continuously increasing demand for plant-based products, the current level of annual yield enhancement in major crop species (varying from 0.8–1.2%) must be doubled [12].

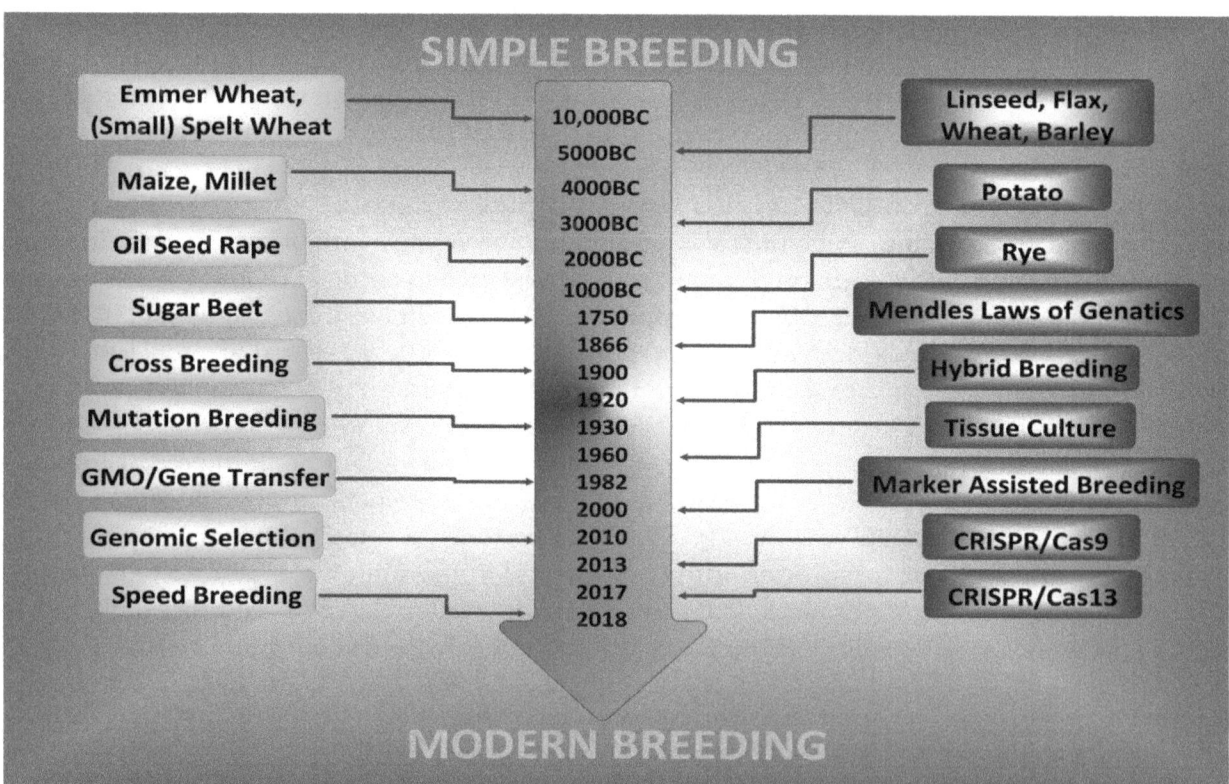

Figure 1. Historical milestones in plant breeding. For 10,000 years, farmers and breeders have been developing and improving crops. Presently, farmers feed 10 times more people using the same amount of land as 100 years ago.

The introduction of Mendelian laws revolutionized the field of crop breeding. Over the last 150 years, crop development has been altered to a great extent as a consequence of contemporary cutting-edge genomics [13]. Different approaches have been used to shorten the duration of plant reproductive cycles. Novel techniques developed in this decade, such as genomic selection, high-throughput phenotyping (HTP), and modern speed breeding, have been shown to accelerate plant breeding. Genetic engineering and molecular methods have also played a role in developing crops with desirable characteristics using gene transformation [14–17]. Other techniques like large-scale sequencing, genomics, rapid gene isolation, and high-throughput molecular markers have also been proposed to improve the breeding of commercially important crop species, such as cisgenesis, intragenesis, polyploidy breeding, and mutation breeding [18–21].

Conventional breeding techniques are inadequate for plant genome enhancement to develop new plant varieties. To overcome this obstacle in plant breeding practices, molecular markers have been used since the 1990s for the selection of superior hybrid lines [22]. Improving plant phenotype for a specific desirable trait involves the artificial selection and breeding of this given trait by the plant breeder. Generally, breeders tend to focus on traits of diploid or diploid-like crops (e.g., maize and tomatoes) rather than polyploid crops (e.g., alfalfa and potatoes), which have more complex genetics. Breeders hence prefer to use crops with shorter reproductive cycles, which allow the production of several generations in a single year and, leading to faster production of the desired phenotypes by artificial breeding compared to crops that only reproduce annually or perennial plants that only reproduce every few years [23–25]. Plant breeding, combined with genome studies, enhances the accuracy of breeding practices and saves time [26]. Compared to other kingdoms, plants are more easily genetically manipulated to obtain desired genetic combinations by selfing, crossbreeding (or both) given their short generation time, and large population size available for analyses [27]. In the early 1980s, NASA partnered with Utah State University to explore the possibility of growing rapid cycling wheat under constant light in space stations. This joint effort resulted in the development of "USU-Apogee", a dwarf wheat line bred for rapid cycling [28,29]. Recently, Lee Hickey and colleagues solved this issue by presenting the idea of "speed breeding", a non-GMO path enabling researcher to turn over many generations and select plants for desired traits between many variations [8,16]. This method uses regulated environmental conditions and prolonged photoperiods to achieve between four and six generations per year of long duration crops (i.e., wheat, barley, and canola) [16,30,31].

Researchers outlined the evolving EU regulatory framework for GMOs and discussed potential ways of regulating plant varieties developed using precision breeding approaches such as clustered regularly interspaced short palindromic repeats (CRISPR), and CRISPR-associated (Cas) proteins CRISPR/Cas9 [32]. Research interest in genetically engineered crops (and more precisely "biotech crops") has been increasing, given the urgent need to ensure food security for the growing human population [33].

Genome editing involves inserting, deleting, or substituting a foreign gene in the organism's DNA. Upon successful transformation, this new sequence is integrated into the host genome [34,35]. Several processes are involved in the fixation of specific DNA sequences, cut with the help of nucleases. Plant breeding alone cannot achieve the required traits, but using the CRISPR-associated (Cas) enzymes (CRISPR/Cas and CRISPR/Cpf1) can help meet the needs for efficient crop research [36,37]. In this review, we discuss the use of conventional and non-conventional plant breeding techniques for different crops, as well as the use of genome editing techniques to change and improve desired phenotypes. Moreover, the potential correlations between these approaches used to develop future strategies for crop improvement will also be explored.

2. Mutation through Traditional or Conventional Breeding

The advantage of conventional plant breeding consists of increasing the availability of genetic resources for crop improvement through introgression of the desired traits. However, some plants are at risk of becoming susceptible to environmental stress and losing genetic diversity [38]. Thus, traditional cultivation methods are not sufficient to resolve global food security issues. Combining multiple phenotypic characters within a single plant variety would successfully increase yield and has been widely used, however, new breeding techniques are less expensive and will enable faster production of genetically improved crops [39].

In recent times, improvements in traditional plant breeding have been introduced, such as wide crosses, introgression of traits from wild relatives by hybrid breeding, mutagenesis, double haploid technology, and some tissue culture-based approaches such as embryo and ovule rescue (to achieve maximum plant regeneration) and protoplast fusion [40–42]. Food and feed crops developed by conventional plant breeding have specific natural phenotypic and agronomic properties. To improve crop quality, researchers have introgressed many beneficial traits through plant breeding with wild relatives, such as higher yield, abiotic and biotic stress resistance, and increased nutritional value [39,43,44]. The identification and combination of traits in familiar genotypes and the selection of high-performing varieties can establish a crop lineage with the desired properties. That being said, this approach can have potentially adverse impacts on food and environmental safety as it occasionally gives rise to safety concerns through unpredictable effects [9,45].

A trait (e.g., stress tolerance) can be improved by selecting the best hybrid progeny with the desired trait using cross breeding [46] (Figure 2a). Desired traits can also be introduced into a chosen 'best' recipient line through backcrossing of the selected progeny with the recipient line for several generations to reduce unwanted phenotype combinations [47]. Genetic variability can be reduced by the use of long-term traditional breeding methods; thus, the introduction of new genes is required for the improvement of desired traits by speed breeding, mutation breeding, and rapid generation advance (RGA) [16,31,48]. From this point of view, mutations could be useful in plant breeding programs and all these precision breeding tools can contribute to the improvement of specific features during the breeding cycle. Plant breeding is always approached holistically by analyzing all applicable agricultural functionality (Figure 3).

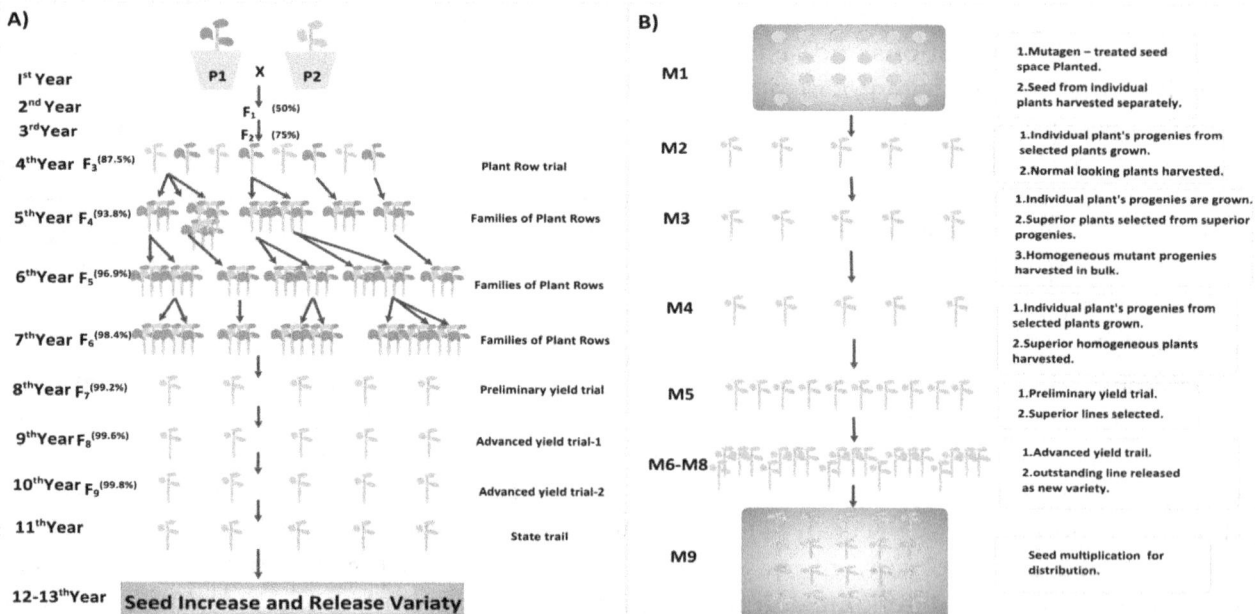

Figure 2. Improvement of agronomic traits using traditional breeding and chemical or physical mutagenic approaches. (**A**) Improving a trait (e.g., disease resistance) by the traditional breeding and for the introduction of the desired donor trait into the 'chosen' recipient line by selecting the progeny with the desired traits from the recipient line and crossing it with the donor line. (**B**) This process uses chemical or physical mutagens to generate mutants via random mutagenesis.

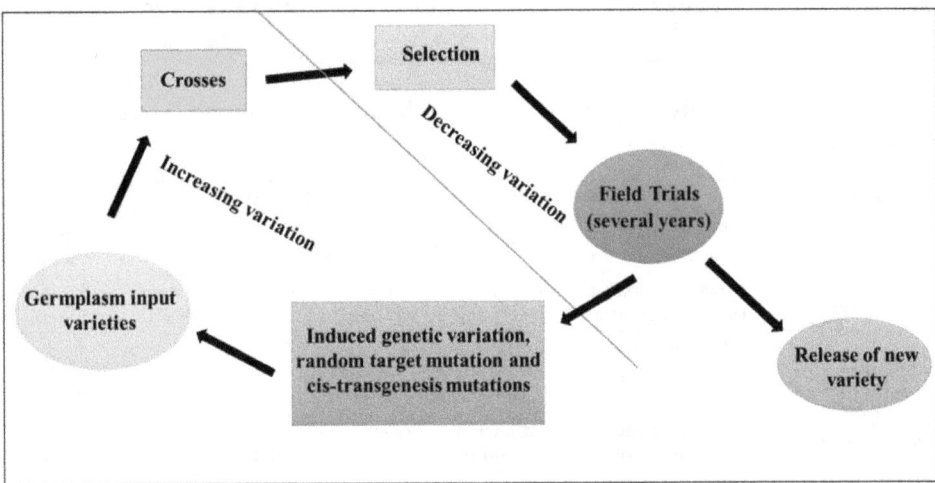

Figure 3. The plant breeding innovation cycle.

Identifying plants with desirable traits among existing plant varieties (or developing new phenotypes if these are not found naturally) is the initial and most important step in plant breeding. It would be impossible to develop new varieties or improve existing ones without natural genetic variation determined by spontaneous mutations. Ossowski et al. [49] concluded that the de novo spontaneous mutation rate was 7×10^{-9} base replacements per site per generation in all the nuclear genomes of five *Arabidopsis thaliana* accumulation lines sustained by single seed descent (SSD) over 30 generations. [50]. This is expected to be true for the genomes of most other plant species: for example, about 20 billion mutations occur each year in a one-hectare wheat field (personal communication with Professor Detlef Weigel, Max Planck Institute for Developmental Biology, Germany).

Another technique to improve plant varieties by conventional breeding is through mutation breeding. Mutagenesis is the phenomenon in which sudden heritable changes occur in the genetic material of an organism. It can occur spontaneously in nature or can be a result of exposure to different chemical, physical, or biological agents [51]. Mutation breeding is classified based on the three known types of mutagenesis. The first is radiation-induced mutagenesis in which mutations occur as a result of exposure to radiation (gamma rays, X-rays, or ion beams.); second is chemically induced mutagenesis; while the third is insertional mutagenesis, a consequence of DNA insertions either through the genetic transformation and insertion of T-DNA or the activation of transposable elements (i.e., site-directed mutagenesis; Table 1) [50,52]. According to Van Harten (Professor Agricultural University, Wageningen, The Netherlands), the history of plant mutation spans back to 300 BC, while the term mutation was first used in 1901 by Hugo De Vries, who reported during the final year of his studies that heredity might be changed by another mechanism, different from recombination and segregation [53]. He examined genomic variations and described them as heritable changes arising from this unique mechanism [54]. Numerous steps are required in any mutation breeding strategy: first, reducing the number of potential variants among the mutagenized seeds or other propagules for close evaluation of the first (M1) plant generation [51,54]. The benefits of mutation breeding over other breeding methods rely on the ability to select useful variant mutants in the second (M2) or third (M3) generations (Figure 2b).

Artificial mutation-causing agents are called mutagens; they are generally classified into two categories: physical and chemical mutagens [55]. They can induce mutations in almost any planting materials, including in vitro cultured cells, seedlings, and whole plants. Seeds are the most frequently used plant material for this specific purpose, but recently, various forms of plant propagules, such as tubers, bulbs, rhizomes, and mutation-induced vegetative propagated plants, are being used more frequently, as scientists take advantage of totipotency in single cells [56]. For example, with the use of ethyl methanesulfonate (EMS) and fast neutrons, collections of M82 tomato mutants were produced and more than 3000 phenotype alterations were classified [57]. An EMS-induced mutation library for the miniature dwarf tomato cultivar Micro-Tom has also been created, creating another resource for tomato genetic studies [58].

Table 1. Examples of commonly used physical and chemical mutagens, their characteristics, and hazard impacts.

Types	Mutagens	Characteristics (Sources and Description)	Hazards	References
Physical Mutagens	X-rays	Electromagnetic radiation; penetrates tissues from just a few millimeters to many centimeters.	Dangerous, penetrating	[59]
	Gamma rays	60Co (Cobalt-60) and 137Cs (Caesium-137); electric magnet radiation generated with radiation isotope and nuclear reactors.	Dangerous, penetrating	[59,60]
	Neutron	235U; there are fast, slow, thermal types; formed in nuclear reactors; unloaded particles; penetrate tissues up to large numbers centimeter;	Very dangerous	[59,60]
	Beta particles	32P and 14C; reduced particle accelerators or radioisotopes; electrons; ionizing and penetrating tissues shallowly	Maybe dangerous	[60]
	Alpha particles	Sources originating from radiological isotopes; helium nucleus able to penetrate tissues heavily	Very dangerous	[59]
	Proton	Present in nuclear reactors and accelerators; derived from the nucleus of hydrogen; penetrate tissues up to several inches.	Very dangerous	[59,60]
	Ion beam	Positively charged ions are accelerated at a high speed and used to irradiate living materials, including plant seeds and tissue culture.	Dangerous	[60]
Chemical Mutagens	Alkylating agents	The alkylated base can then degrade with bases to create a primary site which is mutagenic or recombinogenic or mispairs in DNA replication mutations, depending on the atom concerned.	Dangerous	[59]
	Azide	Just like alkylating agents.	Dangerous	[59]
	Hydroxylamine	Just like alkylating agents.	Dangerous	[56,59]
	Nitrous acid	Acts through deamination, replacing cytosine with uracil, which can pair with adenine and thus result in transitions via subsequent replication cycles.	Very Hazard	[56]
	Acridines	Interspersing between the DNA bases, thus distorting the DNA double helix and the DNA polymerase, recognizes the new basis for this expanded (intercalated) molecule and inserts a frameshift in front of it.	Dangerous	[56]
	Base analog	Comprises the transformations (purine to purine and pyrimidine to pyrimidine) into DNA in place of the regular bases during DNA replication and tautomerizing (existent in two forms, which interconvert into one another such that guanine may be present in keto and enol forms).	Some may be dangerous	[56]

Despite considerable success during the last century, the advances in yields of major crops (e.g., wheat) stabilized or even declined in many regions of the world [61,62]. Restrictions on phenotyping efficiency are increasingly being perceived as key constraints to genetic enhancements in breeding practices [63,64]. Specifically, HTP may cause a bottleneck in traditional breeding, marker-assisted selection (MAS), or genomic selection, where phenotyping is important to establish the accuracy of statistical models [63,65]. Accurate phenotyping is also required to replicate the outcomes of mutagenesis (i.e., GMOs) [66]. Deery et al. [67] and White and Conley [68] reviewed in great detail the benefits and challenges of potential phenotyping platforms, such as HTP.

Furthermore, SSD can be accelerated through the use of HTP [69,70]. SSD is most suitable for handling large segregating populations; while HTP tools are used in breeding programs [71]. Without undermining genetic variability and genetic development, SSD optimizes resource distribution, reducing the time spent growing crops and lowering costs associated with earlier generations' progress [72]. SSD has been successfully used in the groundnut breeding program, with the implementation of an inbreeding cycle producing multiple generations annually to advance fixed lines to multisite evaluation tests [73]. This speed breeding approach is ideal for SSD programs, particularly in cereal crops, allowing for the rapid cycling of multiple lines with healthy plants and viable seeds [30].

3. Mutagens for Molecular Breeding

One of the principal goals in the field of molecular biology is to identify and manipulate genes involved in human, animal, and plant disorders. Genomic tools used in such studies include restriction enzymes, biomarkers, molecular glue (ligases), as well as transcription and post-translational modification machinery [74]. Furthermore, molecular biological approaches are widely used to develop biofortified crops and plant varieties with high yield, new traits, and resistance to insect pests and diseases [75,76]. Globally, about 40 million hectares have been assigned to transgenic cultivars, which were commercialized after testing their biosafety level in 1999 [75]. Plant breeding was then reformed when researchers started to combine traditional practices with molecular tools to address phenotypic changes concerning the genotype of plant traits [77]. Accurate genome sequencing is essential before molecular tools can be used, and next-generation sequencing (NGS) allows researchers to decipher entire genomes and produce vast gene libraries for bioinformatics studies [78]. NGS opens new possibilities in phylogenetic and evolutionary studies, enabling the discovery of novel regulatory sequences and molecular markers [79]. Molecular biology is also facilitating the identification of diverse cytoplasmic male sterility sources in hybrid breeding. Some fertility restorer genes have been cloned in maize, rice, and sorghum [80]. Mutations in the target gene can be screened using target-induced local lesions in the genome (TILLING) and Eco-TILLING, which can directly identify allelic variations in the genome [81]. The most recent studies have determined the structure of plant germplasm using bulked segregant analysis [82], association mapping, genome resequencing [83,84], and fine gene mapping. This allows for the identification of single base-pair polymorphisms based on single sequence repeats, single nucleotide polymorphisms (SNPs), and unique biomarkers linked to quantitative trait loci (QTL) for genome manipulation, germplasm enhancement, and creating high-density gene libraries [85]. Traditional mutagenesis has certain limitations, as it can produce undesirable knockout mutations. It is also time-consuming and requires large-scale screening [86]. However, MAS is a direct approach for tracking mutations that improve backcrossing efficiency (or "breeding by design") [87] and determining the homogeneity of the progeny phenotypes.

In principle, all genome cleavage techniques produce double-stranded breaks (DSBs), blunt ends, or overhangs of the target nucleotide fragment, whether by homologous recombination, site-directed insertion/substitution of genes, or knockout mutations [88]. These DSBs, produced as a result of the action of sequence-specific nucleases (SSNs), are repaired by the non-homologous end joining (NHEJ) mechanism, which adds or removes nucleotides by the homology-directed repair pathway, directing DNA substitutions at target sites [89]. Various literature reviews report three primary SSN systems for genome editing. The first involves zinc finger nucleases (ZFNs), which form the basis for DNA manipulation. The second system involves transcription activator-like effector nucleases (TALENs), while the third system, the most important revolution in cutting-edge genomics, is a clustered regularly interspaced short palindromic repeats/associated protein 9 (CRISPR/Cas9) system [88,90–92]. The use of ZFNs has certain limitations: the constructs are not easy to design and transform, even in plants, and it is an expensive approach. Moreover, some researchers have reported non-specific nucleotide recognition because of their origin from eukaryotic transcription motifs, making this approach less reliable for genome editing [93,94]. Most restriction nucleases are derived from bacteria and TALENs were isolated from the prokaryotic plant pathogen *Xanthomonas* [95]. However, TALENs comprise large and repetitive constructs that require a lot of time and precision to edit the target sequence [96]. Soon after the discovery of TALENs, another promising nuclease (CRISPR/Cas9) was found in a bacterial immune system [97]. This system has been widely used in recent plant genome editing studies and has started replacing the TALEN and ZFN systems due to its high efficiency and accuracy in inducing site-directed breaks in double-stranded DNA [98]. Recently, a CRISPR-associated endonuclease from *Prevotella* and *Francisella* (Cpf1) has emerged as a replacement tool for precise genome editing, including DNA-free dissection of plant material, with higher potency, specificity, and enormous possibilities of wider application [99,100]. The base-editing approach using CRISPR/nCas9 (Cas9 nickase) or dCas9 (deactivated Cas9) fused with cytidine deaminase is a powerful tool to create point mutations. In this study, we point out the remarkable *G. hirsutum*-base editor 3 (GhBE3) base enhancing system developed to create single base mutations in the allotetraploid genome of cotton (*Gossypium hirsutum*) [101].

4. CRISPR/Cas9 and CRISPR/Cpf1 as Genetic Dissection Tools

The CRISPR/Cas9 system is a high-throughput discovery system in cutting-edge genomics, with recent studies reporting extensive use of Cas9 in gene transformation, drug delivery, and knockout mutations based on NHEJ-mediated DSBs [102]. Several studies investigated the mode of action of this potent nuclease and discovered the presence of a CRISPR loci, a cluster of repeating nucleotides in bacterial and archaeal immune systems [103]. These loci have a unique sequence, comprising of Cas9-encoding operons, transcription machinery, and consecutive repeats originating from various viral genomes separated by spacer sequences. These repeats were incorporated into the bacterial genome either by a virus or another foreign invader following an immune reaction [17].

Yin et al. (2017) found that Cas9 can be 'tricked' by supplementing any foreign nucleotide sequence that is digested and inserted in the bacterial genome. To knock a gene out, the CRISPR/Cas9 system is designed accordingly and transformed to explants via *Agrobacterium*, electroporation or the biolistic method. The regenerated plantlets' grown from the transformed callus are then transferred to planting soil [104]. The CRISPR/Cas9 gene knockout system has four significant features: (a) synthetic guide RNA (about 18–20 nucleotides) binding to target DNA, (b) Cas9 cleavage at 3–4 nucleotides after the adjacent proto spacer motif (PAM) (generally, 50 NGG identifies the PAM sequence) [105], (c) selection of a suitable binary vector and sgRNA cloning, and (d) transforming the construct in explants via *Agrobacterium* or microprojectile gene bombardment (Figure 4). *Agrobacterium*-mediated transformation is preferred in most studies given its efficiency and secure delivery [106]. The transformants are raised in growth chambers and examined for mutation studies using PCR, western blotting, ELISA, genotyping, sequencing, and other molecular techniques.

With recent advances in molecular biology and the discovery of sequences in the microbial immune system, biotechnologists can manipulate the organism's genome in a specific and precise way with the aid of CRISPR and its associated Cas proteins. This remarkable genome editing technique is categorized into two broad classes and six types: class 1 with types I, III, and IV and class 2 with types II, type V, and type VI [107]. Type II is the most widely used system in genome editing, while CRISPR/Cas9 from the *Streptococcus* pyogenes is the most commonly used method in the genome editing process. CRISPR class II has a type V effector named Cpf1, which can be designed with highly specific CRISPR RNA to cleave corresponding DNA sequences [108,109]. Cpf1 was recently developed as a substitute to Cas9, because of its unique ability to target T-rich motives through staggered DSBs without the need to trans-activate crRNA. Cpf1 can also process RNA and the DNA nuclease operation. Studies have been conducted to examine the Cpf1 mechanism, aiming to achieve more precise DNA editing and to address it the crystal structure of Cas12b homologous [110]. Another study reported that small molecular compounds can enhance Cpf1 efficiency as they are directly involved in activating or suppressing signaling pathways for cellular repair. Thus, small-molecule–mediated DNA repair aids in useful CRISPR mediated knock-outs [111].

Unfortunately, the development of new crop varieties by genome editing has been delayed in many countries by strict GMO regulations across the globe. This is particularly true for areas obeying a process rather than a regulatory framework based on the product, like in the EU, where authorizations for new varieties developed by genome editing techniques are subject to time- and cost-intensive verification procedures [112]. A recent decision by the European Court of Justice announced the enforcement of strict GMO legislation on target genome editing tools, even if the product is entirely free of transgenes [113]. Process-based regulations were also introduced in at least 15 countries, such as Brazil, India, China, and Australia, while 14 countries, including Canada, Argentina, and the Philippines adopted a product-based regulation. Several countries still have no specific regulatory system in operation, including Paraguay, Myanmar, Chile, and Vietnam. One of the most interesting aspects of regulation is Argentina's adoption, which is more versatile as it allows recent developments in genome editing to be taken into account. In the EU, genome-edited plants are typically listed as GMOs in compliance with the current legislation [114]. The control of genetically modified (GM) crops in the United States is authorized on a case-by-case basis, as set out in the structured framework for the control of biotechnology [115].

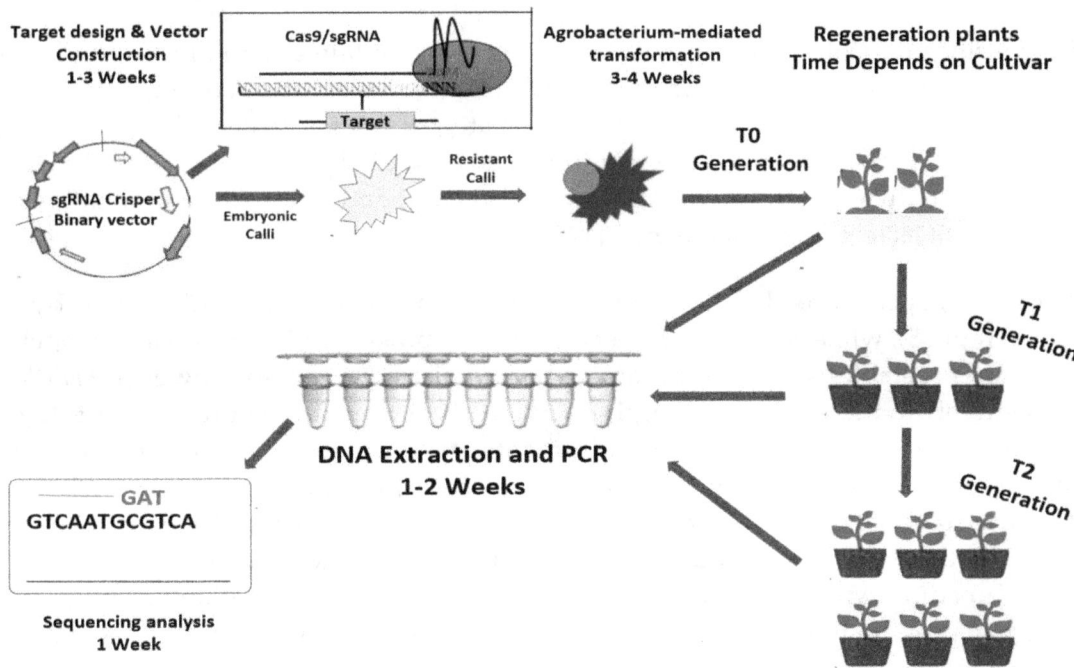

Figure 4. CRISPR–Cas9-based genome editing. CRISPR/Cas9 system uses Cas9 and sgRNA to cleave foreign DNA. It works in three steps: (1) the expression of the nuclear-localized Cas9 protein, (2) the generation of gRNA containing first 20-nt complementary to the target gene, and (3) the NGG PAM site recognition located nearly at the 3′ end of the target site. This process is followed by three additional steps: (1) design target and construction of a gene-specific sgRNA (vector), (2) CRISPR–Cas9 sgRNA can be transfected into the plant protoplast through *Agrobacterium*-mediated transformation, and (3) regenerated plants are screened for mutation via PCR-assay and sequencing. The estimated time needed is indicated for most steps.

5. Speed Breeding (Time-Saving Tools) for Accelerating Plant Breeding

Most plant species create a bottleneck in their applied research and breeding programs, generating the need for technologies to accelerate up plant growth and generation turnover. In the early 1980s, NASA's work was an inspiration for all plant scientists. In 2003, researchers at the University of Queensland coined the term "speed breeding" as a combination of methods developed to accelerate the speed of wheat breeding. Speed breeding protocols are currently being developed for several crops [16,30]. Speed breeding is suitable for diverse germplasm and does not require specific equipment for in vitro culturing, unlike doubled haploid (DH) technology, in which haploid embryos are produced to yield completely homozygous lines [116]. The principle behind speed breeding is to use optimum light intensity, temperature, and daytime length control (22 h light, 22 °C day/17 °C night, and high light intensity) to increase the rate of photosynthesis, which directly stimulates early flowering, coupled with annual seed harvesting to shorten the generation time [16,117]. Light intensity and wavelength plays a key in the regulation of flowering [118,119]. Croser et al. [120] developed early- and late-flowering genotypes for peas, chickpeas, faba beans, and lupins under controlled conditions using various parts of the light spectrum (blue and far red-improved LED lights and metal halide). These species showed a positive correlation to the diminishing red:far red-red proportion (R:FR). Accordingly, light with the most elevated power in the FR area is the most inductive [121,122]. In general, light with high R:FR (e.g., from fluorescent lamps) reduces stem enlargement and increases lateral branching, whereas light with a low R:FR (e.g., from incandescent lamps) strongly enhances stem elongation but inhibits lateral branching and flowering. This process is regulated by FR, while blue light mediates phytochrome FR (Pfr). Furthermore, the effect of R light on flowering repression is mediated by phytochrome R (Pr) [122,123].

Species-specific protocols to induce early flowering using certain environmental signals have been developed, such as short days or vernalization like RGA [48]. Greenhouse strategies under controlled conditions were compared with in vitro plus in vivo strategies and fast generation cycling by extended photoperiod [124–126]. The cost and space requirements associated with developing a large number of inbred lines can be reduced by implementing these practices in the breeding of small grain cereals grown at high densities (e.g., 1000 plants/m^2) [127].

Until recently, speed breeding had been reported to shorten generation time by extending photoperiods (Figure 5), while certain crop species, such as radish (*Raphanus sativus*), pepper (*Capsicum annum*), and leafy vegetables such as Amaranth (*Amaranthus* spp.) and sunflower (*Helianthus annuus*) responded positively to increased day length [27,30,117,128]. Speed breeding of short-day crops has been limited because of their flowering requirements. Nevertheless, recently, Lee Hickey and his research team worked on developing protocols for short-day crop like sorghum, millet and pigeon pea with the International Crop Research Institute for the Semi-Arid Tropics (ICRISAT) as part of a project funded by the Bill and Melinda Gates Foundation (https://geneticliteracyproject.org/2020/03/02/how-speed-breeding-will-help-us-expand-crop-diversity-to-feed-10-billion-people/). Sorghum, millet, and pigeon peas are important plants for many smallholder farmers in Africa and Asia, refining protocols targeted for these types of users has significant implications for global subsistence agriculture.

This goal involves improving the protocols and conditions required for the induction of early flowering and rapid crop development [117]. O'Connor et al. (2014) already reported successful results in the speed breeding of peanuts (*Arachis hypogaea*). Increased day length helped amaranth (*Amaranthus* spp.) to achieve more generations annually [129]. In staple food crops requiring shorter photoperiods to initiate the reproductive phases, such as rice (*Oryza sativa*) and maize (*Zea mays*), speed breeding can accelerate vegetative growth [130]. Using speed breeding, it is possible to develop successive generations of improved crops for field examination via SSD, which is cheaper compared to the production of DHs. Speed breeding is also favorable to gene insertion (common haplotypes) of distinct phenotypes followed by MAS of elite hybrid lines [31,131].

In conclusion, recent advances in plant breeding and genomics have contributed to the development of qualitatively and quantitatively improved cultivars. Innovative agronomic strategies, in addition to the usual practices, have led to remarkable agricultural outcomes. However, sustainable crop development to ensure global food security can only be achieved with the combined investments of private firms, extension workers, and the public sector.

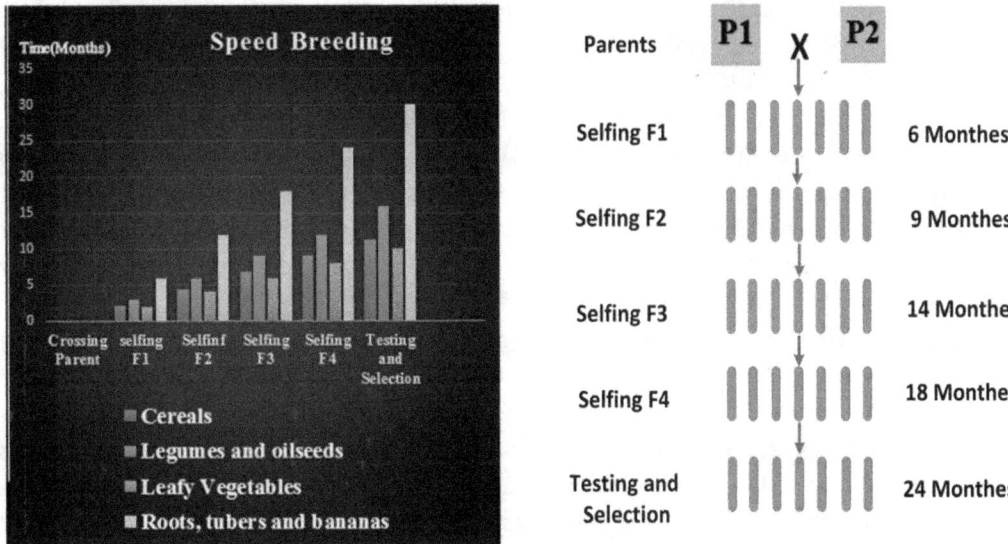

Figure 5. Graphical presentation of the elite line development procedure. Comparison of time (in months) required to develop elite lines from selected parents of some crops. Extended photoperiods induced earlier flowering and created 4 generations annually. The optimal temperature regime (maximum and minimum temperatures) should be applied for each crop. A higher temperature should be maintained during the photoperiod, whereas a fall in temperature during the dark period can aid in stress recovery. At the University of Queensland; (UQ), a 12-h 22 °C/17 °C temperature cycling regime with 2 h of darkness occurring within 12 h of 17 °C has proven successful. The figure is briefly modified from Watson et al. (2018).

6. Contribution of Plant Breeding to Crop Improvement

Molecular plant breeding was revolutionized in the 21st century, leading to crop improvement based on genomics, molecular marker selection, and conventional plant breeding practices [10,39]. For instance, the average yield of wheat (*Triticum spp.*), maize (*Zea mays*), and soybean (*Glycine max*), all significant crops in the United States, showed a positive linear increase from 1930 to 2012 [132,133]. The introduction of recessive genes in off-season nurseries was commercialized by pioneering plant breeder Norman Borlaug (among others), which helped to reduce the time needed to develop new cultivars. For example, the time for developing a new wheat cultivar was reduced from 10–12 years to only 5–6 years [134].

In hybrid and pure line crop breeding, developing similar and homozygous lines is a time-consuming process. Cycle time has been reduced from five to two generations by producing homogeneous and homozygous lines using DHs in diverse crops [135,136]. The maize DH system is one of the most common, it uses the R1-NJ color marker. However, the DH system has various genotypic and biological limitations [136,137]. Different crop species show dependence on the genotype for haploid induction [138], adapting tissue culture (e.g., in case of anther culture), and chromosome doubling by colchicine [139]. Breeders using the DH system unintentionally practice many selections for loci, increasing the success rate of this approach [140], but this might limit genetic variation in the breeding populations in responsive genome regions. Another approach is the RNAi suppression of plant genes (for instance, the MutS HOMOLOG1 (*MSH1*) gene) in multiple plant species, which produces a variety of developmental modifications accompanied by adjustments in plant defense, phytohormones, and abiotic stress response pathways combined with methylome repatterning [141].

Although the evaluation and production of GM crops is an active area of research, this technology is currently restricted because of political and ethical concerns. Nevertheless, GMO technologies make use of the variations that are present in deliberately mutated or naturally occurring populations [39,142,143]. GMOs have a variety of practical applications, for instance, they can be used to produce plant proteins that are toxic to insect pests, create herbicide tolerance genes for weed control, and create "golden rice" biofortified with vitamin A [144]. The characterization and discovery of genes and promoters can offer precise and effective temporal and spatial control of the expression of different genes, which is crucial for the future use of GM crops [145].

The availability of published genome sequences for different crops is increasing every year, facilitated by the use of sequencing technologies that improve sequencing speed and cost [146,147]. Current sequencing technologies, such as the NGS technique, can sequence multiple cultivars with both small and large genomes at a reasonable cost [148]. Although various published genomes are considered to be incomplete, they remain a valuable tool to evaluate important crop traits such as grain traits, fruit ripening, and flowering time adaptation [83,149].

Modern plant breeding programs have engaged interdisciplinary teams with expertise in the fields of statistics, biochemistry, physiology, bioinformatics, molecular biology, agronomy, and economics [150]. Crop breeding has been revolutionized and research on the advancement of DNA sequencing technologies has started the "genomics era" of crop improvement [151]. The genomes of most of the essential crops have been sequenced, creating a much cheaper genotyping platform for DNA fingerprinting. SNPs are ubiquitous DNA markers in crop genomes, they are also cost-efficient and easy to handle. Therefore, in today's crop improvement practices, genotyping large populations with a large number of markers is standard practice. Even whole-genome resequencing data are becoming easily available, giving unprecedented access to the structural diversity of crop genomes [65,83,152].

Currently, researchers are also using molecular genetic mapping of QTL of many complex traits vital in plant breeding. The detection and molecular cloning of genes underlying QTL enable the investigation of naturally occurring allelic variations for specific complex traits [85,153]. Plant productivity can be improved by identifying novel alleles through functional genomics or haplotype analysis. Advances in cereal genomics research in recent years have enabled scientists to improve the prediction of phenotypes from genotypes in cereal breeding [11,46].

Recently developed DNA-free CRISPR/Cas9 system delivery methods, different Cas9 variants, and RNA-guided nucleases offer new possibilities for crop genomic engineering [154]. The need to increase food security makes boosting crop production the primary objective of gene editing (Table 2). Crop yield is a complex trait that depends on several factors. The required phenotypes were found in plants with the loss of function mutations in yield related genes, highlighting the usefulness of CRISPR/Cas9 in improving yield-related traits by knocking out negative regulators affecting yield-determination factors, such as OsGS3 for grain size, OsGn1a for grain number; OsGW5, TaGW2, TaGASR7, and OsGLW2 for grain weight; TaDEP1 and OsDEP1 for panicle size, and OsAAP3 for tiller numbers. [155,156]. Similarly, three rice weight-related genes (GW5, GW2, and TGW6) were knocked out, causing pyramiding and increased weight [157]. The knockout of the *Waxy* gene using CRISPR/Cas9 resulted in the development of rice cultivars with higher nutritional quality [158]. DuPont Pioneer introduced a CRISPR/Cas9 knockout waxy corn line with high yields, ideal for commercial use [155]. A knockout of the *MLO* gene in tomato using CRISPR/Cas9 resulted in resistance to powdery mildew [159]. CRISPR has also been used to mutate the *OsERF922* transcription factor, resulting in resistance to rice blast, a destructive fungal disease [160].

By adopting a 22 h photoperiod and a temperature-controlled regime, generation times were considerably reduced in durum wheat (*T. durum*), spring bread wheat (*T. aestivum*), chickpea (*Cicer arietinum*), pea (*Pisum sativum*), barley (*Hordeum vulgare*), stiff brome (*Brachypodium distachyon*), canola (*Brassica napus*), and barrel clover (*Medicago truncatula*), compared with plants grown in a greenhouse with no supplementary light or those grown in the field. Under rapid growth conditions, plant development was normal, plants (such as wheat and barley) could be crossed easily, and seed germination rates were high [31,161–163].

Table 2. Application of breeding techniques toward crop improvement.

Sr.no.	Species	Method	Traits	References
1	Rice	Cross Breeding	Increased spikelet number per panicle	[164]
2	Rice	Cross Breeding	Yield Increases	[165]
3	Wheat	Cross Breeding	Increase Grain Yield	[166]
4	Tomato	Mutation Breeding	Resistance to bacterial wilt (*Ralstonia solanacearum*)	[167]
5	Rapeseed	Mutation Breeding	Resistance to stem rot (*Sclerotinia sclerotiorum*)	[168]
6	Cotton	Mutation Breeding	Resistance to bacterial blight, cotton leaf curl virus	[169]
7	Barley	Mutation Breeding	Salinity tolerance	[170]
8	Sunflower	Mutation Breeding	Semi-dwarf cultivar/dwarf	
9	Cassava	Mutation Breeding	High-amylose content preferred by diabetes patients because it lowers the insulin level, which prevents quick spikes in glucose contents	[171]
10	Groundnut	Mutation Breeding	Dark green, obovate leaf pod; increased seed size, higher yield, moderately resistant to diseases, increased oil and protein content	[172]
11	Maize	Transgenic Breeding	increased vitamin content (vitamins C, E, or provitamin A)	[173]
12	Tomato	Transgenic Breeding	Dry Matter Increases	[174]
13	Soybean	Transgenic Breeding	Altered carbohydrates metabolism	[174]
14	Barley	Molecular Marker	Adult resistance to stripe rust	[175]
15	Maize	Molecular Marker	Development of quality protein maize	[22]
16	Watermelon	Marker-Assisted Selection	Early Flowering	[176]
17	Canola	QTL	Dynamic growth QTL	[153]
18	Alfalfa	Intragenesis	Lignin content	[129]
19	Apple	Cisgenesis, Intragenesis	Scab resistance	[177,178]
20	Barley	Cisgenesis	Grain phytase activity	[179]
21	Durum wheat	Cisgenesis	Baking quality	[180]
22	Perennial ryegrass	Intragenesis	Drought tolerance	[181]
23	Poplar	Cisgenesis	Plant growth and stature, wood properties	[181]
24	Potato	Cisgenesis	Late blight resistance	[182]
25	Strawberry	Intragenesis	Gray mold resistance	[183]
26	Tomato	Gene editing/ZFN	Reduction of cholesterol and steroidal glycoalkaloids, such as toxic α-solanine and α- chaconine	[184]
27	Wheat	Gene editing/TALEN	Heritable Modification	[185]
28	Rice	Gene knockout/ CRISPR/Cas9	Fragrance	[186]
29	Bread Wheat and Maize	Gene knockout/ CRISPR/Cas9	Leaf development; Male fertility, Herbicide resistance	[187]
30	Poplar	Gene knockout/ CRISPR/Cas9	Lignin content; Condensed tannin content	[188]
31	Tomato	Gene editing/ CRISPR/Cas9	Leaf development	[189]
32	Soybean	Gene replacement/ CRISPR/Cas9	Herbicide resistance	[190]
33	Maize	Gene replacement/ CRISPR/Cas9	Herbicide resistance	[187]
34	Cotton	Genome Editing/ CRISPR/Cas9	Produce transgenic seeds without regeneration	[191]
35	Soybean	Genome Editing/ CRISPR/Cas9	Early Flowering	[192]
36	Rice	Genome Editing/ CRISPR/Cas9	Increased grain weight	[157]
37	Tomato	Genome Editing/ CRISPR/Cas9	Resistance to powdery mildew	[159]
38	Wheat	Gene knockout/ CRISPR/Cas9	low-gluten foodstuff	[193]
39	Rice	Gene knockout/ CRISPR/Cas9	Generate mutant plants which is sensitive to salt stress	[194]
40	Rapeseed	Gene knockout/ CRISPR/Cas9	Controlling pod shattering resistance in oilseed rape	[195]
41	Tomato, Potato	CRISPR/Cas9 Cytidine Base Editor	Transgene-free plants in the first generation in tomato and potato	[196]
42	Tobacco	Genome Editing /CRISPR/Cpf1	Plants harboring	[197]
43	Rice	Genome Editing /CRISPR/Cpf1	Regulate the stomatal density in leaf	[198]
44	Rice	Genome Editing /CRISPR/Cpf1	Stable mRNA equal	[100,199]
45	Maize	Genome Editing /CRISPR/Cpf1	Mutation frequencies doubled	[199]

Table 2. *Cont.*

Sr.no.	Species	Method	Traits	References
46	Chickpea	Rapid generation advance (RGA)	Seven generations per year and enable speed breeding	[48]
47	Pea	Greenhouse strategy	6 Generation/year	[124]
48	Chickpea	Speed Breeding	4-6 Generation/year	[200]
49	Barley	Speed Breeding	Resistance to Leaf Rust	[16]
50	Spring wheat	Speed Breeding	Resistance to Stem Rust	[201]
51	Spring wheat	Speed Breeding	4-6 Generation/year	[16]
52	Barley	Speed Breeding	4-6 Generation/year	[16]
53	Peanut	Speed Breeding	2-3 Generation/year	[200]
54	Canola	Speed Breeding	4-6 Generation/year	[16]
55	Wheat	High-throughput phenotyping (HTP)	Development of improved, high-yielding crop varieties	[202]
56	Tomato	High-throughput phenotyping (HTP)	Using biostimulants to increase the plant capacity of using water	[203]

7. Future Outlook

Although modern plant breeding relies on traditional techniques, the emergence of new approaches will undoubtedly increase its efficiency and effectiveness. In the future, we can expect a wide range of techniques to be developed using interdisciplinary principles to increase their benefits. Strategies for crop production, breeding methods, approaches to field testing, genotyping technologies, even equipment and facilities need to be implemented across crop species to keep our food, fiber and biobased economy diverse. The discovery of CRISPR/Cas9, CRISPR/Cpf1, base-editing, and RGA has revolutionized molecular biology and its innovative applications in agriculture, setting a turning point in plant breeding and cultivation. GMOs can positively disseminate a selectable gene across wild populations in a gene drive process.

Altogether, CRISPR-based gene drive systems will prove—in time—to be beneficial for mankind. They will, for instance, prevent epidemics, improve agricultural practices, and control the spread of invasive species as plant cultivars resistant to insect pests and pathogens and tolerant to herbicides are developed. Existing genome editing techniques can be improved with the help of speed breeding (e.g., genes responsible for late flowering could be knocked out using CRISPR/Cas9). After the successful transfer of Cas9 into the plant, the transgenic plant can then be grown under speed breeding conditions rather than the usual glasshouse conditions to obtain transgenic seeds as early as possible. Using this method, it is possible to obtain stable homozygous phenotypes in less than a year. Furthermore, this method also decreases generation time, as it normally takes several years to develop a GMO crop. However, more efficient breeding strategies combining these technologies could lead to a step-change in the rate of genetic gain. Therefore, CRISPR/Cas9, primarily based on genome editing and speed breeding, will likely gain in popularity. It will be a crucial technique to obtain plants with specific desirable traits and contribute to reaching our objectives for zero-hunger globally. The development of innovations is often applied to a few important economic crops, which require specific adaptation to the reproduction and propagation method and the "process" of the new line development for the various crops of interest. Likewise, the transition of technology usually originates in developed countries, mostly in the private sector. It should be transferred to the public sector and into the developing world, given the significant financial investment required for groundbreaking work.

8. Conclusions

The primary methods for crop improvement in modern agriculture are cross breeding, mutation breeding, and transgenic breeding. Such time-consuming, laborious, and untargeted breeding programs cannot satisfy the increasing global food demand. To deal with this challenge and to enhance crop selection efficiency, marker-assisted breeding, and transgenic approaches have been adopted, generating desired traits via exogenous transformation into elite varieties. These genome editing systems are excellent tools that provide rapid, targeted mutagenesis and can identify the specific plant molecular mechanisms for crop improvement. Crop breeding was revolutionized by the development of next-generation breeding techniques. Genome editing technologies have many advantages over traditional agricultural methods, given their simplicity, efficiency, high specificity, and amenability to multiplexing. We conclude that speed breeding, combined with genetic tools and resources, enable plant biologists to scale up their research in the field of crop improvement.

Author Contributions: S.A., R.A.G., K.-H.J., and W.Z. designed the study; S.A., R.A.G., and M.U.Q. wrote the MS; S.A., A.F., M.M., R.A.G., W.Z., and K.H.J. revised the MS. In this study, S.A. and R.A.G. contributed equally. All authors have read and agreed to the published version of the manuscript.

Funding: This work was supported by the National Key Research and Development Program (2018YFD0100601), the National Natural Science Foundation of China (31650110476), the Jiangsu Collaborative Innovation Center for Modern Crop Production, the Sino-German Research Project (GZ 1362), the Science and Technology Department of Zhejiang Province (2016C02050–8), the Next-Generation Bio Green 21 Program (PJ01325901 and PJ01366401 to KHJ), Republic of Korea, and the Collaborative Genome Program of the Korea Institute of Marine Science and Technology Promotion (KIMST) funded by the Ministry of Oceans and Fisheries (MOF) (No. 20180430 to JKH).

Acknowledgments: Authors are thankful to an anonymous reviewer for their comments and critical reading of the manuscript.

Abbreviations

CRISPR	Clustered Regularly Interspaced Short Palindromic Repeats (CRISPR);
Cas9	CRISPR-associated Proteins;
Cpf1	CRISPR-associated endonuclease in Prevotella and Francisella;
DSB	Double Strand Breaks;
HTP	High-Throughput Phenotyping;
NASA	National Aeronautics and Space Administration;
USU	Utah State University;
GMO	Genetically Modified Organism;
EU	European Union;
Non-GMO	Non-Genetically modified Organism;
SSD	Single Seed Descent;
SB	Speed Breeding;
RGA	Rapid Generation Advance;
UQ	University of Queensland;
DH	Double Haploid;
ICRISAT	International Crop Research Institute for the Semi-Arid Tropics.

References

1. Tester, M.; Langridge, P. Breeding technologies to increase crop production in a changing world. *Science (80-.).* **2010**, *327*, 818–822. [CrossRef] [PubMed]
2. Shiferaw, B.; Smale, M.; Braun, H.-J.; Duveiller, E.; Reynolds, M.; Muricho, G. Crops that feed the world 10. Past successes and future challenges to the role played by wheat in global food security. *Food Secur.* **2013**, *5*, 291–317. [CrossRef]
3. Von Braun, J.; Rosegrant, M.W.; Pandya-Lorch, R.; Cohen, M.J.; Cline, S.A.; Brown, M.A.; Bos, M.S. *New Risks and Opportunities for Food Security Scenario Analyses for 2015 and 2050*; IFPRI: Washington, DC, USA, 2005.
4. Ristaino, J.B. Tracking historic migrations of the Irish potato famine pathogen, Phytophthora infestans. *Microbes Infect.* **2002**, *4*, 1369–1377. [CrossRef]
5. Tatum, L.A. The Southern Corn Leaf Blight Epidemic. *Science* **1971**, *171*, 1113–1116. [CrossRef]
6. UN World Population Projected to Reach 9.8 Billion in 2050, and 11.2 Billion in 2100. Available online: https://www.un.org/development/desa/en/news/population/world-population-prospects-2017.html.
7. FAO How to Feed the World in 2050. In *Insights from an Expert Meet*; FAO: Roma, Italy, 2009; Volume 2050, pp. 1–35.

8.   Voss-Fels, K.P.; Stahl, A.; Hickey, L.T. Q&A: Modern crop breeding for future food security. *BMC Biol.* **2019**, *17*, 18.

9.   Cheema, K.S. K. Plant Breeding its Applications and Future Prospects. *Int. J. Eng. Technol. Sci. Res.* **2018**, *5*, 88–94.

10. Moose, S.P.; Mumm, R.H. Molecular Plant Breeding as the Foundation for 21st Century Crop Improvement. *Plant Physiol.* **2008**, *147*, 969–977. [CrossRef]

11. Varshney, R.K.; Hoisington, D.A.; Tyagi, A.K. Advances in cereal genomics and applications in crop breeding. *Trends Biotechnol.* **2006**, *24*, 490–499. [CrossRef]

12. Li, H.; Rasheed, A.; Hickey, L.T.; He, Z. Fast-forwarding genetic gain. *Trends Plant Sci.* **2018**, *23*, 184–186. [CrossRef] [PubMed]

13. Collins, F.S.; Green, E.D.; Guttmacher, A.E.; Guyer, M.S. A vision for the future of genomics research. *Nature* **2003**, *431*, 835–847.

14. Majid, A.; Parray, G.A.; Wani, S.H.; Kordostami, M.; Sofi, N.R.; Waza, S.A.; Shikari, A.B.; Gulzar, S. Genome Editing and its Necessity in Agriculture. *Int. J. Curr. Microbiol. Appl. Sci.* **2017**, *6*, 5435–5443. [CrossRef]

15. Araus, J.L.; Kefauver, S.C.; Zaman-Allah, M.; Olsen, M.S.; Cairns, J.E. Translating High-Throughput Phenotyping into Genetic Gain. *Trends Plant Sci.* **2018**, *23*, 451–466. [CrossRef] [PubMed]

16. Watson, A.; Ghosh, S.; Williams, M.J.; Cuddy, W.S.; Simmonds, J.; Rey, M.D.; Asyraf Md Hatta, M.; Hinchliffe, A.; Steed, A.; Reynolds, D.; et al. Speed breeding is a powerful tool to accelerate crop research and breeding. *Nat. Plants* **2018**, *4*, 23–29. [CrossRef] [PubMed]

17. Zhang, F.; Wen, Y.; Guo, X. CRISPR/Cas9 for genome editing: Progress, implications and challenges. *Hum. Mol. Genet.* **2014**, *23*, R40–R46. [CrossRef] [PubMed]

18. Murovec, J.; Pirc, Ž.; Yang, B. New variants of CRISPR RNA-guided genome editing enzymes. *Plant Biotechnol. J.* **2017**, *15*, 917–926. [CrossRef]

19. Acquaah, G. Polyploidy in Plant Breeding. In *Principles of Plant Genetics and Breeding*; John Wiley & Sons: Hoboken, NJ, USA, 2012; pp. 452–469.

20. Muth, J.; Hartje, S.; Twyman, R.M.; Hofferbert, H.R.; Tacke, E.; Prüfer, D. Precision breeding for novel starch variants in potato. *Plant Biotechnol. J.* **2008**, *6*, 576–584. [CrossRef]

21. Mujjassim, N.E.; Mallik, M.; Rathod, N.K.K.; Nitesh, S.D. Cisgenesis and intragenesis a new tool for conventional plant breeding: A review. *J. Pharmacogn. Phytochem.* **2019**, *8*, 2485–2489.

22. Dreher, K.; Morris, M.; Khairallah, M.; Ribaut, J.M.; Shivaji, P.; Ganesan, S. Is marker-assisted selection cost-effective compared with conventional plant breeding methods? The case of quality protein Maize. *Econ. Soc. Issues Agric. Biotechnol.* **2009**, 203–236.

23. Abreu, G.B.; Ramalho, M.A.P.; Toledo, F.H.R.B.; De Souza, J.C. Strategies to improve mass selection in maize. *Maydica* **2010**, *55*, 219–225.

24. Kandemir, N.; Saygili, İ. Apomixis: New horizons in plant breeding. *Turkish J. Agric. For.* **2015**, *39*, 549–556. [CrossRef]

25. Leifert, C.; Tamm, L.; Lammerts van Bueren, E.T.; Jones, S.S.; Murphy, K.M.; Myers, J.R.; Messmer, M.M. The need to breed crop varieties suitable for organic farming, using wheat, tomato and broccoli as examples: A review. *NJAS - Wageningen J. Life Sci.* **2011**, *58*, 193–205.

26. Doust, A.; Diao, X. Plant Genetics and Genomics: Crops and Models Volume 19. *Genet. Genom. Setaria Ser.* **2017**, *19*, 377.

27. Stetter, M.G.; Zeitler, L.; Steinhaus, A.; Kroener, K.; Biljecki, M.; Schmid, K.J. Crossing Methods and Cultivation Conditions for Rapid Production of Segregating Populations in Three Grain Amaranth Species. *Front. Plant Sci.* **2016**, *7*, 816. [CrossRef] [PubMed]

28. Bugbee, B.; Koerner, G. Yield comparisons and unique characteristics of the dwarf wheat cultivar "USU-Apogee". *Adv. Sp. Res.* **1997**, *20*, 1891–1894. [CrossRef]

29. Bula, R.J.; Morrow, R.C.; Tibbitts, T.W.; Barta, D.J.; Ignatius, R.W.; Martin, T.S. Light-emitting diodes as a radiation source for plants. *HortScience* **1991**, *26*, 203–205. [CrossRef]

30. Ghosh, S.; Watson, A.; Gonzalez-Navarro, O.E.; Ramirez-Gonzalez, R.H.; Yanes, L.; Mendoza-Suárez, M.; Simmonds, J.; Wells, R.; Rayner, T.; Green, P.; et al. Speed breeding in growth chambers and glasshouses for crop breeding and model plant research. *Nat. Protoc.* **2018**, *13*, 2944–2963. [CrossRef]

31. Hickey, L.T.; Germa, S.E.; Diaz, J.E.; Ziems, L.A.; Fowler, R.A.; Platz, G.J.; Franckowiak, J.D.; Dieters, M.J. *Speed Breeding for Multiple Disease Resistance in Barley*; Springer: New York, NY, USA, 2017.

32. Chen, K.; Wang, Y.; Zhang, R.; Zhang, H.; Gao, C. CRISPR/Cas Genome Editing and Precision Plant Breeding in Agriculture. *Annu. Rev. Plant Biol.* **2019**, *70*, annurev. [CrossRef]

33. Godwin, I.D.; Rutkoski, J.; Varshney, R.K.; Hickey, L.T. Technological perspectives for plant breeding. *Theor. Appl. Genet.* **2019**, *132*, 555–557. [CrossRef]

34. Lee, J.; Chung, J.H.; Kim, H.M.; Kim, D.W.; Kim, H. Designed nucleases for targeted genome editing. *Plant Biotechnol. J.* **2016**, *14*, 448–462. [CrossRef] [PubMed]

35. Zhang, H.; Zhang, J.; Lang, Z.; Botella, J.R.; Zhu, J.K. Genome Editing—Principles and Applications for Functional Genomics Research and Crop Improvement. *Crit. Rev. Plant Sci.* **2017**, *36*, 291–309. [CrossRef]

36. Hsu, P.D.; Scott, D.A.; Weinstein, J.A.; Ran, F.A.; Konermann, S.; Agarwala, V.; Li, Y.; Fine, E.J.; Wu, X.; Shalem, O.; et al. DNA targeting specificity of RNA-guided Cas9 nucleases. *Nat. Biotechnol.* **2013**, *3*, 827. [CrossRef] [PubMed]

37. Zetsche, B.; Gootenberg, J.S.; Abudayyeh, O.O.; Slaymaker, I.M.; Makarova, K.S.; Essletzbichler, P.; Volz, S.E.; Joung, J.; Van Der Oost, J.; Regev, A.; et al. Cpf1 Is a Single RNA-Guided Endonuclease of a Class 2 CRISPR-Cas System. *Cell* **2015**. [CrossRef]

38. Basey, A.C.; Fant, J.B.; Kramer, A.T. Producing native plant materials for restoration: 10 rules to collect and maintain genetic diversity. *Nativ. Plants J.* **2015**, *16*, 37–53. [CrossRef]

39. Krimsky, S. Traditional Plant Breeding. In *GMOs Decoded*; MIT Press: Cambridge, MA, USA, 2019.

40. Shepard, J.F.; Bidney, D.; Barsby, T.; Kemble, R. Fusion of Protoplasts. *Biotechnol. Biol. Front.* **2019**.

41. Marthe, F. Tissue culture approaches in relation to medicinal plant improvement. In *Biotechnologies of Crop Improvement*; Research Gate: Berlin, Germany, 2018; Volume 1, pp. 487–497. ISBN 9783319782836.

42. Germana, M.A. Anther culture for haploid and doubled haploid production. *Plant Cell Tissue Organ Cult.* **2011**, *104*, 283–300. [CrossRef]

43. Hajjar, R.; Hodgkin, T. The use of wild relatives in crop improvement: A survey of developments over the last 20 years. *Euphytica* **2007**, *156*, 1–13. [CrossRef]

44. Ceccarelli, S.; Guimaraes, E.P.; Weltzien, E. *Plant breeding and farmer participation*; NHBS: Devon, UK, 2009; ISBN 9789251063828.

45. Cellini, F.; Chesson, A.; Colquhoun, I.; Constable, A.; Davies, H.V.; Engel, K.H.; Gatehouse, A.M.R.; Kärenlampi, S.; Kok, E.J.; Leguay, J.-J. Unintended effects and their detection in genetically modified crops. *Food Chem. Toxicol.* **2004**, *42*, 1089–1125. [CrossRef]

46. Dolferus, R.; Ji, X.; Richards, R.A. Abiotic stress and control of grain number in cereals. *Plant Sci.* **2011**, *181*, 331–341. [CrossRef]

47. Caligari, P.D.S.; Brown, J. Plant Breeding, Practice. In *Encyclopedia of Applied Plant Sciences*; Academic Press: Cambridge, MA, USA, 2016; Volume 2, pp. 229–235. ISBN 9780123948083.

48. Samineni, S.; Sen, M.; Sajja, S.B.; Gaur, P.M. Rapid generation advance (RGA) in chickpea to produce up to seven generations per year and enable speed breeding. *Crop J.* **2019**. [CrossRef]

49. Ossowski, S.; Schneeberger, K.; Lucas-Lledó, J.I.; Warthmann, N.; Clark, R.M.; Shaw, R.G.; Weigel, D.; Lynch, M. The rate and molecular spectrum of spontaneous mutations in Arabidopsis thaliana. *Science (80-.).* **2010**, *327*, 92–94. [CrossRef] [PubMed]

50. Oladosu, Y.; Rafii, M.Y.; Abdullah, N.; Hussin, G.; Ramli, A.; Rahim, H.A.; Miah, G.; Usman, M. Principle and application of plant mutagenesis in crop improvement: A review. *Biotechnol. Biotechnol. Equip.* **2016**, *30*, 1–16. [CrossRef]

51. Roychowdhury, R.; Tah, J. Mutagenesis—A potential approach for crop improvement. In *Crop Improvement*; Springer: New York, NY, USA, 2013; pp. 149–187.

52. Forster, B.P.; Shu, Q.Y.; Nakagawa, H. Plant mutagenesis in crop improvement: Basic terms and applications. *Plant Mutat. Breed. Biotechnol.* **2012**, 9–20.

53. Van Harten, A.M. *Mutation Breeding: Theory and Practical Applications*; Cambridge University Press: Cambridge, MA, USA, 1998; ISBN 0521470749.

54. Kharkwal, M.C. A brief history of plant mutagenesis. *Plant Mutat. Breed. Biotechnol.* **2012**, 21–30.

55. Mba, C.; Afza, R.; Bado, S.; Jain, S.M. Induced Mutagenesis in Plants. *Plant Cell Cult. Essent. Methods* **2010**, 111–130.

56. Mba, C. Induced Mutations Unleash the Potentials of Plant Genetic Resources for Food and Agriculture. *Agronomy* **2013**, *3*, 200–231. [CrossRef]

57. Menda, N.; Semel, Y.; Peled, D.; Eshed, Y.; Zamir, D. In silico screening of a saturated mutation library of tomato. *Plant J.* **2004**, *38*, 861–872. [CrossRef]
58. Watanabe, S.; Mizoguchi, T.; Aoki, K.; Kubo, Y.; Mori, H.; Imanishi, S.; Yamazaki, Y.; Shibata, D.; Ezura, H. Ethylmethanesulfonate (EMS) mutagenesis of Solanum lycopersicum cv. Micro-Tom for large-scale mutant screens. *Plant Biotechnol.* **2007**, *24*, 33–38. [CrossRef]
59. Wani, M.R.; Kozgar, M.I.; Tomlekova, N.; Khan, S.; Kazi, A.G.; Sheikh, S.A.; Ahmad, P. Mutation breeding: A novel technique for genetic improvement of pulse crops particularly Chickpea (*Cicer arietinum* L.). In *Improvement of Crops in the Era of Climatic Changes*; Springer: New York, NY, USA, 2014; pp. 217–248.
60. Mba, C.; Afza, R.; Shu, Q.Y.; Forster, B.P.; Nakagawa, H. Mutagenic radiations: X-rays, ionizing particles and ultraviolet. *Plant Mutat. Breed. Biotechnol.* **2012**, 83–90.
61. Acreche, M.M.; Briceño-Félix, G.; Sánchez, J.A.M.; Slafer, G.A. Physiological bases of genetic gains in Mediterranean bread wheat yield in Spain. *Eur. J. Agron.* **2008**, *28*, 162–170. [CrossRef]
62. Sadras, V.O.; Lawson, C. Genetic gain in yield and associated changes in phenotype, trait plasticity and competitive ability of South Australian wheat varieties released between 1958 and 2007. *Crop Pasture Sci.* **2011**, *62*, 533–549. [CrossRef]
63. Araus, J.L.; Cairns, J.E. Field high-throughput phenotyping: The new crop breeding frontier. *Trends Plant Sci.* **2014**, *19*, 52–61. [CrossRef] [PubMed]
64. Tardieu, F.; Cabrera-Bosquet, L.; Pridmore, T.; Bennett, M. Plant Phenomics, From Sensors to Knowledge. *Curr. Biol.* **2017**, *27*, R770–R783. [CrossRef] [PubMed]
65. Crossa, J.; Pérez, P.; de los Campos, G.; Mahuku, G.; Dreisigacker, S.; Magorokosho, C. Genomic selection and prediction in plant breeding. *J. Crop Improv.* **2011**, *25*, 239–261. [CrossRef]
66. Blum, A. Genomics for drought resistance-getting down to earth. In *Functional Plant Biology*; CSIRO Publishing: Melbourne, Australia, 2014; Volume 41, pp. 1191–1198.
67. Deery, D.; Jimenez-Berni, J.; Jones, H.; Sirault, X.; Furbank, R. Proximal remote sensing buggies and potential applications for field-based phenotyping. *Agronomy* **2014**, *4*, 349–379. [CrossRef]
68. White, J.W.; Conley, M.M. A flexible, low-cost cart for proximal sensing. *Crop Sci.* **2013**, *53*, 1646–1649. [CrossRef]
69. Saxena, K.; Saxena, R.K.; Varshney, R.K. Use of immature seed germination and single seed descent for rapid genetic gains in pigeonpea. *Plant Breed.* **2017**, *136*, 954–957. [CrossRef]
70. Shakoor, N.; Lee, S.; Mockler, T.C. High throughput phenotyping to accelerate crop breeding and monitoring of diseases in the field. *Curr. Opin. Plant Biol.* **2017**, *38*, 184–192. [CrossRef]
71. Janila, P.; Variath, M.T.; Pandey, M.K.; Desmae, H.; Motagi, B.N.; Okori, P.; Manohar, S.S.; Rathnakumar, A.L.; Radhakrishnan, T.; Liao, B.; et al. Genomic tools in groundnut breeding program: Status and perspectives. *Front. Plant Sci.* **2016**, *7*. [CrossRef]
72. Sarutayophat, T.; Nualsri, C. The efficiency of pedigree and single seed descent selections for yield improvement at generation 4 (F4) of two yardlong bean populations. *Kasetsart J. Nat. Sci.* **2010**, *44*, 343–352.
73. Holbrook, C.C.; Timper, P.; Culbreath, A.K.; Kvien, C.K. Registration of "Tifguard" Peanut. *J. Plant Regist.* **2008**, *2*, 92. [CrossRef]
74. Huang, X. From Genetic Mapping to Molecular Breeding: Genomics Have Paved the Highway. *Mol. Plant* **2016**. [CrossRef] [PubMed]
75. Jung, C. *Chapter 3 Molecular Tools for Plant Breeding*; Springer: New York, NY, USA, 2000; pp. 25–37.
76. Schaart, J.G.; van de Wiel, C.C.M.; Lotz, L.A.P.; Smulders, M.J.M. Opportunities for Products of New Plant Breeding Techniques. *Trends Plant Sci.* **2016**. [CrossRef] [PubMed]
77. Vilanova, S.; Cañizares, J.; Pascual, L.; Blanca, J.M.; Díez, M.J.; Prohens, J.; Picó, B. Application of Genomic Tools in Plant Breeding. *Curr. Genomics* **2012**, *13*, 179–195.
78. Wendler, N.; Mascher, M.; Nöh, C.; Himmelbach, A.; Scholz, U.; Ruge-Wehling, B.; Stein, N. Unlocking the secondary gene-pool of barley with next-generation sequencing. *Plant Biotechnol. J.* **2014**, *12*, 1122–1131. [CrossRef]
79. Metzker, M.L. Sequencing technologies—The next generation. *Nat. Rev. Genet.* **2010**, *11*, 31. [CrossRef]
80. Dwivedi, S.; Perotti, E.; Ortiz, R. Towards molecular breeding of reproductive traits in cereal crops. *Plant Biotechnol. J.* **2008**, *6*, 529–559. [CrossRef]
81. Wang, T.L.; Uauy, C.; Robson, F.; Till, B. TILLING in extremis. *Plant Biotechnol. J.* **2012**, *10*, 761–772. [CrossRef]

82. Zou, C.; Wang, P.; Xu, Y. Bulked sample analysis in genetics, genomics and crop improvement. *Plant Biotechnol. J.* **2016**, *14*, 1941–1955. [CrossRef]

83. Bolger, M.E.; Weisshaar, B.; Scholz, U.; Stein, N.; Usadel, B.; Mayer, K.F.X. Plant genome sequencing—Applications for crop improvement. *Curr. Opin. Biotechnol.* **2014**, *26*, 31–37. [CrossRef] [PubMed]

84. Edwards, D.; Batley, J. Plant genome sequencing: Applications for crop improvement. *Plant Biotechnol. J.* **2010**, *8*, 2–9. [CrossRef] [PubMed]

85. Dhingani, R.M.; Umrania, V.V.; Tomar, R.S.; Parakhia, M.V.; Golakiya, B. Introduction to QTL mapping in plants. *Ann. Plant Sci* **2015**, *4*, 1072–1079.

86. McCallum, C.M.; Comai, L.; Greene, E.A.; Henikoff, S. Targeted screening for induced mutations. *Nat. Biotechnol.* **2000**, *18*, 455. [CrossRef] [PubMed]

87. Nadeem, M.A.; Nawaz, M.A.; Shahid, M.Q.; Doğan, Y.; Comertpay, G.; Yıldız, M.; Hatipoğlu, R.; Ahmad, F.; Alsaleh, A.; Labhane, N.; et al. DNA molecular markers in plant breeding: Current status and recent advancements in genomic selection and genome editing. *Biotechnol. Biotechnol. Equip.* **2018**, *32*, 261–285. [CrossRef]

88. Lloyd, A.; Plaisier, C.L.; Carroll, D.; Drews, G.N. Targeted mutagenesis using zinc-finger nucleases in Arabidopsis. *Proc. Natl. Acad. Sci. USA* **2005**. [CrossRef]

89. Symington, L.S.; Gautier, J. Double-strand break end resection and repair pathway choice. *Annu. Rev. Genet.* **2011**, *45*, 247–271. [CrossRef]

90. Wood, A.J.; Lo, T.W.; Zeitler, B.; Pickle, C.S.; Ralston, E.J.; Lee, A.H.; Amora, R.; Miller, J.C.; Leung, E.; Meng, X.; et al. Targeted genome editing across species using ZFNs and TALENs. *Science* **2011**, *333*, 307. [CrossRef]

91. Sprink, T.; Metje, J.; Hartung, F. Plant genome editing by novel tools: TALEN and other sequence specific nucleases. *Curr. Opin. Biotechnol.* **2015**. [CrossRef]

92. Mao, Y.; Zhang, H.; Xu, N.; Zhang, B.; Gou, F.; Zhu, J.K. Application of the CRISPR-Cas system for efficient genome engineering in plants. *Mol. Plant* **2013**. [CrossRef]

93. Schneider, K.; Schiermeyer, A.; Dolls, A.; Koch, N.; Herwartz, D.; Kirchhoff, J.; Fischer, R.; Russell, S.M.; Cao, Z.; Corbin, D.R. Targeted gene exchange in plant cells mediated by a zinc finger nuclease double cut. *Plant Biotechnol. J.* **2016**, *14*, 1151–1160. [CrossRef] [PubMed]

94. De Pater, S.; Pinas, J.E.; Hooykaas, P.J.J.; van der Zaal, B.J. ZFN-mediated gene targeting of the Arabidopsis protoporphyrinogen oxidase gene through Agrobacterium-mediated floral dip transformation. *Plant Biotechnol. J.* **2013**, *11*, 510–515. [CrossRef] [PubMed]

95. Li, T.; Huang, S.; Jiang, W.Z.; Wright, D.; Spalding, M.H.; Weeks, D.P.; Yang, B. TAL nucleases (TALNs): Hybrid proteins composed of TAL effectors and FokI DNA-cleavage domain. *Nucleic Acids Res.* **2010**, *39*, 359–372. [CrossRef] [PubMed]

96. Char, S.N.; Unger-Wallace, E.; Frame, B.; Briggs, S.A.; Main, M.; Spalding, M.H.; Vollbrecht, E.; Wang, K.; Yang, B. Heritable site-specific mutagenesis using TALENs in maize. *Plant Biotechnol. J.* **2015**, *13*, 1002–1010. [CrossRef]

97. Mahfouz, M.M.; Piatek, A.; Stewart, C.N. Genome engineering via TALENs and CRISPR/Cas9 systems: Challenges and perspectives. *Plant Biotechnol. J.* **2014**, *12*, 1006–1014. [CrossRef]

98. Zhang, Y.; Xie, X.; Liu, Y.G.; Zhang, Y.; Xie, X.; Liu, Y.G.; Ma, X. *CRISPR/Cas9-Based Genome Editing in Plants*, 1st ed; Elsevier Inc.: Amsterdam, The Netherlands, 2017; Volume 149.

99. Zaidi, S.S.-e.-A.; Mahfouz, M.M.; Mansoor, S. CRISPR-Cpf1: A New Tool for Plant Genome Editing. *Trends Plant Sci.* **2017**, *22*, 550–553. [CrossRef]

100. Li, S.; Zhang, X.; Wang, W.; Guo, X.; Wu, Z.; Du, W.; Zhao, Y.; Xia, L. Expanding the scope of CRISPR/Cpf1-mediated genome editing in rice. *Mol. Plant* **2018**, *11*, 995–998. [CrossRef]

101. Qin, L.; Li, J.; Wang, Q.; Xu, Z.; Sun, L.; Alariqi, M.; Manghwar, H.; Wang, G.; Li, B.; Ding, X.; et al. High Efficient and Precise Base Editing of C•G to T•A in the Allotetraploid Cotton (*Gossypium hirsutum*) Genome Using a Modified CRISPR /Cas9 System. *Plant Biotechnol. J.* **2020**, *18*, 45–56. [CrossRef]

102. Bortesi, L.; Zhu, C.; Zischewski, J.; Perez, L.; Bassié, L.; Nadi, R.; Forni, G.; Lade, S.B.; Soto, E.; Jin, X.; et al. Patterns of CRISPR/Cas9 activity in plants, animals and microbes. *Plant Biotechnol. J.* **2016**, *14*, 2203–2216. [CrossRef]

103. Lee, K.; Zhang, Y.; Kleinstiver, B.P.; Guo, J.A.; Aryee, M.J.; Miller, J.; Malzahn, A.; Zarecor, S.; Lawrence-Dill, C.J.; Joung, J.K.; et al. Activities and specificities of CRISPR/Cas9 and Cas12a nucleases for targeted mutagenesis in maize. *Plant Biotechnol. J.* **2019**, *17*, 362–372. [CrossRef]

104. Zhang, Y.; Liang, Z.; Zong, Y.; Wang, Y.; Liu, J.; Chen, K.; Qiu, J.L.; Gao, C. Efficient and transgene-free genome editing in wheat through transient expression of CRISPR/Cas9 DNA or RNA. *Nat. Commun.* **2016**. [CrossRef] [PubMed]

105. Jinek, M.; Chylinski, K.; Fonfara, I.; Hauer, M.; Doudna, J.A.; Charpentier, E. A programmable dual-RNA–guided DNA endonuclease in adaptive bacterial immunity. *Science (80-.).* **2012**, *337*, 816–821. [CrossRef]

106. Wang, X.; Tu, M.; Wang, D.; Liu, J.; Li, Y.; Li, Z.; Wang, Y.; Wang, X. CRISPR/Cas9-mediated efficient targeted mutagenesis in grape in the first generation. *Plant Biotechnol. J.* **2018**, *16*, 844–855. [CrossRef] [PubMed]

107. Murugan, K.; Babu, K.; Sundaresan, R.; Rajan, R.; Sashital, D.G. The Revolution Continues: Newly Discovered Systems Expand the CRISPR-Cas Toolkit. *Mol. Cell* **2017**, *68*, 15–25. [CrossRef]

108. Ma, X.; Chen, X.; Jin, Y.; Ge, W.; Wang, W.; Kong, L.; Ji, J.; Guo, X.; Huang, J.; Feng, X.H.; et al. Small molecules promote CRISPR-Cpf1-mediated genome editing in human pluripotent stem cells. *Nat. Commun.* **2018**. [CrossRef]

109. Riesenberg, S.; Maricic, T. Targeting repair pathways with small molecules increases precise genome editing in pluripotent stem cells. *Nat. Commun.* **2018**. [CrossRef]

110. Yang, F.; Li, Y. The new generation tool for CRISPR genome editing: CRISPR/Cpf1. *Sheng wu gong cheng xue bao= Chinese J. Biotechnol.* **2017**, *33*, 361–371.

111. Maruyama, T.; Dougan, S.K.; Truttmann, M.C.; Bilate, A.M.; Ingram, J.R.; Ploegh, H.L. Corrigendum: Increasing the efficiency of precise genome editing with CRISPR-Cas9 by inhibition of nonhomologous end joining. *Nat. Biotechnol.* **2016**, *34*, 210. [CrossRef] [PubMed]

112. Ishii, T.; Araki, M. A future scenario of the global regulatory landscape regarding genome-edited crops. *GM Crop. Food* **2017**, *8*, 44–56. [CrossRef] [PubMed]

113. Wolt, J.D.; Wang, K.; Yang, B. The Regulatory Status of Genome-edited Crops. *Plant Biotechnol. J.* **2016**, *14*, 510–518. [CrossRef]

114. Callaway, E. CRISPR plants now subject to tough GM laws in European Union. *Nature* **2018**, *560*, 16. [CrossRef] [PubMed]

115. Sprink, T.; Eriksson, D.; Schiemann, J.; Hartung, F. Regulatory hurdles for genome editing: Process- vs. product-based approaches in different regulatory contexts. *Plant Cell Rep.* **2016**, *35*, 1493–1506. [CrossRef] [PubMed]

116. Slama-Ayed, O.; Bouhaouel, I.; Ayed, S.; De Buyser, J.; Picard, E.; Amara, H.S. Efficiency of three haplomethods in durum wheat (Triticum turgidum subsp. durum Desf.): Isolated microspore culture, gynogenesis and wheat× maize crosses. *Czech J. Genet. Plant Breed.* **2019**, *55*, 101–109. [CrossRef]

117. Chiurugwi, T.; Kemp, S.; Powell, W.; Hickey, L.T.; Powell, W. Speed breeding orphan crops. *Theor. Appl. Genet.* **2018**. [CrossRef] [PubMed]

118. Weller, J.L.; Beauchamp, N.; Kerckhoffs, L.H.J.; Platten, J.D.; Reid, J.B. Interaction of phytochromes A and B in the control of de-etiolation and flowering in pea. *Plant J.* **2001**, *26*, 283–294. [CrossRef] [PubMed]

119. Giliberto, L.; Perrotta, G.; Pallara, P.; Weller, J.L.; Fraser, P.D.; Bramley, P.M.; Fiore, A.; Tavazza, M.; Giuliano, G. Manipulation of the blue light photoreceptor cryptochrome 2 in tomato affects vegetative development, flowering time, and fruit antioxidant content. *Plant Physiol.* **2005**, *137*, 199–208. [CrossRef]

120. Croser, J.S.; Pazos-Navarro, M.; Bennett, R.G.; Tschirren, S.; Edwards, K.; Erskine, W.; Creasy, R.; Ribalta, F.M. Time to flowering of temperate pulses in vivo and generation turnover in vivo–in vitro of narrow-leaf lupin accelerated by low red to far-red ratio and high intensity in the far-red region. *Plant Cell. Tissue Organ Cult.* **2016**, *127*, 591–599. [CrossRef]

121. Ribalta, F.M.; Pazos-Navarro, M.; Nelson, K.; Edwards, K.; Ross, J.J.; Bennett, R.G.; Munday, C.; Erskine, W.; Ochatt, S.J.; Croser, J.S. Precocious floral initiation and identification of exact timing of embryo physiological maturity facilitate germination of immature seeds to truncate the lifecycle of pea. *Plant Growth Regul.* **2017**, *81*, 345–353. [CrossRef]

122. Moe, R.; Heins, R. Control of plant morphogenesis and flowering by light quality and temperature. *Acta Hortic.* **1990**, 81–90. [CrossRef]

123. Ausín, I.; Alonso-Blanco, C.; Martínez-Zapater, J.M. Environmental regulation of flowering. *Int. J. Dev. Biol.* **2005**, *49*, 689–705. [CrossRef]

124. Ochatt, S.J.; Sangwan, R.S.; Marget, P.; Assoumou Ndong, Y.; Rancillac, M.; Perney, P. New approaches towards the shortening of generation cycles for faster breeding of protein legumes. *Plant Breed.* **2002**, *121*, 436–440. [CrossRef]

125. Ochatt, S.J.; Sangwan, R.S. In vitro shortening of generation time in Arabidopsis thaliana. *Plant Cell. Tissue Organ Cult.* **2008**, *93*, 133–137. [CrossRef]

126. Heuschele, D.J.; Case, A.; Smith, K.P. Evaluation of Fast Generation Cycling in Oat (*Avena sativa*). *Cereal Res. Commun.* **2019**, *47*, 626–635. [CrossRef]

127. Yao, Y.; Zhang, P.; Liu, H.; Lu, Z.; Yan, G. A fully in vitro protocol towards large scale production of recombinant inbred lines in wheat (*Triticum aestivum* L.). *Plant Cell. Tissue Organ Cult.* **2017**, *128*, 655–661. [CrossRef]

128. Sysoeva, M.I.; Markovskaya, E.F.; Shibaeva, T.G. Plants under continuous light: A review. *Plant Stress* **2010**, *4*, 5–17.

129. Achigan-Dako, E.G.; Sogbohossou, O.E.; Maundu, P. Current knowledge on *Amaranthus* spp.: research avenues for improved nutritional value and yield in leafy amaranths in sub-Saharan Africa. *Euphytica* **2014**, *197*, 303–317. [CrossRef]

130. Collard, B.C.Y.; Beredo, J.C.; Lenaerts, B.; Mendoza, R.; Santelices, R.; Lopena, V.; Verdeprado, H.; Raghavan, C.; Gregorio, G.B.; Vial, L.; et al. Revisiting rice breeding methods–evaluating the use of rapid generation advance (RGA) for routine rice breeding. *Plant Prod. Sci.* **2017**, *20*, 337–352. [CrossRef]

131. Wolter, F.; Schindele, P.; Puchta, H. Plant breeding at the speed of light: The power of CRISPR/Cas to generate directed genetic diversity at multiple sites. *BMC Plant Biol.* **2019**, *19*, 1–8. [CrossRef]

132. Bartley, G. Wheat (*Triticum aestivum*) residue management before growing soybean (*Glycine max*) in Manitoba. Master's Thesis, Department of Plant Science, University of Manitoba, Winnipeg, Manitoba, 2019.

133. Wilton, M. A Broad-Scale Characterization of Corn (*Zea mays*)-Soybean (*Glycine max*) Intercropping as a Sustainable-Intensive Cropping Practice. Ph.D. Thesis, University of Waterloo, Ontario, Canada, 2019.

134. Borlaug, N.E. Sixty-two years of fighting hunger: Personal recollections. *Euphytica* **2007**, *157*, 287–297. [CrossRef]

135. Ferrie, A.M.R.; Möllers, C. Haploids and doubled haploids in Brassica spp. for genetic and genomic research. *Plant Cell Tissue Organ. Cult.* **2011**, *104*, 375–386. [CrossRef]

136. Lübberstedt, T.; Frei, U.K. Application of doubled haploids for target gene fixation in backcross programmes of maize. *Plant Breed.* **2012**, *131*, 449–452. [CrossRef]

137. Dirks, R.; Van Dun, K.; De Snoo, C.B.; Van Den Berg, M.; Lelivelt, C.L.C.; Voermans, W.; Woudenberg, L.; De Wit, J.P.C.; Reinink, K.; Schut, J.W.; et al. Reverse breeding: A novel breeding approach based on engineered meiosis. *Plant Biotechnol. J.* **2009**, *7*, 837–845. [CrossRef] [PubMed]

138. Kebede, A.Z.; Dhillon, B.S.; Schipprack, W.; Araus, J.L.; Bänziger, M.; Semagn, K.; Alvarado, G.; Melchinger, A.E. Effect of source germplasm and season on the in vivo haploid induction rate in tropical maize. *Euphytica* **2011**, *180*, 219–226. [CrossRef]

139. Castillo, A.M.; Cistué, L.; Vallés, M.P.; Soriano, M. Chromosome Doubling in Monocots. In *Advances in Haploid Production in Higher Plants*; Springer: New York, NY, USA, 2009; pp. 329–338.

140. Prigge, V.; Melchinger, A.E. Production of haploids and doubled haploids in maize. In *Plant Cell Culture Protocols*; Springer: New York, NY, USA, 2012; pp. 161–172.

141. Raju, S.K.K.; Shao, M.R.; Sanchez, R.; Xu, Y.Z.; Sandhu, A.; Graef, G.; Mackenzie, S. An epigenetic breeding system in soybean for increased yield and stability. *Plant Biotechnol. J.* **2018**, *16*, 1836–1847. [CrossRef]

142. Halpin, C. Gene stacking in transgenic plants—The challenge for 21st century plant biotechnology. *Plant Biotechnol. J.* **2005**, *3*, 141–155. [CrossRef]

143. Belhaj, K.; Chaparro-Garcia, A.; Kamoun, S.; Nekrasov, V. Plant genome editing made easy: Targeted mutagenesis in model and crop plants using the CRISPR/Cas system. *Plant Methods* **2013**. [CrossRef]

144. Low, J.W.; Mwanga, R.O.M.; Andrade, M.; Carey, E.; Ball, A.-M. Tackling vitamin A deficiency with biofortified sweetpotato in sub-Saharan Africa. *Glob. Food Sec.* **2017**, *14*, 23–30. [CrossRef]

145. Møller, I.S.; Gilliham, M.; Jha, D.; Mayo, G.M.; Roy, S.J.; Coates, J.C.; Haseloff, J.; Tester, M. Shoot Na+ exclusion and increased salinity tolerance engineered by cell type–specific alteration of Na+ transport in Arabidopsis. *Plant Cell* **2009**, *21*, 2163–2178. [CrossRef]

146. Singh, N.K.; Gupta, D.K.; Jayaswal, P.K.; Mahato, A.K.; Dutta, S.; Singh, S.; Bhutani, S.; Dogra, V.; Singh, B.P.; Kumawat, G.; et al. The first draft of the pigeonpea genome sequence. *J. Plant Biochem. Biotechnol.* **2012**, *21*, 98–112. [CrossRef]

147. Jackson, S.A. Rice: The First Crop Genome. *Rice* **2016**, 9. [CrossRef]

148. Egan, A.N.; Schlueter, J.; Spooner, D.M. Applications of next-generation sequencing in plant biology. *Am. J. Bot.* **2012**, *99*, 175–185. [CrossRef] [PubMed]

149. Bernier, G.; Périlleux, C. A physiological overview of the genetics of flowering time control. *Plant Biotechnol. J.* **2005**, *3*, 3–16. [CrossRef] [PubMed]

150. Kondić-špika, A.; Kobiljski, B. Biotechnology in Modern Breeding and Agriculture. In Proceedings of the International Conference on BioScience: Biotechnology and Biodiversity-Step in the Future. The Fourth Joint UNS-PSU Conference, Novi Sad, Serbia, 18–20 June 2012.

151. Liang, Z.; Chen, K.; Li, T.; Zhang, Y.; Wang, Y.; Zhao, Q.; Liu, J.; Zhang, H.; Liu, C.; Ran, Y.; et al. Efficient DNA-free genome editing of bread wheat using CRISPR/Cas9 ribonucleoprotein complexes. *Nat. Commun.* **2017**. [CrossRef] [PubMed]

152. Hedden, P. The genes of the Green Revolution. *TRENDS Genet.* **2003**, *19*, 5–9. [CrossRef]

153. Knoch, D.; Abbadi, A.; Grandke, F.; Meyer, R.C.; Samans, B.; Werner, C.R.; Snowdon, R.J.; Altmann, T. Strong temporal dynamics of QTL action on plant growth progression revealed through high-throughput phenotyping in canola. *Plant Biotechnol. J.* **2020**, *18*, 68–82. [CrossRef]

154. Yin, K.; Gao, C.; Qiu, J.L. Progress and prospects in plant genome editing. *Nat. Plants* **2017**, *3*, 1–6. [CrossRef]

155. Waltz, E. *CRISPR-Edited Crops Free to Enter Market, Skip Regulation*; Nature Publishing Group: Berlin, Germany, 2016.

156. Eş, I.; Gavahian, M.; Marti-Quijal, F.J.; Lorenzo, J.M.; Mousavi Khaneghah, A.; Tsatsanis, C.; Kampranis, S.C.; Barba, F.J. The application of the CRISPR-Cas9 genome editing machinery in food and agricultural science: Current status, future perspectives, and associated challenges. *Biotechnol. Adv.* **2019**. [CrossRef]

157. Xu, R.; Yang, Y.; Qin, R.; Li, H.; Qiu, C.; Li, L.; Wei, P.; Yang, J. Rapid improvement of grain weight via highly efficient CRISPR/Cas9-mediated multiplex genome editing in rice. *J. Genet. Genomics* **2016**, *43*, 529–532. [CrossRef]

158. Zhang, J.; Zhang, H.; Botella, J.R.; Zhu, J. Generation of new glutinous rice by CRISPR/Cas9-targeted mutagenesis of the Waxy gene in elite rice varieties. *J. Integr. Plant Biol.* **2018**, *60*, 369–375. [CrossRef]

159. Nekrasov, V.; Wang, C.; Win, J.; Lanz, C.; Weigel, D.; Kamoun, S. Rapid generation of a transgene-free powdery mildew resistant tomato by genome deletion. *Sci. Rep.* **2017**. [CrossRef]

160. Wang, F.; Wang, C.; Liu, P.; Lei, C.; Hao, W.; Gao, Y.; Liu, Y.G.; Zhao, K. Enhanced rice blast resistance by CRISPR/ Cas9-Targeted mutagenesis of the ERF transcription factor gene OsERF922. *PLoS ONE* **2016**. [CrossRef] [PubMed]

161. Kumar, M.; Aslam, M.; Manisha, Y.; Manoj, N. An Update on Genetic Modification of Chickpea for Increased Yield and Stress Tolerance. *Mol. Biotechnol.* **2018**, *60*, 651–663. [CrossRef] [PubMed]

162. Domoney, C.; Knox, M.; Moreau, C.; Ambrose, M.; Palmer, S.; Smith, P.; Christodoulou, V.; Isaac, P.G.; Hegarty, M.; Blackmore, T.; et al. Exploiting a fast neutron mutant genetic resource in Pisum sativum (pea) for functional genomics. *Funct. Plant Biol.* **2013**, *40*, 1261. [CrossRef]

163. Raman, H.; Raman, R.; Kilian, A.; Detering, F.; Carling, J.; Coombes, N.; Diffey, S.; Kadkol, G.; Edwards, D.; Mccully, M.; et al. Genome-Wide Delineation of Natural Variation for Pod Shatter Resistance in Brassica napus. *PLoS ONE* **2014**, *9*, e101673. [CrossRef] [PubMed]

164. Panigrahi, R.; Kariali, E.; Panda, B.; Lafarge, T.; Mohapatra, P.K. Controlling the trade-off between spikelet number and grain filling; the hierarchy of starch synthesis in spikelets of rice panicle in relation to hormone dynamics. *Funct. Plant Biol.* **2019**, *46*, 507–523. [CrossRef] [PubMed]

165. Witcombe, J.R.; Gyawali, S.; Subedi, M.; Virk, D.S.; Joshi, K.D. Plant breeding can be made more efficient by having fewer, better crosses. *BMC Plant Biol.* **2013**, *13*, 22. [CrossRef]

166. Basnet, B.R.; Crossa, J.; Dreisigacker, S.; Perez-Rodriguez, P.; Manes, Y.; Singh, R.P.; Rosyara, U.R.; Camarillo-Castillo, F.; Murua, M. Hybrid Wheat Prediction Using Genomic, Pedigree, and Environmental Covariables Interaction Models. *Plant Genome* **2019**, *12*. [CrossRef]

167. Xu, Y.; Babu, R.; Skinner, D.J.; Vivek, B.S.; Crouch, J.H. Maize Mutant opaque2 and the Improvement of Protein Quality through Conventional and Molecular Approaches. In Proceedings of the International Symposium on Induced Mutation in Plants, Vienna, Austria, 2–15 August 2008.

168. Shuwen, S.; Lianghong, L.; Jiangsheng, W.; Yongming, Z. In vitro screening stem rot resistant (tolerant) materials in Brassica napus L. *Chin. J. Oil Crop Sci.* **2003**, *25*, 5–8.

169. Pathirana, R. Plant mutation breeding in agriculture. *Plant Sci. Rev.* **2011**, 107–126. [CrossRef]

170. International Atomic Energy Agency. Proceedings of the International Symposium on Plant Mutation Breeding and Biotechnology, Vienna, Austria, 27–31 August 2018.

171. Ceballos, H.; Sanchez, T.; Denyer, K.; Tofino, A.P.; Rosero, E.A.; Dufour, D.; Smith, A.; Morante, N.; Perez, J.C.; Fahy, B. Induction and identification of a small-granule, high-amylose mutant in cassava (Manihot esculenta Crantz). *J. Agric. Food Chem.* **2008**, *56*, 7215–7222. [CrossRef]

172. Hamid, M.A.; Azad, M.A.K.; Howelider, M.A.R. Development of Three Groundnut Varieties with Improved Quantitative and Qualitative Traits through Induced Mutation. *Plant Mutat. reports* **2006**, *1*, 14–16.

173. Newell-McGloughlin, M. Nutritionally improved agricultural crops. *Plant Physiol.* **2008**, *147*, 939–953. [CrossRef] [PubMed]

174. Dunwell, J.M. Transgenic approaches to crop improvement. *J. Exp. Bot.* **2000**, *51*, 487–496. [CrossRef] [PubMed]

175. Toojinda, T.; Baird, E.; Booth, A.; Broers, L.; Hayes, P.; Powell, W.; Thomas, W.; Vivar, H.; Young, G. Introgression of quantitative trait loci (QTLs) determining stripe rust resistance in barley: An example of marker-assisted line development. *Theor. Appl. Genet.* **1998**, *96*, 123–131. [CrossRef]

176. Gimode, W.; Clevenger, J.; McGregor, C. Fine-mapping of a major quantitative trait locus Qdff3-1 controlling flowering time in watermelon. *Mol. Breed.* **2020**, *40*, 1–12. [CrossRef]

177. Joshi, S.G.; Schaart, J.G.; Groenwold, R.; Jacobsen, E.; Schouten, H.J.; Krens, F.A. Functional analysis and expression profiling of HcrVf1 and HcrVf2 for development of scab resistant cisgenic and intragenic apples. *Plant Mol. Biol.* **2011**, *75*, 579–591. [CrossRef]

178. Würdig, J.; Flachowsky, H.; Saß, A.; Peil, A.; Hanke, M.V. Improving resistance of different apple cultivars using the Rvi6 scab resistance gene in a cisgenic approach based on the Flp/FRT recombinase system. *Mol. Breed.* **2015**, *35*. [CrossRef]

179. Holme, I.B.; Wendt, T.; Holm, P.B. Intragenesis and cisgenesis as alternatives to transgenic crop development. *Plant Biotechnol. J.* **2013**, *11*, 395–407. [CrossRef]

180. Gadaleta, A.; Giancaspro, A.; Blechl, A.E.; Blanco, A. A transgenic durum wheat line that is free of marker genes and expresses 1Dy10. *J. Cereal Sci.* **2008**, *48*, 439–445. [CrossRef]

181. Cardi, T. Cisgenesis and genome editing: Combining concepts and efforts for a smarter use of genetic resources in crop breeding. *Plant Breed.* **2016**, *135*, 139–147. [CrossRef]

182. Jo, K.-R.; Kim, C.-J.; Kim, S.-J.; Kim, T.-Y.; Bergervoet, M.; Jongsma, M.A.; Visser, R.G.F.; Jacobsen, E.; Vossen, J.H. Development of late blight resistant potatoes by cisgene stacking. *BMC Biotechnol.* **2014**, *14*, 50. [CrossRef] [PubMed]

183. Schaart, J.G. *Towards Consumer-Friendly Cisgenic Strawberries which are Less Susceptible to Botrytis Cinerea*; Research Gate: Berlin, Germany, 2004; ISBN 908504104X.

184. Sawai, S.; Ohyama, K.; Yasumoto, S.; Seki, H.; Sakuma, T.; Yamamoto, T.; Takebayashi, Y.; Kojima, M.; Sakakibara, H.; Aoki, T.; et al. Sterol side chain reductase 2 is a key enzyme in the biosynthesis of cholesterol, the common precursor of toxic steroidal glycoalkaloids in potato. *Plant Cell* **2014**, *26*, 3763–3774. [CrossRef] [PubMed]

185. Luo, M.; Li, H.; Chakraborty, S.; Morbitzer, R.; Rinaldo, A.; Upadhyaya, N.; Bhatt, D.; Louis, S.; Richardson, T.; Lahaye, T.; et al. Efficient TALEN mediated gene editing in wheat. *Plant Biotechnol. J.* **2019**, *17*, 2026–2028. [CrossRef] [PubMed]

186. Zhou, J.; Peng, Z.; Long, J.; Sosso, D.; Liu, B.; Eom, J.-S.; Huang, S.; Liu, S.; Vera Cruz, C.; Frommer, W.B.; et al. Gene targeting by the TAL effector PthXo2 reveals cryptic resistance gene for bacterial blight of rice. *Plant J.* **2015**, *82*, 632–643. [CrossRef] [PubMed]

187. Svitashev, S.; Young, J.K.; Schwartz, C.; Gao, H.; Falco, S.C.; Cigan, A.M. Targeted Mutagenesis, Precise Gene Editing, and Site-Specific Gene Insertion in Maize Using Cas9 and Guide RNA. *Plant Physiol.* **2015**, *169*, 931–945. [CrossRef] [PubMed]

188. Zhou, X.; Jacobs, T.B.; Xue, L.-J.; Harding, S.A.; Tsai, C.-J. Exploiting SNPs for biallelic CRISPR mutations in the outcrossing woody perennial Populus reveals 4-coumarate:CoA ligase specificity and redundancy. *New Phytol.* **2015**, *208*, 298–301. [CrossRef] [PubMed]

189. Brooks, C.; Nekrasov, V.; Lippman, Z.B.; Van Eck, J. Efficient gene editing in tomato in the first generation using the clustered regularly interspaced short palindromic repeats/CRISPR-associated9 system. *Plant Physiol.* **2014**, *166*, 1292–1297. [CrossRef]

190. Li, Z.; Liu, Z.-B.; Xing, A.; Moon, B.P.; Koellhoffer, J.P.; Huang, L.; Ward, R.T.; Clifton, E.; Falco, S.C.; Cigan, A.M. Cas9-Guide RNA Directed Genome Editing in Soybean. *Plant Physiol.* **2015**, *169*, 960–970. [CrossRef]

191. Zhao, X.; Meng, Z.; Wang, Y.; Chen, W.; Sun, C.; Cui, B.; Cui, J.; Yu, M.; Zeng, Z.; Guo, S.; et al. Pollen magnetofection for genetic modification with magnetic nanoparticles as gene carriers. *Nat. Plants* **2017**, *3*, 956–964. [CrossRef]

192. Han, J.; Guo, B.; Guo, Y.; Zhang, B.; Wang, X.; Qiu, L.-J. Creation of Early Flowering Germplasm of Soybean by CRISPR/Cas9 Technology. *Front. Plant Sci.* **2019**, *10*, 1–10. [CrossRef]

193. Sánchez-León, S.; Gil-Humanes, J.; Ozuna, C.V.; Giménez, M.J.; Sousa, C.; Voytas, D.F.; Barro, F. Low-gluten, nontransgenic wheat engineered with CRISPR/Cas9. *Plant Biotechnol. J.* **2018**, *16*, 902–910. [CrossRef] [PubMed]

194. Farhat, S.; Jain, N.; Singh, N.; Sreevathsa, R.; Das, P.K.; Rai, R.; Yadav, S.; Kumar, P.; Sarkar, A.; Jain, A. CRISPR-cas 9 directed genome engineering for enhancing salt stress tolerance in rice. In *Proceedings of the Seminars in Cell & Developmental Biology*; Elsevier: Amsterdam, The Netherlands, 2019.

195. Zaman, Q.U.; Chu, W.; Hao, M.; Shi, Y.; Sun, M.; Sang, S.-F.; Mei, D.; Cheng, H.; Liu, J.; Li, C. CRISPR/Cas9-Mediated Multiplex Genome Editing of JAGGED Gene in Brassica napus L. *Biomolecules* **2019**, *9*, 725. [CrossRef] [PubMed]

196. Veillet, F.; Perrot, L.; Chauvin, L.; Kermarrec, M.-P.; Guyon-Debast, A.; Chauvin, J.-E.; Nogué, F.; Mazier, M. Transgene-free genome editing in tomato and potato plants using agrobacterium-mediated delivery of a CRISPR/Cas9 cytidine base editor. *Int. J. Mol. Sci.* **2019**, *20*, 402. [CrossRef] [PubMed]

197. Hsu, C.-T.; Cheng, Y.-J.; Yuan, Y.-H.; Hung, W.-F.; Cheng, Q.-W.; Wu, F.-H.; Lee, L.-Y.; Gelvin, S.B.; Lin, C.-S. Application of Cas12a and nCas9-activation-induced cytidine deaminase for genome editing and as a non-sexual strategy to generate homozygous/multiplex edited plants in the allotetraploid genome of tobacco. *Plant Mol. Biol.* **2019**. [CrossRef]

198. Yin, X.; Anand, A.; Quick, P.; Bandyopadhyay, A. Editing a Stomatal Developmental Gene in Rice with CRISPR/Cpf1. In *Plant Genome Editing with CRISPR Systems*; Springer: New York, NY, USA, 2019; pp. 257–268.

199. Malzahn, A.A.; Tang, X.; Lee, K.; Ren, Q.; Sretenovic, S.; Zhang, Y.; Chen, H.; Kang, M.; Bao, Y.; Zheng, X.; et al. Application of CRISPR-Cas12a temperature sensitivity for improved genome editing in rice, maize, and Arabidopsis. *BMC Biol.* **2019**, *17*, 1–14. [CrossRef]

200. Oconnor, D.; Wright, G.; George, D.; Hunter, M. Development and Application of Speed Breeding Technologies in a Commercial Peanut Breeding Program Development and Application of Speed Breeding Technologies in a Commercial Peanut Breeding Program. *Peanut Sci.* **2013**, *40*, 107–114. [CrossRef]

201. Riaz, A. Unlocking new sources of adult plant resistance to wheat leaf rust. Ph.D. Thesis, The University of Queensland, Queensland, Australia, 2018; pp. 1–241.

202. Wang, X.; Xuan, H.; Evers, B.; Shrestha, S.; Pless, R.; Poland, J. High-throughput phenotyping with deep learning gives insight into the genetic architecture of flowering time in wheat. *bioRxiv* **2019**, 527911.

203. Danzi, D.; Briglia, N.; Petrozza, A.; Summerer, S.; Povero, G.; Stivaletta, A.; Cellini, F.; Pignone, D.; de Paola, D.; Janni, M. Can high throughput phenotyping help food security in the mediterranean area? *Front. Plant. Sci.* **2019**, *10*, 15.

2

Linkage Map of a Gene Controlling Zero Tannins (*zt-1*) in Faba Bean (*Vicia faba* L.) with SSR and ISSR Markers

Wanwei Hou [1,†], Xiaojuan Zhang [2,†], Qingbiao Yan [1], Ping Li [1], Weichao Sha [1], Yingying Tian [1] and Yujiao Liu [1,*]

[1] Qinghai Academy of Agricultural and Forestry Sciences, State Key Laboratory of Plateau Ecology and Agriculture and Qinghai Research Station of Crop Gene Resource & Germplasm Enhancement, Qinghai University, Xining 810016, Qinghai, China; houwanwei333@163.com (W.H.); Yqb2001@hotmail.com (Q.Y.); pinger44@km169.net (P.L.); shaweichao@126.com (W.S.); 17809711447@163.com (Y.T.)

[2] State Key Laboratory of Plateau Ecology and Agriculture and College of Eco-Environmental Engineering, Qinghai University, Xining 810016, Qinghai, China; xiaojuan830136@163.com

* Correspondence: 13997058356@163.com

† These authors contributed equally to this work.

Abstract: Faba bean (*Vicia faba* L.), a partially allogamous species, is rich in protein. Condensed tannins limit the use of faba beans as food and feed. Two recessive genes, *zt-1* and *zt-2*, control the zero tannin content in faba bean and promote a white flower phenotype. To determine the inheritance and develop a linkage map for the *zt-1* gene in the faba bean germplasm M3290, F_2 and F_3 progenies were derived from the purple flower and high tannin content genotypes Qinghai12 and *zt-1* line M3290, respectively. Genetic analysis verified a single recessive gene for zero tannin content and flower colour. In total, 596 SSR markers and 100 ISSR markers were used to test the polymorphisms between the parents and bulks for the contrasting flower colour via Bulked Segregant Analysis (BSA). Subsequently, six SSR markers and seven ISSR markers were used to genotype the entire 413 F_2 population. Linkage analysis showed that the *zt-1* gene was closely linked to the SSR markers SSR84 and M78, with genetic distances of 2.9 and 5.8 cM, respectively. The two flanked SSR markers were used to test 34 faba bean genotypes with different flower colours. The closely linked SSR marker SSR84 predicted the *zt-1* genotypes with absolute accuracy. The results from the marker-assisted selection (MAS) from this study could provide a solid foundation for further faba bean breeding programmes.

Keywords: faba bean; *zt-1*; linkage map; SSR; ISSR

1. Introduction

Faba bean (*Vicia faba* L.), one of the most important temperate food crops, is widely grown for human consumption in China, Ethiopia, Egypt and the Andean States of South America and for livestock feed in Europe and Australia [1]. To date, the average faba bean cultivation area is close to 2.5 million hectares annually, which ranks fourth among cool-season crops [2]. It has been demonstrated that growing faba bean is the most effective strategy for managing soil fertility through crop rotation, which contributes to sustainable agriculture [3].

Faba bean seeds together with other relative beans, have high nutritional values as they are excellent sources of protein, carbohydrates, minerals and fibre [4]. Nevertheless, faba bean also suffer from both biotic and abiotic factors that constrain their productivity and digestibility. Previous studies have demonstrated that condensed tannins are responsible for low-protein seeds and may decrease feed consumption due to their astringent taste [5]. Condensed tannins from faba bean may also decrease the efficiency of food utilization [6,7]. Although several methods, such as cooking and autoclaving, have been used to remove condensed tannins, these processes may also promote other changes in the seed compounds. Meanwhile, a significant problem in tannin removal is the high cost [8]. Compared to traditional methods, growing cultivars with low-tannin and zero-tannin content are the most effective, economic and environment-friendly strategy.

A previous study first reported the absence of tannins in the white flowers of faba bean varieties; this served an important role in the in vitro digestibility of nutrients in monogastric animals [9]. According to Picard [9], there are two inherited recessive genes, zt-1 and zt-2, that control the zero-tannin characteristic in faba bean and promote a white flower characteristic in the plant. Genetic studies also discovered that the genes in faba bean that control white-flowered plants actually block anthocyanin synthesis [10,11]. Breeders usually use crosses between intergeneric and interspecific plants to improve their characteristics. However, faba bean genotype hybrids carrying different zero tannin genes generally give rise to segregating progenies. Therefore, identifying varieties with zero tannins will be helpful for both choosing appropriate crosses for breeders [12] and representing a reservoir of genes for tannin-free plants. To date, great progress has been made in developing faba beans with zero tannins. Several markers have been mapped to the zt-1 region. The number of markers is still limited, and more markers are needed to fill the gaps for more efficient marker-assisted selection, further fine mapping and map-based cloning of the gene.

Faba bean, a partially allogamous and genetically isolated plant, tolerates no exchange of genes with any other species, including its close relative *Vicia narbonensis* [13]. The perception is that genetic mapping and marker-assisted selection (MAS) in faba bean faces enormous challenges because of its huge genome size (13,000 Mb) [14,15], even though faba bean is diploid and has fewer chromosomes ($2n = 2x = 12$) than other species in the genus *Vicia* L. [16].

Various molecular markers have been widely used in faba bean, especially in genetic diversity and relationships among germplasm collections. For example, amplified fragment length polymorphism (AFLP) markers were used to assess the genetic diversity in 22 recent faba bean elite cultivars [17]. Zong et al. [18] subsequently analysed winter and spring [19] faba bean accessions worldwide using AFLP markers. Linkage maps of the gene controlling zero tannin in faba bean with SCAR markers developed from linked RAPD markers has been published [20]. Compared with other molecular markers, simple sequence repeats (SSR) markers are based on the amplification of sequence repetitions. It is a simple and repeatable method that can produce abundant polymorphic fragments. Therefore, SSR markers have the advantage of being a valuable tool for constructing genetic linkage maps and marker-assisted trait selection in faba bean breeding efforts.

As mentioned above, Picard [9] and Bond [10] first reported that the seed coat of all white flowered varieties of faba bean was free of tannins. The faba bean germplasm M3290, which is originally from the Mediterranean region, is a tannin-free cultivar with white flowers and the zt-1 gene [9]. The variety was collected from the International Center for Agricultural Research in the Dry Areas (ICARDA) Syria [1]. It was then developed by the Qinghai Academy of Agricultural Science and has been widely used in faba bean breeding programmes in China in the past few years (unpublished data).

The objective of this study was to (1) construct a linkage map of the temporary named gene zt-1 the controls the zero-tannin trait in the M3290 variety and (2) identify closely linked markers that could be useful for marker-assisted selection (MAS) in faba bean and further cloning of the gene.

2. Materials and Methods

2.1. Plant Materials

The Qinghai12 variety has coloured flowers (purple) and high tannin content, while the genotype zt-1 M3290 [1] produces white flowers and has a tannin-free seed coat. M3290 was used as the male parent and Qinghai12 was used as the female parent to develop the population lines. An F_2 population with 413 plants and the derived $F_{2:3}$ families with 8–10 plants each that were derived from the cross between the tannin-free cultivar, M3290, and a condensed tannin line, Qinghai12, were used for mapping the tannin-free gene *zt-1*. The parents and populations used in this study were grown in the fields of the Qinghai Agriculture and Forestry Academy. A total of 413 F_2 progenies were grown during the 2015 growing season on the experimental farm and all F_3 lines with ten or twenty plants each were grown in the next spring in 2016. All the parents and progenies were carefully characterized for their phenotypes by their colour in the field and tested for tannin content in the laboratory during the flowering period. The colours of the offspring were classified as two types, e.g., "white" and "purple", those with the same flower colour as M3290 were scored as "white" and the remainder with the same flower colour as Qinghai12 were scored as "purple".

A representative collection of 34 elite faba bean accessions from the major faba bean production regions in China, including ten main spring varieties from Qinghai province, four spring varieties from Gansu province, four Yunnan germplasms, six winter varieties from Sichuan province, seven Jiangsu winter cultivars and three Zhejiang winter cultivars, were used to validate the molecular markers identified to be linked to the *zt-1* gene. Two main types in China, the spring and winter faba bean, were both selected in this study to compare different ecotypes of faba bean germplasm. All the representative faba bean samples from different areas were grown on the experimental farm at the Qinghai Agriculture and Forestry Academy.

2.2. Tannin Measurement

To enhance the phenotype accuracy and to confirm the tannin content in the genotypes with different colours, the tannin content was determined in the parents and F_2 individuals with different flower colours. The Folin-Donis (F-D) method was used to measure the tannin content with a few modifications [21].

2.3. DNA Extraction

After measuring the condensed tannin content, the newly expanded faba bean leaves were used to extract genomic DNA. Genomic DNA was extracted using the DS (Sodium Lauroylsarcosine) protocol [22,23]. Tannin-free and condensed tannin bulks were established from 20 free (white flower) and 20 condensed (purple flower) tannin content F_2 plants, respectively. Bulked segregant analysis (BSA) [24] was used to identify whether the markers were linked to the gene controlling the zero-tannin characteristic.

2.4. Marker Analysis

A total of 596 pairs of SSR primers were screened between the two parents and bulks. Among them, 128 SSR markers were referred to in Ma et al. [25], 236 SSR markers were selected from a linkage map developed by El-Rodeny et al. [26], and the remaining SSR markers (unpublished data) were kindly provided by the Institute of Crop Science, Chinese Academy of Agricultural Sciences (ICP, CAAS). The ISSR markers used in this study were according to Zietkiewicz et al. [27]. All the primers used in this study were synthesized by Shanghai Sangon Biological Engineering Technology and Services Company Ltd., Shanghai, China.

SSR reactions were performed in a 20 μL reaction volume containing 1 unit of *Taq* DNA polymerase (TaKaRa), 2 μL of 10× buffer (50 mmol KCl (TaKaRa), 10 mmol Tris-HCl (TaKaRa, pH 8.3), and 1.5 mmol $MgCl_2$ (TaKaRa), 200 μmol of each dNTP (Roche, Basel, Switzerland), 6 pmol of each primer and 50–100 ng of template DNA. The PCR conditions were as follows: denaturation at 94 °C for 4 min; 35 cycles of 94 °C for 1 min, 50–61 °C (depending on primers) for 1 min, and 72 °C for 1 min; and a final extension at 72 °C for 10 min. ISSR-PCR amplifications were performed in 25 μL reaction volumes with 80 ng of genomic template DNA, 2 μL of 10 mM Tris–HCl, 50 mM KCl, 15 mM $MgCl_2$, 0.2 mM of each dNTP, 120 nM of each primer, and 1 U of Taq DNA polymerase. ISSR-PCR reactions were performed with the following conditions: denaturation at 95 °C for 5 min; 35 cycles of 95 °C for 30 s, annealing at optimal temperature for 1 min, and 72 °C for 1 min; and a final elongation step at 72 °C for 10 min.

PCR reactions were performed in a PTC200 Peltier Thermal Cycler. PCR products were then mixed with 4 μL of the formamide loading buffer (98% formamide, 10 mM EDTA, 0.25% bromophenol blue, and 0.25% xylene cyanol, pH 8.0, Shanghai Sangon) and heated at 94 °C for 5 min. The PCR products were separated on 6% denaturing polyacrylamide gels, 8% non-denaturing polyacrylamide gels or 1.5% agarose gels. Each 5–7 μL sample was loaded and then resolved using the silver staining method as described by Bassam et al. [28] or ethidium bromide and then photographed.

2.5. Statistical Analysis and Genetic Mapping

The Chi-square tests (χ^2) were used to determine the theoretical expectation based on the assumption of a single Mendelian gene controlling *zt-1*. Recombination fractions were converted to centiMorgans (cM) and the genetic distances of closely linked markers were calculated with software JOINMAP version 4.0 using the Kosambi mapping function [29]. A LOD score of 3.0 was used as a threshold for grouping and a maximum recombination fraction of 0.5 were employed as linkage criteria to establish the linkage group [30].

3. Results

3.1. Phenotypic and Genetic Analyses

In the flower testing in the field, M3290 had white flowers, whereas Qinghai12 had purple flowers. In the F_2 population, there were 95 white flowers and 318 purple flowers in the flower test, which fits a 1:3 ratio ($\chi^2_{1:3} = 0.879$, $p = 0.348$), consistent with the *zt-1* gene behaving as a single recessive gene in this population. Simultaneously, the tannin content also segregated in a 1:3 ratio after measuring the F_2 population, as the 95 plants with white flowers were all tannin-free (0 mg/mL), and the 318 plants with purple flowers presented tannin content ranging from 0.1 to 0.4 mg/mL. The histogram for tannin content was drawn to see the distribution of the trait (Figure 1). When the flower colours of F_3 families were tested during the same period in the next year, the segregation of these families conformed to a 1:2:1 ratio ($\chi^2_{1:2:1} = 1.59$, $p = 0.451$) as expected for a single gene (Table 1).

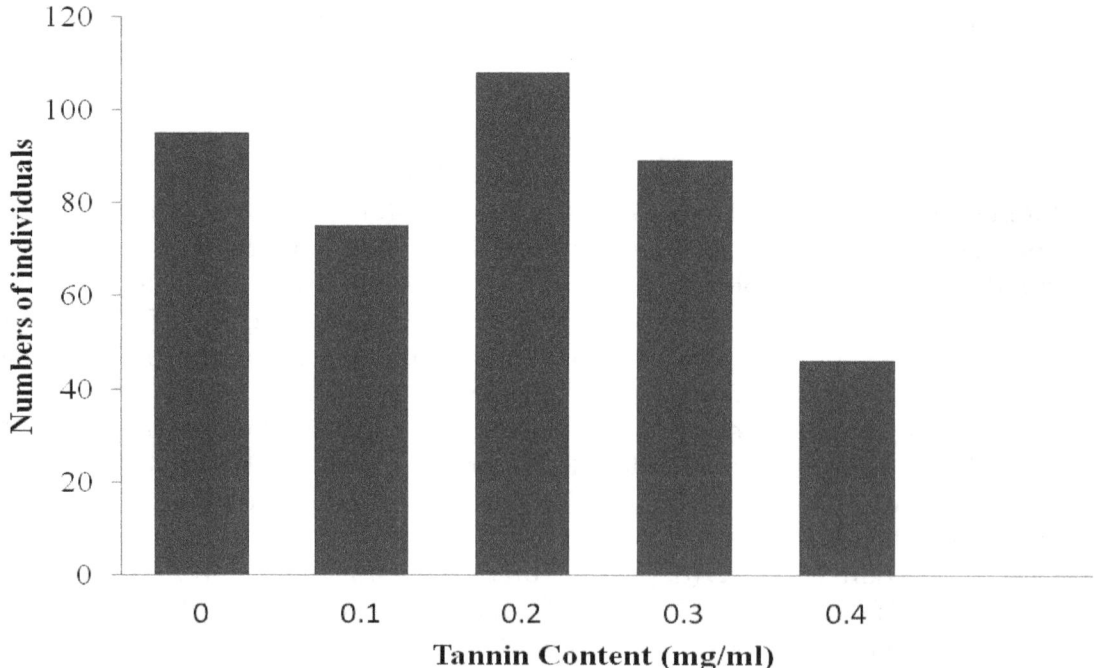

Figure 1. Frequency distributions of tannin content in 413 plants of F_2 population.

Table 1. Segregation for flower colours in the M3290/Qinghai12 F_1, F_2 and $F_{2:3}$ progenies.

CrossProgeny	Observed Number of Plants or Lines			Expected Ratio	χ^2	P
	W	Seg	P			
M3290	15	0	0			
Qinghai12	0	0	15			
F_1	15	-	-	1:0		
F_2	95	-	318	1:3	0.879	0.348
F_3	95	205	113	1:2:1	1.59	0.451

W, white flowers; P, purple flowers; Seg, segregation.

3.2. Identification of the SSR and ISSR Markers

Of the tested SSR primers, fifteen SSR markers, were polymorphic and contrasted between the purple and white flower bulks as well as the two parents (Figure 2). The selected polymorphic SSR markers were all co-dominant and could only be tested by 6% denaturing polyacrylamide gels. For example, the association between the SSR marker SSR84 in part of the F_2 population segregating for the *zt-1* gene is shown in Figure 2. The results shown in Figure 2 indicated that there were five genotypes with white flowers which showed the 900 bp bands with M3290, five genotypes with purple flowers which showed the same bands as Qinghai12 (1100 bp) and the other ten genotypes which showed heterozygous bands. For the tested 100 ISSR markers, seven markers, namely ISSR7, ISSR9, ISSR10, ISSR12, ISSR25, ISSR36 and ISSR48, were identified as polymorphic between the parents and bulks in this study (Figure S1). Among the seven ISSR markers, ISSR12 and ISSR25 were co-dominant and the other five were dominant (Figure 3).

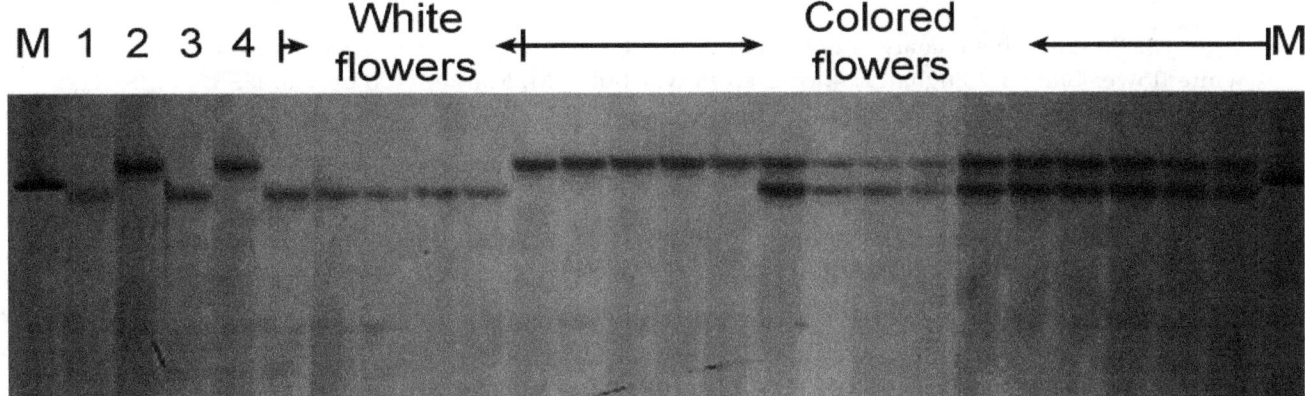

Figure 2. PCR amplification results of the SSR marker SSR 84 in part of the F_2 population segregating for *zt-1*. 1, M3290 (900 bp); 2, Qinghai12 (1100 bp); 3, white flower bulk (900 bp); 4, purple flower bulk (1100 bp); M, Marker (100 bp). The F_2 population includes 5 white flower genotypes (zz, 900 bp), 5 purple flower genotypes (ZZ, 1100 bp) and 10 heterozygous genotypes (Zz, 1100 bp); This is a composite picture of several different gel picture.

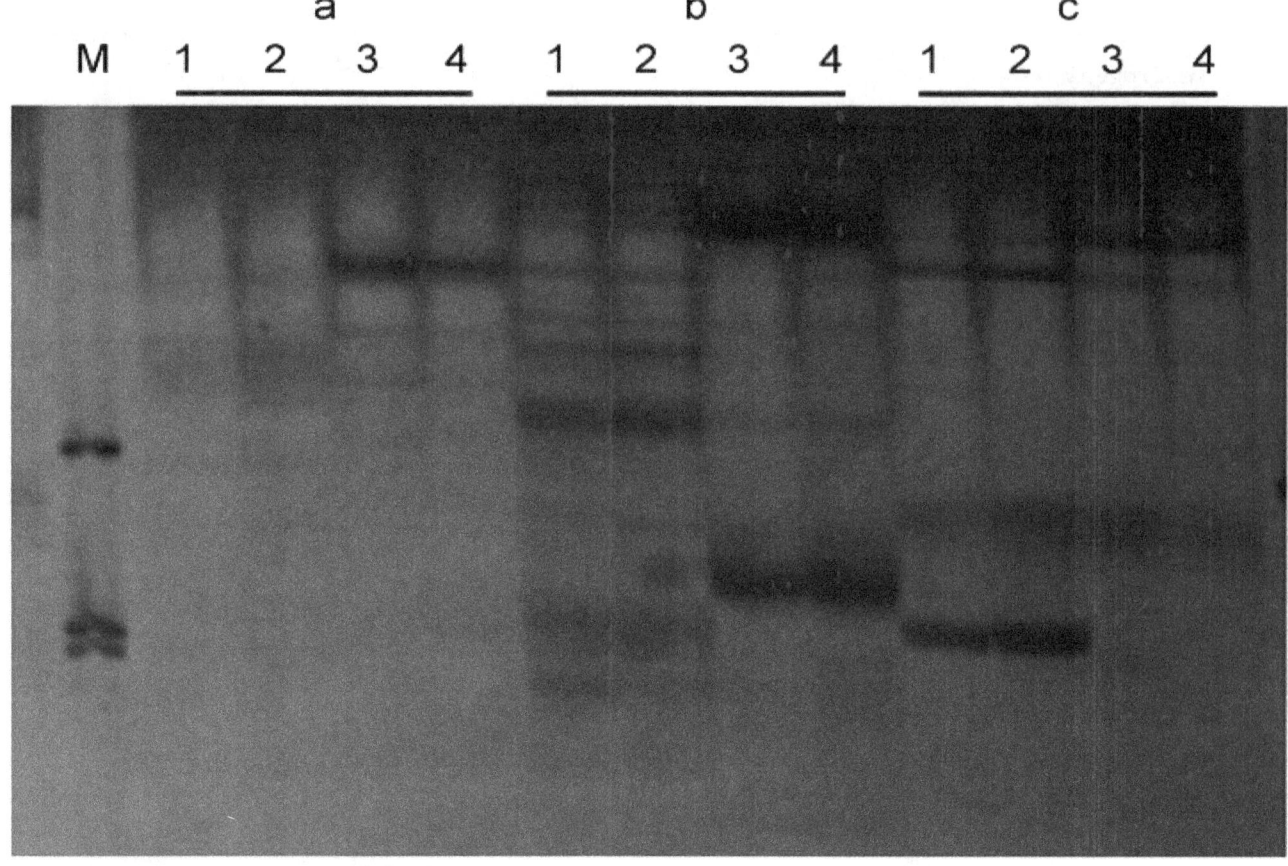

Figure 3. Polymorphism analysis of the partial ISSR markers in the parents and bulks. 1, M3290; 2, white flower bulk; 3, Qinghai12; 4, colored flower bulk. M, Marker (100 bp). a, ISSR7; b, ISSR12; c, ISSR36.

3.3. Mapping the SSR and ISSR Markers

The selected fifteen SSR markers and seven ISSR markers were then used to genotype the 413 F_2 plants to construct the linkage map. After testing the 413 plants in the F_2 population, it was evident that six SSR markers, M78, ssi85H, M233, SSR84, M81 and M38 (Table 2) and three ISSR markers were linked to *zt-1* (Table 2). The results of the *zt-1* linkage estimates with the nine polymorphic markers based on the phenotype and genotype data are shown in Figure 4.

Table 2. Molecular markers mapped at or close to the *zt-1* locus.

Name	Marker Type	Forward Primer (5′-3′)	Reverse Primer (5′-3′)	Annealing Temperature (°C)
SSR84	SSR	TCTGAAAACGAGTTCAGTGGA	CTGGTGCCGAACTAACCAGT	52
M38	SSR	GCTACTGGAGGAGGCTTTCA	GCCTTCTACACAACGGCTTC	53
M78	SSR	GTCAAATCGAGTGGCGAAAG	TTGGGATATGGAAGTAGCTTCAG	52
M81	SSR	CCTCATGCCATTCCTCTGAT	TTCCGCGTGGTAAATTCTATG	55
M233	SSR	CATCCCAACAATATACCGGC	CTGGGGTACCACCGTAACTC	51
ssi85H	SSR	AACAACTACGTAATGCCAGAC	ACATGAGGGGCCAAGTAT	52
ISSR7	ISSR	AGA GAG AGA GAG AGA GT		53
ISSR9	ISSR	AGA GAG AGA GAG AGA GG		50
ISSR10	ISSR	GAG AGA GAG AGA GAG AT		51

The selected nine markers were mapped within a genetic interval of 20.6 cM flanking *zt-1* (Figure 4). The flanking markers, *SSR84* and *M78*, were closely linked with the *zt-1* gene with genetic distances of 2.9 cM and 6.2 cM, respectively (Figure 4).

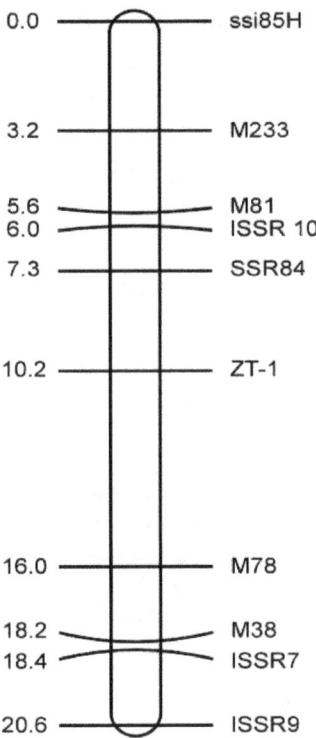

Figure 4. Linkage map of the zero tannins *zt-1* gene flanked by six SSR and three ISSR markers. The locus name and corresponding locations are indicated on the right side and the genetic distances between them are indicated on the left side.

3.4. Closely Linked Markers for the Marker-Assisted Selection of zt-1

The two flanked markers (Figure 4), SSR84 and M78, with genetic distances of 2.9 cM and 6.2 cM, respectively, were first used to identify representative varieties from different areas of China to assess their potential use in the MAS. The results showed that when SSR84 was used for MAS, only M3290 and *zt-1*-carrying genotypes (white flowers) produced the expected band of 900 bp and the genotypes with purple flowers (without the *zt-1* gene) produced 1100 bp bands. Nevertheless, the results of molecular detection with M78 was not helpful in the selection of the *zt-1* gene in faba bean breeding programs (Table 3).

Table 3. The 34 Chinese faba bean genotypes used for validation of the closely linked markers.

No.	Variety	Province	Flower [a]	SSR84 [b]		M78 [b]	
				900 bp	1100 bp	400 bp	420 bp
1	Qinghai11	Qinghai	Purple	− [c]	+ [c]	−	+
2	Qinghai12	Qinghai	Purple	−	+	−	+
3	Qinghai13	Qinghai	Purple	−	+	−	+
4	Qingcan14	Qinghai	Purple	−	+	−	+
5	Qingcan15	Qinghai	Purple	−	+	−	+
6	M3290	Qinghai	White	+	−	+	−
7	TF26	Qinghai	White	+	−	+	−
8	TF29	Qinghai	White	+	−	−	+
9	TF34	Qinghai	White	+	−	−	+
10	2005-00	Qinghai	White	+	−	−	+
11	Lincan6	Gansu	Purple	−	+	−	+
12	Lincan7	Gansu	Purple	−	+	−	+
13	Lincan8	Gansu	Purple	−	+	−	+

Table 3. *Cont.*

No.	Variety	Province	Flower [a]	SSR84 [b]		M78 [b]	
				900 bp	1100 bp	400 bp	420 bp
14	Yangyandou	Gansu	Purple	-	+	-	+
15	Yundou7	Yunnan	Purple	-	+	-	+
16	Yundou8	Yunnan	Purple	-	+	-	+
17	Yundou9	Yunnan	Purple	-	+	-	+
18	Touxinlv	Yunnan	Purple	-	+	-	+
19	Dahudou	Sichuan	Purple	-	+	-	+
20	Xiaohudou	Sichuan	Purple	-	+	-	+
21	Honghudou	Sichuan	Purple	-	+	-	+
22	Chenghu9	Sichuan	Purple	-	+	-	+
23	Chenghu10	Sichuan	Purple	-	+	-	+
24	Chenghu11	Sichuan	Purple	-	+	-	+
25	Tongcanxian7	Jiangsu	Purple	-	+	-	+
26	Tongcanxian8	Jiangsu	Purple	-	+	-	+
27	Qidongbaipi	Jiangsu	Purple	-	+	-	+
28	Haimendabaipi	Jiangsu	Purple	-	+	-	+
29	Tongcan5	Jiangsu	Purple	-	+	-	+
30	Nantongsanbai	Jiangsu	Purple	-	+	-	+
31	DAqingpi	Jiangsu	Purple	-	+	-	+
32	Lvpidou	Zhejiang	Purple	-	+	-	+
33	Luohandou	Zhejiang	Purple	-	+	-	+
34	Xiaoqingdou	Zhejiang	Purple	-	+	-	+

[a] Flower colour. Purple: genotype with tannin content and without the *zt-1* gene.; white: genotype with zero tannin content carrying the *zt-1* gene. [b] Closely linked markers. [c] '+' and '-' indicate the presence and absence of the specific alleles of the SSR markers, respectively.

The other different PCR fragments amplified with other linked markers also could not distinguish lines with white flowers from coloured varieties. Therefore, it was verified that only SSR84 is helpful for selecting the *zt-1* gene in faba bean programmes for tannin content in this study.

4. Discussion

4.1. Zt-1 Gene in Faba Bean Variety M3290

In this study, we identified markers to the single recessive gene *zt-1* in the faba bean variety M3290 and mapped it with six SSR markers and three ISSR markers. We also tested the elite faba bean germplasms with the closely linked markers and the results provided a sound basis for further MAS in faba bean.

The absence of tannin content in faba bean is determined by two recessive genes, *zt-1* and *zt-2*. In this study, *zt-1* was inherited as a single recessive gene in the M3290/Qinghai12 population. This result is in accordance with that of Gutierrez et al. [20], who used a segregated F_2 population derived from Vf6 and a *zt-1* line. The segregation for both the flower colour and tannin content fit the expected 1:3 and 1:2:1 ratios, respectively, which is consistent with a single recessive gene that controls zero tannin content in faba bean. The linkage map of Vf6 × *zt-1* F_2 populations showed that the *zt-1* gene was flanked with two SCAR markers with genetic distances of 3.6 cM (OPAF20$_{776}$) and 9.7 cM (SCC5$_{551}$).

4.2. SSR and ISSR Markers

A total of 596 SSR markers and 100 ISSR markers were used to screen the polymorphisms between parents as well as bulks in this study. The SSR markers were firstly randomly selected from each linkage group (LG) according to Ma et al. [25] and El-Rodeny et al. [26]. Also subsequently, unpublished SSR markers (including EST sequences) were kindly provided by the Institute of Crop Science,

Chinese Academy of Agricultural Sciences (ICP, CAAS). Finally, ISSR markers were downloaded according to Zietkiewicz et al. [27] and used to test the polymorphisms between parents and bulks. The polymorphism selection results indicated that six SSR markers and seven ISSR markers showed clear and repeatable bands between parents and bulks. The linkage analysis showed that the six SSR markers were all linked with the zt-1 gene after genotyping the 413 F_2 plants, but only three ISSR markers indicated linkage correlation with the zt-1 gene. The polymorphism tests verified the application of SSR markers in faba bean, and also provided us with a sound basis for further fine mapping of the zt-1 gene. Nine markers, including six SSR markers and three ISSR markers were mapped at zt-1 locus in this study. The flanking markers, SSR84 and M78, were closely linked with the zt-1 gene with genetic distances of 2.9 cM and 6.2 cM, respectively. Although several markers have been mapped to the zt-1 region, the number of the markers is still limited, and more are needed for more efficient marker-assisted selection, fine mapping and map-based cloning of the zt-1 gene.

Compared to previous studies, a linkage map of the zt-1 region with nine markers, including six SSR markers and three ISSR markers were constructed with a F_2 population.

4.3. Closely Linked Markers and Their Application in MAS

Molecular markers closely linked to the target gene are considered important tools for MAS in plant breeding programmes [31]. However, faba bean possess a large genome size and limited molecular markers. Therefore, MAS progress in faba bean breeding faces enormous challenges [1,32]. In this regard, it is urgent to develop more valuable and closely linked markers for faba bean breeding.

Previous studies allowed the prediction of the zt-1 genotypes with a 95% accuracy [20]. In this study, two flanking markers were used to test their suitability in MAS, and the closely linked marker SSR84 was verified to be a powerful tool (100% accuracy of the selection of the zt-1 gene) for further faba bean breeding. Owing to the recessive nature of the faba bean flower and tannin content traits, crops are often segregated by crosses that cause devaluation a few years after being grown for commercial production [33]. We present a convenient marker in this study that is closely linked with the zt-1 gene and might resolve this problem.

4.4. Faba Bean Breeding Program of Variety M3290

White flowers and zero tannin content are controlled by a single recessive gene. These and many other good agronomic traits make M3290 a desirable donor for faba bean breeding programmes. In fact, M3290 was used in breeding programmes a few years ago in China. The faba bean variety 'TF26' and some other lines were developed with M3290 by the Qinghai Academy of Agriculture and Forestry Sciences; these varieties showed pure white flowers in fields and zero tannin contents. However, the limited genetic background and linked molecular markers for the zt-1 gene still hampers the use of the gene in breeding programs. The demonstration of the zt-1 gene in the germplasm M3290 and the closely linked markers identified in this study should accelerate its application in breeding programmes and SSR84 closely linked with the zt-1 gene could exactly distinguish flowers with different genotypes.

5. Conclusions

F_2 and F_3 progenies derived from M3290 and Qinghai12 were used for phenotypic and genetic analyses, and the results indicated that the zt-1 gene in this population behaved as a single recessive gene. Selected SSR markers and ISSR markers were used to genotype the entire 413 F_2 population, and linkage analysis showed that the zt-1 gene was closely linked to the SSR markers SSR84 and M78, with genetic distances of 2.9 and 5.8 cM, respectively. SSR marker SSR84 could predict the zt-1 genotypes in faba bean breeding.

Supplementary Materials: The following are available online at Figure S1: PCR amplification results of the ISSR marker ISSR10 in part of the F_2 population segregating for zt-1.

Author Contributions: X.Z., W.H., Y.L. conceived and designed the experiments; Q.Y. and W.S. performed the experiments; P.L. and Y.T. analyzed the data; W.H. contributed reagents/materials/analysis tools; X.Z. and W.H. wrote the paper.

Acknowledgments: This research was supported by the National Natural Science Foundation of China (NSFC-31360363) and the China Agriculture Research System (CARS-08). We are also grateful to Zong at CAAS in Beijing, China for the valuable advice and for providing SSR markers for our study.

References

1. Duc, G.; Bao, S.; Baumc, M.; Redden, B.; Sadiki, M. Diversity maintenance and use of *Vicia faba* L. genetic resources. *Field Crop. Res.* **2010**, *115*, 270–278. [CrossRef]

2. FAO. United Nations. 2016. Available online: http://faostat3.fao.org/download/Q/QC/E (accessed on 10 October 2016).

3. Stoddard, F.L.; Hovinen, S.; Kontturi, M.; Lindström, K.; Nykänen, A. Legumes in Finnish agriculture: history, present status and future prospects. *Agric. Food Sci.* **2009**, *18*, 191–205. [CrossRef]

4. Friedman, M. Nutritional value of proteins from different food sources. A review. *J. Agric. Food Chem.* **1996**, *44*, 6–29. [CrossRef]

5. Wiseman, J.; Cole, D.J.A. European legumes in diets for non-ruminants. In *Recent Advances in Animal Nutrition*; Haresign, W., Cole, D.J.A., Eds.; Butterworths: London, UK, 1988; pp. 13–37.

6. Martin-Tanguy, H.; Guillaume, J.; Kossa, A. Condensed tannins in horse bean seeds: Chemical structure and apparent effects on poultry. *J. Sci. Food Agric.* **1977**, *28*, 757–765. [CrossRef]

7. Cansfield, P.E.; Marquardt, R.R.; Campbell, L.D. Condensed proanthocyanidins of faba beans. *J. Sci. Food Agric.* **1980**, *31*, 802–812. [CrossRef] [PubMed]

8. Vander Poel, A.F.B.; Dellaert, L.M.W.; Van Norel, A.; Helsper, J.P.F.G. The digestibility in piglets of faba bean (*Vicia faba* L.) as affected by breeding towards the absence of condensed tannins. *Br. J. Nutr.* **1992**, *68*, 793–800. [CrossRef]

9. Picard, J. Apercu sur l'hérédité du caractére absence de tannins dans les graines de féverole (*Vicia faba* L.). *Ann. Amelior. Plant.* **1976**, *26*, 101–106.

10. Bond, D.A. In vitro digestibility of the testa in tannin-free field beans (*Vicia faba* L.). *J. Agric. Sci. Camb.* **1976**, *86*, 561–566. [CrossRef]

11. Cabrera, A.; Martin, A. Genetics of tannin content and its relationship with flower and testa colours in *Vicia faba* L. *J. Agric. Sci. Camb.* **1989**, *113*, 93–98. [CrossRef]

12. Crofton, G.R.A.; Bond, D.A.; Duc, G. Potential seed multiplication problems arising from the existence of two genes for the absence of tannin in *Vicia faba* L. *Plant Var. Seeds* **2000**, *13*, 131–139.

13. Alghamdi, S.S.; Migdadi, H.M.; Ammar, M.H.; Paull, J.G.; Siddique, K.H.M. Faba bean genomics: Current status and future prospects. *Euphytica* **2012**, *186*, 609–624. [CrossRef]

14. Bennett, M.D.; Smith, J.B. Nuclear DNA amounts in angiosperms. *Proc. R. Soc. Lond. B Biol. Sci.* **1982**, *216*, 179–199. [CrossRef]

15. Johnston, J.S.; Bennett, M.D.; Rayburn, A.L.; Galbraith, D.W.; Price, H.J. Reference standards for determination of DNA content of plant nuclei. *Am. J. Bot.* **1999**, *86*, 609–613. [CrossRef] [PubMed]

16. Raina, S.N.; Ogihara, Y. Ribosomal DNA repeat unit polymorphism in 49 *Vicia* species. *Theor. Appl. Genet.* **1995**, *90*, 477–486. [CrossRef] [PubMed]

17. Zeid, M.; Schon, C.C.; Lin, W. Genetic diversity in recent elite faba bean lines using AFLP markers. *Theor. Appl. Genet.* **2003**, *107*, 1304–1314. [CrossRef] [PubMed]

18. Zong, X.; Liu, X.; Guan, J.; Wang, S.; Liu, Q.; Paull, J.G.; Redden, R. Molecular variation among Chinese and global winter faba bean germplasm. *Theor. Appl. Genet.* **2009**, *118*, 971–978. [CrossRef] [PubMed]

19. Zong, X.; Ren, J.; Guan, J.; Wang, S.; Liu, Q.; Paull, J.G.; Redden, R. Molecular variation among Chinese and global germplasm in spring faba bean areas. *Plant Breed.* **2010**, *129*, 508–513. [CrossRef]

20. Gutierrez, N.; Avila, C.; Rodriguez-Suarez, C.; Moreno, M.; Torres, A. Development of SCAR markers linked to a gene controlling absence of tannins in faba bean. *Mol. Breed.* **2007**, *19*, 305–314. [CrossRef]

21. Hagerman, A.E.; Butler, L.G. Protein precipitation method for the quantitative determination of tannins. *J. Agric. Food Chem.* **1978**, *26*, 809–812. [CrossRef]

22. Song, W.N.; Langridge, P. Identification and mapping polymorphism in cereals based on polymerase chain reaction. *Theor. Appl. Genet.* **1991**, *82*, 209–213.

23. Song, W.N.; Henry, R. Polymorphisms in the a-amy1 gene of wild and cultivated barley revealed by the polymerase chain reaction. *Theor. Appl. Genet.* **1994**, *89*, 509–512.

24. Michelmore, R.W.; Paran, I.; Kesseli, R.V. Identification of markers linked to disease-resistance genes by bulked segregant analysis: A rapid method to detect markers in specific genomic regions by using segregating populations. *Proc. Natl. Acad. Sci. USA* **1991**, *88*, 9828–9832. [CrossRef] [PubMed]

25. Ma, Y.; Yang, T.; Guan, S.; Wang, H.; Wang, X.; Zong, X. Development and characterization of 21 EST-derived microsatellite markers in *Vicia faba* (faba bean). *Am. J. Bot.* **2011**, *98*, 22–24. [CrossRef] [PubMed]

26. El-Rodeny, W.; Kimura, M.; Hirakawa, H.; Sabah, A.; Shirasawa, K.; Sato, S. Development of EST-SSR markers and construction of a linkage map in faba bean (*Vicia faba*). *Breed. Sci.* **2014**, *64*, 252–263. [CrossRef] [PubMed]

27. Zietkiewicz, E.; Rafalski, A.; Labuda, D. Genome fingerprinting by simple sequence repeat (SSR)-anchored polymerase chain reaction amplification. *Genomics* **1994**, *20*, 176–183. [CrossRef] [PubMed]

28. Bassam, B.J.; Anolles, G.C.; Gresshoff, P.M. Fast and sensitive silver staining of DNA in polyacrylamide gels. *Anal. Biochem.* **1991**, *196*, 80–83. [CrossRef]

29. Van Ooijen, J.W. *JoinMap4, Software for the Calculation of Genetic Linkage Maps in Experimental Populations*; Kyazma BV: Wageningen, The Netherlands, 2006.

30. Kosambi, D.D. The estimation of map distances from recombination values. *Ann. Eugen.* **1944**, *12*, 172–175. [CrossRef]

31. Kelly, J.D. Use of random amplified polymorphic DNA markers in breeding for major gene resistance to plant pathogens. *HortScience* **1995**, *30*, 461–465.

32. Terzopoulos, P.J.; Bebeli, P.J. Genetic diversity analysis of Mediterranean faba bean (*Vicia faba* L.) with ISSR markers. *Field Crop. Res.* **2008**, *108*, 39–44. [CrossRef]

33. Link, W.; Ederer, W.; Metz, P.; Buiel, H.; Melchinger, A.E. Genotypic and environmental variation for degree of cross-fertilization in faba bean. *Crop Sci.* **1994**, *34*, 960–964. [CrossRef]

3

Development of High Yielding Glutinous Cytoplasmic Male Sterile Rice (*Oryza sativa* L.) Lines through CRISPR/Cas9 based Mutagenesis of *Wx* and *TGW6* and Proteomic Analysis of Anther

Yue Han [1,†], Dengjie Luo [1,†], Babar Usman [1], Gul Nawaz [1], Neng Zhao [1], Fang Liu [1] and Rongbai Li [1,2,*]

[1] College of Agriculture, State Key Laboratory for Conservation and Utilization of Subtropical Agro-Bioresources, Guangxi University, Nanning 530004, China; 17739899868@163.com (Y.H.); luodengjie01@126.com (D.L.); babarusman119@gmail.com (B.U.); gulnawazmalik@yahoo.com (G.N.); nengzhao_gxu@163.com (N.Z.); liufang1975@163.com (F.L.)

[2] Guangxi Academy of Agricultural Sciences, Guangxi, Nanning 530007, China

* Correspondence: lirongbai@126.com

† These authors contributed equally to this work.

Abstract: Development of high yielding and more palatable glutinous rice is an important goal in breeding and long-standing cultural interaction in Asia. In this study, the *TGW6* and *Wx*, major genes conferring 1000 grain weight (GW) and amylose content (AC), were edited in a maintainer line by CRISPR/Cas9 technology. Four targets were assembled in pYLCRISPR/Cas9Pubi-H vector and T_0 mutant plants were obtained through *Agrobacterium* mediated transformation with 90% mutation frequency having 28% homozygous mutations without off-target effects in three most likely sites of each target and expression level of target genes in mutant lines was significantly decreased ($P < 0.01$), the GW and gel consistency (GC) were increased, and the AC and gelatinization temperature (GT) were decreased significantly and grain appearance was opaque, while there was no change in starch content (SC) and other agronomic traits. Mutations were inheritable and some T_1 plants were re-edited but T_2 generation was completely stable. The pollen fertility status was randomly distributed, and the mutant maintainer lines were hybridized with Cytoplasmic Male Sterile (CMS) line 209A and after subsequent backcrossing the two glutinous CMS lines were obtained in BC_2F_1. The identified proteins from anthers of CMS and maintainer line were closely associated with transcription, metabolism, signal transduction, and protein biosynthesis. Putative mitochondrial NAD^+-dependent malic enzyme was absent in CMS line which caused the pollen sterility because of insufficient energy, while upregulation of putative acetyl-CoA synthetase and Isoamylase in both lines might have strong relationship with CMS and amylose content. High yielding glutinous CMS lines will facilitate hybrid rice breeding and investigations of proteins linked to male sterility will provide the insights to complicated metabolic network in anther development.

Keywords: rice; CRISPR/Cas9; *Wx*; *TGW6*; mutations; maintainer; cytoplasmic male sterile; amylose content; anther; protein

1. Introduction

The rice (*Oryza sativa* L.) is an important widely adapted food crop and 20% of the world's dietary energy supply which is feeding more than half of the world's population and 3 billion people uptake rice daily [1,2]. Due to the fast-growing population, the global rice consumption is projected to increase from 450 million tons in 2011 to about 490 million tons in 2020 and 40% more rice is needed to be produced by 2050 to meet people's demand for food [3,4]. The cytoplasmic male sterility (CMS) is the foundation to exploit the heterosis of hybrid rice which uses a three-line system consisting of a cytoplasmic male sterile (CMS) line (A line), a maintainer (B), and a restorer (R line) for hybrid seed production [5]. China is the pioneer of hybrid rice production and with the development of latest breeding tools the yield of rice has been increased more than 20% and newly developed genotypes performing better than conventional verities and now accounts more than half of the annual rice planting area in China [6]. The development of new CMS has become the main interest of breeders because very few genotypes exhibit a strong restoration ability as effective restorer for CMS in the development of hybrid rice [7,8]. Yield and quality are typical quantitative traits governed by multiple genomic loci, while yield is directly depends on grain weight (GW) which is mainly determined by the synthesis and accumulation of starch in the endosperm of the grain [9,10]. To solve this problem, we must resort to new technologies and new genetic improvement strategies. Starch is one of the important indicators for evaluating rice quality and 90% of rice endosperm is starch [11].

Rice waxy gene Wx-encoded granular bound starch synthase I (GBSSI), also known as Waxy protein is the major gene controlling amylose synthesis in endosperm [12]. Wx gene differentiates into alleles Wx^a and Wx^b, indica rice is dominated by Wx^a which confers higher amylose content by producing 10-fold higher mRNA and protein level than Wx^b while japonica rice is dominated by Wx^b with lower amylose content [13,14]. Wx exon or intron structural change would affect Wx expression by affecting messenger RNA (mRNA) stability [12,15]. Several studies have reported that mutations in the functional site of the Wx gene led to 14.6 to 2.6% reduced amylose content (AC) in rice transgenic lines and hybrids obtained with mutant lines [15–23], while Wx overexpression lines showed increased AC by 6–11% [24].

At present, the GW related genes that have been cloned including $qSW5/GW5$ [25,26], $TGW6$ [27], $GS3$ [28], $GS5$ [29], $GW2$ [30], $GW8/OsSPL16$ [31], $qGL3/qGL3-1/GL3.1$ [32–34], $GW7$ [35], and $OsSPL13$ [36]. Among them, $TGW6$ is one of the most important genes regulating rice GW traits, which encodes a purine acetic acid-glucose hydrolase. Its loss-of-function mutation causes a decrease in the content of indoleacetic acid in the endosperm resulted in increased cell numbers which finally resulted with increased grain length and GW with 15% enhanced production of rice [27]. Rice genes including $DEP1$, $GS3$, $GW2$, $GS5$, $Gn1a$, and $TGW6$, that are negative regulators of grain size and number and grain weight has been knocked-out to improve yield [37,38], and CRISPR/Cas9 based simultaneous mutations of $GW2$, $GW5$, and $TGW6$ resulted in 29.3% increase in GW [39]. This suggests that generation of mutation in major yield related genes in a single cultivar would be helpful to increase large scale production of rice.

With the development of some new molecular biology techniques such as CRISPR/Cas9 (clustered regulatory interspersed short palindromic repeat/CRISPR associated proteins) a lot of achievements has been made in plants and animals. CRISPR/Cas9 technology is widely used to study the gene function and regarded as the third-generation genome-editing tool established after zinc finger nucleases (ZFNs) and transcription activator-like effector nucleases (TALEN), based on guided RNA (gRNA) engineered nucleases, which is most applicable due to their simplicity, efficiency, and versatility [40,41]. CRISPR/Cas9 make a double-stranded break (DSB) in the target DNA which is subsequently repaired by natural repair mechanism of homologous recombination (HR) precise pathway or non-homologous end joining (NHEJ) [42], which creates random insertions and deletions and results in targeted gene knockouts or gene replacement [40,43,44]. CRISPR/Cas9 is the most advanced genome editing tool in plant biology [45,46] and has been widely used in animals, yeast, human non-human cell lines [42,47,48], as well as in the model species *A. thaliana* and *N. benthamiana* [43,49], as well as crops such as rice [50–52], wheat [53], maize [54], potato [55], and tomato [56].

Conventional plant breeding techniques are effective but laborious and time consuming, therefore we used CRISPR/Cas9-mediated gene editing to introduce a loss-of- function mutations into the Wx and $TGW6$ genes associated with lower AC and increase yield in rice maintainer line 209B. Our results show that mutations in the *Wx* and *TGW6* gene produce decreased AC and enhanced yield in rice CMS line offering an effective strategy of accelerating the hybrid rice breeding program. Through one generation of hybridization and two generations of backcrossing with mutant maintainer lines as the male parent and 209A as female parent, the glutinous cytoplasmic male sterile lines (CMS) were successfully achieved. The protein of CMS line pollen and mutant maintainer line were separated by two-dimensional electrophoresis and sodium dodecyl sulphate-polyacrylamide gel electrophoresis (SDS-PAGE) and differentially expressed spots were analyzed. This study gave new insights into the mechanism of CMS and maintainer lines and demonstrated the power of proteomic in plant biology. Present study showed that the CRISPR/Cas9 technology provides the tool set to fasten the rice breeding program to achieve desired agronomic characters and improved yield.

2. Materials and Methods

2.1. Rice Material, CRISPR/Cas9, and gRNA Vectors

The cytoplasmic male sterile line 209A and its maintainer line 209B developed by Professor Li Rongbai, were collected from Rice Research Institute of Guangxi University. The maintainer line 209B was used for genetic transformation to which have the characteristics of resistance to drought and blast with compact plant type. Plants were grown in the experimental field of Guangxi University during normal rice growing season and maintained regularly. The Cas9 vector pYLCRISPR/Cas9-MT(I) and the gRNA vectors (OsU6a, OsU6b, OsU6c, and OsU3m) (Figure 1) were provided by Professor Liu Yaoguang, South China Agricultural University, Guangzhou, China.

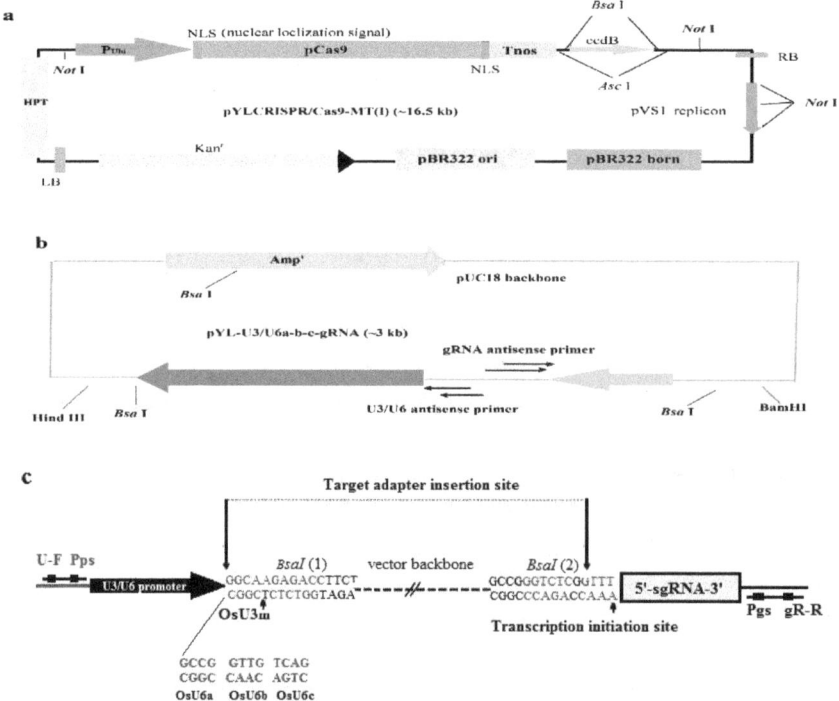

Figure 1. Maps of pYLCRISPR/Cas9-MT(I) and pYL-U3/U6a-b-gRNA vectors. (**a**) The binary vector with *Cas9p* driven by maize ubiquitin promoter (P_{Ubi}). The key sequences and restriction sites for Golden Gate ligation are shown. The expression of the sgRNA scaffold is driven by the rice U6a/U3 small nuclear RNA promoters; the expression of hygromycin (HPT) is driven by 2 CaMV35S promoters. NLS: nuclear localization signal; Tnos: gene terminator; LB and RB: left border and right border, respectively. (**b**) The physical map of the sgRNA intermediate plasmids. U3/U6 promoters from rice used for preparation of multiple sgRNA expression cassettes in single binary constructs. (**c**) *BsaI* sites (1, 2) in the sgRNA plasmids and their sequence information. These *BsaI*-cutting (small arrows) sites of the plasmids makes compatible sites for ligation by generating distinct non-palindromic ends to the U3/U6 promoters and a common end to the sgRNA sequence. Modified from Ma et al. (2015a) [57].

2.2. gRNA Target Selection and Synthesis of Oligonucleotide Strands

The gRNA target sequences were designed according to the exon sequence of *Wx* (LOC_Os06g04200) and *TGW6* (LOC_Os06g41850) provided by the Rice Genomics Annotation website (http://rice.plantbiology.msu.edu/) (Figure 2). The targets were 20 bp long gRNA sequences followed by the protospacer adjacent motif (PAM) NGG. The targets were selected with high GC%, low off-target score (Table S1) in exon regions by using online toolkit CRISPR-GE (http: //skl.scau.edu.cn/) and sgRNA structures (Figure S1) were developed by online tool CRISPR-P 2.0 (http://crispr.hzau.edu.cn/CRISPR2/). The CRISPR/Cas9 constructs that we designed to target *Wx* were in the first exon (WxT1: bases 1522–1541; WxT2: bases 2011–2030), with expected targeted mutations. The both targets for *TGW6* were also designed in the exon region (TGW6T1: bases 184–203; TGW6T2: bases 751–770) and expected mutations were in the coding region (Figure 2). The gRNA sequences were aligned and validated by using National Center for Biotechnology Information (NCBI) (https://blast.ncbi.nlm.nih.gov/Blast.cgi) and non-specific targets were excluded. Oligonucleotide sequences were synthesized by Beijing Genomics Institute (BGI) and shown in Table S2.

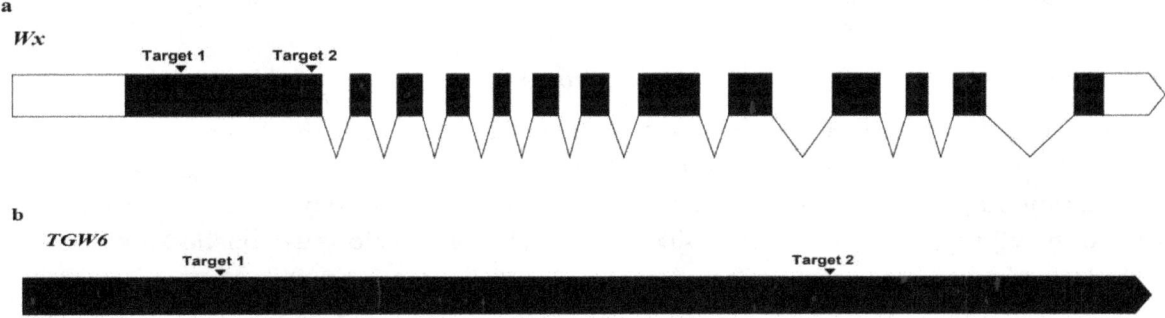

Figure 2. Schematic diagram of gene structures and target sites of the gRNA in the genes locus. (**a**) Position of two targets in *Wx* gene and (**b**) both targets in *TGW6* gene locus.

2.3. Vector Construction

The selected binary plasmids were isolated from *E. coli* (*Escherichia coli*) TOP10F' strains according to the previous established methods [57,58] with some modifications (Figure S2). The target site containing sequence primers WxgRT1/OsU6aWxT1 and WxgRT2/OsU6bWxT2 TGgRT1/OsU6cTGT1 and TGgRT2/OsU3TGT2, (Table S2) were combined by annealing, and then the target site sequence-containing chimeric primers were cloned into the sgRNA expression cassettes pYLsgRNA-U6a, pYLsgRNA-U6b, pYLsgRNAU6c and pYLsgRNA-U3m at a *Bsa*I site (Figure S3). The integrated sgRNA expression cassettes were then amplified by nested polymerase chain reaction (PCR) using U-F/Reverse adapter primers and gR-R for the first round, and the corresponding site-specific primers Pps-R/Pgs-2, Pps-2/Pgs-3, Pps-3/Pgs-4, and Pps-4/Pgs-L (Table S2) for the overlapping PCR (Figure S4) to ligate four-target-sgRNAs expression cassettes into the pYLCRISPR/Cas9Pubi-H vector and ligation product was transformed into DH5α competent cells according to the established protocol [20] with some modifications (Figure S5). The monoclonal inoculation culture was picked and were amplified by using SPL1 and SP-R primers (Table S2) and the clones confirming the product length ware sent to Beijing Genomics Institute (BGI) for sequencing.

2.4. Agrobacterium-Mediated Transformation of Rice Callus

The constructed plasmid was transformed into *Agrobacterium* EHA105 by electroporation according to the established method [59], and positive clones were used for rice callus transformation and transformed plants were obtained by hygromycin screening.

2.5. T_0 Genotypinng

The genomic DNA of T_0 mutant lines was extracted by cetyl trimethylammonium bromide (CTAB) method and PCR amplification was performed by using target specific primers for *Wx* gene Target1 OsWaxyT1F/OsWaxyT1R, and for Target2 OsWaxyT2F/OsWaxyT2R, and for *TGW6* Target1 Tgw6-T1F/Tgw6-T1R, for Target2 Tgw6-T2F/Tgw6-T2R (Table S2). The amplified products were visualized by 1% agarose gel electrophoresis and then sent to BGI for sequencing and mutations were decoded by using online tool DSDecodeM (http://skl.scau.edu.cn/dsdecode/). The multiple amino-acid sequence alignment was performed by using Clustal Omega online tool (https://www.ebi.ac.uk/Tools/msa/clustalo/). The off-target regions were selected from CRISPR-GE online tool (http://skl.scau.edu.cn/offtarget/).

2.6. Identification of T-DNA Free Mutant Lines and Cross Section Analysis of Grain Endosperm

The genomic DNA of T_1 and T_2 generations was extracted and amplified by using Cas9-F and Cas9-R specific primers (Table S2). The amplified product was subjected to 1% agarose gel to check the T-DNA free mutant plants. Scanning electron microscopy was used to observe the cross section of the mutant and its wild type (WT) mature grain according to the previous established method [60].

2.7. Expression Analysis

Total RNA was extracted from WT and T_1 mutant plants by using TaKaRa MiniBEST Plant RNA Extraction Kit according to manufacturer instructions. The specific primers for *Wx* gene W-F/W-R and TG-F/TG-R were used for *TGW6* (Table S2). The rice *OsActin* gene was used as internal control and 20 µL reaction was prepared with 2 µL cDNA, 0.4 µL each of the forward and reverse primers, 10 µL, Synergy Brands, Incorporated (SYBR) Green Master Mix, 7.2 µL ddH$_2$O. PCR Amplification procedure was as followed, 30s at 94 °C, 5s at 94 °C, and 30s at 60 °C with 45 cycles. The relative expression of genes was calculated from three biological replicates per sample according to the $2^{-\Delta\Delta Ct}$ method [61].

2.8. Determination of AC, SC, GC, and GT

The AC of T_0, T_1, and T_2 generations were measured after 3 months of harvesting. The total AC and SC were measured by using Megazyme Amylose Assay Kit (KAMYL), Guangzhou, China and Total Starch Assay Kit (AA/AMG), Guangzhou, China. The GC was evaluated for random five samples [62] and alkali digestion test was used to estimate GT [63].

2.9. Phenotyping

The data was recorded in five plants per line for GW (g) in T_0, T_1, and T_2 generation while the data for main agronomic traits was recorded in T_2 generation, such as plant height (PH) (cm), number of panicles (NOP), flag leaf length (FLL) (cm), flag leaf width (FLW) (cm), panicle length (PL) (cm), grains per spike (GPS), and seed setting rate (SSR) (%), as described previously [28].

2.10. Backcrossing and Observation of Pollen and Spikelet Fertility

The mutant maintainer lines T2-4-2, T2-7-1, T2-14-4, T2-19-3, T2-23-5, and 209A sterile line were crossed with two CMS lines MX-G1 and MX-G2 and during next season the 12 testcrosses along with respective male parents were transplanted and fertility test was performed at flowering stage. The pollen and spikelet fertility test were done by 1% (m/v) I_2-KI solution at flowering stage to evaluate the fertility restoration ability of restorer lines. The young spikelets were collected in the early morning to determine pollen fertility and kept in the jar about 2 h for opening the spikelets. The pollen was stained with KI solution and observed under a compound microscope. The stained pollens with round shape and well developed were considered as viable and irregular shaped and non-stained pollens were counted as sterile pollens. The criteria for classifying the parental lines as maintainers and restorers were followed as proposed previously [64]. Subsequent backcrosses were made between mutant lines as male and 209A as female to develop glutinous CMS lines.

2.11. Pollen Protein Analysis

Anthers of CMS line and maintainer line were taken from upper part of one panicle inside in the spikelet and located in the middle of panicles were collected and protein was extracted according to previous established method [65]. The protein separation was performed through SDS-PAGE gel electrophoresis [66] and proteins were identified by peptide mass fingerprinting (PMF) [67]. The analysis was performed with mass to charge ratio (m/z) formula to identify the monoisotopic masses and search in the NCBInr database using MASCOT (Matrix Science) software.

2.12. Statistical Analysis

The data were analyzed using SPSS 16.0 Statistical Software Program. The graphs were developed by GraphPad Prism (version 7.0, GraphPad Software Inc., San Diego, CA, USA).

3. Results

3.1. gRNA Design and Vector Construction

The 20bp long target sites were chosen in *Wx* and *TGW6* codon region and predictive cleavage site for the *TGW6* was 5pb upstream from initiation codon (ATG) and 72bp downstream of *Wx*. The sgRNA expression cassette was generated by overlapping PCR (Figure 3a) and ligating of the target adaptors to the *BsaI*-digested sgRNA intermediate plasmid and amplified products were successfully assembled in to pYLCRISPR/Cas9-MT(I) binary plasmid with Golden Gate ligation method (Ma et al. 2015). The ligation product was transformed to *E. coli* and positive clones were verified. The CRISPR/Cas9 binary plasmid was constructed and sequencing peaks confirmed that four targets were assembled in the plasmid successfully (Figure 3d,e).

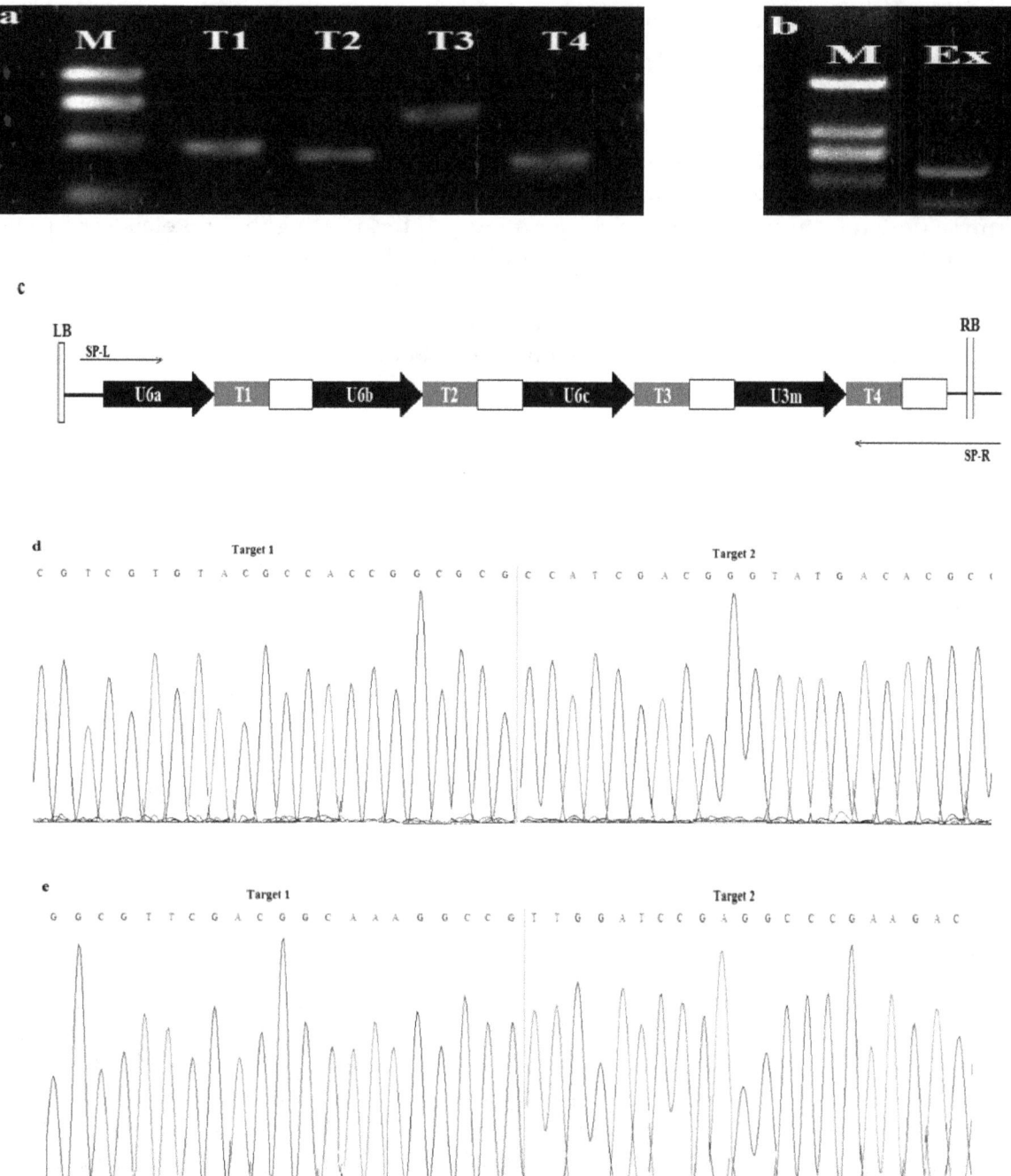

Figure 3. (**a**) sgRNA expression cassette after second round of PCR, M:2000, T1 (OsU6a-gRNA): 629 bp, T2 (OsU6b-sgRNA): 564 bp, T3 (OsU6c-sgRNA): 767 bp, and T4 (OsU3m-sgRNA): 515 bp, M: molecular marker. (**b**) sgRNA expression cassette after mixing and purification, Ex: expression cassette. (**c**) Illustration of the assembly of four sgRNA expression cassette into a pYLCRISPR/Cas9-MT(I) vector by single Golden Gate ligation. SP-L1 and SP-R are the flanking primers used to amplify the ligated sgRNA expression cassettes. (**d**) Sequencing results for the two target sequences of *Wx* gene and (**e**) and two targets of *TGW6*.

3.2. T_0 Genotyping and Off-Target Analysis

Total 55 positive mutant plants were obtained, and the DNA of 25 plants was extracted by CTAB method to analyze the mutations in target sites. The sequencing results showed that for Target1 of *Wx* gene there were 10 homozygous mutant plants, 5 heterozygous, 7 bi-allelic, and 1 chimeric and 2 were WT; while, for the Target2 of *Wx* gene, there were 4 homozygous mutant plants, 6 heterozygous, 10 bi-allelic and, 1 chimeric and 4 were WT. The mutation frequency for Target1 of TGW6 gene was as; 9 homozygous, 7 heterozygous, 6 bi-allelic, 1 chimeric, and 2 WT and for Target2 of TGW6 there were 5 homozygous, 8 heterozygous, 9 bi-allelic, 1 chimeric and 2 WT (Table 1). Five plants with good phenotype (4, 7, 14, 19, 23) were selected and identified with deletions of DNA fragments between distinct target sites (Figure 4a,b). In two of these homozygous mutant plants (4 and 7) were found with large fragment deletion at target sites of both genes. The mutant plant (14) with bi-alleic mutations in both genes was also found. The DSBs occurred either specifically in the upstream of the PAM (Figure 4), or at imprecise sites thus eliminating the genomic sequence beyond the PAMs. However, we analyzed the limited number of amplicons per transformant so maybe the actual number of mutations might be higher. In 25 of the plants the WT sequences of both genes were also detected, suggesting that Cas9 was inactive in these events. The average mutation types for both genes were 32% bi-allelic, 28% homozygous, 26% heterozygous, 4% chimeric, and 10.0% were WT (Table 1). The mutation rate for both genes was 90% and based on allele mutation types, for *Wx* gene 80% (4/5) of the mutations were simultaneous nucleotide deletions and insertions, 20% (1/5) of the mutations were only deletions with no insertions, and there were no mutations with only insertions (Figure 4a). The allelic mutation types for *TGW6* were 60% deletions, 40.0% simultaneous deletions and insertions and there was no mutation with only insertions (Figure 4b). As for the deletion mutations, the 40% mutations were large fragment deletions ranging from −11 to −120, while 60% were short (≤ 10 bp) deletions ranging from −1 to −6 and as for the insertion mutations, 90% (9/10) were 1 bp insertions and 10% (1/5) were +2 insertions (Figure 4). Comparison of WT and mutant's deduced amino acid sequences revealed that mutations resulted in changed conserved amino acid sequences (Figure 5).

The off-target predictions by CRISPR-P tool were analyzed and three off-target sites were selected for each target and examined by PCR based sanger sequencing in T_0 generation. The results showed that there were no off-target effects in the selected putative loci and targeted mutation were easily detected (Table S3).

Table 1. Mutation rate of T_0 generation

Gene	Target Site		Bi-Allelic	Homozygous	Heterozygous	Chimeric	WT	Total
				Mutation Type				
Wx	Target 1	No. of plants	7	10	5	1	2	25
		Mutation rate (%)	28	40	20	4	8	100
	Target 2	No. of plants	10	4	6	1	4	25
		Mutation rate (%)	40	16	24	4	16	100
TGW6	Target 1	No. of plants	6	9	7	1	2	25
		Mutation rate (%)	24	36	28	4	8	100
	Target 2	No. of plants	9	5	8	1	2	25
		Mutation rate (%)	36	20	32	4	8	100

WT: wild-type.

a

Mutant Plants	Target1 Sequence	PAM	Target2 Sequence	PAM	Mutation type
WT	GTCGTGTACGCCACCGGCGC_ _ _ _	CGG _ _ _ _	CATCGACGGGTATGACACGC_ _ _ _	_CGG	WT
Wx-4	GTCGTGT————————— _ _ _ _	CGG _ _ _ _	CATCGACGGTG——————	_CGG	−13, −10/+1
Wx-7	GTCGTGTACGGC————— _ _ _	CGG _ _ _ _	CATCTCGC———————— _ _ _	_CGG	−19/+1, −13/+1
Wx-14	GTCGTGTA——AACCGGCGC_ _ _ _	CGG _ _ _ _	CATCGAC——GTATGACACTGC_ _ _ _	CGG	−4/+1, −2/+1
			CATCGACGGAG——GACACGC_ _ _	_CGG	−3/+1
Wx-19	GTCGTGTACGCCACC——— _ _ _ _	CGG _ _ _ _	CATCGACGGGTAT—ACACGC_ _ _ _	_CGG	−4, −1
Wx-23	GTCGTGTACGCCA——CGC_ _ _ _	CGG _ _ _ _	CATCGAC—GGTATGTACACGC_ _ _ _	_CGG	−3, −1/+1

b

Mutant Plants	Target1 Sequence	PAM	Target2 Sequence	PAM	Mutation type
WT	GCGTTCGACGGCAAAGGCCG_ _ _ _	CGG _ _ _ _	TGGATCCGAGGCCCGAAGAC_ _ _ _	_TGG	WT
GW-4	GCGTT——————————— _ _ _	CGG _ _ _ _	TGGAT———————————	_TGG	−120, −42
GW-7	GCGTTCGACGG————— _ _ _	CGG _ _ _ _	TGGATCCGAGG—————— _ _ _	_TGG	−95, −55
GW-14	GCGTTCG———AAAGGCCG_ _ _ _	CGG _ _ _ _	TGGATCCGCGAGGC———AC_ _ _	_TGG	−5, −6/+2
			TGGA————CCCGAGAGAC_ _ _ _	_TGG	−7/+1
GW-19	GCGTT————————GCCG_ _ _ _	CGG _ _ _ _	TGGAT——GGCCCGAAGAC_ _ _	_TGG	−11, −4
GW-23	GCGTTCG——GCAAAGGCCG_ _ _ _	CGG _ _ _	TGGATCCGAGGTCCCGAAG——_ _ _	_TGG	−3, −2/+1

Figure 4. Nucleotide sequences at the target site in the 5 T_0 mutant rice plants. (**a**) Mutations induced at *Wx* and (**b**) *TGW6* target sites. The recovered mutated alleles are shown below the wild-type (WT) sequence. The target sites nucleotides are shown in black capital letters and black dashes. The red capital letters indicate inserted nucleotides and the apostrophe followed to red capital letters indicates inserted nucleotides are not shown in. The Protospacer Adjacent Motif (PAM) site nucleotides are shown in yellow background letters. The red dashes indicate the deleted nucleotides. − and + indicate deletion and insertion of the indicated number of nucleotides, respectively −/+ indicates simultaneous deletion and insertion of the indicated number of nucleotides; GW: grain weight.

a

```
WT      ------MNVVFVGAE---------MAPWSKTGGLGDVLGGLPPAMAANGHRVMVISPRYD
Wx-4    MVDEHTVDSTLVAMEKPHPLVPVCNLDLSHNAGIPSVL-------------VLVVPRRD
Wx-7    MVDEHTVDSTLVAMEKPHPLVPVCNLDLSHNAGIPSVL-------------VLVVPRRD
Wx-14   MVDEHTVDSTLVAMEKPHPLVPVCNLDLSHNAGIPSVL-------------VLVVPRRD
Wx-19   MVDEHTVDSTLVAMEKPHPLVPVCNLDLSHNAGIPSVL-------------VLVVPRRD
Wx-23   MVDEHTVDSTLVAMEKPHPLVPVCNLDLSHNAGIPSVL-------------VLVVPRRD
            ::.:*. *           *:.:*. .**

WT      QYKDAWDTSVVAEIKVADRYERVRFFHCYKRGVDRVFIDHPSFLEKVWGKTGEKIYGPDT
Wx-4    HHDPV----A---IR-----------SHGRGEAT--------EDVTEAAG-LAPGGHL
Wx-7    HHDPV----A---IR-----------SHGRGEAT--------EDVTEAAG-LAPGGHL
Wx-14   HHDPV----A---IR-----------SHGRGEAT--------EDVTEAAG-LAPGGHL
Wx-19   HHDPV----A---IR-----------SHGRGEAT--------EDVTEAAG-LAPGGHL
Wx-23   HHDPV----A---IR-----------SHGRGEAT--------EDVTEAAG-LAPGGHL
        ::.,.        *:           .: **     .: * .:*   * *

WT      GVDYKDNQMRFSLLCQAALEAPRILNLNNNPYFKGTYGEDVVFVCNDWHTGPLASYLKNN
Wx-4    GADEHDVH----------------------------------------------
Wx-7    GADEQGGV----------------------------------------------
Wx-14   GADEHDVH----------------------------------------------
Wx-19   GADEHDVH----------------------------------------------
Wx-23   GADEHDVH----------------------------------------------
        *.* :.

WT      YQPNGIYRNAKVAFCIHNISYQGRFAFEDYPELNLSERFRSSFDFIDGYDTPVEGR
Wx-4    -ADT----------------------------------------------
Wx-7    -HD-----------------------------------------------
Wx-14   -AGAGL---------MD---------------------------------
Wx-19   -AAVAY---------T----------------------------------
Wx-23   -AGGGV---------HD---------------------------------
```

b

```
WT      MR----WNGEAAGWSTYTYSPSYTKNKCAASTLPTVQTESKCGRPLGLRFHYKTGNLYIA
GW-4    MRSTARQAATAAAFA-------------------------------------------L
GW-7    ---------------------------------------------------------
GW-14   MR----WNGEAAGWSTYTYSPSYTKNKCAASTLPTVQTESKCGRPLGLRFHYKTGNLYIA
GW-19   MR----WNGEAAGWSTYTYSPSYTKNKCAASTLPTVQTESKCGRPLGLRFHYKTGNLYIA
GW-23   MR----WNGEAAGWSTYTYSPSYTKNKCAASTLPTVQTESKCGRPLGLRFHYKTGNLYIA

WT      DAYMGLMRVGPKGPKGGEATVLAMKADGVPLRFTNGVDIDQVTGDVYFTDSSMNYQRSQHEQV
GW-4    IVFLVLLSPSPTAAATATT---------RM--------------F--KTIDARRSQHLDL
GW-7    ---MGLMRVGPKGPKGGEATVLAMKADGVPLRFTNGVDIDQVTGDVYFTDSSMNYQRSQHEQV
GW-14   DAYMGLMRVGPKGPKGGEATVLAMKADGVPLRFTNGVDIDQVTGDVYFTDSSMNYQRSQHEQV
GW-19   DAYMGLMRVGPKGPKGGEATVLAMKADGVPLRFTNGVDIDQVTGDVYFTDSSMNYQRSQHEQV
GW-23   DAYMGLMRVGPKGPKGGEATVLAMKADGVPLRFTNGVDIDQVTGDVYFTDSSMNYQRSQHEQV
        : *:  .*...:;.              *:         *   .:::  :****  ::

WT      TATKDSTGRLMKYDPRTNQVTVLQSNITYPNGVAMSADRTHL--IVALTG-------PC
GW-4    G------GSLVG--P----ESVALRGIDSPHGPDREQMRPPVRPTVSLQNRQPVHRRRLH
GW-7    TATKDSTGRLMKYDPRTNQVTVLQSNITYPNGVAMSADRTHL--IVALTG-------PC
GW-14   TATKDSTGRLMKYDPRTNQVTVLQSNITYPNGVAMSADRTHL--IVALTG-------PC
GW-19   TATKDSTGRLMKYDPRTNQVTVLQSNITYPNGVAMSADRTHL--IVALTG-------PC
GW-23   TATKDSTGRLMKYDPRTNQVTVLQSNITYPNGVAMSADRTHL--IVALTG-------PC
        *. *   .   :.* .* *;.*    . *  :    *;.*.

WT      KLMRHWIRGPK--------
GW-4    GIDASWSKRRGGNRASHEG
GW-7    KLMRHWIR----------
GW-14   KLMRHWIRQA--------
GW-19   KLMRHWMARR--------
GW-23   KLMRHWIRGPE--------
        :    *
```

Figure 5. Amino acid sequence alignment for WT and five transformants in T_0 generation. (**a**) Amino acid sequences showing the alignment about *Wx* gene, and (**b**) *TGW6* mutant plants and WT. The deleted amino acids are shown by black hyphens, the translation was terminated earlier in mutants 4 and 7. Highly conserved and partially conserved amino-acid sequences are indicated with asterisks (*) and dash (-) signs, respectively.

3.3. Expression Level of Target Genes in WT and Mutant Lines

qPCR was used to detect the relative expression of TGW6 and *Wx* gene of T_0 plants. The expression of WT was not altered, and the expression of mutant plants was substantially downregulated in the *Wx* and *TGW6* mutants compared with WT ($P < 0.01$, Figure 6) indicating that mutations have successfully affected the target genes expression. The homozygous mutant lines (4,7) with large fragment deletions showed lower expression level of *Wx* and *TGW6* gene.

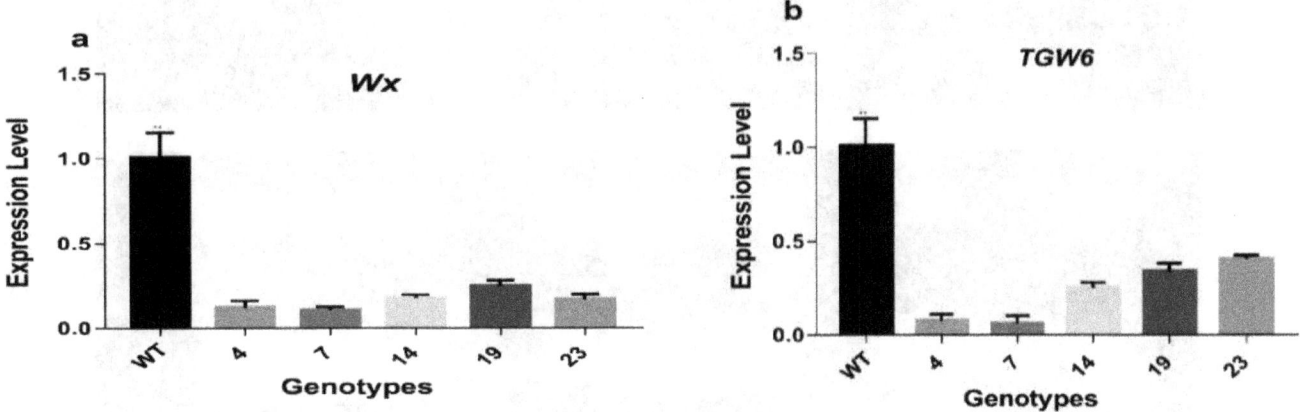

Figure 6. Relative expression analysis of target genes (**a**) *Wx* and (**b**) *TGW6* in wild type (WT) and mutant plants. Transcripts level was determined by Q-RT-PCR with cDNA generated from leaves of four-week-old plants. The expression values of the individual genes were normalized by using expression level of rice *Actin* gene as an internal standard. The data represents the mean values of three independent samples [Mean ± SD (standard deviation)]. ** indicates significant difference at $P < 0.01$ (*t* test).

3.4. Screening of T-DNA Free T_1 Generation and Seed Cross-Section Analysis

We addressed the genetically modified (GM) related regulations and issue of social acceptance of GM foods and to avoid public controversy by removing the transgenes from CRISPR/Cas9- edited waxy rice lines by self-pollination in the T_1 generation which allowed to produce non-GM lines containing the desired mutations. The DNA of T_1 and T_2 generations was extracted to investigate the possibility of obtaining rice lines harboring the desired modifications in target genes but without transferred DNA (T-DNA), the Cas9 gene specific primers Cas9-F and Cas9-R were used and amplified by PCR and T-DNA negative lines were selected for sequencing of the target regions. Notably, 13 T_1 plants were failed to generate the Cas9-specific amplicon (Figure 7a). Similarly, the PCR assay also failed to detect the Cas9 specific amplicon in the same 13 mutant lines of T_2 generation (Figure 7b). These results showed that T-DNA-free plants carrying the desired gene modifications can be acquired through genetic segregation in later generations.

Scanning electron microscopy of cross section of grain endosperm revealed that CRISPR/Cas9 mutant line showed opaque appearance compared to WT endosperm. The starch granules in the cross-section of WT grain were packed like polyhedral structure but in the mutant line a greater number of small and irregularly arranged starch granule structures were observed (Figure 7d–g), indicating a change in the mature grain of mutant line. The change in structure may cause the scattering of the light as it passes and resulted in opaque appearance.

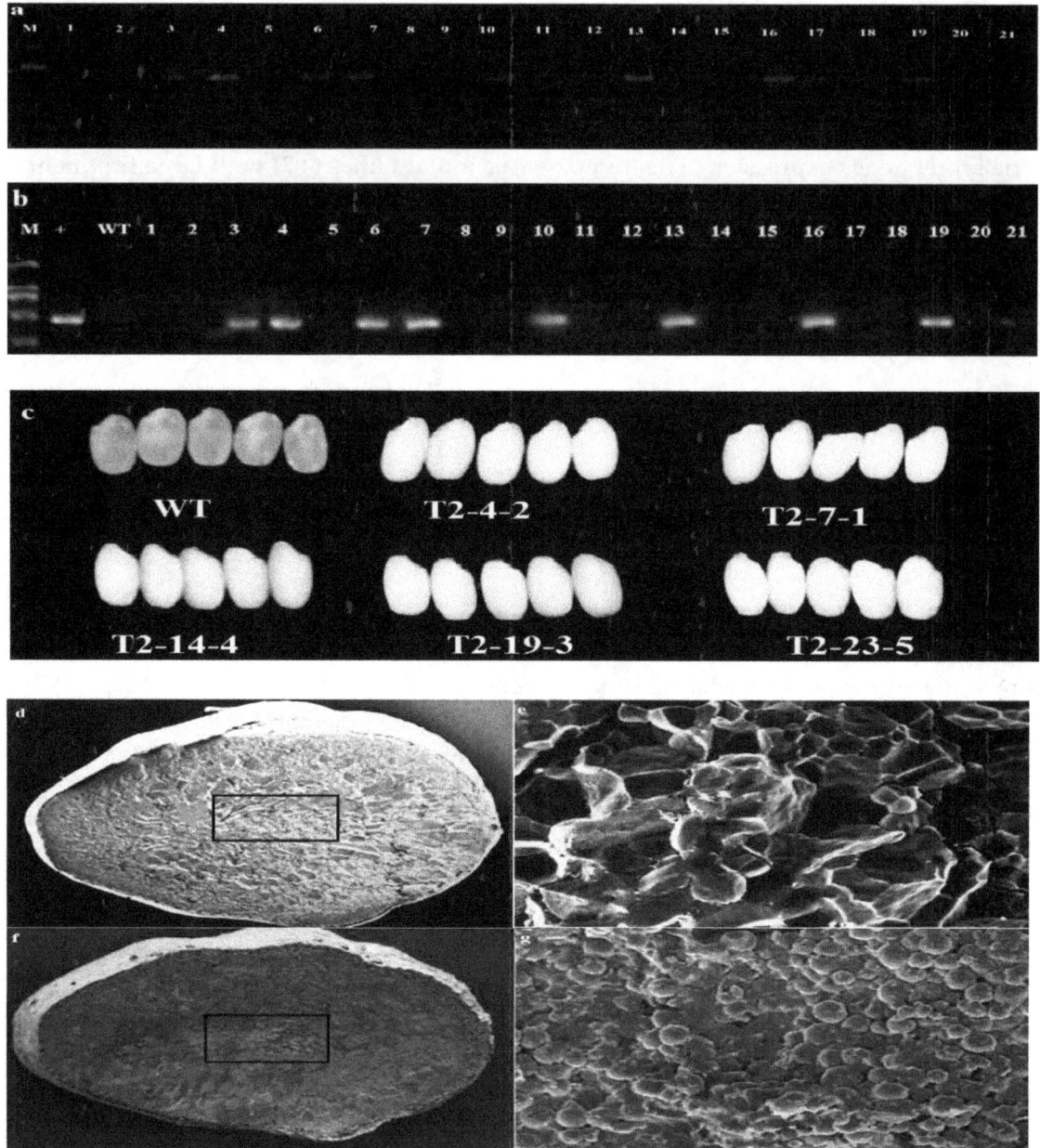

Figure 7. PCR-based identification of T-DNA-free rice mutant plants and seed analysis of WT and mutant line (T2-4-2). PCR products amplified from the progenies of (**a**) T_1 and (**b**) T_2 mutant lines genomic DNA using the Cas9 specific primers Cas9-F and Cas9-R. WT: wild-type, M: DNA molecular marker, +: positive control (**c**) Grain phenotype of WT and mutant lines. Cross-section analysis of endosperm in WT (**d,e**) and mutant line (**f,g**).

3.5. Transmission of Mutations in T_1 and T_2 Generations

The sequencing results showed that mutations in T_0 generation were not stable in some mutant plants but T_2 generation was completely stable. The mutant plants of 7-1, 14-4, and 19-3 lines in T_1 generation exhibited insertions and deletions in the Wx target regions, while 4-2 and 23-5 showed consistent mutations in T_0 and T_1 (Figure 8a; Figure 4a). The mutant plants of 4-2, 7-1, and 23-5 showed insertions and deletions at $TGW6$ target sites in T_1 generation, while mutations in 23-5 and 14-4 were similar in T_0 and T_1 generation (Figure 8b; Figure 4b). The transmission of mutations from T_1 to T_2 generation were investigated and sequencing results of T_2 plants showed that the mutations were consistent with the T_1 generation without any insertions or deletions, indicating that the T_2 generation has been stabilized (Figure 8a,b; Figure 4a,b).

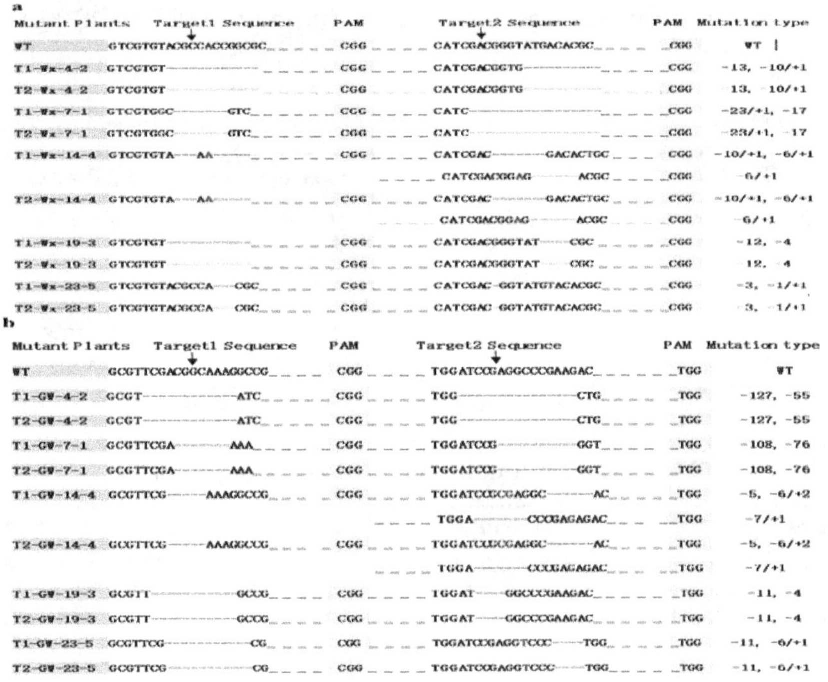

Figure 8. Sequence alignment for transmission of mutations at (**a**) *Wx* and (**b**) *TGW6* target sites in T_1 and subsequent T_2 generations. The targeted sequence is shown in capital black letters and the PAM sequence in yellow background. Insertion is represented by red letters, and deletion by red hyphens.

3.6. AC, GC, GT, and SC

AC of T_0, T_1, and T_2 generations were determined while the GC and GT was recorded for T_2 generation. The AC of mutant lines were significantly decreased ($P < 0.01$) as 18.2% to 1.7%. The homozygous mutant lines with long fragment deletion (4, 7) showed more decreased AC than heterozygous and bi-allelic mutants (Table 2). Grains of mutant plants were white and fully opaque in contrast with the typical non-waxy WT (Figure 7c). There was no effect on other grain quality traits, as result showed that total SC was unchanged in mutant and WT plants (Table 2). Another trait related to eating and cooking, GT, was also greatly decreased in the mutant plants and GC was increased as compared to the WT (Table 2).

Table 2. Amylose content, GC, and GT of WT and mutant plants.

T_0 plant	AC (%)	T_1 plants	AC (%)	T_2 Plants	AC (%)	GC (mm)	GT (ASV)	SC (%)
4	2.6 ± 0.5 **	4-2	1.7 ± 0.1 **	T2-4-2	1.8 ± 0.1 **	138.62 ± 2.8 **	3.12 ± 0.9 **	62.5 ± 1.4 ns
7	3.6 ± 0.3 **	7-1	2.2 ± 0.5 **	T2-7-1	2.1 ± 0.3 **	129.65 ± 3.9 **	3.09 ± 1.1 **	63.2 ± 2.3 ns
14	10.5 ± 0.2 **	14-4	2.4 ± 0.1 **	T2-14-4	2.6 ± 0.2 **	125.32 ± 4.6 **	3.21 ± 0.2 **	63.2 ± 1.9 ns
19	12.2 ± 0.1 **	19-3	2.8 ± 0.3 **	T2-19-3	2.7 ± 0.2 **	114.22 ± 2.6 **	3.19 ± 0.8 **	62.8 ± 2.9 ns
23	9.5 ± 0.6 **	23-5	3.2 ± 0.5 **	T2-23-5	3.1 ± 0.1 **	111.56 ± 5.2 **	3.24 ± 0.6 **	64.5 ± 3.4 ns
WT	18.2 ± 1.2	WT	17.6 ± 1.3	WT	18.1 ± 2.1	58.65 ± 3.7	5.67 ± 1.4	62.97 ± 2.7

Note: Data is shown the average of three independent samples, ** indicate significant difference; ns indicate non-significant difference, mm, millimeters; ASV: alkali spreading value. Data listed in table are presented as means ± SD, ($P < 0.01$). AC: amylose content; GC: gel consistency; GT: gelatinization temperature; SC: starch content.

3.7. Yield and Yield Contributing Traits

The results of the GW in T_0, T_1, and T_2 generation was recorded (Table 3), and results showed that the mutant plants significantly increased the GW (>5%). As expected, the GW of mutant line (T2-4-2) was 24.0 g maximum, whereas T2-7-1 presented a GW of 23.1 g (Table 3). The GW of T2-14-4, T2-19-3, and T2-23-5 was 23.4 g, 23.1 g and 23.7 g respectively, which was greater than the value of 21.1 g recorded in WT (Table 3). However, there were no significant differences detected in the other main agronomic traits between mutant lines and WT, including the PH, NOP, FLL, FLW, PL, GPS, and SSR (Table 4).

Table 3. 1000 grain weight (g) of mutant lines and WT in T_0, T_1, and T_2 generations.

T_0 plants	GW (g)	T_1 Plants	GW (g)	T_2 Plants	GW (g)
4	23.6 ± 0.50 *	4-2	23.9 ± 0.50 *	T2-4-2	24.0 ± 0.50 *
7	23.1 ± 0.57 *	7-1	23.8 ± 0.57 *	T2-7-1	23.1 ± 0.57 *
14	23.4 ± 0.50 *	14-4	23.7 ± 0.50 *	T2-14-4	23.4 ± 0.50 *
19	23.1 ± 0.47 *	19-3	23.3 ± 0.47 *	T2-19-3	23.1 ± 0.47 *
23	23.7 ± 0.38 *	23-5	23.7 ± 0.38 *	T2-23-5	23.7 ± 0.38 *
WT	21.0 ± 0.35	WT	21.3 ± 0.35	WT	21.1 ± 0.35

The data listed in the table are mean ± standard error. * indicate significant difference ($P < 0.01$). GW: grain weight.

Table 4. Main agronomic characters in T_2 generation.

T_2 Plants	PH (cm)	NOP	FLL (cm)	FLW (cm)	PL (cm)	GPS	SSR (%)
T2-4-2	84 ± 3.4 [ns]	9.5 ± 5.4 [ns]	42.3 ± 4.3 [ns]	1.7 ± 0.2 [ns]	25.1 ± 1.2 [ns]	191 ± 6.7 [ns]	87.8 ± 2.3 [ns]
T2-7-1	83 ± 4.5 [ns]	8.8 ± 2.6 [ns]	39.5 ± 5.2 [ns]	1.6 ± 0.1 [ns]	24.6 ± 2.3 [ns]	195 ± 6.9 [ns]	87.9 ± 4.5 [ns]
T2-14-4	85 ± 2.7 [ns]	9.5 ± 1.6 [ns]	43.2 ± 2.1 [ns]	1.5 ± 0.3 [ns]	23.9 ± 1.5 [ns]	196 ± 4.5 [ns]	88.9 ± 6.2 [ns]
T2-19-3	86 ± 3.6 [ns]	9.3 ± 3.4 [ns]	44.6 ± 1.9 [ns]	1.8 ± 0.4 [ns]	25.1 ± 2.4 [ns]	188 ± 7.6 [ns]	88.4 ± 1.3 [ns]
T2-23-5	85 ± 2.8 [ns]	9.2 ± 4.6 [ns]	43.3 ± 2.5 [ns]	1.6 ± 0.5 [ns]	23.9 ± 3.1 [ns]	193 ± 5.8 [ns]	86.9 ± 2.8 [ns]
WT	83 ± 4.6	9.5 ± 2.9	44.5 ± 3.6	1.7 ± 0.2	25.4 ± 1.9	192 ± 4.9	87.6 ± 4.6

PH: plant height; NOP: number of panicles; FLL: flag leaf length; FLW: flag leaf width; PL: panicle length; GPS: grains per spike; SSR: seed setting rate; The data listed in the table are mean ± standard error. [ns] indicate non-significant difference.

3.8. Pollen Fertility Status

In the T_3 generation, a total 16 lines (4-2A, 4-1A, 4-4B, 4-3A, 4-5C, 4-7A, 4-8A, 7-5B, 7-3A, 14-4A, 14-4C, 19-3C, 19-5A, 19-3B, 23-5B, 23-7A) were assessed for pollen fertility status and pollen fertility rate was randomly distributed, which showed that CRISPR/Cas9 mutations did not affect the fertility status of maintainer lines. Among 16 mutant lines, the 3 genotypes (4-2A, 19-5A, 19-3B) were completely sterile (CS), 3 (4-8A, 7-5B, 7-3A) were sterile (S) and 2 (4-1A, 4-4B) were partially sterile (PS) which was considered as male sterile lines or A line. Six genotypes were found sterile (CS and S) having pollen fertility 0–9% while two genotypes showed 10–29% pollen fertility. Two genotypes were recorded PS which is 12.5% of total (Table 5). Two genotypes were found partially fertile (PF) which is also 12.5% of total. Two genotypes were identified as fertile (F) and four were fully fertile (FF) as these genotypes had above 70% and 80% pollen and spikelet fertility respectively, which is 37.50% of the total genotypes (Table S4).

Table 5. Classification of mutant lines based on pollen fertility status.

S. No	Symbol	Fertility Status	Genotypes
1	CS	Completely Sterile	4-2A, 19-5A, 19-3B
2	S	Sterile	4-8A, 7-5B, 7-3A
3	PS	Partially Sterile	4-1A, 4-4B
4	PF	Partially fertile	23-5B, 23-7A
5	F	Fertile	4-5C, 4-7A,
6	FF	Highly/fully Fertile	4-3A, 14-4C, 19-3C, 14-4A

Note: Pollen sterility status was classified as, CS: (0%), S: (1–9%), PS: (10–29%), PF: (30–69%): F: (70–79%), FF: (≥80% and above).

To reduce the breeding cycles to develop glutinous rice lines, the developed mutant glutinous maintainer lines (as the male parent) were used to hybridize with CMS line 209A (as female parent) to produce F_1 hybrids, and then the F_1 hybrids were backcrossed with mutant lines. Molecular marker-assisted selection (MAS) was used to select the homozygous plants in the BC_1F_1 and CMS plants were selected and again backcrossed with mutant lines and two homozygous BC_2F_1 CMS lines (GX-B1, GX-B6) were obtained with no genetic segregation and increased yield and waxy-grain phenotype (Figure 7c), which will be used for the further breeding of hybrid glutinous rice.

3.9. Pollen Protein Identification

A total 25 spots in both genotypes were exercised and ultimately 16 spots were successfully identified, and proteins associated with pollen development (Table 6). The important proteins Putative acetyl-CoA synthetase and isoamylase were upregulated in both lines which clearly showed that these proteins have important role in CMS and control of AC in rice (Table 6).

Table 6. Rice pollen protein identified by peptide mass fingerprinting.

Sr. No.	Matched Protein	Organism	Accession No.	Mr/pI *	Spot Regulation	
					GX-B1	4-2A
1	20S proteasome beta 4 subunit	O. sativa	Q9LST6	23.6/5.42	+	−
2	Putative RNA-binding protein	O. sativa	Q852C0	97.3/9.34	+	−
3	Putative berberine bridge enzyme	O. sativa	Q84pv5	60.10/6.0	+	−
4	Putative mitochondrial NAD$^+$ -dependent malic enzyme	O. sativa	Q9FVY8	57.34/8.2	−	+
5	Putative calcium-binding protein annexin	O. sativa	Q84Q48	35.5/9.44	+	−
6	UDP-glucuronic acid decarboxylase	O. sativa	Q8W3J0	39.5/7.16	−	+
7	Putative phosphoribosyl pyrophosphate synthase	O. sativa	Q8S2E5	44.17/6.9	+	++
8	Putative RNA binding protein	O. sativa	Q7XC34	48.4/5.21	−	+
9	H$^+$ -transporting two-sector ATPase alpha chain–rice mitochondria	O. sativa	P15998	55.53/7.9	+	-
10	Glucose-1-phosphate adenylyltransferase large subunit 3	O. sativa	Q688T8	56.2/6.48	−	+
11	Putative membrane-associated salt-inducible protein	O. sativa	Q8W2V6	78.02/9.2	+	−
12	Putative leucine-rich repeat protein	O. sativa	Q6I5I5	29.58/9.6	++	+
13	Putative acetyl-CoA synthetase	O. sativa	Q6H798	78.5/5.69	+	++
14	Putative lipoamide dehydrogenase	O. sativa	Q94GU7	58.8/6.35	−	+
15	Isoamylase (fragrant)	O. sativa	D0TZF0	82.1/5.46	++	+
16	DNA binding protein	O. sativa	Q40691	33.0/8.96	+	+

* Molecular weight (Mr) and isoelectric point (pI) of matched proteins, + Indicates that the protein spot is present, − Indicates that the protein spot is absent, ++ Indicates more than a 2-fold increase. UDP: uridine diphosphate; NAD$^+$: nicotinamide adenine dinucleotide.

The identified proteins have various biological functions based on known functions from known functions from the European Bioinformatics Institute (EMBL-EBI) and literature. The protein spots related to GX-B1 were cellular protein catabolic process, RNA-binding, oxidoreductase activity, calcium-dependent phospholipid binding, nucleoside metabolic process, photosystem I assembly, acetyl-CoA biosynthetic process from acetate and amylopectin and their beta-limit dextrins (highly expressed). The protein spots of 4-2A were related to malate metabolic process (highly expressed), NAD$^+$ binding, nucleoside metabolic process, RNA binding, starch biosynthetic process, acetyl-CoA biosynthetic process from acetate (highly expressed) and amylopectin and their beta-limit dextrins.

4. Discussion

CRISPR/Cas9 is an emerging genome editing technology developed in past few years with high specificity and editing efficiency. Relative to ZFNs [68] and TALENs [69,70], CRISPR/Cas9 is simple and flexible and only one gRNA and one nuclease (Cas9) are needed to achieve the mutations in the DNA sequence of the target gene. Current research focuses on the development of CRISPR/Cas9 technology and specific gene knockouts.

Breeding for consumer-preferred grain yield and quality have thus become a major goal for breeding programs and in the last few decades, the classical, mutational, and molecular breeding approaches have brought about tremendous increase in rice productivity with the development of novel rice varieties for food security considerations. The improved living standards and fast economic growth are shifting public attention toward quality characteristics such as, nutrition, flavor, appearance, and cooking which are linked to starch physical properties. With the development of latest gene editing technologies such as CRISPR/Cas9, many yield related quantitative trait loci (QTLs) has been edited and their functions have been explored in different verities [37]. In the rice grain endosperm, starch is the major component consisting of a linear polysaccharide amylose which determines the cooking and eating quality of rice. Cooking of high AC (25–33%) verities results in separated, dry and firm rice grains, becoming hard after cooling while glutinous rice with low AC (5–20%) is especially sticky and soft when cooked [71]. The improvement of maintainer line in hybrid rice breeding system is most inevitable to achieve target traits. In China, the *indica* hybrids considered low quality owing to high AC that makes them hard and dry when cooked. The breeding for low AC and improved grain quality and yield is a major objective of breeders.

In this study the CRISPR/Cas9 construct with 20-nt target sequence for the sgRNA was carefully designed with high GC content and low off-target score and the *Wx* and *TGW6* gene with expectation to produce a null mutation were edited. The schematic representation of whole procedure of generation and analysis of targeted mutated plants was described in Figure S6. The goal of this study was to develop a high yielding CMS rice line with low amylose content to facilitate the hybrid rice breeding program and proteins from anthers of maintainer and CMS line were also identified. We sequenced the Wx and TGW6 gene and confirmed that 209B contains both genes (Figure S7). Four targets were designed in the exon regions of *Wx* and *TGW6* gene, the corresponding promoters were OsU6a, OsU6b, OsU6c, and OsU3m and *Agrobacterium tumefaciens* based transformations was successfully achieved with the CRISPR/Cas9 cassette and mutations in the target regions were analyzed by sanger sequencing by decoding it using online DSDecodeM tool. The results of this study indicate that the CRISPR/Cas9 gene editing technology can successfully edit rice targeting DNA sequences with high efficiency and multiple mutations can be generated at the same target site, and base deletion or insertion occurs before the target site PAM.

The total mutation frequency was up to 90%, wherein homozygous mutations were about 28%, which indicate that the CRISPR/Cas9 editing facilitates homozygous mutations in the T_0 generation (Table 1). The previous studies showed that the CRISPR/Cas9 induced the homozygous mutations in T_0 generation and mutations mainly take place in transformed calli cells [72]. The expression level of targeted genes was lower in mutant lines than WT (Figure 6). The off-target mutations were not detected for all targets (Table S3). The comparison of T_0 and T_1 generations showed that the mutation frequency of homozygotes was stably inherited regardless of whether T-DNA is present. The conserved amino acid sequence was totally changed in mutant plants and mutant plants showed divergence to WT in amino acid sequence alignment (Figure 5).

The glutinous rice lines were obtained, and all mutant lines seeds showed low AC decreased from 18.2% to 1.7 % and homozygous mutant lines showed less percentage of AC than heterozygous and bi-allelic mutants (Table 2). It is reported that the *Wx* gene also affect the GC and GT of rice [22,73] and our results showed an increase in GC from 58.65 mm to 138.62 mm and decrease in GT from 5.67 ASV to 3.12 ASV, while there was no effect on total SC (Table 2). The GW was increased from 21.1 g to 240.8 g (Table 3), while there was no effect on PH, NOP, FLL, FLW, PL, GPS, and SSR (Table 4). The cross-section analysis by electron microscope showed that endosperm of mutant grains was shrunken corresponding to their WT. The T-DNA free lines were obtained to address the social values of laws about genetically modified (GM) foods by selecting the transgene free lines by self-pollination in the T_1 and T_2 generations (Figure 7). Our results showed that the T_1 mutant lines were re-edited while mutations were inherited and stable in T_2 generation (Figure 8). The T_0 lines are frequently difficult to predict which suggests that the mutations in T_0 generations are not stable but the mutations in T_1 generations transmitted stably to later generations. These results are consistent with previous reports that the editing site of the T_1 generation mutant plant target sequence may also have a sequence recognized by the gRNA target, resulting in re-editing, which makes the T_1 generation unpredicted which can stabilize in later generations [74]. Together, these results clearly demonstrate that CRISPR/Cas9-induced gene mutations can be stably transmitted to subsequent generations.

The shape of pollen grains and staining patterns in male sterility inducing cytoplasm and sterility maintaining nuclear genes are influenced by the pollen abortion stage related to nuclear stage [75]. Mutant maintainer lines were assessed for pollen fertility status and results showed that pollen fertility rate was randomly distributed and six genotypes were found sterile having pollen fertility 0–9%, two genotypes PS with 10–29% pollen fertility, two genotypes were recorded PF which is 12.5% of total and four genotypes were identified as CF and two were F as these genotypes had above 80% pollen and spikelet fertility which is 37.50% of the total genotypes (Table 5). The developed maintainer lines were crossed with CMS line to develop F_1 and after subsequent backcrossing glutinous CMS line was achieved.

The CMS lines has been widely used in hybrid rice production, but the molecular mechanism of CMS remains poor understood. The protein identification tool is a powerful tool to study anther development and pollen production in plants [76–79]. The CMS is different plant species are cause by a specific ORF containing chimeric genes in mitochondrial genome [80] with rare similarity but sharing same male gamete abortion phenomenon [81]. The mitochondrial amplification events suggest an increased demand for energy during pollen development [82] but lowered ATP production was also observed in some CMS flowers [81,83]. In this study the proteins identified in maintainer and CMS line helped to understand the molecular mechanism of rice male sterility. Sixteen proteins were identified between sterile and maintainer anthers (Table 6). The identified proteins have potential roles in anther and pollen development and may help to clarify the mechanism of male sterility in rice.

The proteins of CMS line and GX4-2 maintainer mutant line anthers were separated by two-dimensional electrophoresis and SDS-PAGE as the second. The silver stained proteins were analyzed using Image Master 2D software. The identified proteins were, 20S proteasome beta 4 subunit, putative RNA-binding protein, Putative berberine bridge enzyme (BBE), putative mitochondrial NAD$^+$-dependent malic enzyme, Putative calcium-binding protein annexin, UDP-glucuronic acid decarboxylase, putative phosphoribosyl pyrophosphate synthase, putative RNA binding protein (RBP), H$^+$-transporting two-sector ATPase alpha chain–rice mitochondria, glucose-1-phosphate adenylyltransferase large subunit 3, putative membrane-associated salt-inducible protein, putative leucine-rich repeat protein, putative acetyl-CoA synthetase (ACOS), putative lipoamide dehydrogenase, Isoamylase (fragrant), and DNA binding protein (Table 6). These proteins are closely associated with metabolism, protein biosynthesis, transcription, signal transduction and many other activities which are important in cell activities and essential to pollen development.

Dysfunctions of mitochondria in the pollen caused CMS in plants and several other mitochondrion regions have been identified associated with CMS [84]. ATP synthase β-subunit helps to fulfill the demand of energy for respiratory function and cellular energy to develop male gametophyte also observed in mitochondria [85], and defective β-subunit resulted non-functional pollens and abnormal anther development [84]. The 20S proteasome is the proteolytic complex actively involved in removing abnormal proteins with several biological functions [86], while RBP is involved to regulate transcriptional and post-transcriptional levels to control the gene expression. Plants respond to pathogen infection with rapid reprogramming of gene expression and loss of function of RBP showed enhanced resistance to pathogens [77]. Biochemical and biological function of BBE are unexplored [87]. Plant annexins regulate diverse aspects of plant development, stress responses and growth [88]. ACOS played role in plastids and in several metabolic pathways [89] and has significant role in anther development [90,91]. The ACOS in anther prevent the conversion of pyruvate into acetyl-CoA which leads to pollen sterility. The degeneration and formation of various tissues during pollen development needs high energy for key biosynthetic intermediates. Isoamylase in combination with pullulanase plays a predominant role in amylopectin synthesis and also essential for the construction of the amylopectin multiple-cluster structure by removing the excessive branches to avoid interference with the formation of double helices of the cluster chains of amylopectin and crystallization of starch in the endosperm. These proteins or enzymes are involved in multiple physiological and biochemical reactions such as carbon metabolism and starch synthesis, as well as signal transduction and protein expression regulation [92].

In short, the increase yield and reduction of AC are valuable parameters in crop breeding and CRISPR/Cas9 is excellent technology to achieve targeted mutations in genes. In this study the rice maintainer line and new CMS lines were developed with increased yield and improved quality while maintaining all agronomic traits. We also took precautionary approach and produced T-DNA-free plants to avoid foreign bacterial DNA integration and bypass GMO rules. The most likely off-target effects were analyzed and Cas9 free plants were selected for food safety assessments and it was ensured that the other plant traits were not affected. In our work, we improved existing traits by directly rewriting the plant genetic code without any cutting and pasting genes from animals or bacteria into rice plants. Our study provides some insights to study the gene functions and generation of new rice CMS lines with increased yield and improved quality without compromising on nutritional value to facilitate the hybrid breeding programs of rice to develop elite crop verities. This study is the first example to develop rice CMS lines with increased yield and low AC and the protein identification in mutant rice maintainer and CMS line which will be the source material for further breeding of hybrid glutinous rice verities in short period. The identified proteins in anther of maintainer and CMS lines provide the insights to the actual mechanism underlying in sterility of rice lines. The study showed the genetic mutations are not only helpful to improve the plant characteristics, they also help in understanding the mechanisms underlying the biochemical behavior changes in cell of the plants.

5. Conclusions

The CRISPR/Cas9 technology induces fastest changes to plant genome than other molecular approaches and mutations passed to the next generations without any rewriting or emendations. Different types of mutations were achieved for both genes and a mutant library was generated which laid an important material basis for further high-yield and stable hybrid breeding of rice. This study provides an important theoretical and practical significance and reference for the rapid creation of excellent rice germplasm with important application value such as rice quality, and male sterility, and is expected to provide a safe and efficient new way for rice germplasm resources innovation. This study applied a proteomic approach to identify the regulating proteins of a CMS and mutant maintainer rice line and it is concluded that pollen development in different genetic material is associated with the differential expression of several proteins. These results collectively suggested that the knowledge of these parameters in rice breeding may be further applied as criteria to develop rice CMS lines. The new germplasm with important application value was obtained which laid an important material basis for further breeding program to facilitate the rice breeding to improve yield and quality.

Supplementary Materials: Figure S1: Schematic representation of secondary structures of both sgRNAs used in this experiment. (a) structure both sgRNA's for Wx targets and (b) structure of both sgRNA's for *TGW6* targets. The stem loop sgRNA secondary structure was predicted by online tool (http://crispr.hzau.edu.cn/cgi-bin/CRISPR/ CRISPR); Figure S2: Isolation of the binary plasmids and sgRNA intermediate plasmids; Figure S3: Sequences of the sgRNA vectors and those of the expression cassettes; Figure S4: sgRNA expression cassette procedure by overlapping PCR containing a target sequence. The chimeric primers with target sequence strands are given in additional file 3. The first PCR is carried out in two separated reactions with U-F/U#T#- and gRT#+/gR-R primer pair, U# indicates a given promoter, and T#+ and T#- indicate forward and reverse strands of a target sequence; Figure S5: Illustration for transformation of *E. coli*; Figure S6: Schematic diagram of the procedure for CRISPR/Cas9 based generation of mutant plants and analysis of target regions. The targets were selected using CRISPR-GE online web-based tool and expression cassette was constructed by using overlapping PCR and inserted into a binary vector. Agrobacterium mediated transformation was performed and T_0 plants were regenerated and sequencing was performed, and later generations were produced by self-pollination and genotyping was performed by using target specific primers in T_1 and T_2 generations. The phenotypic data of mutant and wild type plants were recorded and further analyzed. Pollen fertility analysis and protein identification was also performed; Figure S7: Sequence alignment of the (a) Wx and (b) TGW6 gene in reference genome and 209B maintainer line. The SNPs between reference genome and 209B are indicated in red box; Table S1: Efficiency score and positions of four targets; Table S2: List of primers used in the study; Table S3: Detection of mutations on the putative off-target sites; Table S4: Pollen fertility status of F_1 lines.

Author Contributions: Conceptualization, Y.H. and D.L.; Data curation, B.U., G.N. and F.L.; Formal analysis, Y.H., D.L., B.U. and N.Z.; Funding acquisition, R.L.; Investigation, D.L., B.U. and N.Z.; Methodology, H.Y., D.L. and B.U.; Project administration, R.L.; Resources, R.L.; Software, B.U.; Supervision, R.L.; Validation, F.L. and R.L.; Visualization, G.N., F.L. and R.L.; Writing—original draft. D.L., B.U. and G.N.; Writing—review & editing, R.L.

Acknowledgments: We would like to thank Li Zhihua and Qin Baoxiang for the helpful discussion and invaluable comments to make this research meaningful. We are highly grateful to Liu Yaoguang for providing us the vector and promoters for the experiment.

Accession Numbers: Sequence data from this article can be found in the GenBank data library under accession numbers GenBank: KR029105, KR029107, KR029108 and KR559259 for the sgRNA intermediate plasmids and GenBank: KR029109 for the CRISPR/Cas9 binary vector.

References

1. Birla, D.S.; Malik, K.; Sainger, M.; Chaudhary, D.; Jaiwal, R.; Jaiwal, P.K. Progress and challenges in improving the nutritional quality of rice (*Oryza sativa* L.). *Crit. Rev. Food Sci. Nutr.* **2017**, *57*, 2455–2481. [CrossRef] [PubMed]

2. Pérez-Montaño, F.; Alías-Villegas, C.; Bellogín, R.; Del Cerro, P.; Espuny, M.; Jiménez-Guerrero, I.; López-Baena, F.J.; Ollero, F.; Cubo, T. Plant growth promotion in cereal and leguminous agricultural important plants: From microorganism capacities to crop production. *Microbiol. Res.* **2014**, *169*, 325–336. [CrossRef] [PubMed]

3. Milovanovic, V.; Smutka, L. Asian Countries in the Global Rice Market. *Acta Univ. Agric. Silvic. Mendel. Brun.* **2017**, *65*, 679–688. [CrossRef]

4. Hsu, Y.C.; Tseng, M.C.; Wu, Y.P.; Lin, M.Y.; Wei, F.J.; Hwu, K.K.; Hsing, Y.I.; Lin, Y.R. Genetic factors responsible for eating and cooking qualities of rice grains in a recombinant inbred population of an inter-subspecific cross. *Mol. Breed.* **2014**, *34*, 655–673. [CrossRef] [PubMed]

5. Yuan, L.; Tang, C. Retrospect, current status and prospect of hybrid rice. *Rice in China* **1999**, *4*, 3–6.

6. Cheng, S.H.; Zhuang, J.Y.; Fan, Y.Y.; Du, J.H.; Cao, L.Y. Progress in research and development on hybrid rice: a super-domesticate in China. *Ann. Bot.* **2007**, *100*, 959–966. [CrossRef] [PubMed]

7. Sharma, S.; Singh, S.; Nandan, R.; Kumar, M. Identification of restorers and maintainers for CMS lines of rice (*Oryza sativa* L.). *Ind. J. Plant Gen. Res.* **2012**, *25*, 186–188. [CrossRef]

8. Umesh, S.; Lal, J.; Kumar, H. Isolation and evaluation of restorers and examining possibility of developing new version of CMS lines for upland rainfed rice hybrids. *Environ. Ecol.* **2012**, *30*, 872–876.

9. Xing, Y.; Zhang, Q. Genetic and molecular bases of rice yield. *Annu. Rev. Plant Biol.* **2010**, *61*, 421–442. [CrossRef]

10. You, A.; Lu, X.; Jin, H.; Ren, X.; Liu, K.; Yang, G.; Yang, H.; Zhu, L.; He, G. Identification of quantitative trait loci across recombinant inbred lines and testcross populations for traits of agronomic importance in rice. *Genetics* **2006**, *172*, 1287–1300. [CrossRef]

11. Wani, A.A.; Singh, P.; Shah, M.A.; Schweiggert-Weisz, U.; Gul, K.; Wani, I.A. Rice starch diversity: Effects on structural, morphological, thermal, and physicochemical properties—A review. *Compr. Rev. Food Sci. Food Saf.* **2012**, *11*, 417–436. [CrossRef]

12. Wang, Z.Y.; Zheng, F.Q.; Shen, G.Z.; Gao, J.P.; Snustad, D.P.; Li, M.G.; Zhang, J.L.; Hong, M.M. The amylose content in rice endosperm is related to the post-transcriptional regulation of the waxy gene. *Plant J.* **1995**, *7*, 613–622. [CrossRef] [PubMed]

13. Isshiki, M.; Morino, K.; Nakajima, M.; Okagaki, R.J.; Wessler, S.R.; Izawa, T.; Shimamoto, K. A naturally occurring functional allele of the rice waxy locus has a GT to TT mutation at the 5′ splice site of the first intron. *Plant J.* **1998**, *15*, 133–138. [CrossRef] [PubMed]

14. Sano, Y. Differential regulation of waxy gene expression in rice endosperm. *Theor. Appl. Genet.* **1984**, *68*, 467–473. [CrossRef] [PubMed]

15. Larkin, P.D.; Park, W.D. Association of waxy gene single nucleotide polymorphisms with starch characteristics in rice (*Oryza sativa* L.). *Mol. Breed.* **2003**, *12*, 335–339. [CrossRef]

16. Bligh, H.F.J.; Larkin, P.D.; Roach, P.S.; Jones, C.A.; Fu, H.; Park, W.D. Use of alternate splice sites in granule-bound starch synthase mRNA from low-amylose rice varieties. *Plant Mol. Biol.* **1998**, *38*, 407–415. [CrossRef]

17. Sato, H.; Suzuki, Y.; Sakai, M.; Imbe, T. Molecular characterization of Wx-mq, a novel mutant gene for low-amylose content in endosperm of rice (*Oryza sativa* L.). *Breed. Sci.* **2002**, *52*, 131–135. [CrossRef]

18. Bergman, C.; Delgado, J.; McClung, A.; Fjellstrom, R. An improved method for using a microsatellite in the rice waxy gene to determine amylose class. *Cereal Chem.* **2001**, *78*, 257–260. [CrossRef]

19. Inukai, T.; Sako, A.; Hirano, H.Y.; Sano, Y. Analysis of intragenic recombination at wx in rice: correlation between the molecular and genetic maps within the locus. *Genome* **2000**, *43*, 589–596. [CrossRef]

20. Ma, X.; Zhang, Q.; Zhu, Q.; Liu, W.; Chen, Y.; Qiu, R.; Wang, B.; Yang, Z.; Li, H.; Lin, Y. A robust CRISPR/Cas9 system for convenient, high-efficiency multiplex genome editing in monocot and dicot plants. *Mol. Plant.* **2015**, *8*, 1274–1284. [CrossRef]

21. Terada, R.; Nakajima, M.; Isshiki, M.; Okagaki, R.J.; Wessler, S.R.; Shimamoto, K. Antisense waxy genes with highly active promoters effectively suppress waxy gene expression in transgenic rice. *Plant Cell Physiol.* **2000**, *41*, 881–888. [CrossRef] [PubMed]

22. Liu, Q.; Wang, Z.; Chen, X.; Cai, X.; Tang, S.; Yu, H.; Zhang, J.; Hong, M.; Gu, M. Stable inheritance of the antisense Waxy gene in transgenic rice with reduced amylose level and improved quality. *Transgenic Res.* **2003**, *12*, 71–82. [CrossRef] [PubMed]

23. Liu, Q.; Yu, H.; Chen, X.; Cai, X.; Tang, S.; Wang, Z.; Gu, M. Field performance of transgenic indica hybrid rice with improved cooking and eating quality by down-regulation of Wx gene expression. *Mol. Breed.* **2005**, *16*, 199–208. [CrossRef]

24. Itoh, K.; Ozaki, H.; Okada, K.; Hori, H.; Takeda, Y.; Mitsui, T. Introduction of Wx transgene into rice wx mutants leads to both high-and low-amylose rice. *Plant Cell Physiol.* **2003**, *44*, 473–480. [CrossRef] [PubMed]

25. Shomura, A.; Izawa, T.; Ebana, K.; Ebitani, T.; Kanegae, H.; Konishi, S.; Yano, M. Deletion in a gene associated with grain size increased yields during rice domestication. *Nat. Genet.* **2008**, *40*, 1023. [CrossRef] [PubMed]

26. Weng, J.; Gu, S.; Wan, X.; Gao, H.; Guo, T.; Su, N.; Lei, C.; Zhang, X.; Cheng, Z.; Guo, X. Isolation and initial characterization of GW5, a major QTL associated with rice grain width and weight. *Cell Res.* **2008**, *18*, 1199. [CrossRef]

27. Ishimaru, K.; Hirotsu, N.; Madoka, Y.; Murakami, N.; Hara, N.; Onodera, H.; Kashiwagi, T.; Ujiie, K.; Shimizu, B.-i.; Onishi, A. Loss of function of the IAA-glucose hydrolase gene TGW6 enhances rice grain weight and increases yield. *Nat. Genet.* **2013**, *45*, 707. [CrossRef]

28. Fan, C.; Xing, Y.; Mao, H.; Lu, T.; Han, B.; Xu, C.; Li, X.; Zhang, Q. GS3, a major QTL for grain length and weight and minor QTL for grain width and thickness in rice, encodes a putative transmembrane protein. *Theor. Appl. Genet.* **2006**, *112*, 1164–1171. [CrossRef]

29. Li, Y.; Fan, C.; Xing, Y.; Jiang, Y.; Luo, L.; Sun, L.; Shao, D.; Xu, C.; Li, X.; Xiao, J. Natural variation in GS5 plays an important role in regulating grain size and yield in rice. *Nat. Genet.* **2011**, *43*, 1266–1269. [CrossRef]

30. Song, X.J.; Huang, W.; Shi, M.; Zhu, M.Z.; Lin, H.X. A QTL for rice grain width and weight encodes a previously unknown RING-type E3 ubiquitin ligase. *Nat. Genet.* **2007**, *39*, 623. [CrossRef]

31. Wang, S.; Wu, K.; Yuan, Q.; Liu, X.; Liu, Z.; Lin, X.; Zeng, R.; Zhu, H.; Dong, G.; Qian, Q. Control of grain size, shape and quality by *OsSPL16* in rice. *Nat. Genet.* **2012**, *44*, 950. [CrossRef] [PubMed]

32. Hu, Z.; He, H.; Zhang, S.; Sun, F.; Xin, X.; Wang, W.; Qian, X.; Yang, J.; Luo, X. A Kelch motif-containing serine/threonine protein phosphatase determines the large grain QTL trait in rice. *J. Integr. Plant Biol.* **2012**, *54*, 979–990. [CrossRef] [PubMed]

33. Qi, P.; Lin, Y.S.; Song, X.J.; Shen, J.B.; Huang, W.; Shan, J.X.; Zhu, M.Z.; Jiang, L.; Gao, J.P.; Lin, H.X. The novel quantitative trait locus *GL3. 1* controls rice grain size and yield by regulating Cyclin-T1; 3. *Cell Res.* **2012**, *22*, 1666–1680. [CrossRef] [PubMed]

34. Zhang, X.; Wang, J.; Huang, J.; Lan, H.; Wang, C.; Yin, C.; Wu, Y.; Tang, H.; Qian, Q.; Li, J. Rare allele of *OsPPKL1* associated with grain length causes extra-large grain and a significant yield increase in rice. *Proc. Natl. Acad. Sci. USA* **2012**, *109*, 21534–21539. [CrossRef] [PubMed]

35. Wang, S.; Li, S.; Liu, Q.; Wu, K.; Zhang, J.; Wang, S.; Wang, Y.; Chen, X.; Zhang, Y.; Gao, C. The *OsSPL16-GW7* regulatory module determines grain shape and simultaneously improves rice yield and grain quality. *Nat. Genet.* **2015**, *47*, 949. [CrossRef] [PubMed]

36. Si, L.; Chen, J.; Huang, X.; Gong, H.; Luo, J.; Hou, Q.; Zhou, T.; Lu, T.; Zhu, J.; Shangguan, Y. OsSPL13 controls grain size in cultivated rice. *Nat. Genet.* **2015**, *47*, 447. [CrossRef] [PubMed]

37. Zhang, H.; Zhang, J.; Lang, Z.; Botella, J.R.; Zhu, J.K. Genome editing—Principles and applications for functional genomics research and crop improvement. *Crit. Rev. Plant Sci.* **2017**, *36*, 291–309. [CrossRef]

38. Li, M.; Li, X.; Zhou, Z.; Wu, P.; Fang, M.; Pan, X.; Lin, Q.; Luo, W.; Wu, G.; Li, H. Reassessment of the four yield-related genes Gn1a, DEP1, GS3, and IPA1 in rice using a CRISPR/Cas9 system. *Front. Plant Sci.* **2016**, *7*, 377. [CrossRef]

39. Xu, R.; Yang, Y.; Qin, R.; Li, H.; Qiu, C.; Li, L.; Wei, P.; Yang, J. Rapid improvement of grain weight via highly efficient CRISPR/Cas9-mediated multiplex genome editing in rice. *J. Genet. Genom.* **2016**, *43*, 529–532. [CrossRef]

40. Bortesi, L.; Fischer, R. The CRISPR/Cas9 system for plant genome editing and beyond. *Biotechnol. Adv.* **2015**, *33*, 41–52. [CrossRef]

41. Baltes, N.J.; Gil-Humanes, J.; Cermak, T.; Atkins, P.A.; Voytas, D.F. DNA replicons for plant genome engineering. *Plant Cell.* **2014**, *26*, 161–163. [CrossRef] [PubMed]

42. Miglani, G.S. Genome editing in crop improvement: Present scenario and future prospects. *J. Crop Improv.* **2017**, *31*, 453–559. [CrossRef]

43. Feng, Z.; Zhang, B.; Ding, W.; Liu, X.; Yang, D.L.; Wei, P.; Cao, F.; Zhu, S.; Zhang, F.; Mao, Y. Efficient genome editing in plants using a CRISPR/Cas system. *Cell Res.* **2013**, *23*, 1229–1232. [CrossRef] [PubMed]

44. Voytas, D.F.; Gao, C. Precision genome engineering and agriculture: opportunities and regulatory challenges. *PLoS Biol.* **2014**, *12*, e1001877. [CrossRef] [PubMed]

45. Belhaj, K.; Chaparro-Garcia, A.; Kamoun, S.; Patron, N.J.; Nekrasov, V. Editing plant genomes with CRISPR/Cas9. *Curr. Opin. Biotechnol.* **2015**, *32*, 76–84. [CrossRef] [PubMed]

46. Weeks, D.P.; Spalding, M.H.; Yang, B. Use of designer nucleases for targeted gene and genome editing in plants. *Plant Biotechnol. J.* **2016**, *14*, 483–495. [CrossRef] [PubMed]

47. Doudna, J.A.; Charpentier, E. The new frontier of genome engineering with CRISPR-Cas9. *Science* **2014**, *346*, 1258096. [CrossRef]

48. Khatodia, S.; Bhatotia, K.; Passricha, N.; Khurana, S.; Tuteja, N. The CRISPR/Cas genome-editing tool: Application in improvement of crops. *Front. Recent Dev. Plant Sci.* **2016**, *7*, 506. [CrossRef]

49. Nekrasov, V.; Staskawicz, B.; Weigel, D.; Jones, J.D.; Kamoun, S. Targeted mutagenesis in the model plant Nicotiana benthamiana using Cas9 RNA-guided endonuclease. *Nat. Biotechnol.* **2013**, *31*, 691. [CrossRef]

50. Jiang, W.; Zhou, H.; Bi, H.; Fromm, M.; Yang, B.; Weeks, D.P. Demonstration of CRISPR/Cas9/sgRNA-mediated targeted gene modification in Arabidopsis, tobacco, sorghum and rice. *Nucleic Acids Res.* **2013**, *41*, e188. [CrossRef]

51. Miao, J.; Guo, D.; Zhang, J.; Huang, Q.; Qin, G.; Zhang, X.; Wan, J.; Gu, H.; Qu, L.J. Targeted mutagenesis in rice using CRISPR-Cas system. *Cell Res.* **2013**, *23*, 1233–1236. [CrossRef]

52. Shan, Q.; Wang, Y.; Li, J.; Zhang, Y.; Chen, K.; Liang, Z.; Zhang, K.; Liu, J.; Xi, J.J.; Qiu, J.L. Targeted genome modification of crop plants using a CRISPR-Cas system. *Nat. Biotechnol.* **2013**, *31*, 686–688. [CrossRef]

53. Wang, Y.; Cheng, X.; Shan, Q.; Zhang, Y.; Liu, J.; Gao, C.; Qiu, J.L. Simultaneous editing of three homoeoalleles in hexaploid bread wheat confers heritable resistance to powdery mildew. *Nat. Biotechnol.* **2014**, *32*, 947. [CrossRef] [PubMed]

54. Char, S.N.; Neelakandan, A.K.; Nahampun, H.; Frame, B.; Main, M.; Spalding, M.H.; Becraft, P.W.; Meyers, B.C.; Walbot, V.; Wang, K.; et al. An Agrobacterium-delivered CRISPR/Cas9 system for high-frequency targeted mutagenesis in maize. *Plant Biotech. J.* **2017**, *15*, 257–268. [CrossRef]

55. Butler, N.M.; Atkins, P.A.; Voytas, D.F.; Douches, D.S. Generation and inheritance of targeted mutations in potato (*Solanum tuberosum* L.) using the CRISPR/Cas system. *PLoS ONE* **2015**, *10*, e0144591. [CrossRef] [PubMed]

56. Pan, C.; Ye, L.; Qin, L.; Liu, X.; He, Y.; Wang, J.; Chen, L.; Lu, G. CRISPR/Cas9-mediated efficient and heritable targeted mutagenesis in tomato plants in the first and later generations. *Sci. Rep.* **2016**, *6*, 24765. [CrossRef] [PubMed]

57. Ma, X.; Liu, Y.G. CRISPR/Cas9-based multiplex genome editing in monocot and dicot plants. *Curr. Protoc. Mol. Biol.* **2016**, *115*, 31.6.1–31.6.21. [CrossRef]

58. Moore, D.; Dowhan, D. Purification and concentration of DNA from aqueous solutions. *Curr. Protoc. Mol. Biol.* **2002**, *59*, 2.1.1–2.1.10. [CrossRef]

59. Hiei, Y.; Ohta, S.; Komari, T.; Kumashiro, T. Efficient transformation of rice (*Oryza sativa* L.) mediated by Agrobacterium and sequence analysis of the boundaries of the T-DNA. *Plant J.* **1994**, *6*, 271–282. [CrossRef]

60. Kang, H.G.; Park, S.; Matsuoka, M.; An, G. White-core endosperm floury endosperm-4 in rice is generated by knockout mutations in the C4-type pyruvate orthophosphate dikinase gene (OsPPDKB). *Plant J.* **2005**, *42*, 901–911. [CrossRef]

61. Livak, K.J.; Schmittgen, T.D. Analysis of relative gene expression data using real-time quantitative PCR and the $2^{-\Delta\Delta CT}$ method. *Methods* **2001**, *25*, 402–408. [CrossRef] [PubMed]

62. Cagampang, G.B.; Perez, C.M.; Juliano, B.O. A gel consistency test for eating quality of rice. *J. Sci. Food Agric.* **1973**, *24*, 1589–1594. [CrossRef] [PubMed]

63. Little, R.R. Differential effect of dilute alkali on 25 varieties of milled white rice. *Cereal Chem.* **1958**, *35*, 111–126.

64. Virmani, S.S. *Hybrid Rice Breeding Manual*; International Rice Research Institute: Laguna, Philippines, 1997.

65. Sarhadi, E.; Bazargani, M.M.; Sajise, A.G.; Abdolahi, S.; Vispo, N.A.; Arceta, M.; Nejad, G.M.; Singh, R.K.; Salekdeh, G.H. Proteomic analysis of rice anthers under salt stress. *Plant Physiol. Biochem.* **2012**, *58*, 280–287. [CrossRef] [PubMed]

66. Laemmli, U.K. Cleavage of structural proteins during the assembly of the head of bacteriophage T4. *Nature* **1970**, *227*, 680–685. [CrossRef] [PubMed]

67. Mathesius, U.; Imin, N.; Chen, H.; Djordjevic, M.A.; Weinman, J.J.; Natera, S.H.; Morris, A.C.; Kerim, T.; Paul, S.; Menzel, C.; et al. Evaluation of proteome reference maps for cross-species identification of proteins by peptide mass fingerprinting. *Proteomics* **2002**, *2*, 1288–1303. [CrossRef]

68. Dreier, B.; Beerli, R.R.; Segal, D.J.; Flippin, J.D.; Barbas, I. Development of zinc finger domains for recognition of the 5'-ANN-3'family of DNA sequences and their use in the construction of artificial transcription factors. *J. Biol. Chem.* **2001**, *276*, 29466–29478. [CrossRef] [PubMed]

69. Tesson, L.; Usal, C.; Ménoret, S.; Leung, E.; Niles, B.J.; Remy, S.; Santiago, Y.; Vincent, A.I.; Meng, X.; Zhang, L.; et al. Knockout rats generated by embryo microinjection of TALENs. *Nat. Biotechnol.* **2011**, *29*, 695. [CrossRef]

70. Huang, P.; Xiao, A.; Zhou, M.; Zhu, Z.; Lin, S.; Zhang, B. Heritable gene targeting in zebrafish using customized TALENs. *Nat. Biotechnol.* **2011**, *29*, 699. [CrossRef]

71. Juliano, B. Varietal impact on rice quality. *Cereal Foods World* **1998**, *43*, 207–222.

72. Zhang, H.; Zhang, J.; Wei, P.; Zhang, B.; Gou, F.; Feng, Z.; Mao, Y.; Yang, L.; Zhang, H.; Xu, N.; et al. The CRISPR/C as9 system produces specific and homozygous targeted gene editing in rice in one generation. *Plant Biotechnol. J.* **2014**, *12*, 797–807. [CrossRef] [PubMed]

73. Tian, Z.; Qian, Q.; Liu, Q.; Yan, M.; Liu, X.; Yan, C.; Liu, G.; Gao, Z.; Tang, S.; Zeng, D.; et al. Allelic diversities in rice starch biosynthesis lead to a diverse array of rice eating and cooking qualities. *Proc. Natl. Acad. Sci. USA* **2009**, *106*, 21760–21765. [CrossRef] [PubMed]

74. Xu, R.F.; Li, H.; Qin, R.Y.; Li, J.; Qiu, C.H.; Yang, Y.C.; Ma, H.; Li, L.; Wei, P.C.; Yang, J.B. Generation of inheritable and "transgene clean" targeted genome-modified rice in later generations using the CRISPR/Cas9 system. *Sci. Rep.* **2015**, *5*, 11491. [CrossRef] [PubMed]

75. Elkonin, L.A.; Tsvetova, M.I. Heritable effect of plant water availability conditions on restoration of male fertility in the "9E" CMS-inducing cytoplasm of sorghum. *Front. Plant Sci.* **2012**, *3*, 91. [CrossRef] [PubMed]

76. Wen, L.; Liu, G.; Li, S.Q.; Wan, C.X.; Tao, J.; Xu, K.Y.; Zhang, Z.J.; Zhu, Y.G. Proteomic analysis of anthers from Honglian cytoplasmic male sterility line rice and its corresponding maintainer and hybrid. *Bot. Stud.* **2007**, *48*, 293–309.

77. Qi, J.; Ma, H.; Xu, J.; Chen, M.; Zhou, D.; Wang, T.; Chen, S. Proteomic analysis of bud differentiation between cytoplasmic male-sterile line and maintainer in tobacco. *Acta Agron. Sin.* **2012**, *38*, 1232–1239. [CrossRef]

78. Sheoran, I.S.; Ross, A.R.; Olson, D.J.; Sawhney, V.K. Differential expression of proteins in the wild type and 7B-1 male-sterile mutant anthers of tomato (*Solanum lycopersicum*): A proteomic analysis. *J. Proteomics* **2009**, *71*, 624–636. [CrossRef]

79. Zheng, R.; Yue, S.; Xu, X.; Liu, J.; Xu, Q.; Wang, X.; Han, L.; Yu, D. Proteome analysis of the wild and YX-1 male sterile mutant anthers of wolfberry (*Lycium barbarum* L.). *PLoS ONE* **2012**, *7*, e41861. [CrossRef]

80. Ivanov, M.; Dymshits, G. Cytoplasmic male sterility and restoration of pollen fertility in higher plants. *Russ. J. Genet.* **2007**, *43*, 354–368. [CrossRef]

81. Teixeira, R.T.; Knorpp, C.; Glimelius, K. Modified sucrose, starch, and ATP levels in two alloplasmic male-sterile lines of B. napus. *J. Exp. Bot.* **2005**, *56*, 1245–1253. [CrossRef]

82. Warmke, H.; Lee, S.L.J. Pollen abortion in T cytoplasmic male-sterile corn (*Zea mays*): A suggested mechanism. *Science* **1978**, *200*, 561–563. [CrossRef] [PubMed]

83. Bergman, P.; Edqvist, J.; Farbos, I.; Glimelius, K. Male-sterile tobacco displays abnormal mitochondrial atp1 transcript accumulation and reduced floral ATP/ADP ratio. *Plant Mol. Biol.* **2000**, *42*, 531–544. [CrossRef] [PubMed]

84. Hanson, M.R.; Bentolila, S. Interactions of mitochondrial and nuclear genes that affect male gametophyte development. *Plant Cell.* **2004**, *16*, S154–S169. [CrossRef] [PubMed]

85. De Paepe, R.; Forchioni, A.; Chetrit, P.; Vedel, F. Specific mitochondrial proteins in pollen: Presence of an additional ATP synthase beta subunit. *Proc. Natl. Acad. Sci. USA* **1993**, *90*, 5934–5938. [CrossRef]

86. Sassa, H.; Oguchi, S.; Inoue, T.; Hirano, H. Primary structural features of the 20S proteasome subunits of rice (*Oryza sativa*). *Gene* **2000**, *250*, 61–66. [CrossRef]

87. Benedetti, M.; Verrascina, I.; Pontiggia, D.; Locci, F.; Mattei, B.; De Lorenzo, G.; Cervone, F. Four Arabidopsis berberine bridge enzyme-like proteins are specific oxidases that inactivate the elicitor-active oligogalacturonides. *Plant J.* **2018**, *94*, 260–273. [CrossRef] [PubMed]

88. Clark, G.B.; Morgan, R.O.; Fernandez, M.P.; Roux, S.J. Evolutionary adaptation of plant annexins has diversified their molecular structures, interactions and functional roles. *New Phytol.* **2012**, *196*, 695–712. [CrossRef]

89. Schnurr, J.; Shockey, J. The acyl-CoA synthetase encoded by LACS2 is essential for normal cuticle development in Arabidopsis. *Plant Cell.* **2004**, *16*, 629–642. [CrossRef]

90. Lallemand, B.; Erhardt, M.; Heitz, T.; Legrand, M. Sporopollenin biosynthetic enzymes interact and constitute a metabolon localized to the endoplasmic reticulum of tapetum cells. *Plant Physiol.* **2013**, *162*, 616–625. [CrossRef]

91. Shi, J.; Cui, M.; Yang, L.; Kim, Y.J.; Zhang, D. Genetic and biochemical mechanisms of pollen wall development. *Trends Plant Sci.* **2015**, *20*, 741–753. [CrossRef]

92. Kubo, A.; Fujita, N.; Harada, K.; Matsuda, T.; Satoh, H.; Nakamura, Y. The starch–debranching enzymes isoamylase and pullulanase are both involved in amylopectin biosynthesis in rice endosperm. *Plant Physiol.* **1999**, *121*, 399–410. [CrossRef]

4

Comparative Analysis of the Transcriptional Response of Tolerant and Sensitive Wheat Genotypes to Drought Stress in Field Conditions

Shuzuo Lv [1,2,*], Kewei Feng [3], Shaofeng Peng [1,2], Jieqiong Wang [1], Yuanfei Zhang [1], Jianxin Bian [3] and Xiaojun Nie [3,*]

[1] Luoyang Academy of Agriculture and Forestry Science, Luoyang Key Laboratory of Crop Molecular Biology and Germplasm Enhancement, Luoyang 471000, Henan, China; pengsf@gmail.com (S.P.); fkwyc@sina.com (J.W.); yfzhang@gmail.com (Y.Z.)
[2] BGI Luoyang Agricultural Innovation Center, Luoyang 471023, Henan, China
[3] State Key Laboratory of Crop Stress Biology in Arid Areas, College of Agronomy and Yangling Branch of China Wheat Improvement Center, Northwest A&F University, Yangling 712100, Shaanxi, China; fkwyc@hotmail.com (K.F.); brian1791@nwsuaf.edu.cn (J.B.)
* Correspondence: lvshuzuo@gmail.com (S.L.); small@nwsuaf.edu.cn (X.N.)

Abstract: Drought stress is one of the most adverse environmental limiting factors for wheat (*Triticum aestivum* L.) productivity worldwide. For better understanding of the molecular mechanism of wheat in response to drought, a comparative transcriptome approach was applied to investigate the gene expression change of two wheat cultivars, Jimai No. 47 (drought-tolerant) and Yanzhan No. 4110 (drought-sensitive) in the field under irrigated and drought-stressed conditions. A total of 3754 and 2325 differential expressed genes (DEGs) were found in Jimai No. 47 and Yanzhan No. 4110, respectively, of which 377 genes were overlapped, which could be considered to be the potential drought-responsive genes. GO (Gene Ontology) analysis showed that these DEGs of tolerant genotype were significantly enriched in signaling transduction and MAP (mitogen-activated protein) kinase activity, while that of sensitive genotype was involved in photosynthesis, membrane protein complex, and guard cell differentiation. Furthermore, 32 and 2 RNA editing sites were identified in drought-tolerant and sensitive genotypes under drought compared to irrigation, demonstrating that RNA editing also plays an important role in response to drought in wheat. This study investigated the gene expression pattern and RNA editing sites of two wheat cultivars with contrasting tolerance in field condition, which will contribute to a better understanding of the molecular mechanism of drought tolerance in wheat and beyond.

Keywords: differentially expressed genes; drought; RNA-seq; RNA editing; wheat

1. Introduction

Drought is one of the most hazardous environmental stresses limiting plant growth and development, which gradually becomes a major threat to the world's agricultural production nowadays and leads to huge yield losses of major crops annually [1–3]. According to statistics, only 20% of cropland worldwide is available for irrigation and it provides approximately 40% of global food production, whereas rain-fed agriculture provides the remaining 60% [4]. Understanding of the molecular mechanism of drought response in crops is crucial for genetic improvement and breeding for drought tolerance, which could meet the challenge of population boom and food security in the 21st century [5]. To date, extensive studies have been conducted to uncover the complexity of mechanisms of plants in response to drought at the morphological, physiological, and molecular level [6–9]. Generally, plants rapidly close their stomata when subjected to drought stress to decrease water losses from leaves, and then a series of downstream response processes are triggered [10–12]. On a cellular level, an osmotic adjustment is first activated and the osmolytes such as proline, glutamate, and mannitol as well as sorbitol and trehalose are accumulated, which could prevent the plant cell from dehydration by increasing the osmotic stress to keep cell membrane integration and enzyme function under drought stress [13,14]. In addition, these substances have been used as the physiological indictors to assess the drought tolerance in plants. On molecular level, several genes and proteins have been reported to be induced in response to drought tolerance, such as dehydration-responsive element binding protein (DREB), C-repeat-binding factor (CBF) and myeloblastosis oncogene (MYB) [15–17]. These drought-responsive genes are mostly transcription factors, which play the hub role in drought-signaling transduction and regulation pathway, such as ROS (reactive oxygen species) [18,19]. It is well known that drought is a complex quantitative trait, which is affected by various factors including environmental condition, genotype, developmental stage, drought severity, and duration [16,20–22]. Thus, an increasing number of studies are further needed to uncover the complex regulatory mechanism of drought tolerance.

Wheat is one of the most important cereal crops all over the world, occupying 17% of cultivated lands and providing the main food source for 30% of the global population [23,24]. Furthermore, wheat is widely grown in a large range of lands under both irrigated and rain-fed conditions. Due to global warming, drought has become the most serious environmental constraint to wheat production and has caused about 5.5% average loss annually [25,26]. Therefore, mining and using drought-tolerant genes to improve wheat varieties with enhanced drought tolerance is urgently needed to meet the challenge of global climate change and food security [27]. Recently, great progress has been made in revealing the molecular mechanism of drought response and many drought-responsive genes have been identified in wheat [28–31]. It demonstrated that over-expression of *TaNAC69* could enhance the drought tolerance in bread wheat [32], and *TaSAP5* could alter the drought stress responses by promoting the degradation of DRIP (DREB interacting protein) proteins [33]. In addition, the wheat MYB gene *TaMYBsdu1* is found to be up-regulated under drought stress but showed differential expression between tolerant and sensitive genotypes, suggesting it plays a crucial role in regulating drought tolerance [34]. With the advent of high-throughput sequencing technology, RNA deep sequencing has been widely used to investigate the gene differential expression profiles involved in drought response in wheat at the transcriptome level, which provided a direct and effective method to identify drought-inducible genes and also contribute to better understanding of drought-signaling pathways [35,36]. It has been demonstrated that different wheat cultivars had diverse molecular basis for drought response and adaptation and the performance of wheat under controlled conditions showed less correlation with field performance [20,37,38]. However, most current studies were conducted using the limited genotypes (drought-tolerant or sensitive) under the controlled conditions. The transcriptional difference of drought-tolerant and drought-sensitive wheat varieties under irrigated and drought-stressed field conditions are not well understood up to now. Here, we investigated the gene expression profiles of two elite wheat cultivars with contrasting drought tolerance, namely Jimai No. 47 (drought-tolerant) and Yanzhan No. 4110 (drought-sensitive) in the field under irrigated and drought-stressed conditions to provide more information for better understanding of the molecular mechanism of drought tolerance in wheat and beyond.

2. Materials and Methods

2.1. Plant Samples and RNA Isolation

Two wheat elite varieties, Jimai No. 47 and Yanzhan No. 4110, were used in this study, of which Jimai No. 47 is a widely grown in the arid area of northern China with excellent drought tolerance and Yanzhan No. 4110 is a high-yield but drought-sensitive variety. These two varieties were grown in field plots of Luoyang A&F institute (Luoyang, Henan, China) with the same plant density in the 2017–2018 crop season. Each plot was 6 m^2 with 2 m in width and 3 m in length. The normal field management was used but differed in water condition. The irrigated treatment was applied three times in November 2017, January 2018, and March 2018 while the drought-stressed treatment was rain-fed with 280 mm of rain in overall crop season. Six replication and randomized block design were used. At the grain-filling stage, the flag leave of 10 wheat plants in each plot were randomly collected for RNA isolation.

Total RNA of the above prepared samples was firstly isolated using TRIzol reagent (Invitrogen, Carlsbad, CA, USA) and treated with RNase-free DNase I to remove any contaminating genomic DNA according to the manufacturer's instructions. Then, the quality of RNA was checked by agarose gel electrophoresis and the quantity was measured by NanoDropND-1000 Spectrophotometer (NanoDrop Technologies, Waltham, MA, USA). Finally, the equal quantity RNA of three replications for the same treatment was pooled and used for RNA sequencing.

2.2. RNA Deep Sequencing and Data Analysis

The four pooled RNA samples were used to construct the RNA sequencing libraries following Illumina's standard pipeline (Illumina, San Diego, CA, USA). High-throughput sequencing was performed on an Illumina HiSeq3000 platform following the standard protocol at Sangong Bio-Technology Corporation (Shanghai, China). The RNA-seq data have been deposited into the genome sequence archive (GSA) database in BIG Data Center, Beijing Institute of Genomics (BIG), Chinese Academy of Sciences, with the accession number CRA001148, and are publicly accessible at http://bigd.big.ac.cn/gsa [39].

The quality of raw data obtained for sequencing was tested by FastQC and then quality filtered by FASTX-toolkit. The adaptor contamination, low-quality reads (quality scores < 20), reads with ambiguous "N" bases more than 10% bases, as well as reads less than 20 bases were removed to obtain the clean data. Then, all the clean data was aligned to wheat reference genome The International Wheat Genome Sequencing Consortium (IWGSC) RefSeq v1.0 [40] by HISAT2 with the default parameters (version 2.0.5) [41]. Then, gene expression levels were calculated based on the FPKM (fragments per kilobaseof exon per million fragments mapped) method and only the high-confidence gene models annotated in wheat RefSeq v1.0 were used. Pearson correlations between biological replicates were also calculated based on the FPKM values of all expressed genes, to assess the reliability.

2.3. Identification of Differentially Expressed Genes (DEGs)

Differential expression genes were identified using DESeq2.0 [42]. The adjusted p-value < 0.05 and fold-changes > 2 (Log_2 $^{(treatment/control)}$ | $\geq$ 1) were used as thresholds for differentially expressed analysis. Then, a K-means clustering was used to extract the fundamental patterns of gene expression [43]. GO terms that are significantly over-represented in each cluster were determined by the AgriGO [44]. Singular Enrichment Analysis (SEA) in AgriGO was used to detect over-represented GO categories in each cluster compared to the whole genes. GO terms with corrected FDR of less than 0.05 were taken as significant ones. KEGG (Kyoto Encyclopedia of Genes and Genomes) pathways were obtained by searching against KEGG database [45].

2.4. Analysis of RNA Editing Sites

To identify the RNA editing sites of these two wheat varieties between irrigated and drought-stressed, all the clean reads of RNA-seq were mapped to the wheat genome IWGSC RefSeq v1.0 using SPRINT (SnP-free RNA editing IdeNtification Toolkit) software with default parameters [46]. To avoid any errors, the independencies mapping software BWA (http://biobwa.sourceforge.net) [47] and SNP nomenclature tools Samtools (http://www.htslib.org) [48] were integrated to identify the SNP between RNA reads and reference genomic DNA. The overlapping results obtained by both methods were retained for further analysis. Then, using the irrigation samples as background, the SNPs at the same position in drought-stressed sample which were different from the background were considered as the potential RNA editing sites. Finally, the editing sites were further filtered using the following parameters: (1) the edited sites having more than 5 mapped reads; (2) the ratio of editing reads and total mapped reads more than 50%; (3) the editing site identified in both biological replications.

3. Results and Discussion

3.1. RNA-seq Analysis

To enrich the knowledge of drought tolerance mechanism in wheat, we investigated the gene expression profiles of wheat cultivar Jimai No. 47 (drought-tolerant, JM) and Yanzhan No. 4110 (drought-sensitive, YZ) in the field under irrigated (I) and rain-fed (R) conditions by RNA-seq technology. A total of 8 pooled RNA samples were sequenced, and about 61.43 Gb of raw data was obtained, with an average of about 53 million pair-end reads with 150 bp in size for each sample (Table 1). After a quality filter, 55.81 Gb of clean data remained, representing 90.8% of the raw data. In addition, the clean reads of these samples ranged from 41,653,874 to 51,913,178, with an average of 48,402,376 reads representing 6.97 Gb. Then, all the clean reads were mapped to the wheat genotype Chinese Spring reference genome IWGSC_V1 (accessed from URGI (Unité de Recherche Génomique Info) database on 6 June 2018). Results showed that on average, about 80% of clean reads could be mapped to the wheat reference genome, and approximately 77% were uniquely mapped, which was similar to previous studies [9,35]. The unmapped 20% reads might be due to genotype specifics or incompletion of the reference genome. Furthermore, most of RNA-seq reads were mapped onto the exon region of the reference genome, although they were also mapped to intergenic and intron regions with very low frequency. Based on the mapping result, the expression level of the annotated genes in these mapped regions were obtained. Then, we used the gene expression level to calculate the correlating coefficient of two biological replicates of all samples to examine the repeatability and reliability. Results showed the biological replicates of all samples showed strong correlation relationships with the coefficient (R^2) of more than 0.90.

Furthermore, we compared the abundance of expressed gene in these samples. Out of 110,790 high-confidence genes annotated in wheat genome, 58,498 (52.8%) genes were detected to be expressed in these samples, of which 45,258 genes were found to be expressed in all tested samples, representing the core gene set of wheat (Figure 1). A total of 50,981, 51,140, 52,622 and 52,961 genes were expressed in JM_I, JM_R and YZ_I and YZ_R, respectively. This means that compared to irrigation condition, more genes were induced to express by drought stress in both drought-tolerant and sensitive genotypes, which is consistent with previous studies [20,36]. From the point of view of the genotype, the variety YZ has more abundant expressed genes than the JM variety. Additionally, 1167, 903, 1189 and 1223 genes were found to be specifically expressed in JM_I, JM_R and YZ_I and YZ_R respectively (Figure 1).

Table 1. Summary of RNA-seq and mapping data.

	Total Reads	Raw Base (bp)	Clean Reads	Clean Bases (bp)	Total Mapped to Wheat Genome	Multiple Mapped	Uniquely Mapped	Reads Mapped in Proper Pairs
JM-I2	54,939,280	7,913,976,131	50,542,262	7,278,085,728	40,962,623 (81.05%)	1,908,065 (3.78%)	39,054,558 (77.27%)	34,136,584 (67.54%)
JM-I1	49,544,428	7,152,858,915	47,469,372	6,835,589,568	38,093,160 (80.25%)	1,458,216 (3.07%)	36,634,944 (77.18%)	32,711,652 (68.91%)
YZ-R1	52,387,108	7,582,519,380	51,913,178	7,475,497,632	41,618,432 (80.17%)	1,576,802 (3.04%)	40,041,630 (77.13%)	36,267,234 (69.86%)
YZ-I1	53,867,378	7,742,552,474	51,768,216	7,454,623,104	41,280,310 (79.74%)	1,858,852 (3.59%)	39,421,458 (76.15%)	34,636,332 (66.91%)
YZ-I2	53,900,538	7,810,800,148	51,550,574	7,474,833,230	41,415,518 (80.34%)	1,488,853 (2.89%)	39,926,665 (77.45%)	35,443,404 (68.75%)
YZ-R2	53,939,098	7,789,354,276	47,698,318	6,868,557,792	38,634,459 (81.00%)	1,698,950 (3.56%)	36,935,509 (77.44%)	32,060,558 (67.22%)
JM-R2	55,766,550	8,024,037,795	44,623,214	6,425,742,816	36,373,597 (81.51%)	2,175,078 (4.87%)	34,198,519 (76.64%)	28,892,192 (64.75%)
JM-R1	51,561,402	7,422,811,668	41,653,874	5,998,157,856	34,227,750 (82.17%)	2,045,728 (4.91%)	32,182,022 (77.26%)	27,307,370 (65.56%)
Average	53,238,223	7,679,863,848	48,402,376	6,976,385,966				
Total	425,905,782	61,438,910,787	387,219,008	55,811,087,726				

Note: JM represents the Jimai No. 47 (drought-tolerant) and YZ represents Yanzhan No. 4110 (drought-sensitive); I1 and I2 represent the two samples under irrigation condition whereas R1 and R2 represent the two samples under rain-fed condition.

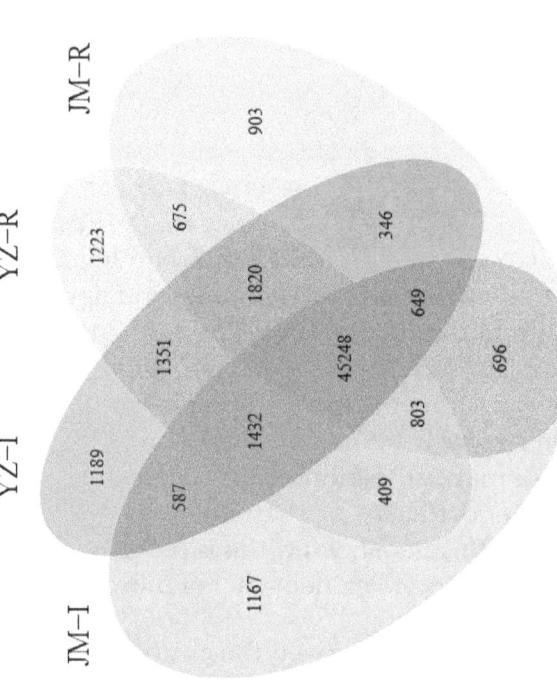

Figure 1. Venn map of the expressed genes identified in different samples. JM = the Jimai No. 47 (drought-tolerant) and YZ = Yanzhan No. 4110 (drought-sensitive); I = irrigation; R = rain-fed (drought).

3.2. Identification of the Differentially Expressed Genes

To identify the drought-responsive genes, the differentially expressed gene (DEGs) were analyzed between different treatments and different genotypes using padj < 0.05 and |log2Ratio| ≥1 as thresholds. In drought-tolerant genotypes (JM), 2754 DEGs were identified between irrigation and drought condition, of which 1152 gene were up-regulated, and 1612 gene were down-regulated (Figure 2). In drought-sensitive genotype (YZ), there were 2325 DEGs, and 1075 and 1250 were up-regulated and down-regulated, respectively (Figure 2). GO analysis found that the DEGs of drought-tolerant genotype JM were enriched into signal transducer activity (GO:0005057), MAP kinase activity (GO:0004707), intracellular signal transduction (GO:0035556) and cellular response to abiotic stimulus (GO:0071214) while those of drought-sensitive genotype YZ were mainly involved in photosynthesis (GO:0015979), membrane protein complex (GO:0098796), ER membrane protein complex (GO:0072546), guard cell differentiation (GO:0010052), and positive regulation of response to oxidative stress (GO:1902884) (Figure 3). These results indicated that the drought-tolerant genotype activated a series of signaling pathways in response to drought stress and made adaptive adjustment, so the cell process and membrane activity were not negatively affected by drought stress, while the drought-sensitive genotype did not rapidly activate signaling transduction but activated photosynthesis and cellular process to cope with drought stress. The enriched KEGG pathway of the DEGs also showed the drought-tolerant genotype as mainly involved in signaling pathway while the drought-sensitive genotype was involved in photosynthesis and cellular activity (Figure 4). The different mechanisms of tolerant and sensitive genotypes responding to drought will contribute to develop an effective method to cope with drought stress and to assess the ability of stress tolerance. Furthermore, it is found that 377 DEGs were overlapped in drought-tolerant and sensitive genotypes, which could play fundamental roles in the regulation of drought response in wheat, including AP2/ERF (APETALA2/ethylene responsive factor), MYB and WRKY transcription factors. Additionally, 2377 and 1948 DEGs were specifically identified in genotypes JM and YZ respectively, which might be involved in genotype-specific regulatory pathway in response to drought.

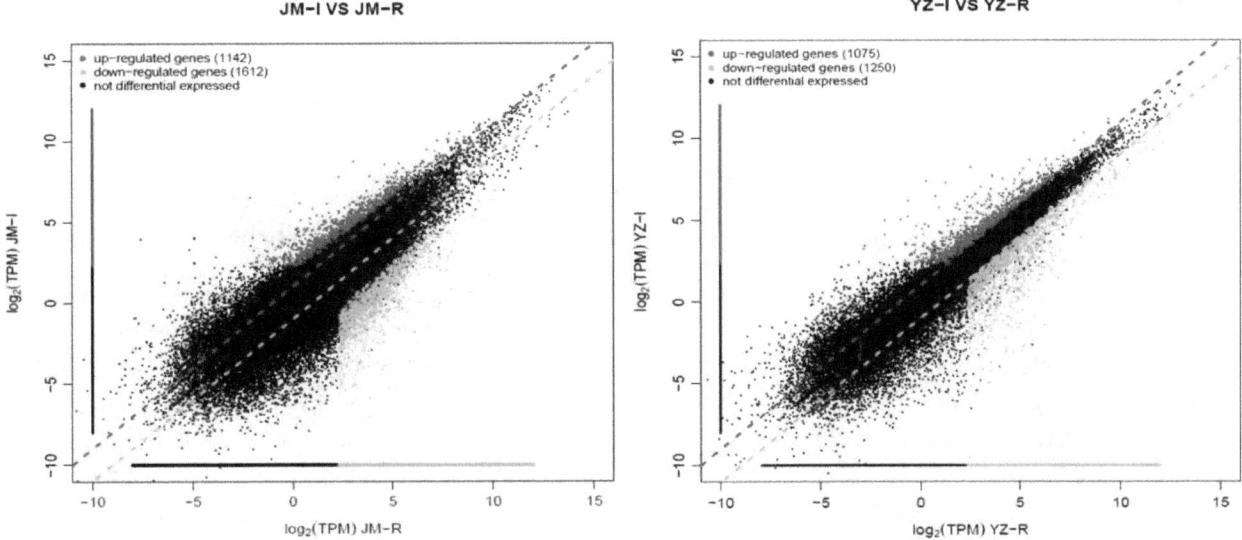

Figure 2. The differentially expressed genes between irrigation and rain-fed condition identified in drought-tolerant genotype Jimai No. 47 and drought-sensitive genotype Yanzhan No. 4110. JM = the Jimai No. 47 (drought-tolerant) and YZ = Yanzhan No. 4110 (drought-sensitive); I = irrigation; R = rain-fed (drought).

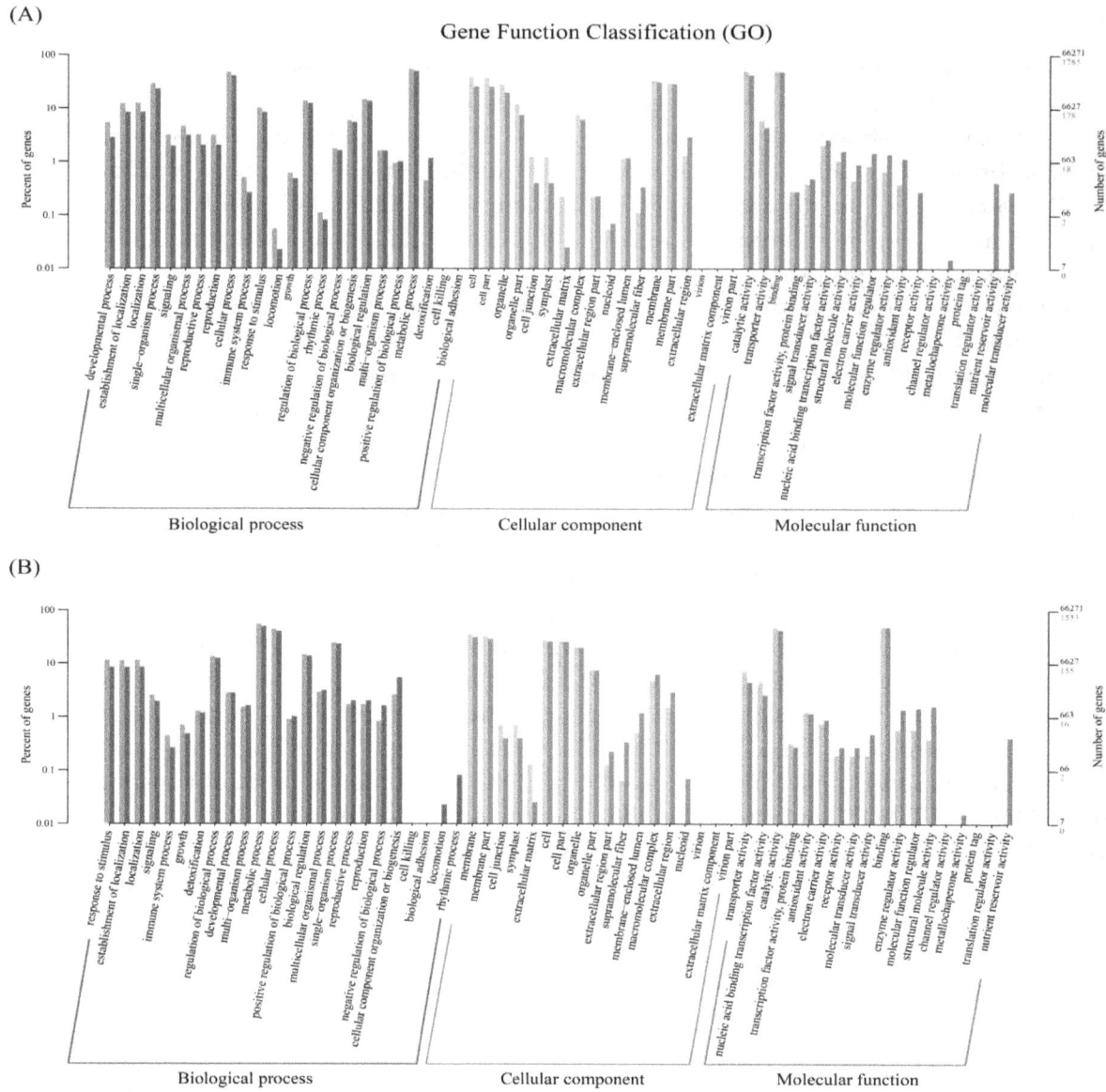

Figure 3. GO enrichment analysis of the DEGs in drought-tolerant and drought-sensitive genotypes. (**A**): drought-tolerant genotype Jimai No. 47; (**B**) drought-sensitive genotype Yanzhan No. 4110.

The expression patterns of genes will provide the crucial information for understanding their biological function [49]. Thus, all the identified DEGs were further used to define clusters based on their specific expression patterns in each sample. A K-mean clustering method was conducted with the squared Euclidean distance measure, and all DEGs could be classified into 10 categories (Figure 5). The cluster I-V showed the relatively higher expression in sensitive genotype than tolerant genotype under both well-watered and drought conditions. Among them, Cluster I comprised 84 genes showing high expression in YZ_R but showed almost similar expression in other samples, suggesting they might be specifically induced by drought in sensitive genotypes. Cluster 2 with 647 genes showed low expression in JM_I and showed high expression in other samples. These genes might be involved in response to drought and GO enrichment analysis found they mainly functioned in cellular response to stress (GO:0033554), plant organ senescence (GO:0090693) and ARF (Auxin response factor) protein signal transduction (GO:0032011) and regulation of ARF protein signal transduction (GO:0032012). The cluster VI-X showed relatively high expression in the tolerant genotype compared to the sensitive genotype. Cluster 9 with 169 genes showed specifically high expression in JM_R, suggesting these genes play a crucial role in regulating tolerance to drought. Then, GO analysis of these 169 genes found that they mainly enriched the positive regulation of cellular response to oxidative stress (GO:1900409), positive regulation of response to ROS (GO:1901033), stomatal movement (GO:0010118) as well as auxin-activated signaling pathway (GO:0009734), which proved their roles in regulating wheat defense systems to tolerate drought stress.

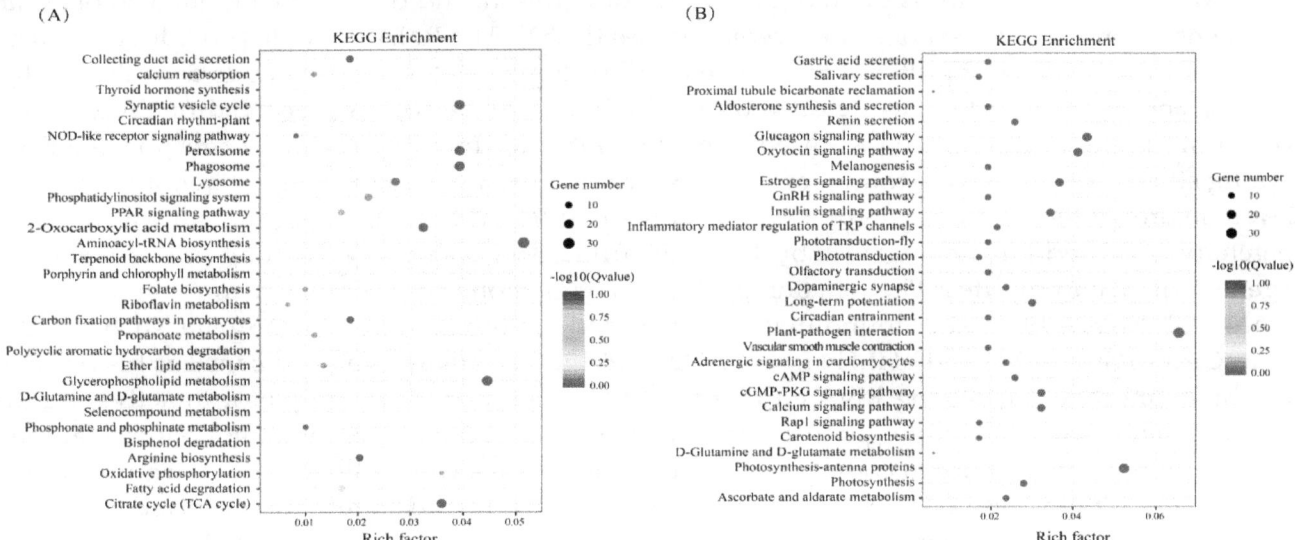

Figure 4. Comparison of KEGG enrichment analysis of the DEGs in drought-tolerant and drought-sensitive genotypes. (**A**): drought-tolerant genotype Jimai No. 47; (**B**) drought-sensitive genotype Yanzhan No. 4110.

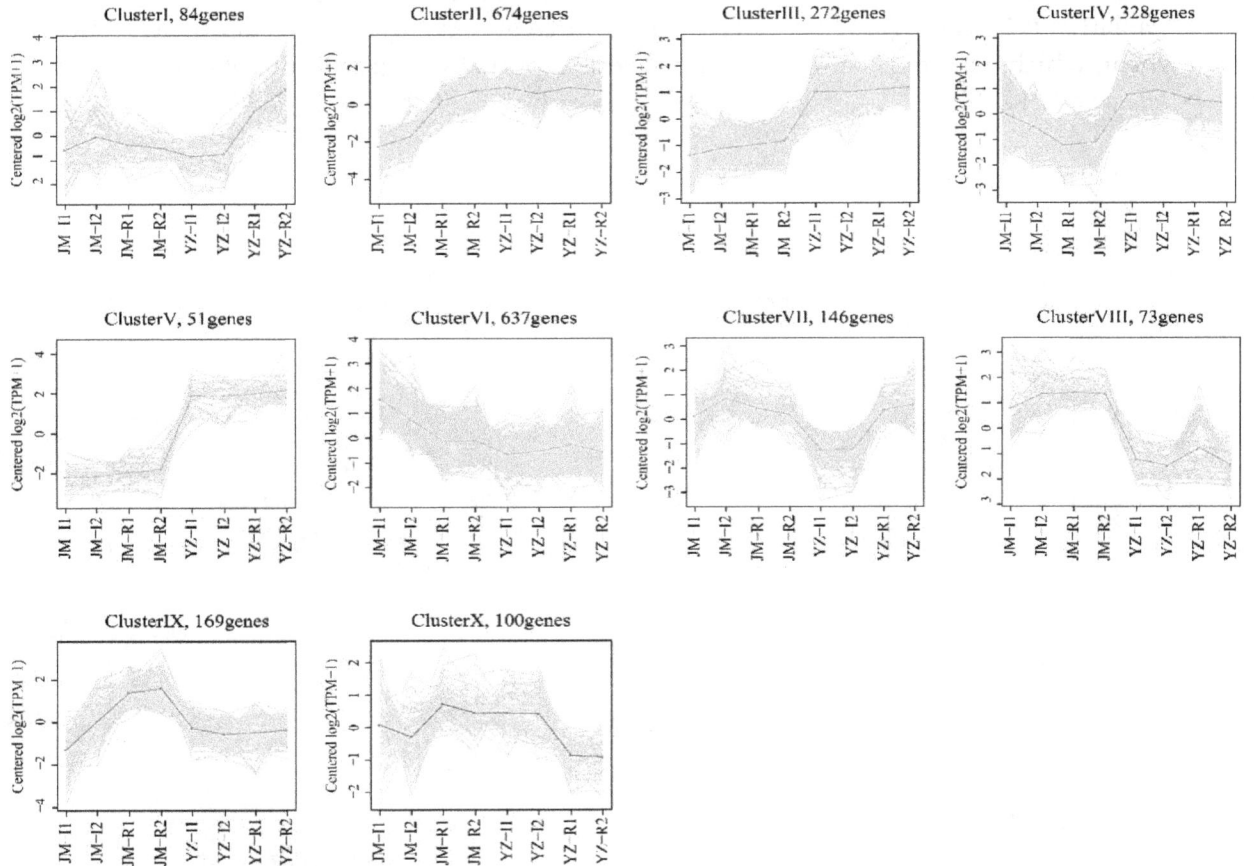

Figure 5. Cluster analysis of the differentially expressed genes. JM = the Jimai No. 47 (drought-tolerant) and YZ = Yanzhan No. 4110 (drought-sensitive); I = irrigation; R = rain-fed (drought).

3.3. Analysis of RNA Editing Sites

RNA editing is a process that occurs in the RNA molecular base change or modification when transcribed, which is one of most important mechanisms regulating gene expression and enriching genetic information at the post-transcription level [50]. A larger number of studies have reported that RNA editing not only controls plant organ formation, growth, and development, but also plays an indispensable role in the response to diverse stresses [51,52]. The RNA-seq data provides a resource to identify the RNA editing sites at the whole transcriptome level. Using methods described in the Material and Methods section, we detected the RNA editing sites between water and drought condition to identify the drought-induced RNA editing sites (Table 2). In total, 32 drought-responsive RNA editing sites in 22 genes were found, of which 30 were found in drought-tolerant genotype JM and 2 were in drought-sensitive genotype YZ. TraesCS6B01G079400.1 have 4 editing sties, of which 3 were in genotype JM and 1 in YZ, followed by TraesCS4B01G293600.3 showing 3 editing sites and another 6 genes with 2 editing sites. The remaining 13 genes owned one sites. Function annotation of these genes with RNA editing sites found that they included the transcription factor gene, such as MYB (TraesCS6B01G012800.1) and bHLH (TraesCS5A01G279200.1), kinase proteins as well as histone and plasma membrane, suggesting RNA editing might function as the key regulator in activating the processes and pathways of drought tolerance. Furthermore, 13 sites were found in UTR regions and the remaining 19 sites were in coding regions, of which 9 sites were edited at the third position of the codon, 8 at the second position and 2 at the first position. The 10 edited sites occurred at the first and second position of the codon caused the alteration of amino acid, which could be considered as candidates for further functional study. Finally, a total of 7 types of base change were introduced by RNA editing in these 32 sites, of which C to T mutation is the most abundant type with the value of 10 times, followed by T to C and G to A with the number of 5, C to A with the number of 4, G to T and G to C with the number of 3 as well as A to G with the number of 2. This result showed that the transition (22) was significantly higher than transversion (10) in these identified RNA editing sites in wheat, which was consistent with the previous reports in the plastomes of einkorn wheat and *Aegilops tauschii*. L. [51,52].

Table 2. The drought-responsive RNA editing sites identified in wheat.

Gene ID	Genotype [a]	Position [b]	Samples [c]				Editing Site [d]	NR_annotation
			I1	I2	R1	R2		
TraesCS3D01G002100.1	JM	912	A/7	A/5	G/5	G/9	acA/acG	Receptor kinase-like protein
TraesCS3D01G002100.1	JM	507	G/9	G/5	C/5	C/6	gtG/gtC	Receptor kinase-like protein
TraesCS3D01G517100.1	JM	660	G/122	G/99	T/76	T/144	gcC/gcA	Abscisic stress ripening protein
TraesCS4B01G220200.1	JM	843	C/5	C/7	T/5	T/6	-	Phosphoenolpyruvate carboxykinase (ATP)-like
TraesCS4B01G280800.1	JM	417	G/6	G/8	C/5	C/11	-	40S ribosomal protein S27
TraesCS4B01G287800.1	JM	1356	C/13	C/24	A/16	A/21	ggC/ggA	Sucrose transporter
TraesCS4B01G292500.1	JM	735	G/9	G/7	A/10	A/13	-	Transcription initiation factor IIE subunit beta
TraesCS4B01G293600.3	JM	1893	G/11	G/25	A/19	A/18	-	Copper-transporting ATPase
TraesCS4B01G293600.3	JM	1221	C/53	C/78	T/37	T/89	gCa/gTa	Copper-transporting ATPase
TraesCS4B01G293600.3	JM	135	C/64	C/86	T/46	T/85	Cca/Tca	Copper-transporting ATPase
TraesCS4D01G268600.1	JM	1215	C/9	C/21	A/6	A/35	agG/agT	Ribosomal large subunit pseudouridine synthase B
TraesCS5A01G232200.1	JM	498	G/5	G/6	A/4	A/7	tCg/tTg	RING/U-box superfamily protein, putative
TraesCS5A01G279200.1	JM	1437	C/5	C/13	A/5	A/8	-	Basic helix-loop-helix transcription factor
TraesCS5A01G501900.1	JM	819	C/5	C/32	A/6	A/35	Gcc/Tcc	ATP phosphoribosyltransferase
TraesCS5A01G509700.1	JM	795	C/8	C/10	T/6	T/13	cCg/cTg	Thyroid adenoma-associated protein-like protein
TraesCS5A01G518300.1	JM	213	G/49	G/142	A/50	A/158	ggC/ggT	Aquaporin-like protein
TraesCS5A01G518300.1	JM	1071	G/5	G/12	A/5	A/15	-	Aquaporin-like protein
TraesCS5D01G043500.1	JM	909	C/5	C/5	T/5	T/11	gCg/gTg	Receptor-like protein kinase
TraesCS5D01G045300.1	JM	618	T/10	T/45	C/9	C/48	aTg/aCg	NRT1/PTR family protein 2.2
TraesCS6B01G012800.1	JM	6282	T/5	T/9	C/5	C/5	taC/taT	MYB family transcription factor-like protein
TraesCS6B01G069800.1	JM	1533	C/17	C/5	T/3	T/6	acG/acA	Kinase family protein
TraesCS6B01G069800.1	JM	1185	T/8	T/5	C/8	C/16	-	Kinase family protein
TraesCS6B01G078200.1	JM	1473	G/5	G/9	A/5	A/31	-	Histone H4
TraesCS6B01G078200.1	JM	834	T/8	T/16	C/5	C/50	-	Histone H4
TraesCS6B01G079400.1	JM	969	A/9	A/23	G/5	G/36	gTg/gCg	Syntaxin, putative
TraesCS6B01G079400.1	JM	957	G/5	G/23	T/5	T/20	-	Syntaxin, putative
TraesCS6B01G079400.1	JM	2106	T/5	T/21	C/5	C/19	-	Syntaxin, putative
TraesCS6B01G434200.1	JM	1665	T/6	T/5	C/5	C/7	gAg/gGg	Thioredoxin
TraesCS7A01G437900.1	JM	408	G/5	G/14	C/10	C/5	-	Histone H2A
TraesCS7D01G399500.1	JM	816	C/8	C/7	T/6	T/7	gaC/gaT	Eukaryotic aspartyl protease family protein
TraesCS6B01G079400.1	YZ	1122	C/8	C/12	T/8	T/15	-	Syntaxin, putative
TraesCS5A01G509700.1	YZ	795	C/11	C/5	T/5	T/8	cCg/cTg	Thyroid adenoma-associated protein-like protein

Note: [a] JM = the drought-tolerant genotype Jimai No. 47; YZ = the drought-sensitive genotype Yanzhan No. 4110; [b] the position represents the position of the edited site at the mRNA molecular; [c] I1 and I2 = the two samples under irrigation condition whereas R1 and R2 = the two samples under rain-fed condition. The capital = the abbreviation of nucleotide; the number = the number of mapped reads; [d] "_" = the editing sites located in UTR region.

4. Conclusions

In this study, we investigated the gene expression profiles of drought-tolerant and drought-sensitive genotypes under irrigated and rain-fed conditions to identify the drought-responsive genes in the field environment in wheat. Results showed that the drought-tolerant and drought-sensitive genotype adopted different mechanisms to respond to drought. Furthermore, the drought-responsive RNA editing sites were identified and a total of 34 editing sites were found, demonstrating that RNA editing could play a crucial role in regulating drought response and adjustment in wheat. This is the first study to report drought-responsive RNA editing by RNA-seq data, which will contribute to a better understanding of the molecular mechanism of drought tolerance in wheat and beyond.

Author Contributions: Conceptualization, S.L. and X.N.; Methodology, S.P.; Software, K.F. and J.W.; Formal Analysis, S.L. and K.F.; Resources, Y.Z.; Writing—Original Draft Preparation, S.L. and K.F.; Writing–Review and Editing, J.B. and X.N.; Funding Acquisition, S.L. and X.N.

Acknowledgments: We would like to thank High-Performance Computing (HPC) of Northwest A&F University for providing computing resources.

References

1. Trenberth, K.E.; Dai, A.; Schrier, G.V.; Jones, P.D.; Barichivich, J.; Briffa, K.R.; Sheffield, J. Global warming and changes in drought. *Nat. Clim. Chang.* **2014**, *4*, 17–22. [CrossRef]

2. Lesk, C.; Rowhani, P.; Ramankutty, N. Influence of extreme weather disasters on global crop production. *Nature* **2016**, *529*, 84–87. [CrossRef] [PubMed]

3. Lobell, D.B.; Gourdji, S.M. The influence of climate change on global crop productivity. *Plant Physiol.* **2012**, *160*, 1686–1697. [CrossRef] [PubMed]

4. Cropland data in Food and Agriculture Organization of the United Nations. Available online: http://www.fao.org/faostat/en/#data/LR (accessed on 15 September 2018).

5. Thao, N.P.; Tran, L.S. Enhancement of plant productivity in the post-genomics era. *Curr. Genom.* **2016**, *17*, 295–296. [CrossRef] [PubMed]

6. Wang, W.; Vinocur, B.; Altman, A. Plant responses to drought, salinity and extreme temperatures: Towards genetic engineering for stress tolerance. *Planta* **2003**, *218*, 1–14. [CrossRef] [PubMed]

7. Manoj, K.; Raju, S.; Satoshi, O.; Yusaku, U.; Selvaraj, M.G.; Kagale, S. Drought response in wheat: Key genes and regulatory mechanisms controlling root system architecture and transpiration efficiency. *Front. Chem.* **2017**, *5*, 106.

8. Zhang, X.; Lei, L.; Lai, J.; Zhao, H.; Song, W. Effects of drought stress and water recovery on physiological responses and gene expression in maize seedlings. *BMC Plant Biol.* **2018**, *18*, 68. [CrossRef] [PubMed]

9. Ma, J.; Li, R.; Wang, H.; Li, D.; Wang, X.; Zhang, Y.; Zhen, W.; Duan, H.; Yan, G.; Li, Y. Transcriptomics analyses reveal wheat responses to drought stress during reproductive stages under field conditions. *Front. Plant Sci.* **2017**, *8*, 592. [CrossRef] [PubMed]

10. Bhushan, D.; Pandey, A.; Choudhary, M.K.; Datta, A.; Chakraborty, S.; Chakraborty, N. Comparative proteomics analysis of differentially expressed proteins in chickpea extracellular matrix during dehydration stress. *Mol. Cell. Proteom.* **2007**, *6*, 1868–1884. [CrossRef] [PubMed]

11. Mahajan, S.; Tuteja, N. Cold, salinity and drought stresses: An overview. *Arch. Biochem. Biophys.* **2005**, *444*, 139–158. [CrossRef] [PubMed]

12. Maud, V.; Hazel, K.S.; Cohen, D.; Dewoody, J.; Trewin, H.; Steenackers, M.; Bastien, C.; Taylor, G. Adaptive mechanisms and genomic plasticity for drought tolerance identified in European black poplar (*Populus nigra* L.). *Tree Physiol.* **2016**, *36*, 909–928.

13. Naimat, U.; Meral, Y.Z.; Neslihan, Ö.G.; Budak, H. Comparative metabolite profiling of drought stress in roots and leaves of seven Triticeae species. *BMC Genom.* **2017**, *18*, 969.

14. Ramachandra, R.A.; Chaitanya, K.V.; Vivekanandan, M. Drought-induced responses of photosynthesis and antioxidant metabolism in higher plants. *J. Plant Physiol.* **2004**, *161*, 1189–1202.

15. Hossain, Z.; Nouri, M.Z.; Komatsu, S. Plant cell organelle proteomics in response to abiotic stress. *J. Proteome Res.* **2012**, *11*, 37–48. [CrossRef] [PubMed]

16. Moumeni, A.; Satoh, K.; Kondoh, H.; Asano, T.; Hosaka, A.; Venuprasad, R.; Serraj, R.; Kumar, A.; Leung, H.; Kikuchi, S. Comparative analysis of root transcriptome profiles of two pairs of drought-tolerant and susceptible rice near-isogenic lines under different drought stress. *BMC Plant Biol.* **2011**, *11*, 174. [CrossRef] [PubMed]

17. Ramanjulu, S.; Bartels, D. Drought- and desiccation-induced modulation of gene expression in plants. *Plant Cell Environ.* **2002**, *25*, 141–151. [CrossRef] [PubMed]

18. Nakashima, K.; Ito, Y.; Yamaguchi-Shinozaki, K. Transcriptional regulatory networks in response to abiotic stresses in Arabidopsis and grasses. *Plant Physiol.* **2009**, *149*, 88–95. [CrossRef] [PubMed]

19. Ergen, N.Z.; Thimmapuram, J.; Bohnert, H.J.; Budak, H. Transcriptome pathways unique to dehydration tolerant relatives of modern wheat. *Funct. Integr. Genom.* **2009**, *9*, 377–396. [CrossRef] [PubMed]

20. Rampino, P.; Pataleo, S.; Gerardi, C.; Mita, G.; Perrotta, C. Drought stress response in wheat: Physiological and molecular analysis of resistant and sensitive genotypes. *Plant Cell Environ.* **2006**, *29*, 2143–2152. [CrossRef] [PubMed]

21. Budak, H.; Kantar, M.; Kurtoglu, K.Y. Drought tolerance in modern and wild wheat. *Sci. World J.* **2013**, *2013*, 548246. [CrossRef] [PubMed]

22. Cheng, L.X.; Wang, Y.P.; He, Q.; Li, H.J.; Zhang, X.J.; Zhang, F. Comparative proteomics illustrates the complexity of drought resistance mechanisms in two wheat (*Triticum aestivum* L.) cultivars under dehydration and rehydration. *BMC Plant Biol.* **2016**, *16*, 188. [CrossRef] [PubMed]

23. Gill, B.S.; Appels, R.; Botha-Oberholster, A.M.; Buell, C.R.; Bennetzen, J.L.; Chalhoub, B.; Chumley, F.; Dvorák, J.; Iwanaga, M.; Keller, B.; et al. A workshop report on wheat genome sequencing: International genome research on wheat consortium. *Genetics* **2014**, *168*, 1087–1096. [CrossRef] [PubMed]

24. Mayer, K.F.; Rogers, J.; Doležel, J.; Pozniak, C.; Eversole, K.; Feuillet, C.; Gill, B.; Friebe, B.; Lukaszewski, A.J.; Sourdille, P.; et al. A chromosome-based draft sequence of the hexaploid bread wheat (*Triticum aestivum*) genome. *Science* **2014**, *345*, 1251788.

25. Zampieri, M.; Ceglar, A.; Dentener, F.; Toreti, A. Wheat yield loss attributable to heat waves, drought and water excess at the global, national and subnational scales. *Environ. Res. Lett.* **2017**, *12*, 064008. [CrossRef]

26. Manes, Y.; Gomez, H.; Puhl, L.; Reynolds, M.; Braun, H.; Trethowan, R. Genetic yield gains of the CIMMYT international semi-arid wheat yield trials from 1994 to 2010. *Crop Sci.* **2012**, *52*, 1543–1552. [CrossRef]

27. Reynolds, M.; Foulkes, M.J.; Slafer, G.A.; Berry, P.; Parry, M.A.; Snape, J.W.; Angus, W.J. Raising yield potential in wheat. *J. Exp. Bot.* **2009**, *60*, 1899–1918. [CrossRef] [PubMed]

28. Morran, S.; Eini, O.; Pyvovarenko, T.; Parent, B.; Singh, R.; Ismagul, A.; Eliby, S.; Shirley, N.; Langridge, P.; Lopato, S. Improvement of stress tolerance of wheat and barley by modulation of expression of DREB/CBF factors. *Plant Biotechnol. J.* **2011**, *9*, 230–249. [CrossRef] [PubMed]

29. Gao, H.M.; Wang, Y.F.; Xu, P.; Zhang, Z.B. Overexpression of a WRKY transcription factor TaWRKY2 enhances drought stress tolerance in transgenic wheat. *Front. Plant Sci.* **2018**, *9*, 997. [CrossRef] [PubMed]

30. Akpinar, B.A.; Avsar, B.; Lucas, S.J.; Budak, H. Plant abiotic stress signaling. *Plant Signal. Behav.* **2012**, *7*, 1450–1455. [CrossRef] [PubMed]

31. Wilkinson, S.; Davies, W.J. Drought, ozone, ABA and ethylene: New insights from cell to plant to community. *Plant Cell Environ.* **2010**, *33*, 510–525. [CrossRef] [PubMed]

32. Xue, G.P.; Way, H.M.; Richardson, T.; Drenth, J.; Joyce, P.A.; McIntyre, C.L. Overexpression of TaNAC69 leads to enhanced transcript levels of stress up-regulated genes and dehydration tolerance in bread wheat. *Mol. Plant* **2011**, *4*, 697–712. [CrossRef] [PubMed]

33. Zhang, N.; Yin, Y.J.; Liu, X.Y.; Tong, S.M.; Xing, J.W.; Zhang, Y.; Pudake, R.N.; Izquierdo, E.M.; Peng, H.R.; Xin, M.M.; et al. The E3 ligase TaSAP5 alters drought stress responses by promoting the degradation of DRIP proteins. *Plant Physiol.* **2017**, *175*, 1878–1892. [CrossRef] [PubMed]

34. Rahaie, M.; Xue, G.P.; Naghavi, M.R.; Alizadeh, H.; Schenk, P.M. A MYB gene from wheat (*Triticum aestivum* L.) is up-regulated during salt and drought stresses and differentially regulated between salt-tolerant and sensitive genotypes. *Plant Cell Rep.* **2010**, *29*, 835–844. [CrossRef] [PubMed]

35. Dalal, M.; Sahu, S.; Tiwari, S.; Rao, A.R.; Gaikwad, K. Transcriptome analysis reveals interplay between hormones, ROS metabolism and cell wall biosynthesis for drought-induced root growth in wheat. *Plant Physiol. Biochem.* **2018**, *130*, 482–492. [CrossRef] [PubMed]

36. Hu, L.; Xie, Y.; Fan, S.; Wang, Z.; Wang, F.; Zhang, B.; Li, H.; Song, J.; Kong, L. Comparative analysis of root transcriptome profiles between drought-tolerant and susceptible wheat genotypes in response to water stress. *Plant Sci.* **2018**, *272*, 276–293. [CrossRef] [PubMed]

37. Bowne, J.B.; Erwin, T.A.; Juttner, J.; Schnurbusch, T.; Langridge, P.; Bacic, A.; Roessner, U. Drought responses of leaf tissues from wheat cultivars of differing drought tolerance at the metabolite level. *Mol. Plant* **2012**, *5*, 418–429. [CrossRef] [PubMed]

38. Ford, K.L.; Cassin, A.; Bacic, A. Quantitative proteomic analysis of wheat cultivars with differing drought stress tolerance. *Front. Plant Sci.* **2011**, *2*, 44. [CrossRef] [PubMed]

39. Genome Sequence Archive (GSA) Database. Available online: http://bigd.big.ac.cn/gsa (accessed on 1 October 2018).

40. Appels, R.; Eversole, K.; Feuillet, C.; Keller, B.; Rogers, J.; Stein, N.; Pozniak, C.J.; Choulet, F.; Distelfeld, A.; Poland, J.; et al. Shifting the limits in wheat research and breeding using a fully annotated reference genome. *Science* **2018**, *361*, eaar7191. [PubMed]

41. Kim, D.; Langmead, B.; Salzberg, S.L. HISAT: A fast spliced aligner with low memory requirements. *Nat. Methods* **2015**, *12*, 357–360. [CrossRef] [PubMed]

42. Love, M.I.; Huber, W.; Anders, S. Moderated estimation of fold change and dispersion for RNA-seq data with DESeq2. *Genome Biol.* **2014**, *15*, 550. [CrossRef] [PubMed]

43. Brock, G.; Datta, S.; Pihur, V.; Datta, S. clValid: An R package for cluster validation. *J. Stat. Softw.* **2008**, *25*, 1–22. [CrossRef]

44. Tian, T.; Liu, Y.; Yan, H.Y.; You, Q.; Yi, X.; Du, Z.; Xu, W.Y.; Su, Z. agriGO v2.0: A GO analysis toolkit for the agricultural community. *Nucleic Acids Res.* **2017**, *45*, W122–W129. [CrossRef] [PubMed]

45. Kanehisa, M.; Goto, S. KEGG: Kyoto Encyclopedia of Genes and Genomes. *Nucleic Acids Res.* **2000**, *28*, 27–30. [CrossRef] [PubMed]

46. Zhang, F.; Lu, Y.L.; Yan, S.J.; Xing, Q.H.; Tian, W.D. SPRINT: An SNP-free toolkit for identifying RNA editing sites. *Bioinformatics* **2017**, *3*, 3538–3548. [CrossRef] [PubMed]

47. BWA (Burrows-Wheeler Aligner). Available online: http://bio-bwa.sourceforge.net/ (accessed on 1 September 2018).

48. Samtools. Available online: http://www.htslib.org. (accessed on 1 September 2018).

49. Ramírez-González, R.H.; Borrill, P.; Lang, D.; Harrington, S.A.; Brinton, J.; Venturini, L.; Davey, M.; Jacobs, J.; van Ex, F.; Pasha, A.; et al. The transcriptional landscape of polyploid wheat. *Science* **2018**, *361*, eaar6089. [CrossRef] [PubMed]

50. Takenaka, M.; Zehrmann, A.; Verbitskiy, D.; Härtel, B.; Brennicke, A. RNA editing in plants and its evolution. *Annu. Rev. Genet.* **2013**, *47*, 335–352. [CrossRef] [PubMed]

51. Wang, M.; Liu, H.; Ge, L.; Xing, G.; Wang, M.; Weining, S.; Nie, X. Identification and analysis of RNA editing sites in the chloroplast transcripts of *Aegilops tauschii* L. *Genes* **2016**, *30*, 8. [CrossRef] [PubMed]

52. Kumbhar, F.; Nie, X.; Xing, G.; Zhao, X.; Lin, Y.; Wang, S.; Weining, S. Identification and characterization of RNA editing sites in chloroplast transcripts of einkorn wheat (*Triticum monococcum*). *Ann. Appl. Biol.* **2018**, *172*, 197–207. [CrossRef]

5

Unraveling Molecular and Genetic Studies of Wheat (*Triticum aestivum* L.) Resistance against Factors Causing Pre-Harvest Sprouting

Ahmad Ali [1,2,3], **Jiajia Cao** [1,2,3], **Hao Jiang** [1,2,3], **Cheng Chang** [1,2,3,4], **Hai-Ping Zhang** [1,2,3,4,*], **Salma Waheed Sheikh** [5], **Liaqat Shah** [1,2,3] and **Chuanxi Ma** [1,2,3,4]

[1] College of Agronomy, Anhui Agricultural University, Hefei 230036, China; ahmad.zafar18@yahoo.com (A.A.); caojia.phs@foxmail.com (J.C.); jiangh.1230@foxmail.com (H.J.); changtgw@163.com (C.C.); ahmed_pgmb@yahoo.com (L.S.); machuanxi@ahau.edu.cn (C.M.)

[2] Key Laboratory of Wheat Biology and Genetic Improvement on South Yellow and Huai River Valley, Ministry of Agriculture, Hefei 230036, China

[3] National Engineering Laboratory for Crop Stress Resistance Breeding, Hefei 230036, China

[4] Anhui Key Laboratory of Crop Biology, Hefei 230036, China

[5] School of Life Sciences, Anhui Agricultural University, Hefei 230036, China; salma_bt02@yahoo.com

[*] Correspondence: zhhp20@163.com

Abstract: Pre-harvest sprouting (PHS) is one of the most important factors having adverse effects on yield and grain quality all over the world, particularly in wet harvest conditions. PHS is controlled by both genetic and environmental factors and the interaction of these factors. Breeding varieties with high PHS resistance have important implications for reducing yield loss and improving grain quality. The rapid advancements in the wheat genomic database along with transcriptomic and proteomic technologies have broadened our knowledge for understanding the regulatory mechanism of PHS resistance at transcriptomic and post-transcriptomic levels. In this review, we have described in detail the recent advancements on factors influencing PHS resistance, including grain color, seed dormancy, α-amylase activity, plant hormones (especially abscisic acid and gibberellin), and QTL/genes, which are useful for mining new PHS-resistant genes and developing new molecular markers for multi-gene pyramiding breeding of wheat PHS resistance, and understanding the complicated regulatory mechanism of PHS resistance.

Keywords: wheat; pre-harvest sprouting; seed dormancy; abscisic acid; gibberellin; QTL/genes

1. Introduction

Pre-harvest sprouting (PHS) refers to the germination of grains in mature cereal spikes before harvest under continuous wet weather conditions [1]. PHS has adverse impacts on wheat quality and yield [2,3] and reduces baking quality of dough by making it porous, sticky, and off-color. The price of sprouted grain is decreased by 20–50% and is unacceptable for human food if it contains more than 4% sprouted grains [4]. The decreased bread and noodle quality is due to increased activity of lipases, amylases, and proteases, enzymes which degrade lipids, starch, and proteins in sprouting grains [5,6]. Global yield and quality losses due to PHS have a financial impact estimated at $1 billion annually [7]. PHS occurred frequently in many major wheat producing areas of the world, including China, USA, Japan, Canada, Australia, and also in Europe [8]. In China, PHS is a major problem, especially in the northern spring wheat region, Yangtze River Valley, and northeastern spring wheat region which are characterized by heavy rainfall and high humidity before harvest [9]. In recent years, it has also become a serious problem in the Yellow and Huai Valleys' wheat region due to climate changes. Therefore, improving PHS resistance is a major breeding objective to mitigate the risk of PHS and increase the production of high-quality wheat.

PHS resistance is associated with several developmental, physiological, and morphological features of the spike and seed, which includes seed coat (pericarp) color and permeability, seed dormancy, α-amylase activity, and levels of plant growth hormones (abscisic acid, gibberellin and auxin) [1,10–18]. Other factors, such as waxiness, hairiness, ear morphology, and germination-inhibitory compounds produced in bracts surrounding the grains have also been linked with PHS resistance [19,20]. Among them, seed dormancy is the major genetic factor controlling PHS resistance, therefore, much attention has been paid to understand the molecular mechanism of seed dormancy as a means to improve PHS resistance in wheat breeding programs.

PHS resistance is a typical quantitative trait controlled by numerous QTL/genes. Many quantitative trait loci (QTL) have been identified for PHS resistance in wheat [1,14,18,21–37]. Several candidate genes for PHS resistance have also identified, including *TaSdr*, *TaPHS1*, *TaMFT*, *TaVp-1*, *Tamyb10*, and *TaMKK3-A* [38–46]. These QTL/genes are valuable for gene pyramiding in breeding programs. However, the regulatory mechanisms of PHS remain unclear, which is why progress in improving wheat PHS resistance is limited.

To understand the regulatory mechanism of PHS resistance and provide valuable information for developing PHS resistant wheat varieties, this review summarizes recent advances of several major factors affecting PHS resistance, including grain color, seed dormancy, α-amylase activity, and plant growth hormones.

1.1. Grain Color

Grain color (GC) is an important genetic factor affecting the brightness of flour and is also associated with seed dormancy and PHS resistance. It is controlled by the *R-1* gene series distally located on long arms of chromosomes 3A, 3B, and 3D [47]. Dominant *R-1* alleles confer red grain color and are denoted by *R-A1b*, *R-B1b*, and *R-D1b* whereas the recessive alleles contribute white grain color and are named as *R-A1a*, *R-B1a*, and *R-D1a*, respectively. For dominant *R-A1b*, *R-B1b*, and *R-D1b* alleles, only one allele is enough for red color, while redness increases in a gene dosage-dependent manner [48]. The *R* genes act as transcriptional activators of flavonoid synthesis genes and are positioned in the same region as Myb-type transcription factor loci (*Tamyb10-A1*, *Tamyb10-B1*, and *Tamyb10-D1*) [49]. Himi et al. [40] confirmed the three *Tamyb10-1* genes on chromosomes 3AL, 3BL, and 3DL as candidate genes underlying the *R-1* loci for wheat grain color.

The red pigment in the testa of plant grains is composed of catechin, and proanthocyanidins (PA) that are produced in the flavonoid biosynthesis pathway and synthesized by different enzymes such as dihydroflavonol-4-reductase (DFR), chalcone flavanone isomerase (CHI), flavanone 3-hydroxylase (F3H), and chalcone synthase (CHS) [50–52] (Figure 1). These enzymes are expressed only in immature red grains and are almost completely repressed in the grains of white wheat [49]. The above Myb-type *Tamyb10-1* transcription factors control anthocyanin production and the red pigment of wheat grain by up-regulating the structural genes encoding DFR, CHI, F3H, and CHS in the flavonoid biosynthesis pathway.

In general, red-grained genotypes are more resistant to PHS compared to white-grained genotypes [53,54]. Himi et al. [53] observed the effect of *R* genes on grain dormancy by using near-isogenic red grained ANK lines and white grained mutant (EMS-AUS) lines and found that the level of dormancy conferred by *R* genes decreased rapidly in ANK lines during the after-ripening stage whereas reduction in the white grained mutant (EMS-AUS) line was not large indicating that *R* genes might play a minor role in seed dormancy. Groos et al. [1] detected four QTL for both PHS resistance and GC using a recombinant inbred line (RIL) population from a cross between Renan (red-grained) and Récital (white-grained). Three of these QTLs were close to *R* genes, and one was mapped on chromosome 5AS. Lin et al. [55] reported the genetic architecture of GC and PHS and genetic relationship of these two traits in a panel of 185 U.S. elite breeding lines and cultivars using a genome-wide association study (GWAS). These results showed that GC genes (*Tamyb10-A1* and *Tamyb10-D1*) had a significant effect on PHS resistance, but *Tamyb10-B1* was significant only for GC and not for PHS resistance. In addition, a novel QTL for GC was also identified on chromosome 1B. Zhou et al. [37] identified three main QTLs for PHS resistance by GWAS, including a novel locus on chromosome 5D and two loci co-located with *Tamyb10-1* genes on chromosomes 3A and 3D. Furthermore, 32 GC-related QTLs (GCR-QTL) were also detected, and a strong correlation was observed between the number of GCR-QTL and seed germination rate. The above results imply that GC is significantly associated with PHS resistance, and might be controlled jointly by many QTLs in addition to he *Tamyb10-1* gene. Of these, some QTLs are for both GC and PHS resistance; others are for GC only and not for PHS resistance. Therefore, it should be possible to breed PHS-resistant white wheat by using the gene-editing technology known as CRISPR/Cas9 to alter the GC-related genes keeping in view the other dormancy-related QTLs besides those provided by the *R-1* genes of the red grained parent used for such editing.

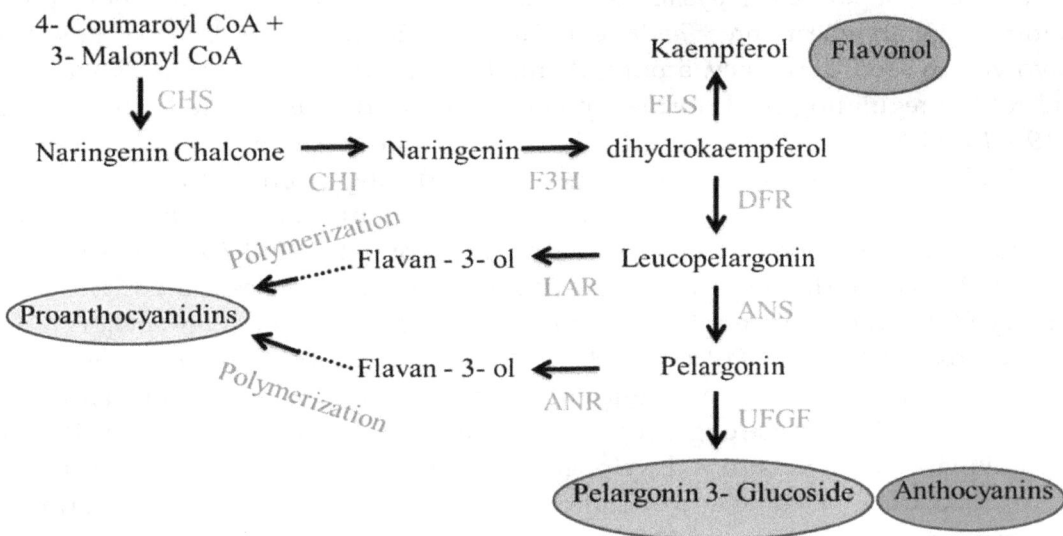

Figure 1. Schematic representation of flavonoid biosynthesis pathway in plants. Enzymes are shown in blue while intermediates are shown in black. End products are indicated in colored shapes. Dotted arrows represent multiple steps. CHS, chalcone synthase; CHI, chalcone isomerase; F3H, flavanone 3-hydroxylase; DFR, dihydroflavonol 4-reductase; ANS, anthocyanidin synthase; UFGT, UDP-glucose flavonoid 3-O glucosyltransferase; FLS, flavonol synthase; LAR, leucoanthocyanidin reductase; ANR, anthocyanidin reductase.

1.2. Seed Dormancy

Dormancy is the inhibition of germination of morphologically ripe and healthy seeds even under optimum conditions of light, moisture, and temperature [56,57]. Initiation and maintenance of dormancy is affected by both genetic and environmental factors [58]. Dormancy is regarded as a major genetic component of PHS resistance [59–61]. Seed dormancy in wheat is a complex phenomenon and can be divided into seed coat-imposed and embryo-imposed dormancy [62,63]. Seed coat inhibitory compounds are associated with seed coat-based dormancy [53], whereas crosstalk of phytohormones, such as abscisic acid (ABA), gibberellin (GA), and auxin, are involved in embryo-imposed dormancy [64,65]. Seed coat-imposed dormancy in particular is involved in the seed survival mechanism of several species [66]. The seed coat exerts its germination-restrictive action by its mechanical resistance to radicle protrusion or being impermeable to water and/or oxygen. These properties are positively correlated with seed coat color due to phenolic compounds in diverse species. In wheat, red-grained genotypes exhibit a wide range of seed dormancy and are more resistant to PHS because they contained dominant alleles in their trigenic series, whereas white-grained cultivars lack seed dormancy at maturity and are susceptible to PHS [63,67–69].

It is widely known that abscisic acid (ABA) is the major mediator for seed dormancy because it plays a significant role in inducing and maintaining dormancy during seed development as well as in imbibed seeds [70,71]. Many genes, like TaPHS1 (a TaMFT-like gene), TaCYP707A1, and TaDOG1, have been identified for seed dormancy and are also involved in ABA synthesis and its signal transduction [41,43,72–74]. Until now, TaPHS1/TaMFT, TaSdr, PM19-A1/A2, and TaMKK3-A are the cloned genes involved in controlling seed dormancy and PHS resistance in wheat. TaMFT (Mother of FT and TFL1) is a homologue of the Arabidopsis MFT gene which controls seed dormancy and also regulates ABA and GA signal transduction. These studies indicated that wheat and Arabidopsis share the same regulatory mechanism of seed dormancy [41,43,72]. An SNP in the promoter region (at position –222) of TaMFT has identified which may increase MFT expression and likely contributes to increase seed dormancy [41]. Another gene TaPHS1 (homolog of wheat MFT gene) involved in the regulation of seed dormancy and PHS resistance was identified on wheat chromosome 3A while the mutations at +646 and +666 positions of the coding region of TaPHS1 gene resulted in PHS susceptibility [42,43]. It has been reported that Sdr gene plays an intermediate role in inhibiting germination and promoting dormancy in rice [75]. In wheat, three TaSdr genes TaSdr-A1, TaSdr-B1, and TaSdr-D1 have been cloned and are involved in seed dormancy, among them; TaSdr-B1 on chromosome 2B was observed to play a vital role in regulating seed dormancy [46,76]. Barrero et al. [44] identified two candidate genes PM19-A1 and PM19-A2 which positively regulate seed dormancy. They also demonstrated that PM19-A1 highly expressed in dormant genotypes during grain maturation while PM19-A2 showed sequence variations between non-dormant and dormant genotypes. In wheat, another gene MKK3-A (mitogen-activated protein kinase kinase 3), also called TaMKK3-A, has been identified on chromosome 4AL as a candidate gene of the Phs-A1 locus which is associated with the length of seed dormancy [45]. Despite the multi-genic control of seed dormancy, a few major loci, including Phs-A1 on chromosome 4AL and TaPHS1/TaMFT on chromosome 3A, have also shown to involve in sprouting resistance and account for a significant proportion of natural variations in diverse mapping populations [77]. Based on the strong effect, Phs-A1 has been identified in at least 15 multi-parent and bi-parental mapping populations developed from diverse germplasm originated in the United Kingdom, Australia, China, Japan, Mexico, Europe, Canada, and Africa [44,78–81]. Shorinola et al. [77] studied the physiological evaluation of Phs-A1 during seed after-ripening and observed that it delayed the rate of loss in dormancy when plants were grown across a wide temperature range of 13–22 °C. In another study, Shorinola et al. [82] conducted a haplotype analysis of the Phs-A1 locus and found that TaMKK3-A, but not PM19, was the causal gene underlying variation in sprouting associated with Phs-A1 in diverse Asian, North American, European, and Australian germplasm.

In addition to the PHS-resistant genes identified in wheat, maize could also act as a model to improve PHS resistance in wheat, e.g., the maize viviparous-1 (Vp1) gene. McCarty et al. [83] reported that Vp1 gene encodes a transcription factor that plays a significant role in the regulation of late embryogenesis in maize and late embryo development in bread wheat. TaVp1 genes were extensively studied in wheat and were linked with seed dormancy and PHS resistance [38,39,61,84–89]. The TaVp1 genes were mapped about 30 cM from R loci on homologous group 3Lchromosomes [86]. Six TaVp-1B alleles, TaVp-1Ba, TaVp-1Bb, TaVp-1Bc, TaVp-1Bd, TaVp-1Be, and TaVp-1B were identified in wheat [38,39,89,90]. Based on this allelic variation, the STS marker (Vp1B3) was developed for seed dormancy and it was observed that TaVp-1Bb and TaVp-1Bc alleles were linked with higher PHS tolerance [38]. Another STS marker (Vp1A3) was also developed for PHS tolerance by observing greater PHS resistance in various combinations of allelic variations, like TaVp-1Agm/TaVp-1Ba, TaVp-1Agm/TaVp-1Bb, TaVp-1Aam/TaVp-1Bb, and TaVp-1Aim/TaVp-1Bb [61]. Moreover, genetic analyses identified other genes such as DOG1 involved in seed germination and dormancy [91–94]. The DELAY OF GERMINATION1 (DOG1) gene was first identified in Arabidopsis as a major QTL involved in increased seed dormancy [95]. The length of seed dormancy was estimated by the amount of expression of DOG1 protein in freshly-harvested seeds, which indicated that the DOG1 gene is a timer for the release from dormancy [96]. In a recent study, Nishimura et al. [97] demonstrated that the

DOG1 and *AHG1* genes interact with multiple environmental factors as well as the PYL/RCAR ABA receptor-mediated regulatory system to establish an important regulatory mechanism for control of seed dormancy and germination. Recent advances in genome sequencing and whole genome assembly of hexaploid wheat will trigger progress in identifying more seed dormancy and PHS resistance genes [98–102].

1.3. α-Amylase Activity

The α-amylase (amy) enzyme is involved in many plant physiological processes such as cold tolerance and germination rate and can hydrolyze α-1,4 -glycosidic bonds in saccharides [103,104]. The expression of amy enzyme is strictly controlled by the phytohormones ABA and gibberellin. ABA inhibits the amy expression during grain development. However, in genetic defect wheat, a high level of high pI amy genes could be expressed, resulting in poor grain quality during late grain development which is referred to as late maturity α-amylase (LMA) [105]. An elevated level of GA promotes amy expression during seed germination [106]. Alpha-amylase activity and PHS resistance are associated with each other possibly due to the fact that increased α-amylase activity upon water absorption promotes seed germination [107,108]. A remarkable difference was found in α-amylase activity between PHS-resistant and -susceptible cultivars in wheat [108]. Of three PHS traits, falling number (FN) [109] was found to indirectly measure the α-amylase enzyme activity that degrades starch in germinating seeds and is an important factor in quantifying PHS [110]. Breakdown of starch due to increased α-amylase activity results in a decreased FN value and is an indirect sign of low seed dormancy and low PHS resistance. Four isozymes of α-amylase affecting PHS have been identified in wheat, including malt-α-amylase (α-amylase-1) on homologous chromosomes 6, green-α-amylase (α-amylase-2) on homologous chromosomes 7, α-amylase-3 on homologous chromosomes 5, and α-amylase-4 has two members on homologous chromosomes 2 and 3 [111,112]. The wheat B genome contains genes for α-amylase-1 and α-amylase-2 among all the three genomes. Promoters of α-amylase-1 gene contains GA responsive complex that consists of a GA-responsive element (CAATAAA), pyrimidine box (CCTTTT), and TATCCAT/C box [112]. GA3 seemed to be involved in regulation of expression level of α-amylase-1 and α-amylase-2 [113]. The α-amylase-1 activity seemed to be significantly correlated with seed dormancy and contributed about 84% to seed germination [111].

In addition to α-amylase variation, α-amylase subtilisin inhibitors (ASI) were also identified in wheat, rice, rye, and barley by limiting α-amylase activity to restrain seed germination [114,115]. Moreover, ten ASI isomerides were identified through monoclonal antibody immune imprinting and isoelectric focusing electrophoresis techniques [116,117]. Yuan et al. [118] reported that PHS tolerance can be increased by reducing the α-amylase activity through combing α-amylase-1 and ASI complex. The α-amylase quantity and activity is very low in dormant seeds and increases after seed germination, therefore, it is necessary to identify the regulatory factors interacting with α-amylase, which can contribute to understand the complicated molecular mechanism of α-amylase regulating PHS tolerance.

1.4. Plant Growth Hormones

Previous studies have described the significance of plant hormones in metabolic and signaling aspects and their probable role in the maintenance and release of dormancy in seeds of cereal crops [43,119,120]. Among plant growth hormones, abscisic acid (ABA) and gibberellin (GA) play important roles in regulation of dormancy and germination, ABA induces dormancy and GA stimulates seed germination [121,122]. A change in balance between ABA and GA levels in seed constitutes a regulatory mechanism that results in maintenance or release of seed dormancy [120,123]. Several studies have reported the regulatory mechanisms of other hormones like ethylene, jasmonate, brassinosteroids, and auxin in controlling seed dormancy, germination and PHS resistance [43,121,124,125]. Environmental factors, such as light and temperature, also affect the dormancy and germination by disturbing the balance between ABA and GA levels in cereal crops [126,127].

Numerous mutants have been developed to understand the regulatory role of plant growth hormones in seed germination and dormancy. Recent advances in the genomics of cereal crops have led to identify many genes involved in metabolic and signaling pathways of plant hormones for regulating seed germination and dormancy. The levels of plant growth hormones are noticeably different in PHS resistant and susceptible varieties; therefore, PHS resistance can be improved by identifying more genes involved in the expression and regulation of plant growth hormones.

1.4.1. Abscisic Acid

Abscisic acid (ABA) is an essential hormone that promotes seed dormancy, seed maturation and tolerance to desiccation [128]. Dormant wheat ABA levels increases by up to 2.5-fold during imbibition but remains unchanged in non-dormant grains [129]. ABA level in seeds/tissues is regulated by its synthesis and catabolism [130]. ABA biosynthesis is catalyzed by numerous enzymes like NCED (9-cis-epoxycarotenoid dioxygenase) that acts as a key regulator of ABA biosynthesis during seed maturation. During ABA biosynthesis, oxidative cleavage of violaxanthin and 9-cis-neoxanthin by NCED is regarded as rate-limiting step [131], whereas ABA catabolism is triggered by ABA 8'-hydroxylase enzyme (ABA8'OH) encoded by *CYP707A* genes that induce ABA hydroxylation at the 8' position [132–134] (Figure 2). Therefore, the *NCED* and *CYP707A* genes play important roles in germination and dormancy by controlling the ABA level in seeds.

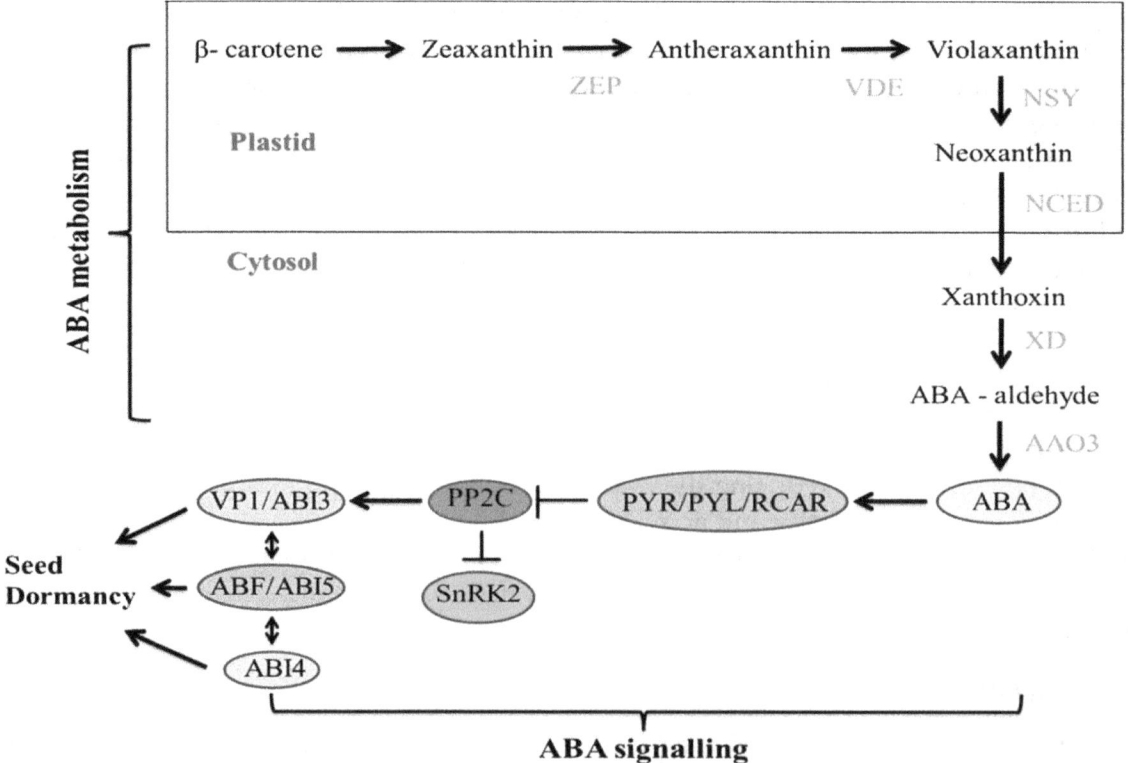

Figure 2. Schematic representation of ABA metabolism and signaling pathway in plants. Enzymes are shown in blue while intermediates are shown in black. The end product is indicated in the yellow circule; ZEP, zeaxanthin epoxidase; VDE, violaxanthin de-epoxidase; NSY, neoxanthin synthase; NCED, 9-cisepoxycarotenoid dioxygenase; XD, xanthoxin dehydrogenase; AAO3, abscisic aldehyde oxidase; PYR/PYL/RCAR, pyrabactin resistance/pyrabactin-like/regulatory components of ABA receptors; PP2C, protein phosphatase 2C; SnRK2, SNF1-related protein kinase2; ABI3, abscisic acid insensitive 3; ABI4, abscisic acid insensitive 4; ABI5, abscisic acid insensitive 5; VP1, viviparous 1; ABF, ABRE binding factor.

Seed development in wheat is characterized by two peaks of ABA accumulation that occur during the mid and late maturation phases. The first peak arises around 25 days after pollination (DAP) while the second peak arises around 35 DAP and extended up to 40 DAP in dormant wheat seeds [135,136], indicating the significance of ABA in inducing embryo dormancy [137]. Moreover, imbibed dormant wheat seeds have shown 3.8-fold higher expression of *TaNCED2* than non-dormant imbibed seed, while non-dormant seeds exhibit 2.5-fold higher expression of *TaABA8'OH1* (a wheat homolog of *CYP707A*) than dormant seeds in both imbibed and dry conditions [138]. Mutational analysis of the two homologs of *TaABA8'OH1* (*TaABA8'OH1A* and *TaABA8'OH1D*) showed an increase in embryonic ABA contents during mid and late stages (40–60 DAA) of seed development resulting in a higher level of seed dormancy [139] and highlighting the importance of higher embryonic ABA levels in inducing seed dormancy during the seed maturation phase in wheat.

ABA has been involved in the regulation of several seed developmental processes like deposition of storage reserves and primary dormancy induction that are evident from the observation of ABA mutants or deficient plants in maize and *Arabidopsis* [140]. A large number of mutants with reduced ability of synthesizing ABA have been developed in various crops, like the *aba1* mutant in *Nicotiana plumbaginifolia*, viviparous (*Vp*) mutants *Vp5, Vp7, Vp10/Vp13, Vp14*, and *Vp15* in maize; *aba1, aba2*, and *aba3* mutants in *Arabidopsis*, and *sit, flc*, and *not* mutants in tomato [91,131]. Several knock-out mutations are available for most wheat genes that provide an invaluable resource for characterizing the gene function. The resource of Targeting Induced Local Lesions In Genome (TILLING) mutants, like Kronos (tetraploid) and Cadenza (hexaploid), have been developed in wheat. The exome sequences of 1535 Kronos and 1200 Cadenza mutants have been resequenced using Illumina next-generation sequencing that can be used to screen for mutations in pre-harvest sprouting and dormancy related genes [141].

ABA biosynthetic mutants failed to induce seed dormancy and revealed a wilty vegetative phenotype, e.g., the *aba1* mutant in *Arabidopsis* and *aba2* mutant in tobacco were not able to produce zeaxanthin epoxidase (ZEP), the first identified ABA biosynthetic enzyme [142]. Another ABA-deficient mutant *aba4* was identified in *Arabidopsis* during a screening of paclobutrazol resistance germination and has known to be impaired in neoxanthin synthase (NSY) enzyme [143]. The *vp14* (*viviparous14*) mutant in maize and the notabilis mutant in tomato have shown impairment for NCED, which acts as a catalyst for oxidative cleavage of 9'-cis neoxanthin and/or xanthophylls, 9-cis-violaxanthin, and produces xanthoxin, as shown in Figure 2 [144,145]. The *vp10* and *vp15* mutants in maize, *sitiens* and *flacca* mutant in tomato and *aba2* and *aba3* mutants in *Arabidopsis* have also shown the impairment in later steps during ABA biosynthetic pathway in the cytosol [146,147].

The role of ABA in seed dormancy of wheat has already been described [16,43]. Nambara et al. [130] reported three core components of ABA signaling in seeds such as protein phosphatase 2Cs (PP2Cs), SNF1-related protein kinase2s (SnRK2s) and pyrabactin resistance/pyrabactin like/regulatory components of ABA receptors (PYR/PYL/RCAR), as shown in Figure 2. ABA forms a complex by binding with its receptor PYR/PYL/RCAR which then interacts with PP2Cs to inhibit its function. The PP2Cs negatively regulate ABA signaling by repressing the SnRK2s activity, which is a positive regulator of downstream targets. Inhibition of PP2Cs causes de-repression of SnRK2s, which in turn phosphorylates and activates down-stream transcriptional factors including ABI3 (B3 type protein), ABI4 (AP2 type transcription factor), ABI5 (abscisic acid insensitive 5), and ABFs (bZIP-type transcription factors). These transcriptional factors are important for the expressional regulation of ABA-responsive genes of seeds [130]. In the absence of ABA, PP2Cs becomes activated and, in turn, dephosphorylate and deactivate the SnRK2s. The molecular components involved in the ABA signaling pathway seem to be conserved in seeds of both monocot and dicot species [148].

The current understanding of signaling elements like ABA that control seed dormancy and germination mainly results from genetic analysis. In wheat, QTL and mutational analysis have revealed the importance of ABA sensitivity in regulating seed dormancy [149,150]. Dormant wheat seeds show more ABA sensitivity than non-dormant seeds [151,152]. *Vp1* was the first gene cloned in maize against ABA response [83]. Expression of *Vp1* in wheat embryos was positively correlated with ABA sensitivity and degree of seed dormancy [87,153]. Splicing of the *Vp1* gene in wheat and rice counterpart resulted in susceptibility to PHS in both species [88,154]. *ABI3* is the ortholog of *Vp1* in *Arabidopsis* and the seeds containing *Vp1* or *ABI3* alleles exhibited similar phenotypes including ABA insensitivity, desiccation intolerance, and premature activation of the shoot apical meristem [155]. Mutational analysis of the *ABI4* and *ABI5* loci in *Arabidopsis* showed similar quantitative effects as *ABI3* on ABA sensitivity and seed development, but *ABI3* null mutations were more destructive than *ABI4* and *ABI5* [156]. These studies may help to explain the role of ABA in inducing wheat seed dormancy and to understand the molecular mechanisms underlying the regulation of ABA metabolism in inducing dormancy during seed maturation.

1.4.2. Gibberellin

Gibberellin (GA) is another major plant hormone that plays an important role in regulation of seed dormancy and germination [64]. GA breaks the seed dormancy and promotes germination by balancing the primitive endogenous inhibitors [157]. It also regulates the expression of α-amylase synthesis genes involved in seed germination and hydrolyzing the starch in the endosperm. In *Arabidopsis*, leafy cotyledon 2 (*lec2*) and fusca3 (*fus3*) could up-regulate GA activity resulting in germination of seeds before maturity [158,159].

The bioactive concentration of GA in plants is regulated by the balance between its synthesis and inactivation, that are mainly controlled by the genes *GA2ox* (encoding GA 2-oxidase), *GA3ox* (encoding GA 3-oxidase), and *GA20ox* (encoding GA 20-oxidase), respectively [160] (Figure 3). Many genes encoding these enzymes have been identified in a range of crop species including wheat, rice, and barley [160,161] and their expression plays significant roles in dormancy and germination by regulating the GA level in seeds.

Variations at the transcriptional level of these genes orthologs due to after-ripening and in non-dormant and dormant cereal crop seeds indicated the role of GA in regulating the seed germination and dormancy. For instance, dormancy loss in imbibed after-ripened barley and wheat seeds has shown to be linked with increased expression of the *TaGA3ox* and *TaGA20ox* genes and a higher level of bioactive GA$_1$ [43,126,162]. Moreover, transcriptional activation of *GA20ox* gene orthologs induced an increase in the level of GA$_4$ in non-dormant embryos of sorghum seeds, whereas up-regulation of *GA2ox* gene orthologs led to a decrease in the GA$_4$ level [163]. Mutational studies of these genes will provide further information regarding the molecular mechanisms of GA in regulating seed germination and dormancy. Genetic studies in rice have identified other candidate genes, such as *OsGA20ox2* and *OsGA2ox3*, responsible for regulating seed germination [164,165], while mutational analysis of *OsGA20ox2* showed greater dormancy due to reduction in the GA levels in seeds [164].

GA signals in plants are perceived by the soluble receptor protein gene *GID1* (Gibberellin insensitive dwarf 1), that was first mapped in rice. Mutational analysis of *GID1* in rice showed repression of α-amylase production and had no inhibitory effect on seed germination [166]. Orthologs of *GID1* protein have also been identified in wheat but further characterization of functions analysis of *GID1* orthologs in seed dormancy is required [167].

DELLA proteins in plants are another important element of GA signaling pathway, which function as a GA repressor and are broken down by ubiquitination induced by GA [168]. GA binds with GIDI and triggers the formation of the GA-GID1-DELLA complex which then interacts with F-box protein (the principal component of SCFSLY 1/GID2 E3 ubiquitin ligase) to degrade the DELLA protein through the ubiquitin-26S proteasome pathway [169–171] (Figure 3). In *Arabidopsis*, five DELLA proteins including RGL1 (RGA like1), RGL2 (RGA like2), RGL3 (RGA like3), GA1 (GA insensitive), and RGA (Repressor of GA1) were identified [172], among them RGL2 is known to be an important seed germination repressor [173]. The DELLA proteins in cereals such as RHT (reduced height) in wheat, SLN1 (slender1) in barley and SLR1 (slender rice1) in rice are transcribed by single genes [174–176]. Chandler [177] observed that DELLA mutant seeds of barley were non-dormant and exhibited higher α-amylase activity in the aleurone layer.

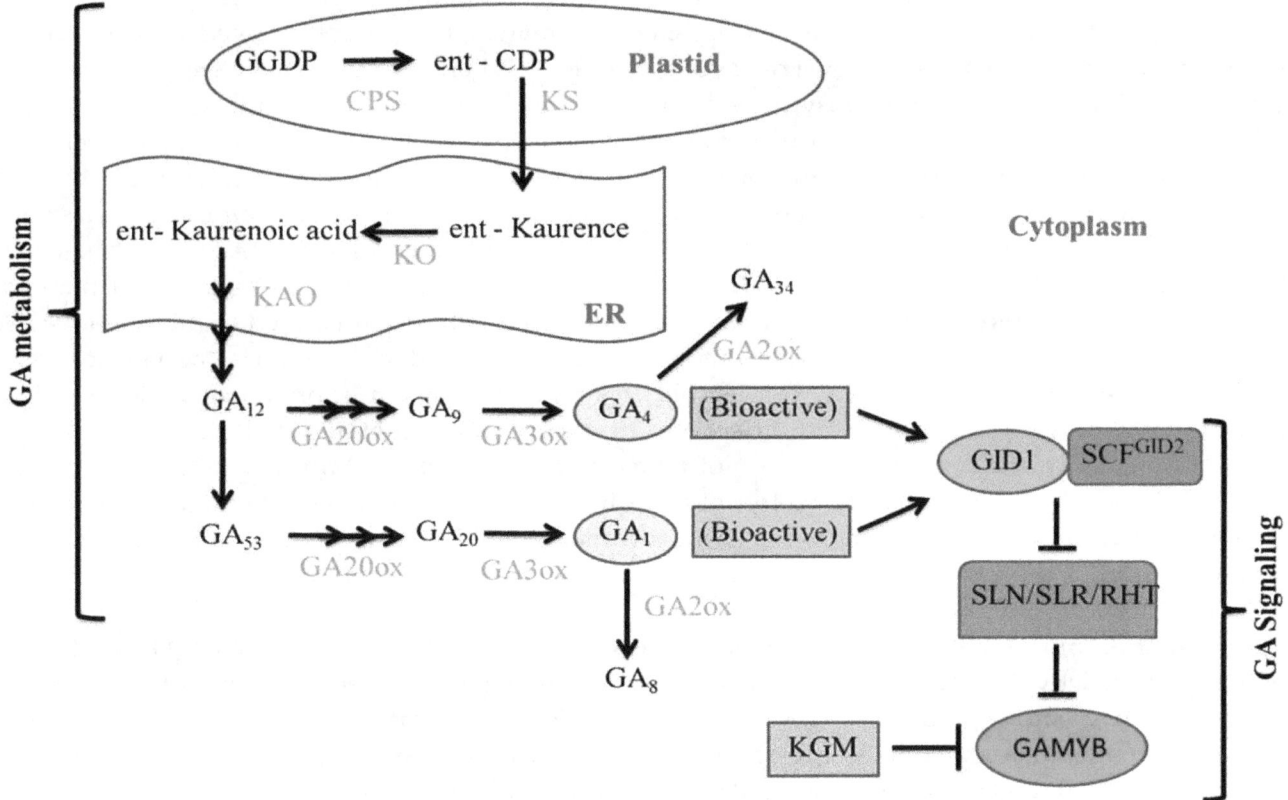

Figure 3. Schematic representation of GA metabolism and signaling pathway in plants. Enzymes are shown in blue while intermediates are shown in black. Multiple arrows represent multiple steps. GGDP, geranylgeraryl diphosphate; CPS, ent- ent- Copalyl diphosphate synthase; KS, ent-Kaurene synthase; KO, ent-Kaurene oxidase; KAO, ent- Kaurenoic acid oxidase; GA20ox,GA20 oxidase; GA3ox,GA3 oxidase; GA2ox,GA2 oxidase; GID1, gibberellin insensitive dwarf 1; GID2, gibberellin insensitive dwarf 2; SLN, slender1 in barley; SLR1, slender rice1; RHT, reduced height; GAMYB, GA regulated MYB transcriptional regulator; KGM, kinase associated with GAMYB.

In addition to DELLA proteins, other GA-regulated MYB transcriptional regulators (GAMYB) also play a significant role in the GA signaling pathway in aleurone cells of cereal crops [178]. In cereal aleurone, GAMYB triggers the transcriptional activation of GA and regulates hydrolytic enzymes especially α-amylase by directly binding to GA responsive elements in promoter regions [178]. The GAMYB function in cereal aleurone cells was repressed by another GA downstream signaling component named KGM1 (KINASE ASSOCIATED WITH GAMYB1) [179]. Mutational analysis of *GAMYB* orthologs in rice showed repression of α-amylase gene expression but had no effect on seed germination [180]. In wheat, whole seed transcriptional analysis showed no transcriptional differences in orthologs of *GID1*, *RHT*, *GAMYB*, and *KGM1* represented in the GeneChip Wheat Genome Array between after-ripened and dormant seeds. However, transcription of GA responsive genes encoding cell wall hydrolases and amalyses are induced in response to after-ripening [43]. These results might indicate that wheat seeds responsiveness to GA is controlled by post-transcriptional mechanisms or functions without these GA signaling elements. Although these studies demonstrate the role of GAMYB in germination and dormancy, but need further studies to identify and characterize more genes that interact with GAMYB and increase our understanding about the role of downstream GA signaling elements in controlling germination and seed dormancy.

1.4.3. Other Plant Hormones

Previous studies have described the importance of some other plant hormones, like ethylene, jasmonate (JA), auxin, and brassinosteroid (BR), in regulating seed dormancy and PHS resistance [121,124,125]. Transcriptomic analysis of dormant wheat seeds showed that imbibition triggered variations in expression level of several orthologous genes encoding key enzymes involved in ethylene, JA, auxin, and BR biosynthesis and their metabolic pathways due to after-ripening, indicating roles in regulating seed germination and dormancy [43,119]. Jacobsen et al. [181] reported that JA promotes dormancy release in dormant wheat seeds and perform antagonistically to ABA. Transcriptional activation of biosynthetic gene orthologs, such as *KAT3* (3-ketoacyl coenzyme a thiolase3), *LOX5* (lipoxygenase5), and *AOS* (allene oxide synthase) involved in JA biosynthesis were detected in imbibed after-ripened seeds. The production of a huge amount of jasmonate-isoleucine along with other orthologs of the biosynthetic pathway during after-ripened seed imbibition revealed the role of JA in controlling wheat seed dormancy. It has been reported that methyl jasmonate decreased the level of seed dormancy in wheat, but its role was regulated by variation in expression levels of the *ABA8′OH* and *NCED1* genes and ABA concentration [181]. Xu et al. [182] studied the role of JA and other hormones in the stratification of wheat dormant seeds and found that JA formation is necessary for seed germination induced by cold stratification. They also concluded that an increase in JA synthesis promoted a reduction in ABA concentration in cold-stratified grain embryos. However, the mechanism of JA in release of seed dormancy to allow germination is still not clear.

Transcriptomic analysis of after-ripened dormant wheat seeds revealed imbibition-mediated activation of BR biosynthetic ortholog genes such as *DET2* (De-etiolated 2) and *DWF4* (Dwarf 4) [119]. Transcriptional activation of these BR biosynthetic and signaling components in imbibed after-ripened wheat seeds have shown to be associated with transcriptional stimulation of BR responsive orthologs, such as *BBE* (BR enhanced expression) and *PRE* (paclobutrazol resistance), that control cell elongation, a process essential for seed germination [183,184]. These results indicated that BR plays a significant role in controlling seed dormancy and germination in wheat.

In *Arabidopsis*, previous studies have revealed that BR plays a significant role in the regulation of seed dormancy along with ABA and also increased the synthesis of ethylene, which has a regulatory role in seed dormancy of monocot species, such as wild oat [185–187]. During ethylene biosynthesis, BR mediates post-transcriptional activation of the ethylene biosynthetic enzymes such as ACC (aminocyclopropane-1- carboxylic acid) synthase (ACS) which acts as a catalyst during the first step [188]. In addition to ACS, ethylene biosynthesis is also mediated by another enzyme ACO (ACC oxidase). In wheat, ACO orthologs are involved in up-regulation in whole imbibed after-ripened seeds compared to dormant seeds [119]. Transcriptional activation of these enzymes and other ethylene receptor orthologs in wheat, such as ERS1 (ethylene response sensor 1), have revealed the role of ethylene in controlling seed germination and dormancy. In other cereal crops like rice and barley, ethylene promotes germination in non-dormant seed but is not involved in the loss of dormancy [189,190]. At present, the role of ethylene in regulating seed dormancy is not clear and needs further investigation of its regulatory role in seed dormancy in different crops.

Recent studies have revealed the role of auxin in maintaining seed dormancy. For instance, exogenous application of auxin increased seed dormancy in wheat [191,192] and *Arabidopsis* [193,194] through ABA activation. Liu et al. [43] studied the temporal expression patterns of metabolic and signaling genes of ABA, GA, IAA and jasmonate in both dormant and after-ripening dry and imbibed wheat seeds and observed that after-ripening mediated developmental switch from dormancy to germination seems to be linked with declines in seed sensitivity to ABA and IAA and repression of auxin signaling. Mutational analysis of wheat ERA8 (Enhance Response To ABA8) mutant showed an increase in dormancy due to increased level of embryonic jasmonate and aleurone IAA [195]. Metabolomic profiling of two water imbibed wheat cultivars Sukang (dormant) and Baegjoong (non-dormant) detected variable amounts of many auxin-related compounds in the 48h samples of Baejoong and found that indoleacetate abundance was not changed in the Sokang sample but showed an abrupt reduction in Baegjoong at 48h water imbibition. Three catabolites of IAA including indole-3-carboxylate were also detected that showed similar of IAA at 48 h but with two other metabolites, such as indoleacetyl-aspartate and 2-oxindole-3-acetate, showed much higher levels at 48h in Baegjoong [196]. It also regulates several plant growths and developmental processes through the auxin signaling pathway mediated by aux/indole-3-acetic acid (IAA), transport inhibitor response1 (TIR1)/additional F box protein (AFB), and auxin response factors (ARFs) [197,198]. The ARFs are responsible for regulating the expression of a large number of auxin-responsive genes by binding with promoters of auxin response elements (AuxREs) [199,200]. Auxin inhibits seed germination and promotes dormancy through ABA-mediated response by regulating seed specific signaling components of ABA like *ABI3*, *ABI4*, and *ABI5*. Among them, *ABI3* is the only transcription factor involved in the regulation of seed dormancy [201–203]. Liu et al. [65] observed that auxin and ARF10/16 are involved in the regulation of *ABI3* expression which in turn inhibits seed germination and promotes seed dormancy in after-imbibed seeds. The function of ARF10 and ARF16 as positive regulators during the ABA signaling pathway contributes in developing a map of integrated hormone signaling during plant growth and development [204,205]. Auxin-induced seed dormancy seems to be an evolutionary mechanism that inhibits seed germination during unfavorable conditions and might be important for conservation of diversity and evolution in seed plant species [65]. Whether auxin is directly involved in seed dormancy is not clear, and its mechanism in controlling seed dormancy is also still unknown.

1.5. Environmental Factors Affecting PHS Resistance

Environmental factors such as rainfall, temperature and high relative humidity during the grain filling and maturation stages play an important role in the regulation of dormancy and sprouting in wheat. Temperature is one of the most important environmental factors for maintaining dormancy during seed development and for inducing dormancy during seed imbibition [206]. During seed development, low temperatures between 10 and 15 °C can induce high and prolonged dormancy while low temperature during germination breaks dormancy of freshly-harvested wheat seeds [207,208]. At low temperatures the MOTHER OF FT AND TFL1 (*MFT*) gene is involved in enhancing the dormancy during seed development in wheat [41]. It has also been reported that rainfall 10–20 days before harvest causes little or no sprouting but may influence the crop to be highly susceptible for sprouting at later rainfall [209]. Lunn et al. [210] studied the relationship between mean temperature and period of seed dormancy during the grain development and found that shorter dormancy periods occur after high mean temperatures.

Environmental factors such as temperature fluctuations, salinity and seed moisture content seem to promote ABA synthesis in plants with consequent effects on seed dormancy [211]. Footitt et al. [212] reported that the depth of seed dormancy and gene expression patterns were linked with seasonal variations in soil temperature. They also explained that ABA signaling was linked with deep dormancy during winter while its repression was linked with relief from dormancy during spring. ABA-signaling repression was concurrent with increased DELLA repression of germination. During winter, the expression of *NCED6* (ABA synthesis) and *GA2ox2* (GA catabolism) genes were found to be increased resulting in increased seed dormancy due to the decrease in soil temperature whereas, during spring, the endogenous ABA biosynthesis was found to be decreased while the expression of *CYP707A2* (ABA catabolism) and *GA3ox1* (GA synthesis) genes was increased resulting in declined seed dormancy in *Arabidopsis* [212]. In another study, Kashiwakura et al. [162] used two PHS-tolerant varieties, OS38 (highly dormant) and Gifu-komugi (Gifu, moderately dormant) to characterize the mechanisms of both dormancy maintenance and breakage at low temperatures. They observed that Gifu grains were germinated after imbibition at 15 °C whereas OS38 grains remained dormant. Imbibition of Gifu grains at low temperature caused a reduction in ABA levels in dormant embryos primarily due to the expression of *TaABA8'OH1* and *TaABA8'OH2* (ABA catabolism) and *TaGA3ox2* (GA synthesis) genes resulting in increased GA levels. On the other hand, imbibition of extremely dormant OS38 grains at a low temperature increased ABA levels by inducing the expression of *TaNCED* (ABA biosynthesis) gene and suppressing *TaGA3ox2* and *TaABA8′OH2* genes. In a recent study, Izydorczyk et al. [127] observed a delay in germination of non-dormant imbibed wheat seeds under supra-optimal and suboptimal temperatures which was due to the expression of ABA signaling genes; ABI3, ABI5, PYL5, and SnRK2 in the embryo tissues resulted in enhanced ABA sensitivity. These studies explained the role of ABA and GA in dormancy and germination but needs further genetic studies to understand the physiological role of metabolic genes of ABA and GA in the regulation of seed dormancy in cereal crops.

1.6. QTL/genes Identified for PHS Resistance

The genetics of PHS resistance is controlled by both epistatic and additive effects which are affected by environmental conditions. The interaction between QTL epistasis (Q × Q) and the environment (Q × E, Q × Q × E) for PHS resistance has been studied to understand the complex genetic structure of QTL [213–215]. In wheat, PHS resistance is controlled jointly by multiple QTLs located an almost 21 chromosomes (1A, 1B, 2A, 2B, 2D, 3A, 3B, 3D, 4A, 4B, 5B, 5D, 6A, 6B, 6D, 7A, 7B and 7D) [1,14,18, 21–37,216–220] (Table 1).

In addition, several candidate genes for PHS resistance have also been identified based on comparative genomics or transcriptomic analysis, such as *TaSdr-A1* and *TaSdr-B1* on 2AS and 2BS [46,76], *TaPHS1* and *TaMFT* on 3AS [16,41,42], *TaVp-1* and *Tamyb10* on group 3 chromosomes [38–40,84,86], and *PM19-A1/A2* and *TaMKK3-A* on 4AL chromosome [44,45] (Table 2).

The interaction of genes with different PHS resistant QTL is different. QTL positioned on chromosome 4A may interact with *R* gene controlling red seed color to affect the PHS tolerance. Another QTL for PHS resistance was identified on wheat chromosome 5D independently of an *R* gene [29]. PHS is a typical quantitative trait controlled by multiple QTL and genes. A major QTL was mapped on the chromosome 4A by using various mapping populations which controlled about 40% of the phenotypic variation in PHS resistance in wheat [3,27,227]. These studies were conducted using SSR markers for the construction of genetic map and QTL mapping. Due to cost-effective and rapid innovations in sequencing technologies, thousands of molecular markers especially SNPs (such as wheat 820K, 660K, and 90K arrays) have been developed in wheat, which are useful for fine-mapping of QTLs and for cloning of candidate genes in the target regions. Moreover, recent advances in genome sequencing and whole genome assembly of hexaploid wheat will provide the bases for rapid identification of various PHS resistance genes [98–102].

Table 1. QTLs for PHS and related traits identified in wheat.

Trait	QTL	Chromosome	Nearest Marker	QTL name	Material	Reference
PHS and GC	5	3AL 3BL 3DL 5AS	Xffb293 Xgwm403, Xbcd131 Xgwm3 Xbcd1871	-	RILs	[1]
PHS and GC	3	1BS 4BL 7AS	Xpsp3000 Xpsp3030-Xpsp3078 Xpsp3050	-	RILs	[22]
PHS and SD	3	3A 3A 3A	Xpsr394-Xgwm5 Xcdo345 Xcdo345-Xbcd141	*taVp1* *QPhs.ocs-3A.1* *QPhs.ocs-3A.2*	RILs	[23]
PHS and SD	2	4AL 3AL	Xksuf8a-Xbcd402b Xpsr903b-XATPased	-	RILs	[217]
PHS	1	3AL	Xwmc153-Xgwm155	*QPhs.ccsu3A.1*	RILs	[24]
SD	1	4A	Xgwm397-Xgwm269-Xbarc170	-	DHLs	[3]
PHS and SD	1	3AL	Xbarc310-Xbcd907	*QPhs.ocs-3A.1*	RILs	[25]
PHS and SD	1	3AS 3AS	Xbarc310 Xbarc321	*QPhs-3AS* *QPhs.pseru-3AS*	RILs	[26]
PHS	3	2B 2B	Xdup398-Xbarc54 Xbarc105-Xbarc334	*QPhs.pseru-2B.1* *QPhs.pseru-2B.2*	RILs	[221]
PHS	1	2DS	Xgwm261-Xgwm484	*Qphs.sau-2D*	F$_2$ and F$_6$	[28]
PHS	4	2B 2D 3D 6D	Xbarc55-Xwmc474 Xwmc111-WxPt-999 7Xbarc1161 Xcfd37-Xbarc196	*QPhs.cnl-2B.1* *QPhs.cnl-2D.1* *QPhs.cnl-3D.1* *QPhs.cnl-6D.1*	DHLs	[14]
PHS	3	2AL 3AL 3BL	Xgwm1045-Xgwm296 Xgwm153-Xgwm155 Xgwm1005-Xgwm980	*QPhs.ccsu-2A.5* *QPhs.ccsu-3A.1* *QPhs.ccsu-3B.6*	RILs	[32]
PHS and GC	5	3B 3D 3A 5D 3D	Xbarc77-Xwmc30 7Xwmc552-Xwmc533 Xcfa2193-Xwmc594 Xgwm469-Xcfd10 Xwmc11-Xcfd223	*QGi.crc-3B* *QGi.crc-3D* *QSi.crc-3A* *QSi.crc-5D* *QCl.crc-3D*	DHLs	[29]

Table 1. *Cont.*

Trait	QTL	Chromosome	Nearest Marker	QTL name	Material	Reference
PHS and SD	1	3BL	Xwmc527-Xgwm77	-	DHLs	[31]
PHS	1	5D	XCFD40-XBARC1097	qPhs5D.1	DHLs	[33]
PHS and SD	5	2A 2B 3A 4A 7B	521-2A 521-2B 521-3A 521-4A 521-7B	-	Single chromosome substitution lines	[218]
PHS	3	1A 2A 7B	Xwmc611-Xwmc333 Xgwm515-Xgwm425 Xgwm297-Xwmc532	QPhsd.spa.-1A.1 QPhsd.spa.-2A.1 QPhsd.spa.-7B.1	RILs	[35]
PHS	4	3B 4A 7B 7D	19 SNPs flanking the QTL 12 SNPs flanking the QTL 10 SNPs flanking the QTL 04 SNPs flanking the QTL	QSi.crc-3B QGi.crc-4A QSi.crc-7B QFn.crc-7D	DHLs	[80]
PHS	5	1A 1B 5B 7A 7B	wPt-6274 Xwmc191 wPt-6910-wPt-7400 Xcfa2174 Xwmc606	QPhs.spa-1A QPhs.spa-1B QPhs.spa-5B QPhs.spa-7A QPhs.spa-7B	DHLs	[222]
PHS and SD	1	2B	Xwmc477-Xbarc55	Sdr2B	RILs	[76]
PHS and SD	1	4A	wsnp_Ex_c66324_64493429 - CD920298	4A-1	RILs	[44]
PHS and SD	4	4A 4B 5A 5B	GBS212432-GBS10994 7Xbarc20-Xwmc238 TTM_199619-TTM_1259 7Xbarc346-2-TTM_62137_50	Qphs.pseru-4A.1 Qphs.pseru-4B.1 Qphs.pseru-5A.1 Qphs.pseru-5B.1	RILs	[223]
PHS and GC	6	3AL 3AL 3AL 3DL 3DL 1A/1D/3A/5B	Xwmc559-1 Tamyb10-A1-66 Tamyb10-A1-74 BS00067163_51 Tamyb10-D1-93 Xbarc148	-	RILs	[55]

Table 1. *Cont.*

Trait	QTL	Chromosome	Nearest Marker	QTL name	Material	Reference
PHS	6	3A	TaMFT	QDor-3A	RILs	[224]
		4A	cfa2256	QDor-4A		
		1B	Xbarc181	QDor-1B		
		7B	UCW99	QHt-7B		
		4A	cfa2256	QAwn-4A		
		6B	Xwmc397	QAwn-6B		
PHS and SD	1	2A	Xgwm95-Xgwm372	Sdr2A	RILs	[46]
PHS and SD	3	2D	Xwmc503	QDor-2D	Back crosspopulation	[225]
		3D	Xcfd22	QDor-3D		
		3D	Vp1-4	TaVp1		
PHS	2	1B	tPt-7980	-	86 Chinesegermplasm	[226]
		1B	wPt-645			
PHS	3	3A	7AX-111578083	QTL1	717 Chinese wheat landraces	[37]
		3D	3 DArT-seq and 5 SNPs	QTL2		
		5D	AX-109028892	QTL3		
PHS	5	1A	wPt-6654-wPt-7030	-	RIL	[220]
		4D	wPt-0710-Rht-D1			
		5A	gwm186-P7560-439			
		5D	P7551-267-wmc574			
		7B	P7455-236-P7553-711			

Note: PHS-pre-harvest sprouting; DHLs-doubled haploid lines; RILs-recombinant inbred lines; GC-grain color; SD-seed dormancy.

Table 2. Genes for PHS and related traits identified in wheat.

Wheat Gene	Chromosomes	Gene Function	Homologs/Orthologs Gene	Experimental Methodology	References
TaSdr-A1	2AS	SD	Rice OsSdr4 orthologs	Comparative genomics approach	[46]
TaSdr-B1	2BS	SD	Rice OsSdr4 orthologs	homologous cloning approach	[76]
TaMFT	3AS	SD	Wheat TaMFT homolog	Transcriptomic approach	[41]
TaPHS1	3AS	SD	Wheat MFT homolog	comparative fine mapping and map-based cloning	[16,42]
TaVp-1	Group 3 Chromosomes	SD and PHS	Maize Vp1 and rice OsVp1 orthologs	Genomic southern analysis	[38,39,84,86]
Tamyb10	Group 3 Chromosomes	GC	Arabidopsis TT2 and Rice OsMYB3 orthologs	Cloning approach	[40]
PM19-A1/A2	4AL	SD	-	Transcriptomic approach	[44]
TaMKK3-A	4AL	SD	-	Map-based approach	[45]

Note: SD-seed dormancy; PHS-pre-harvest sprouting; GC-grain color.

PHS resistance is controlled by genotype, environment and their interaction [58]. The use of molecular markers for marker-assisted selection (MAS) could be helpful for direct identification of favorable or deleterious alleles in diverse groups of genotypes [228]. Iyer-Pascuzzi and McCouch [229] reported that MAS can be used for indirect selection of desired traits with considerable reduction in cost and time. Many gene-specific markers, such as SSRs (Xgwm15, Xgwm894, and Xgwm937), STMS markers (wmc104, Xwmc397, and Xwmc468), and STS markers (Vp1-B2 and Vp1B3), were developed for the *Vp1* gene and can be used for identification of PHS resistance in diverse genotypes [15,27,34,38,230–233]. Ogbonnaya et al. [27] found that Xgwm894 and Xgwm937 markers are significantly associated with PHS resistance and could be used for improving PHS resistance in wheat breeding programs. Liu et al. [16] developed an SNP marker named *TaPHS1-SNP1* that can be used as diagnostic marker for identifying the resistance allele of *TaPHS1* in breeding. Based on SNP flanking sequences on chromosomes 3B, 4A, 7B, and 7D, 18 KASP markers were developed that can be used for PHS resistance in future genetic studies and might also be useful for evaluating the PHS in breeding as well as germplasm materials [80]. In a study, Rasheed et al. [234] developed five KASP based assays of functional markers for four genes, including SDR_SNP for *TaSdr-B1*, TaMFT-1617R for *TaMFT-A1*, TaMFT-721J for *TaMFT-A1*, Vp1B1-83_IND for *TaVp-1B*, and Vp1B1-193_IND for *TaVp-1B*. These four genes may have different pathways to induce PHS tolerance; therefore, these KASP assays could provide an excellent opportunity to combine beneficial alleles for PHS tolerance in breeding programs. Wang et al. [235] developed STS (sequence tagged site) marker Tamyb10D for the *Tamyb10D1* gene and showed that it can be used as an efficient marker for evaluating the depth of seed dormancy in bread wheat genotypes. Moreover, the CAPS (cleaved amplified polymorphism sequence) marker Sdr2A has also been developed which is positioned on 2.9 cM intervals between Xgwm95 and Xgwm372 markers and can be used for identifying PHS resistant genotypes [46]. With the advancements in sequencing technologies, more than 124,000 gene loci have been annotated, [98,101], which provides a huge base for identifying more genes for PHS resistance and for the development of functional markers linked with PHS resistance, that can be used for developing PHS resistant varieties during wheat molecular breeding programs.

2. Conclusions and Future Prospects

PHS is a complex trait and determined by various endogenous and exogenous factors. Development of PHS resistant varieties is desirable in wheat growing areas especially having long wet weather conditions during harvest. Only a few PHS-resistant cultivars are commercially available in the field, and the grain quality of those cultivars needs to be improved. Therefore, selecting and inserting new resources could also be helpful in developing PHS resistant cultivars.

Understanding the genetics of various factors affecting PHS resistance is necessary for improving PHS resistance in wheat cultivars. The combination of both genetic and genomic technologies should be used to deeply study the temporal and spatial transcription of the genes involved directly or indirectly in controlling PHS resistance. Genomic and post-genomic data will broaden our knowledge about various factors affecting PHS resistance.

Construction of mutant libraries is important for future studies. Map-based cloning and mutant analysis of the genes underlying PHS resistance will provide new insights in improving PHS resistance of crop species. It is also necessary to use the available crop genome database that will trigger the progress in this field. Furthermore, rapid advancements in molecular technologies, like next-generation sequencing (NGS) technologies [236], and ongoing chromosomal-based and wheat whole genome sequencing projects (International Wheat Genome Sequencing Consortium, IWGSC) [98–102] will provide new opportunities for identification and functional analysis of the candidate genes controlling PHS resistance.

Author Contributions: A.A. and H.-P.Z. wrote the manuscript. S.W.S. and L.S. drew all the figures in the manuscript. J.C. and H.J. drew all the tables in the manuscript. C.C. and C.M. reviewed and edited the manuscript.

Acknowledgments: This work was supported by National key research and development program "Creation of new breeding material with stress resistance and new wheat variety breeding" (2017YFD0100703); grants from the National Natural Science Foundation of China (31401372); China Agriculture Research System (CARS-03); Wheat Genetics and Breeding Research Platform Innovation Team of Anhui's University, the Promotion Project of High Education of Anhui Province "Team construction of high level teacher of crop discipline"; Jiangsu Collaborative Innovation Center for Modern Crop Production (JCIC-MCP); the Agriculture Research System of Anhui Province (AHCYTX-02); and the Introduced Leading Talent Research Team for Universities in Anhui Province.

References

1. Groos, C.; Gay, G.; Perretant, M.-R.; Gervais, L.; Bernard, M.; Dedryver, F.; Charmet, G. Study of the relationship between pre-harvest sprouting and grain color by quantitative trait loci analysis in a white × red grain bread-wheat cross. *Theor. Appl. Genet.* **2002**, *104*, 39–47. [CrossRef] [PubMed]
2. Kulwal, P.; Ishikawa, G.; Benscher, D.; Feng, Z.; Yu, L.X.; Jadhav, A.; Mehetre, S.; Sorrells, M.E. Association mapping for pre-harvest sprouting resistance in white winter wheat. *Theor. Appl. Genet.* **2012**, *125*, 793–805. [CrossRef] [PubMed]
3. Mares, D.; Mrva, K.; Cheong, J.; Williams, K.; Watson, B.; Storlie, E.; Sutherland, M.; Zou, Y. A QTL located on chromosome 4A associated with dormancy in white- and red-grained wheats of diverse origin. *Theor. Appl. Genet.* **2005**, *111*, 1357–1364. [CrossRef] [PubMed]
4. Sorrells, M.; Sherman, J. Facts: Pre—Harvest Sprouting. MAS Wheat, 2007. Available online: https://maswheat.ucdavis.edu/ (accessed on 2 January 2019).
5. Andreoli, C.; Bassoi, M.C.; Brunetta, D. Genetic control of seed dormancy and pre-harvest sprouting in wheat. *Sci. Agric.* **2006**, *63*, 564–566. [CrossRef]
6. Simsek, S.; Ohm, J.B.; Lu, H.; Rugg, M.; Berzonsky, W.; Alamri, M.S.; Mergoum, M. Effect of pre-Harvest sprouting on physicochemical properties of starch in wheat. *Foods* **2014**, *3*, 194–207. [CrossRef] [PubMed]
7. Bewley, J.D.; Black, M.; Halmer, P. The encyclopedia of seeds science. In *Technology and Uses*; CABI Publishing: Oxfordshire, UK, 2006; p. 528.
8. Rajjou, L.; Duval, M.; Gallardo, K.; Catusse, J.; Bally, J.; Job, C.; Job, D. Seed germination and vigor. *Annu. Rev. Plant Biol.* **2012**, *63*, 507–533. [CrossRef] [PubMed]
9. Xiao, S.-H.; Zhang, X.-Y.; Yan, C.-S.; Lin, H. Germplasm improvement for preharvest sprouting resistance in Chinese white-grained wheat: An overview of the current strategy. *Euphytica* **2002**, *126*, 35–38. [CrossRef]
10. Kruger, J.E. Biochemistry of preharvest sprouting in cereals. In *Preharvest Field Sprouting in Cereals*; Derera, N.F., Ed.; CRC Press, Inc.: Boca Raton, FL, USA, 1989; pp. 61–84.
11. Wahl, T.I.; O'Rourke, A.D. The economics of sprout damage in wheat. In *Preharvest Field Sprouting in Cereals*; Walker-Simmonds, M.K., Ried, J.L., Eds.; CRC Press: Boca Raton, FL, USA; American Association of Cereal Chemists: St. Paul, MN, USA, 1993; pp. 10–17.
12. Lan, X.; Zheng, Y.; Ren, X.; Liu, D.; Wei, Y.; Yan, Z. Utilization of preharvest sprouting tolerance gene of synthetic wheat RSP. *J. Plant Genet. Resour.* **2005**, *6*, 204–209.
13. Lin, R.; Horsley, R.D.; Schwarz, P.B. Associations between caryopsis dormancy, α-amylase activity, and pre-harvest sprouting in barley. *J. Cereal Sci.* **2008**, *48*, 446–456. [CrossRef]
14. Munkvold, J.D.; Tanaka, J.; Benscher, D.; Sorrells, M.E. Mapping quantitative trait loci for preharvest sprouting resistance in white wheat. *Theor. Appl. Genet.* **2009**, *119*, 1223–1235. [CrossRef] [PubMed]
15. Yang, J.H.; Yu, Y.X.; Cheng, J.; Tan, X.L.; Shen, W.P. Study on the Pre-harvest Sprouting Tolerance in *Triticum aestivum* ssp. Yunnanense King. *J. Triticeae Crops* **2011**, *31*, 747–752.
16. Liu, S.; Sehgal, S.K.; Li, J.; Lin, M.; Trick, H.N.; Yu, J.; Gill, B.S.; Bai, G. Cloning and characterization of a critical regulator for preharvest sprouting in wheat. *Genetics* **2013**, *195*, 263–273. [CrossRef] [PubMed]
17. Gao, F.; Ayele, B.T. Functional genomics of seed dormancy in wheat: Advances and prospects. *Front. Plant Sci.* **2014**, *5*, 458. [CrossRef] [PubMed]
18. Mares, D.J.; Mrva, K. Wheat grain preharvest sprouting and late maturity alpha-amylase. *Planta* **2014**, *240*, 1167–1178. [CrossRef] [PubMed]
19. King, R.W.; von Wettstein-Knowles, P. Epicuticular waxes and regulation of ear wetting and pre-harvest sprouting in barley and wheat. *Euphytica* **2000**, *112*, 157–166. [CrossRef]
20. Gatford, K.T.; Eastwood, R.F.; Halloran, G.M. Germination inhibitors in bracts surrounding the grain of *Triticum tauschii. Funct. Plant Biol.* **2002**, *29*, 881–890. [CrossRef]

21. Kato, K.; Nakamura, W.; Tabiki, T.; Miura, H.; Sawada, S. Detection of loci controlling seed dormancy on group 4 chromosomes of wheat and comparative mapping with rice and barley genomes. *Theor. Appl. Genet.* **2001**, *102*, 980–985. [CrossRef]

22. Flintham, J.; Adlam, R.; Bassoi, M.; Holdsworth, M.; Gale, M. Mapping genes for resistance to sprouting damage in wheat. *Euphytica* **2002**, *126*, 39–45. [CrossRef]

23. Osa, M.; Kato, K.; Mori, M.; Shindo, C.; Torada, A.; Miura, H. Mapping QTLs for seed dormancy and the *Vp1* homologue on chromosome 3A in wheat. *Theor. Appl. Genet.* **2003**, *106*, 1491–1496. [CrossRef] [PubMed]

24. Kulwal, P.L.; Kumar, N.; Gaur, A.; Khurana, P.; Khurana, J.P.; Tyagi, A.K.; Balyan, H.S.; Gupta, P.K. Mapping of a major QTL for pre-harvest sprouting tolerance on chromosome 3A in bread wheat. *Theor. Appl. Genet.* **2005**, *111*, 1052–1059. [CrossRef] [PubMed]

25. Mori, M.; Uchino, N.; Chono, M.; Kato, K.; Miura, H. Mapping QTLs for grain dormancy on wheat chromosome 3A and the group 4 chromosomes, and their combined effect. *Theor. Appl. Genet.* **2005**, *110*, 1315–1323. [CrossRef] [PubMed]

26. Kottearachchi, N.S.; Uchino, N.; Kato, K.; Miura, H. Increased grain dormancy in white-grained wheat by introgression of preharvest sprouting tolerance QTLs. *Euphytica* **2006**, *152*, 421–428. [CrossRef]

27. Ogbonnaya, F.C.; Imtiaz, M.; Ye, G.; Hearnden, P.R.; Hernandez, E.; Eastwood, R.F.; van Ginkel, M.; Shorter, S.C.; Winchester, J.M. Genetic and QTL analyses of seed dormancy and preharvest sprouting resistance in the wheat germplasm CN10955. *Theor. Appl. Genet.* **2008**, *116*, 891–902. [CrossRef] [PubMed]

28. Ren, X.B.; Lan, X.J.; Liu, D.C.; Wang, J.L.; Zheng, Y.L. Mapping QTLs for pre-harvest sprouting tolerance on chromosome 2D in a synthetic hexaploid wheat x common wheat cross. *J. Appl. Genet.* **2008**, *49*, 333–341. [PubMed]

29. Fofana, B.; Humphreys, D.G.; Rasul, G.; Cloutier, S.; Brûlé-Babel, A.; Woods, S.; Lukow, O.M.; Somers, D.J. Mapping quantitative trait loci controlling pre-harvest sprouting resistance in a red × white seeded spring wheat cross. *Euphytica* **2009**, *165*, 509–521. [CrossRef]

30. Kumar, A.; Kumar, J.; Singh, R.; Garg, T.; Chhuneja, P.; Balyan, H.S.; Gupta, P.K. QTL analysis for grain colour and pre-harvest sprouting in bread wheat. *Plant Sci.* **2009**, *177*, 114–122. [CrossRef]

31. Mares, D.; Rathjen, J.; Mrva, K.; Cheong, J. Genetic and environmental control of dormancy in white-grained wheat (*Triticum aestivum* L.). *Euphytica* **2009**, *168*, 311–318. [CrossRef]

32. Mohan, A.; Kulwal, P.; Singh, R.; Kumar, V.; Mir, R.R.; Kumar, J.; Prasad, M.; Balyan, H.S.; Gupta, P.K. Genome-wide QTL analysis for pre-harvest sprouting tolerance in bread wheat. *Euphytica* **2009**, *168*, 319–329. [CrossRef]

33. Zhu, Z.L.; Tian, B.; Liu, B.; Xie, Q.G.; Tian, J.C. Quantitative Trait Loci Analysis for Pre-harvest Sprouting Using Intact Spikes in Wheat (*Triticum aestivum* L.). *Shandong Agric. Sci.* **2010**, *6*, 19–23.

34. Zhang, H.P.; Feng, J.M.; Chang, C. Investigation of main loci contributing to strong seed dormancy of Chinese wheat landrace. *J. Agric. Biotechnol.* **2011**, *19*, 270–277.

35. Knox, R.E.; Clarke, F.R.; Clarke, J.M.; Fox, S.L.; DePauw, R.M.; Singh, A.K. Enhancing the identification of genetic loci and transgressive segregants for preharvest sprouting resistance in a durum wheat population. *Euphytica* **2012**, *186*, 193–206. [CrossRef]

36. Kumar, S.; Knox, R.E.; Clarke, F.R.; Pozniak, C.J.; DePauw, R.M.; Cuthbert, R.D.; Fox, S. Maximizing the identification of QTL for pre-harvest sprouting resistance using seed dormancy measures in a white-grained hexaploid wheat population. *Euphytica* **2015**, *205*, 287–309. [CrossRef]

37. Zhou, Y.; Tang, H.; Cheng, M.-P.; Dankwa, K.O.; Chen, Z.-X.; Li, Z.-Y.; Gao, S.; Liu, Y.-X.; Jiang, Q.-T.; Lan, X.-J.; et al. Genome-Wide Association Study for Pre-harvest Sprouting Resistance in a Large Germplasm Collection of Chinese Wheat Landraces. *Front. Plant Sci.* **2017**, *8*, 401. [CrossRef] [PubMed]

38. Yang, Y.; Zhao, X.L.; Xia, L.Q.; Chen, X.M.; Xia, X.C.; Yu, Z.; He, Z.H.; Roder, M. Development and validation of a *Viviparous-1* STS marker for pre-harvest sprouting tolerance in Chinese wheats. *Theor. Appl. Genet.* **2007**, *115*, 971–980. [CrossRef] [PubMed]

39. Chang, C.; Feng, J.M.; Si, H.Q.; Yin, B.; Zhang, H.P.; Ma, C.X. Validating a novel allele of viviparous-1 (*Vp-1Bf*) associated with high seed dormancy of Chinese wheat landrace, Wanxianbaimaizi. *Mol. Breed.* **2010**, *25*, 517–525. [CrossRef]

40. Himi, E.; Maekawa, M.; Miura, H.; Noda, K. Development of PCR markers for *Tamyb10* related to R-1, red grain color gene in wheat. *Theor. Appl. Genet.* **2011**, *122*, 1561–1576. [CrossRef] [PubMed]

41. Nakamura, S.; Abe, F.; Kawahigashi, H.; Nakazono, K.; Tagiri, A.; Matsumoto, T.; Utsugi, S.; Ogawa, T.; Handa, H.; Ishida, H.; et al. Wheat Homolog of MOTHER OF FT AND TFL1 Acts in the Regulation of Germination. *Plant Cell* **2011**, *23*, 3215–3229. [CrossRef] [PubMed]

42. Liu, S.; Sehgal, S.K.; Lin, M.; Li, J.; Trick, H.N.; Gill, B.S.; Bai, G. Independent mis-splicing mutations in *TaPHS1* causing loss of preharvest sprouting (PHS) resistance during wheat domestication. *New Phytol.* **2015**, *208*, 928–935. [CrossRef] [PubMed]

43. Liu, A.; Gao, F.; Kanno, Y.; Jordan, M.C.; Kamiya, Y.; Seo, M.; Ayele, B.T. Regulation of Wheat Seed Dormancy by After-Ripening Is Mediated by Specific Transcriptional Switches That Induce Changes in Seed Hormone Metabolism and Signaling. *PLoS ONE* **2013**, *8*, e56570. [CrossRef] [PubMed]

44. Barrero, J.M.; Cavanagh, C.; Verbyla, K.L.; Tibbits, J.F.G.; Verbyla, A.P.; Huang, B.E.; Rosewarne, G.M.; Stephen, S.; Wang, P.; Whan, A.; et al. Transcriptomic analysis of wheat near-isogenic lines identifies *PM19-A1* and *A2* as candidates for a major dormancy QTL. *Genome Biol.* **2015**, *16*, 93. [CrossRef] [PubMed]

45. Torada, A.; Koike, M.; Ogawa, T.; Takenouchi, Y.; Tadamura, K.; Wu, J.; Matsumoto, T.; Kawaura, K.; Ogihara, Y. A Causal Gene for Seed Dormancy on Wheat Chromosome 4A Encodes a MAP Kinase Kinase. *Curr. Biol.* **2016**, *26*, 782–787. [CrossRef] [PubMed]

46. Zhang, Y.; Xia, X.; He, Z. The seed dormancy allele *TaSdr-A1a* associated with pre-harvest sprouting tolerance is mainly present in Chinese wheat landraces. *Theor. Appl. Genet.* **2017**, *130*, 81–89. [CrossRef] [PubMed]

47. Metzger, R.J.; Silbaugh, B.A. Location of genes for seed coat color in hexaploid wheat (*Triticum aestivum* L.). *Crop Sci.* **1970**, *10*, 495–496. [CrossRef]

48. Wang, D.; Dowell, F.E.; Lacey, R.E. Predicting the Number of Dominant R Alleles in Single Wheat Kernels Using Visible and Near-Infrared Reflectance Spectra. *Cereal Chem.* **1999**, *76*, 6–8. [CrossRef]

49. Himi, E.; Noda, K. Red grain colour gene (R) of wheat is a Myb-type transcription factor. *Euphytica* **2005**, *143*, 239–242. [CrossRef]

50. Chopra, S.; Athma, P.; Peterson, T. Alleles of the maize *P* gene with distinct tissue specificities encode Myb-homologous proteins with C-terminal replacements. *Plant Cell* **1996**, *8*, 1149–1158. [CrossRef] [PubMed]

51. Mol, J.; Grotewold, E.; Koes, R. How genes paint flowers and seeds. *Trends Plant Sci.* **1998**, *3*, 212–217. [CrossRef]

52. Kohyama, N.; Chono, M.; Nakagawa, H.; Matsuo, Y.; Ono, H.; Matsunaka, H. Flavonoid compounds related to seed coat color of wheat. *Biosci. Biotechnol. Biochem.* **2017**, *81*, 2112–2118. [CrossRef] [PubMed]

53. Himi, E.; Mares, D.J.; Yanagisawa, A.; Noda, K. Effect of grain colour gene (*R*) on grain dormancy and sensitivity of the embryo to abscisic acid (ABA) in wheat. *J. Exp. Bot.* **2002**, *53*, 1569–1574. [CrossRef] [PubMed]

54. Warner, R.L.; Kudrna, D.A.; Spaeth, S.C.; Jones, S.S. Dormancy in white-grain mutants of Chinese Spring wheat (*Triticum aestivum* L.). *Seed Sci. Res.* **2007**, *10*, 51–60. [CrossRef]

55. Lin, M.; Zhang, D.; Liu, S.; Zhang, G.; Yu, J.; Fritz, A.K.; Bai, G. Genome-wide association analysis on pre-harvest sprouting resistance and grain color in U.S. winter wheat. *BMC Genom.* **2016**, *17*, 794. [CrossRef] [PubMed]

56. Chouard, P. Vernalization and its Relations to Dormancy. *Annu. Rev. Plant Physiol.* **1960**, *11*, 191–238. [CrossRef]

57. Belderok, B. Seed dormancy problems in cereals. *Field Crop Abstr.* **1968**, *21*, 203–211.

58. Marzougui, S.; Sugimoto, K.; Yamanouchi, U.; Shimono, M.; Hoshino, T.; Hori, K.; Kobayashi, M.; Ishiyama, K.; Yano, M. Mapping and characterization of seed dormancy QTLs using chromosome segment substitution lines in rice. *Theor. Appl. Genet.* **2012**, *124*, 893–902. [CrossRef] [PubMed]

59. Gubler, F.; Millar, A.A.; Jacobsen, J.V. Dormancy release, ABA and pre-harvest sprouting. *Curr. Opin. Plant Biol.* **2005**, *8*, 183–187. [CrossRef] [PubMed]

60. Sun, Y.W.; Jones, H.D.; Yang, Y.; Dreisigacker, S.; Li, S.M.; Chen, X.M.; Shewry, P.R.; Xia, L.Q. Haplotype analysis of *Viviparous-1* gene in CIMMYT elite bread wheat germplasm. *Euphytica* **2012**, *186*, 25–43. [CrossRef]

61. Yang, Y.; Zhang, C.L.; Liu, S.X.; Sun, Y.Q.; Meng, J.Y.; Xia, L.Q. Characterization of the rich haplotypes of *Viviparous-1A* in Chinese wheats and development of a novel sequence-tagged site marker for pre-harvest sprouting resistance. *Mol. Breed.* **2014**, *33*, 75–88. [CrossRef]

62. Belderok, B. Physiological-biochemical aspects of dormancy in wheat. *Cereal Res. Commun.* **1976**, *4*, 133–137.

63. Freed, R.D.; Everson, E.H.; Ringlund, K.; Gullord, M. Seed coat color in wheat and the relationship to seed dormancy at maturity. *Cereal Res. Commun.* **1976**, *4*, 147–149.

64. Finch-Savage, W.E.; Leubner-Metzger, G. Seed dormancy and the control of germination. *New Phytol.* **2006**, *171*, 501–523. [CrossRef] [PubMed]

65. Liu, X.; Zhang, H.; Zhao, Y.; Feng, Z.; Li, Q.; Yang, H.-Q.; Luan, S.; Li, J.; He, Z.-H. Auxin controls seed dormancy through stimulation of abscisic acid signaling by inducing ARF-mediated ABI3 activation in *Arabidopsis*. *Proc. Natl. Acad. Sci. USA* **2013**, *110*, 15485–15490. [CrossRef] [PubMed]

66. Kelly, K.M.; Van Staden, J.; Bell, W.E. Seed coat structure and dormancy. *Plant Growth Regul.* **1992**, *11*, 201–209. [CrossRef]

67. Gfeller, F.; Svejda, F. Inheritance of post-harvest seed dormancy and kernal colour in spring wheat lines. *Can. J. Plant Sci.* **1960**, *40*, 1–6. [CrossRef]

68. Everson, E. Varietal variation for dormancy in mature wheat. *Q. Bull. Mich. St. Univ. Agric. Exp. Stn.* **1961**, *43*, 820–829.

69. McEwan, J.M. The sprouting reaction of stocks with single genes for red grain colour derived from hilgendorf 61 wheat. *Cereal Res. Commun.* **1980**, *8*, 261–264.

70. Jacobsen, J.V.; Pearce, D.W.; Poole, A.T.; Pharis, R.P.; Mander, L.N. Abscisic acid, phaseic acid and gibberellin contents associated with dormancy and germination in barley. *Physiol. Plant.* **2002**, *115*, 428–441. [CrossRef] [PubMed]

71. Koornneef, M.; Bentsink, L.; Hilhorst, H. Seed dormancy and germination. *Curr. Opin. Plant Biol.* **2002**, *5*, 33–36. [CrossRef]

72. Lei, L.; Zhu, X.; Wang, S.; Zhu, M.; Carver, B.F.; Yan, L. *TaMFT-A1* Is Associated with Seed Germination Sensitive to Temperature in Winter Wheat. *PLoS ONE* **2013**, *8*, e73330. [CrossRef] [PubMed]

73. Ashikawa, I.; Mori, M.; Nakamura, S.; Abe, F. A transgenic approach to controlling wheat seed dormancy level by using Triticeae DOG1-like genes. *Transgen. Res.* **2014**, *23*, 621–629. [CrossRef] [PubMed]

74. Murphey, M.; Kovach, K.; Elnacash, T.; He, H.; Bentsink, L.; Donohue, K. DOG1-imposed dormancy mediates germination responses to temperature cues. *Environ. Exp. Bot.* **2015**, *112*, 33–43. [CrossRef]

75. Sugimoto, K.; Takeuchi, Y.; Ebana, K.; Miyao, A.; Hirochika, H.; Hara, N.; Ishiyama, K.; Kobayashi, M.; Ban, Y.; Hattori, T.; et al. Molecular cloning of *Sdr4*, a regulator involved in seed dormancy and domestication of rice. *Proc. Natl. Acad. Sci. USA* **2010**, *107*, 5792–5797. [CrossRef] [PubMed]

76. Zhang, Y.; Miao, X.; Xia, X.; He, Z. Cloning of seed dormancy genes (*TaSdr*) associated with tolerance to pre-harvest sprouting in common wheat and development of a functional marker. *Theor. Appl. Genet.* **2014**, *127*, 855–866. [CrossRef] [PubMed]

77. Shorinola, O.; Bird, N.; Simmonds, J.; Berry, S.; Henriksson, T.; Jack, P.; Werner, P.; Gerjets, T.; Scholefield, D.; Balcárková, B.; et al. The wheat Phs-A1 pre-harvest sprouting resistance locus delays the rate of seed dormancy loss and maps 0.3 cM distal to the *PM-19* genes in UK germplasm. *J. Exp. Bot.* **2016**, *67*, 4169–4178. [CrossRef] [PubMed]

78. Ogbonnaya, F.C.; Imtiaz, M.; Depauw, R.M. Haplotype diversity of preharvest sprouting QTLs in wheat. *Genome Biol.* **2007**, *50*, 107–118. [CrossRef] [PubMed]

79. Torada, A.; Koike, M.; Ikeguchi, S.; Tsutsui, I. Mapping of a major locus controlling seed dormancy using backcrossed progenies in wheat (*Triticum aestivum* L.). *Genome* **2008**, *51*, 426–432. [CrossRef] [PubMed]

80. Cabral, A.L.; Jordan, M.C.; McCartney, C.A.; You, F.M.; Humphreys, D.G.; MacLachlan, R.; Pozniak, C.J. Identification of candidate genes, regions and markers for pre-harvest sprouting resistance in wheat (*Triticum aestivum* L.). *BMC Plant Biol.* **2014**, *14*, 340. [CrossRef] [PubMed]

81. Sydenham, S.L.; Barnard, A. Targeted Haplotype Comparisons between South African Wheat Cultivars Appear Predictive of Pre-harvest Sprouting Tolerance. *Front. Plant Sci.* **2018**, *9*, 63. [CrossRef] [PubMed]

82. Shorinola, O.; Balcárková, B.; Hyles, J.; Tibbits, J.F.G.; Hayden, M.J.; Holušova, K.; Valárik, M.; Distelfeld, A.; Torada, A.; Barrero, J.M.; et al. Haplotype Analysis of the Pre-harvest Sprouting Resistance Locus *Phs-A1* Reveals a Causal Role of *TaMKK3-A* in Global Germplasm. *Front. Plant Sci.* **2017**, *8*, 1555. [CrossRef] [PubMed]

83. McCarty, D.R.; Hattori, T.; Carson, C.B.; Vasil, V.; Lazar, M.; Vasil, I.K. The *Viviparous-1* developmental gene of maize encodes a novel transcriptional activator. *Cell* **1991**, *66*, 895–905. [CrossRef]

84. Chang, C.; Zhang, H.-P.; Zhao, Q.-X.; Feng, J.-M.; Si, H.-Q.; Lu, J.; Ma, C.-X. Rich allelic variations of *Viviparous-1A* and their associations with seed dormancy/pre-harvest sprouting of common wheat. *Euphytica* **2011**, *179*, 343–353. [CrossRef]

85. Hattori, T.; Vasil, V.; Rosenkrans, L.; Hannah, L.C.; McCarty, D.R.; Vasil, I.K. The *Viviparous-1* gene and abscisic acid activate the *C1* regulatory gene for anthocyanin biosynthesis during seed maturation in maize. *Genes. Dev.* **1992**, *6*, 609–618. [CrossRef] [PubMed]

86. Bailey, P.C.; McKibbin, R.S.; Lenton, J.R.; Holdsworth, M.J.; Flintham, J.E.; Gale, M.D. Genetic map locations for orthologous *Vp1* genes in wheat and rice. *Theor. Appl. Genet.* **1999**, *98*, 281–284. [CrossRef]

87. Nakamura, S.; Toyama, T. Isolation of a *VP1* homologue from wheat and analysis of its expression in embryos of dormant and non-dormant cultivars. *J. Exp. Bot.* **2001**, *52*, 875–876. [CrossRef] [PubMed]

88. McKibbin, R.S.; Wilkinson, M.D.; Bailey, P.C.; Flintham, J.E.; Andrew, L.M.; Lazzeri, P.A.; Gale, M.D.; Lenton, J.R.; Holdsworth, M.J. Transcripts of *Vp-1* homeologues are misspliced in modern wheat and ancestral species. *Proc. Natl. Acad. Sci. USA* **2002**, *99*, 10203–10208. [CrossRef] [PubMed]

89. Xia, L.Q.; Ganal, M.W.; Shewry, P.R.; He, Z.H.; Yang, Y.; Röder, M.S. Exploiting the diversity of *Viviparous-1* gene associated with pre-harvest sprouting tolerance in European wheat varieties. *Euphytica* **2008**, *159*, 411–417. [CrossRef]

90. Yang, Y.; Chen, X.; He, Z.; Röder, M.; Xia, L. Distribution of *Vp-1* alleles in Chinese white-grained landraces, historical and current wheat cultivars. *Cereal Res. Commun.* **2009**, *37*, 169–177. [CrossRef]

91. Bentsink, L.; Koornneef, M. Seed Dormancy and Germination. *Am. Soc. Plant Biol.* **2008**, *6*, e0119. [CrossRef] [PubMed]

92. Graeber, K.; Nakabayashi, K.; Miatton, E.; Leubner-Metzger, G.; Soppe, W.J. Molecular mechanisms of seed dormancy. *Plant Cell Environ.* **2012**, *35*, 1769–1786. [CrossRef] [PubMed]

93. Finkelstein, R. Abscisic acid synthesis and response. *Arab Book* **2013**, *11*, e0166. [CrossRef] [PubMed]

94. Nonogaki, H. Seed dormancy and germination-emerging mechanisms and new hypotheses. *Front. Plant Sci.* **2014**, *5*, 233. [CrossRef] [PubMed]

95. Bentsink, L.; Jowett, J.; Hanhart, C.J.; Koornneef, M. Cloning of *DOG1*, a quantitative trait locus controlling seed dormancy in *Arabidopsis*. *Proc. Natl Acad. Sci. USA* **2006**, *103*, 17042–17047. [CrossRef] [PubMed]

96. Nakabayashi, K.; Bartsch, M.; Xiang, Y.; Miatton, E.; Pellengahr, S.; Yano, R.; Seo, M.; Soppe, W.J. The time required for dormancy release in *Arabidopsis* is determined by DELAY OF GERMINATION1 protein levels in freshly harvested seeds. *Plant Cell* **2012**, *24*, 2826–2838. [CrossRef] [PubMed]

97. Nishimura, N.; Tsuchiya, W.; Moresco, J.J.; Hayashi, Y.; Satoh, K.; Kaiwa, N.; Irisa, T.; Kinoshita, T.; Schroeder, J.I.; Yates, J.R.; et al. Control of seed dormancy and germination by DOG1-AHG1 PP2C phosphatase complex via binding to heme. *Nat. Commun.* **2018**, *9*, 2132. [CrossRef] [PubMed]

98. International Wheat Genome Sequencing Consortium (IWGSC). A chromosome-based draft sequence of the hexaploid bread wheat (*Triticum aestivum*) genome. *Science* **2014**, *345*, 1251788. [CrossRef] [PubMed]

99. Chapman, J.A.; Mascher, M.; Buluç, A.; Barry, K.; Georganas, E.; Session, A.; Strnadova, V.; Jenkins, J.; Sehgal, S.; Oliker, L.; et al. A whole-genome shotgun approach for assembling and anchoring the hexaploid bread wheat genome. *Genome Biol.* **2015**, *16*, 26. [CrossRef] [PubMed]

100. Clavijo, B.J.; Venturini, L.; Schudoma, C.; Accinelli, G.G.; Kaithakottil, G.; Wright, J.; Borrill, P.; Kettleborough, G.; Heavens, D.; Chapman, H.; et al. An improved assembly and annotation of the allohexaploid wheat genome identifies complete families of agronomic genes and provides genomic evidence for chromosomal translocations. *Genome Res.* **2017**, *27*, 885–896. [CrossRef] [PubMed]

101. International Wheat Genome Sequencing Consortium (IWGSC); Appels, R.; Eversole, K.; Feuillet, C.; Keller, B.; Rogers, J.; Stein, N.; Pozniak, C.J.; Choulet, F.; Distelfeld, A.; et al. Shifting the limits in wheat research and breeding using a fully annotated reference genome. *Science* **2018**, *361*, eaar7191. [CrossRef] [PubMed]

102. Ramírez-González, R.H.; Borrill, P.; Lang, D.; Harrington, S.A.; Brinton, J.; Venturini, L.; Davey, M.; Jacobs, J.; van Ex, F.; Pasha, A.; et al. The transcriptional landscape of polyploid wheat. *Science* **2018**, *361*, eaar6089. [CrossRef] [PubMed]

103. Autio, K.; Simoinen, T.; Suortti, T.; Salmenkallio-Marttila, M.; Lassila, K.; Wilhelmson, A. Structural and Enzymic Changes in Germinated Barley and Rye. *J. Inst. Brew.* **2001**, *107*, 19–25. [CrossRef]

104. Masojć, P.; Milczarski, P. Relationship between QTLs for preharvest sprouting and alpha-amylase activity in rye grain. *Mol. Breed.* **2009**, *23*, 75–84. [CrossRef]

105. Barrero, J.M.; Mrva, K.; Talbot, M.J.; White, R.G.; Taylor, J.; Gubler, F.; Mares, D.J. Genetic, hormonal and physiological analysis of late maturity alpha-amylase (LMA) in wheat. *Plant Physiol.* **2013**, *161*, 1265–1277. [CrossRef] [PubMed]

106. Woodger, F.; Jacobsen, J.V.; Gubler, F. Gibberellin action in germinated cereal grains. In *Plant Hormones: Biosynthesis, Signal Transduction, Action*; Davies, P.J., Ed.; Springer: Dordrecht, The Netherlands, 2010; pp. 221–240.

107. Wu, Y.; Hu, H.; Wang, G.; Zhang, Y.; Ji, J. Relationship between alpha amylase activity and resistance of pre-harvest sprouting in spring wheat. *J. Jilin Agric. Univ.* **2002**, *24*, 22–25.

108. Wang, X.G.; Ren, J.P.; Yin, J. The mechanism on wheat pre-harvest resistant sprouting. *China Agric. Sci.* **2008**, *24*, 243–250.

109. Paterson, A.H.; Sorrells, M.E.; Obendorf, R.L. Methods of evalution for pre-harvest sprouting resistance in wheat breeding programs. *Can. J. Plant Sci.* **1989**, *69*, 681–689. [CrossRef]

110. Humphreys, D.G.; Noll, J. Methods for characterization of preharvest sprouting resistance in a wheat breeding program. *Euphytica* **2002**, *126*, 61–65. [CrossRef]

111. Gale, M.D.; Ainsworth, C.C. The relationship between alpha-amylase species found in developing and germinating wheat grain. *Biochem. Genet.* **1984**, *22*, 1031–1036. [CrossRef] [PubMed]

112. Zhang, Q.; Li, C. Comparisons of Copy Number, Genomic Structure, and Conserved Motifs for α-Amylase Genes from Barley, Rice, and Wheat. *Front. Plant Sci.* **2017**, *8*, 1727. [CrossRef] [PubMed]

113. Marchylo, B.A.; Kruger, J.E.; Macgregor, A.W. Production of multiple forms of α-amylase in germinated, incubated, whole, de-embryonated wheat kernels. *Cereal Chem.* **1983**, *61*, 305–310.

114. Mundy, J.; Hejgaard, J.; Svendsen, I. Characterization of a bifunctional wheat inhibitor of endogenous α-amylase and subtilisin. *FEBS Lett.* **1984**, *167*, 210–214. [CrossRef]

115. Henry, R.J.; Battershell, V.G.; Brennan, P.S.; Oono, K. Control of wheat α-amylase using inhibitors from cereals. *J. Sci. Food Agric.* **1992**, *58*, 281–284. [CrossRef]

116. Macgregor, A.W.; Marchylo, B.A.; Kruger, J.E. Multiple α-amylase components in germinated cereal grains determined by isoelectric focusing and chromatofocusing. *Cereal Chem.* **1988**, *65*, 326–333.

117. Masojc, P.; Zawistowski, J.; Howes, N.K.; Aung, T.; Gale, M.D. Polymorphism and chromosomal location of endogenous alpha-amylase inhibitor genes in common wheat. *Theor. Appl. Genet.* **1993**, *85*, 1043–1048. [CrossRef] [PubMed]

118. Yuan, Y.P.; Chen, X.; Xiao, S.H.; Zhang, W. Extraction and identification of barley α-amylase/subtilisin inhibitor. *J. Triticeae Crops* **2005**, *25*, 40–43.

119. Chitnis, V.R.; Gao, F.; Yao, Z.; Jordan, M.C.; Park, S.; Ayele, B.T. After-ripening induced transcriptional changes of hormonal genes in wheat seeds: The cases of brassinosteroids, ethylene, cytokinin and salicylic acid. *PLoS ONE* **2014**, *9*, e87543. [CrossRef] [PubMed]

120. Shu, K.; Liu, X.D.; Xie, Q.; He, Z.H. Two faces of one seed: Hormonal regulation of dormancy and germination. *Mol. Plant* **2016**, *9*, 34–45. [CrossRef] [PubMed]

121. Kucera, B.; Cohn, M.A.; Leubner-Metzger, G. Plant hormone interactions during seed dormancy release and germination. *Seed Sci. Res.* **2005**, *15*, 281–307. [CrossRef]

122. Finkelstein, R.; Reeves, W.; Ariizumi, T.; Steber, C. Molecular aspects of seed dormancy. *Ann. Rev. Plant Biol.* **2008**, *59*, 387–415. [CrossRef] [PubMed]

123. Finch-Savage, W.E.; Footitt, S. Seed dormancy cycling and the regulation of dormancy mechanisms to time germination in variable field environments. *J. Exp. Bot.* **2017**, *68*, 843–856. [CrossRef] [PubMed]

124. Matilla, A.J.; Matilla-Vázquez, M.A. Involvement of ethylene in seed physiology. *Plant Sci.* **2008**, *175*, 87–97. [CrossRef]

125. Linkies, A.; Leubner-Metzger, G. Beyond gibberellins and abscisic acid: How ethylene and jasmonates control seed germination. *Plant Cell Rep.* **2012**, *31*, 253–270. [CrossRef] [PubMed]

126. Gubler, F.; Hughes, T.; Waterhouse, P.; Jacobsen, J. Regulation of dormancy in barley by blue light and after-ripening: Effects on abscisic acid and gibberellin metabolism. *Plant Physiol.* **2008**, *147*, 886–896. [CrossRef] [PubMed]

127. Lzydorczyk, C.; Nguyen, T.N.; Jo, S.; Son, S.; Tuan, P.A.; Ayele, B.T. Spatiotemporal modulation of abscisic acid and gibberellin metabolism and signaling mediates the effects of suboptimal and supraoptimal temperatures on seed germination in wheat (*Triticum aestivum* L.). *Plant Cell Environ.* **2017**, *41*, 1022–1037. [CrossRef] [PubMed]

128. Rodriguez, M.V.; Mendiondo, G.M.; Maskin, L.; Gudesblat, G.E.; Iusem, N.D.; Benech-Arnold, R.L. Expression of ABA signalling genes and ABI5 protein levels in imbibed *Sorghum bicolor* caryopses with contrasting dormancy and at different developmental stages. *Ann. Bot.* **2009**, *104*, 975–985. [CrossRef] [PubMed]

129. Ried, J.L.; Walker-Simmons, M.K. Synthesis of abscisic Acid-responsive, heat-stable proteins in embryonic axes of dormant wheat grain. *Plant Physiol.* **1990**, *93*, 662–667. [CrossRef] [PubMed]

130. Nambara, E.; Okamoto, M.; Tatematsu, K.; Yano, R.; Seo, M.; Kamiya, Y. Abscisic acid and the control of seed dormancy and germination. *Seed Sci. Res.* **2010**, *20*, 55–67. [CrossRef]

131. Schwartz, S.H.; Qin, X.; Zeevaart, J.A. Elucidation of the indirect pathway of abscisic acid biosynthesis by mutants, genes, and enzymes. *Plant Physiol.* **2003**, *131*, 1591–1601. [CrossRef] [PubMed]

132. Cutler, A.J.; Krochko, J.E. Formation and breakdown of ABA. *Trends Plant Sci.* **1999**, *4*, 472–478. [CrossRef]

133. Saito, S.; Hirai, N.; Matsumoto, C.; Ohigashi, H.; Ohta, D.; Sakata, K.; Mizutani, M. *Arabidopsis CYP707As* encode (+)-abscisic acid 8′-hydroxylase, a key enzyme in the oxidative catabolism of abscisic acid. *Plant Physiol.* **2004**, *134*, 1439–1449. [CrossRef] [PubMed]

134. Kushiro, T.; Okamoto, M.; Nakabayashi, K.; Yamagishi, K.; Kitamura, S.; Asami, T.; Hirai, N.; Koshiba, T.; Kamiya, Y.; Nambara, E. The *Arabidopsis* cytochrome P450 *CYP707A* encodes ABA 8′-hydroxylases: Key enzymes in ABA catabolism. *EMBO J.* **2004**, *23*, 1647–1656. [CrossRef] [PubMed]

135. Suzuki, T.; Matsuura, T.; Kawakami, N.; Noda, K. Accumulation and leakage of abscisic acid during embryo development and seed dormancy in wheat. *J. Plant Growth Regul.* **2000**, *30*, 253–260. [CrossRef]

136. Chono, M.; Honda, I.; Shinoda, S.; Kushiro, T.; Kamiya, Y.; Nambara, E.; Kawakami, N.; Kaneko, S.; Watanabe, Y. Field studies on the regulation of abscisic acid content and germinability during grain development of barley: Molecular and chemical analysis of pre-harvest sprouting. *J. Exp. Bot.* **2006**, *57*, 2421–2434. [CrossRef] [PubMed]

137. Garello, G.; Le Page-Degivry, M.T. Evidence for the role of abscisic acid in the genetic and environmental control of dormancy in wheat (*Triticum aestivum* L.). *Seed Sci. Res.* **1999**, *9*, 219–226. [CrossRef]

138. Son, S.; Chitnis, V.R.; Liu, A.; Gao, F.; Nguyen, T.N.; Ayele, B.T. Abscisic acid metabolic genes of wheat (*Triticum aestivum* L.): Identification and insights into their functionality in seed dormancy and dehydration tolerance. *Planta* **2016**, *244*, 429–447. [CrossRef] [PubMed]

139. Chono, M.; Matsunak, H.; Seki, M.; Fujita, M.; Kiribuchi-Otobe, C.; Oda, S.; Kojima, H.; Kobayashi, D.; Kawakami, N. Isolation of a wheat (*Triticum aestivum* L.) mutant in ABA 8′-hydroxylase gene: Effect of reduced ABA catabolism on germination inhibition under field condition. *Breed. Sci.* **2013**, *63*, 104–115. [CrossRef] [PubMed]

140. Holdsworth, M.; Kurup, S.; Mkibbin, R. Molecular and genetic mechanisms regulating the transition from embryo development to germination. *Trends Plant Sci.* **1999**, *4*, 275–280. [CrossRef]

141. Ksenia, V.K.; Vasquez-Gross, H.A.; Howell, T.; Bailey, P.; Paraiso, F.; Clissold, L.; Simmonds, J.; Ramirez-Gonzalez, R.H.; Wang, X.; Borrill, P.; et al. Uncovering hidden variation in polyploid wheat. *Proc. Natl. Acad. Sci. USA* **2017**, *114*, 913–921. [CrossRef]

142. Marin, E.; Nussaume, L.; Quesada, A.; Gonneau, M.; Sotta, B.; Hugueney, P.; Frey, A.; Marion-Poll, A. Molecular identification of zeaxanthin epoxidase of *Nicotiana plumbaginifolia*, a gene involved in abscisic acid biosynthesis and corresponding to the ABA locus of Arabidopsis thaliana. *EMBO J.* **1996**, *15*, 2331–2342. [CrossRef] [PubMed]

143. North, H.M.; De Almeida, A.; Boutin, J.P.; Frey, A.; To, A.; Botran, L.; Sotta, B.; Marion-Poll, A. The *Arabidopsis* ABA-deficient mutant aba4 demonstrates that the major route for stress-induced ABA accumulation is via neoxanthin isomers. *Plant J.* **2007**, *50*, 810–824. [CrossRef] [PubMed]

144. Tan, B.C.; Schwartz, S.H.; Zeevaart, J.A.; McCarty, D.R. Genetic control of abscisic acid biosynthesis in maize. *Proc. Natl. Acad. Sci. USA* **1997**, *94*, 12235–12240. [CrossRef] [PubMed]

145. Burbidge, A.; Grieve, T.M.; Jackson, A.; Thompson, A.; McCarty, D.R.; Taylor, I.B. Characterization of the ABA-deficient tomato mutant notabilis and its relationship with maize Vp14. *Plant J.* **1999**, *17*, 427–431. [CrossRef] [PubMed]

146. Schwartz, S.H.; Leon-Kloosterziel, K.M.; Koornneef, M.; Zeevaart, J.A. Biochemical characterization of the aba2 and aba3 mutants in *Arabidopsis thaliana*. *Plant Physiol.* **1997**, *114*, 161–166. [CrossRef] [PubMed]

147. Sagi, M.; Scazzocchio, C.; Fluhr, R. The absence of molybdenum cofactor sulfuration is the primary cause of the flacca phenotype in tomato plants. *Plant J.* **2002**, *31*, 305–317. [CrossRef] [PubMed]

148. Kim, H.; Hwang, H.; Hong, J.W.; Lee, Y.N.; Ahn, I.P.; Yoon, I.S.; Yoo, S.D.; Lee, S.; Lee, S.C.; Kim, B.G. A rice orthologue of the ABA receptor, OsPYL/RCAR5, is a positive regulator of the ABA signal transduction pathway in seed germination and early seedling growth. *J. Exp. Bot.* **2012**, *63*, 1013–1024. [CrossRef] [PubMed]

149. Noda, K.; Matsuura, T.; Maekawa, M.; Taketa, S. Chromosomes responsible for sensitivity of embryo to abscisic acid and dormancy in wheat. *Euphytica* **2002**, *123*, 203–209. [CrossRef]

150. Schramm, E.C.; Nelson, S.K.; Kidwell, K.K.; Steber, C.M. Increased ABA sensitivity results in higher seed dormancy in soft white spring wheat cultivar 'Zak'. *Theor. Appl. Genet.* **2013**, *126*, 791–803. [CrossRef] [PubMed]

151. Morris, C.F.; Moffatt, J.M.; Sears, R.G.; Paulsen, G.M. Seed dormancy and responses of caryopses, embryos, and calli to abscisic Acid in wheat. *Plant Physiol.* **1989**, *90*, 643–647. [CrossRef] [PubMed]

152. Corbineau, F.B.A.; Come, D. Changes in sensitivity to abscisic acid of the developing and maturing embryo of two wheat cultivars with different sprouting susceptibility. *Isr. J. Plant Sci.* **2000**, *48*, 189–197. [CrossRef]

153. De Laethauwer, S.; Reheul, D.; De Riek, J.; Haesaert, G. *Vp1* expression profiles during kernel development in six genotypes of wheat, triticale and rye. *Euphytica* **2012**, *188*, 61–70. [CrossRef]

154. Fan, J.; Niu, X.; Wang, Y.; Ren, G.; Zhuo, T.; Yang, Y.; Lu, B.R.; Liu, Y. Short, direct repeats (SDRs)-mediated post-transcriptional processing of a transcription factor gene *OsVP1* in rice (*Oryza sativa*). *J. Exp. Bot.* **2007**, *58*, 3811–3817. [CrossRef] [PubMed]

155. Giraudat, J.; Hauge, B.M.; Valon, C.; Smalle, J.; Parcy, F.; Goodman, H.M. Isolation of the *Arabidopsis ABI3* gene by positional cloning. *Plant Cell* **1992**, *4*, 1251–1261. [CrossRef] [PubMed]

156. Finkelstein, R.R.; Gampala, S.S.L.; Rock, C.D. Abscisic Acid Signaling in Seeds and Seedlings. *Plant Cell* **2002**, *14*, s15–s45. [CrossRef] [PubMed]

157. McCrate, A.J.; Nielsen, M.T.; Paulsen, G.M.; Heyne, E.G. Relationship between sprouting in wheat and embryo response to endogenous inhibition. *Euphytica* **1982**, *31*, 193–200. [CrossRef]

158. Curaba, J.; Moritz, T.; Blervaque, R.; Parcy, F.; Raz, V.; Herzog, M.; Vachon, G. *AtGA3ox2*, a Key Gene Responsible for Bioactive Gibberellin Biosynthesis, Is Regulated during Embryogenesis by LEAFY COTYLEDON2 and FUSCA3 in *Arabidopsis*. *Plant Physiol.* **2004**, *136*, 3660–3669. [CrossRef] [PubMed]

159. Lu, Q.S.; Paz, J.D.; Pathmanathan, A.; Chiu, R.S.; Tsai, A.Y.; Gazzarrini, S. The C-terminal domain of FUSCA3 negatively regulates mRNA and protein levels, and mediates sensitivity to the hormones abscisic acid and gibberellic acid in *Arabidopsis*. *Plant J.* **2010**, *64*, 100–113. [CrossRef] [PubMed]

160. Yamaguchi, S. Gibberellin metabolism and its regulation. *Ann. Rev. Plant Biol.* **2008**, *59*, 225–251. [CrossRef] [PubMed]

161. Pearce, S.; Huttly, A.K.; Prosser, I.M.; Li, Y.D.; Vaughan, S.P.; Gallova, B.; Patil, A.; Coghill, J.A.; Dubcovsky, J.; Hedden, P.; et al. Heterologous expression and transcript analysis of gibberellin biosynthetic genes of grasses reveals novel functionality in the GA3ox family. *BMC Plant Biol.* **2015**, *15*, 130. [CrossRef] [PubMed]

162. Kashiwakura, Y.-I.; Kobayashi, D.; Jikumaru, Y.; Takebayashi, Y.; Nambara, E.; Seo, M.; Kamiya, Y.; Kushiro, T.; Kawakami, N. Highly Sprouting-Tolerant Wheat Grain Exhibits Extreme Dormancy and Cold Imbibition-Resistant Accumulation of Abscisic Acid. *Plant Cell Physiol.* **2016**, *57*, 715–732. [CrossRef] [PubMed]

163. Rodriguez, M.V.; Mendiondo, G.M.; Cantoro, R.; Auge, G.A.; Luna, V.; Masciarelli, O.; Benech-Arnold, R.L. Expression of seed dormancy in grain sorghum lines with contrasting pre-harvest sprouting behavior involves differential regulation of gibberellin metabolism genes. *Plant Cell Physiol.* **2012**, *53*, 64–80. [CrossRef] [PubMed]

164. Ye, H.; Feng, J.; Zhang, L.; Zhang, J.; Mispan, M.S.; Cao, Z.; Beighley, D.H.; Yang, J.; Gu, X. Map-based cloning of *Seed Dormancy1-2* identified a gibberellin synthesis gene regulating the development of endosperm-imposed dormancy in rice. *Plant Physiol.* **2015**, *169*, 2152–2165. [CrossRef] [PubMed]

165. Magwa, R.A.; Zhao, H.; Xing, Y. Genome-wide association mapping revealed a diverse genetic basis of seed dormancy across subpopulations in rice (*Oryza sativa* L.). *BMC Genet.* **2016**, *17*, 28. [CrossRef] [PubMed]

166. Ueguchi-Tanaka, M.; Ashikari, M.; Nakajima, M.; Itoh, H.; Katoh, E.; Kobayashi, M.; Chow, T.Y.; Hsing, Y.I.; Kitano, H.; Yamaguchi, I.; et al. *GIBBERELLIN INSENSITIVE DWARF1* encodes a soluble receptor for gibberellin. *Nature* **2005**, *437*, 693–698. [CrossRef] [PubMed]

167. Li, A.; Yang, W.; Li, S.; Liu, D.; Guo, X.; Sun, J.; Zhang, A. Molecular characterization of three *GIBBERELLIN-INSENSITIVE DWARF1* homologous genes in hexaploid wheat. *J. Plant Physiol.* **2013**, *170*, 432–443. [CrossRef] [PubMed]

168. Sun, T.P. The molecular mechanism and evolution of the GA-GID1-DELLA signaling module in plants. *Curr. Biol.* **2011**, *21*, 338–345. [CrossRef] [PubMed]

169. Sasaki, A.; Itoh, H.; Gomi, K.; Ueguchi-Tanaka, M.; Ishiyama, K.; Kobayashi, M.; Jeong, D.H.; An, G.; Kitano, H.; Ashikari, M.; et al. Accumulation of phosphorylated repressor for gibberellin signaling in an F-box mutant. *Science* **2003**, *299*, 1896–1898. [CrossRef] [PubMed]

170. McGinnis, K.M.; Thomas, S.G.; Soule, J.D.; Strader, L.C.; Zale, J.M.; Sun, T.P.; Steber, C.M. The *Arabidopsis SLEEPY1* gene encodes a putative F-box subunit of an SCF E3 ubiquitin ligase. *Plant Cell* **2003**, *15*, 1120–1130. [CrossRef] [PubMed]

171. Murase, K.; Hirano, Y.; Sun, T.P.; Hakoshima, T. Gibberellin-induced DELLA recognition by the gibberellin receptor GID1. *Nature* **2008**, *456*, 459–463. [CrossRef] [PubMed]

172. Daviere, J.M.; Achard, P. Gibberellin signaling in plants. *Development* **2013**, *140*, 1147–1151. [CrossRef] [PubMed]

173. Tyler, L.; Thomas, S.G.; Hu, J.; Dill, A.; Alonso, J.M.; Ecker, J.R.; Sun, T.P. Della proteins and gibberellin-regulated seed germination and floral development in *Arabidopsis*. *Plant Physiol.* **2004**, *135*, 1008–1019. [CrossRef] [PubMed]

174. Peng, J.; Richards, D.E.; Hartley, N.M.; Murphy, G.P.; Devos, K.M.; Flintham, J.E.; Beales, J.; Fish, L.J.; Worland, A.J.; Pelica, F.; et al. Green revolution' genes encode mutant gibberellin response modulators. *Nature* **1999**, *400*, 256–261. [CrossRef] [PubMed]

175. Ikeda, A.; Ueguchi-Tanaka, M.; Sonoda, Y.; Kitano, H.; Koshioka, M.; Futsuhara, Y.; Matsuoka, M.; Yamaguchi, J. Slender rice, a constitutive gibberellin response mutant, is caused by a null mutation of the *SLR1* gene, an ortholog of the height-regulating gene *GAI/RGA/RHT/D8*. *Plant Cell* **2001**, *13*, 999–1010. [CrossRef] [PubMed]

176. Chandler, P.M.; Marion-Poll, A.; Ellis, M.; Gubler, F. Mutants at the *Slender1* locus of barley cv Himalaya. Molecular and physiological characterization. *Plant Physiol.* **2002**, *129*, 181–190. [CrossRef] [PubMed]

177. Chandler, P.M. Hormonal regulation of gene expression in the "slender" mutant of barley (*Hordeum vulgare* L.). *Planta* **1988**, *175*, 115–120. [CrossRef] [PubMed]

178. Gubler, F.; Raventos, D.; Keys, M.; Watts, R.; Mundy, J.; Jacobsen, J.V. Target genes and regulatory domains of the GAMYB transcriptional activator in cereal aleurone. *Plant J.* **1999**, *17*, 1–9. [CrossRef] [PubMed]

179. Woodger, F.J.; Gubler, F.; Pogson, B.J.; Jacobsen, J.V. A Mak-like kinase is a repressor of GAMYB in barley aleurone. *Plant J.* **2003**, *33*, 707–717. [CrossRef] [PubMed]

180. Kaneko, M.; Inukai, Y.; Ueguchi-Tanaka, M.; Itoh, H.; Izawa, T.; Kobayashi, Y.; Hattori, T.; Miyao, A.; Hirochika, H.; Ashikari, M.; et al. Loss-of-function mutations of the rice *GAMYB* gene impair alpha-amylase expression in aleurone and flower development. *Plant Cell* **2004**, *16*, 33–44. [CrossRef] [PubMed]

181. Jacobsen, J.V.; Barrero, J.M.; Hughes, T.; Julkowska, M.; Taylor, J.M.; Xu, Q.; Gubler, F. Roles for blue light, jasmonate and nitric oxide in the regulation of dormancy and germination in wheat grain (*Triticum aestivum* L.). *Planta* **2013**, *238*, 121–138. [CrossRef] [PubMed]

182. Xu, Q.; Truong, T.T.; Barrero, J.M.; Jacobsen, J.V.; Hocart, C.H.; Gubler, F. A role for jasmonates in the release of dormancy by cold stratification in wheat. *J. Exp. Bot.* **2016**, *67*, 3497–3508. [CrossRef] [PubMed]

183. Friedrichsen, D.M.; Nemhauser, J.; Muramitsu, T.; Maloof, J.N.; Alonso, J.; Ecker, J.R.; Furuya, M.; Chory, J. Three redundant brassinosteroid early response genes encode putative bHLH transcription factors required for normal growth. *Genetics* **2002**, *162*, 1445–1456. [PubMed]

184. Zhang, L.Y.; Bai, M.Y.; Wu, J.; Zhu, J.Y.; Wang, H.; Zhang, Z.; Wang, W.; Sun, Y.; Zhao, J.; Sun, X.; et al. Antagonistic HLH/bHLH transcription factors mediate brassinosteroid regulation of cell elongation and plant development in rice and *Arabidopsis*. *Plant Cell* **2009**, *21*, 3767–3780. [CrossRef] [PubMed]

185. Adkins, S.W.; Ross, J.D. Studies in Wild Oat Seed Dormancy: The role of ethylene in dormancy breakage and germination of wild oat seeds (*Avena fatua* L.). *Plant Physiol.* **1981**, *67*, 358–362. [CrossRef] [PubMed]

186. Steber, C.M.; McCourt, P. A role for brassinosteroids in germination in *Arabidopsis*. *Plant Physiol.* **2001**, *125*, 763–769. [CrossRef] [PubMed]

187. Divi, U.K.; Krishna, P. Overexpression of the brassinosteroid biosynthetic gene *AtDWF4* in *Arabidopsis* seeds overcomes abscisic acid-induced inhibition of germination and increases cold tolerance in transgenic seedlings. *J. Plant Growth Regul.* **2010**, *29*, 385–393. [CrossRef]

188. Hansen, M.; Chae, H.S.; Kieber, J.J. Regulation of ACS protein stability by cytokinin and brassinosteroid. *Plant J.* **2009**, *57*, 606–614. [CrossRef] [PubMed]

189. Locke, J.M.; Bryce, J.H.; Morris, P.C. Contrasting effects of ethylene perception and biosynthesis inhibitors on germination and seedling growth of barley (*Hordeum vulgare* L.). *J. Exp. Bot.* **2000**, *51*, 1843–1849. [CrossRef] [PubMed]

190. Gianinetti, A.; Laarhoven, L.J.; Persijn, S.T.; Harren, F.J.; Petruzzelli, L. Ethylene production is associated with germination but not seed dormancy in red rice. *Ann. Bot.* **2007**, *99*, 735–745. [CrossRef] [PubMed]

191. Morris, C.F.; Mueller, D.D.; Faubion, J.M.; Paulsen, G.M. Identification of l-Tryptophan as an Endogenous Inhibitor of Embryo Germination in White Wheat. *Plant Physiol.* **1988**, *88*, 435–440. [CrossRef] [PubMed]

192. Ramaih, S.; Guedira, M.; Paulsen, G.M. Relationship of indoleacetic acid and tryptophan to dormancy and preharvest sprouting of wheat. *Funct. Plant Biol.* **2003**, *30*, 939–945. [CrossRef]

193. Brady, S.M.; Sarkar, S.F.; Bonetta, D.; McCourt, P. *The ABSCISIC ACID INSENSITIVE 3 (ABI3)* gene is modulated by farnesylation and is involved in auxin signaling and lateral root development in *Arabidopsis*. *Plant J.* **2003**, *34*, 67–75. [CrossRef] [PubMed]

194. Liu, P.P.; Montgomery, T.A.; Fahlgren, N.; Kasschau, K.D.; Nonogaki, H.; Carrington, J.C. Repression of *AUXIN RESPONSE FACTOR10* by microRNA160 is critical for seed germination and post-germination stages. *Plant J.* **2007**, *52*, 133–146. [CrossRef] [PubMed]

195. Martinez, S.A.; Tuttle, K.M.; Takebayashi, Y.; Seo, M.; Campbell, K.G.; Steber, C.M. The wheat ABA hypersensitive *ERA8* mutant is associated with increased preharvest sprouting tolerance and altered hormone accumulation. *Euphytica* **2016**, *212*, 229–245. [CrossRef]

196. Das, A.; Kim, D.; Khadka, P.; Rakwal, R.; Rohila, J.S. Unraveling Key Metabolomic Alterations in Wheat Embryos Derived from Freshly Harvested and Water-Imbibed Seeds of Two Wheat Cultivars with Contrasting Dormancy Status. *Front. Plant Sci.* **2017**, *8*, 1203. [CrossRef] [PubMed]

197. Chapman, E.J.; Estelle, M. Mechanism of auxin-regulated gene expression in plants. *Ann. Rev. Genet.* **2009**, *43*, 265–285. [CrossRef] [PubMed]

198. Vanneste, S.; Friml, J. Auxin: A trigger for change in plant development. *Cell* **2009**, *136*, 1005–1016. [CrossRef] [PubMed]

199. Guilfoyle, T.J.; Hagen, G. Auxin response factors. *Curr. Opin. Plant Biol.* **2007**, *10*, 453–460. [CrossRef] [PubMed]

200. Mockaitis, K.; Estelle, M. Auxin receptors and plant development: A new signaling paradigm. *Ann. Rev. Cell Dev. Biol.* **2008**, *24*, 55–80. [CrossRef] [PubMed]

201. Ooms, J.; Leon-Kloosterziel, K.M.; Bartels, D.; Koornneef, M.; Karssen, C.M. Acquisition of Desiccation Tolerance and Longevity in Seeds of *Arabidopsis thaliana* (A Comparative Study Using Abscisic Acid-Insensitive abi3 Mutants). *Plant Physiol.* **1993**, *102*, 1185–1191. [CrossRef] [PubMed]

202. Finkelstein, R.R. Mutations at two new *Arabidopsis* ABA response loci are similar to the abi3 mutations. *Plant J.* **1994**, *5*, 765–771. [CrossRef]

203. Brocard-Gifford, I.M.; Lynch, T.J.; Finkelstein, R.R. Regulatory Networks in Seeds Integrating Developmental, Abscisic Acid, Sugar, and Light Signaling. *Plant Physiol.* **2003**, *131*, 78–92. [CrossRef] [PubMed]

204. Nemhauser, J.L.; Hong, F.; Chory, J. Different plant hormones regulate similar processes through largely nonoverlapping transcriptional responses. *Cell* **2006**, *126*, 467–475. [CrossRef] [PubMed]

205. Santner, A.; Estelle, M. Recent advances and emerging trends in plant hormone signalling. *Nature* **2009**, *459*, 1071–1078. [CrossRef] [PubMed]

206. Kendall, S.; Penfield, S. Maternal and zygotic temperature signalling in the control of seed dormancy and germination. *Seed Sci. Res.* **2012**, *22*, S23–S29. [CrossRef]

207. Reddy, L.V.; Metzger, R.J.; Ching, T.M. Effect of Temperature on Seed Dormancy of Wheat. *Crop Sci.* **1985**, *25*, 455–458. [CrossRef]

208. Nyachiro, J.; Clarke, F.R.; Depauw, R.; Knox, R.; Armstrong, K.C. Temperature effects on seed germination and expression of seed dormancy in wheat. *Euphytica* **2002**, *126*, 123–127. [CrossRef]

209. Mares, D. Pre-harvest sprouting in wheat. I. Influence of cultivar, rainfall and temperature during grain ripening. *Aust. J. Agric. Res.* **1993**, *44*, 1259–1272. [CrossRef]

210. Lunn, G.D.; Kettlewell, P.; Major, B.J.; Scott, R.K. Variation in dormancy duration of the U.K. wheat cultivar Hornet due to environmental conditions during grain development. *Euphytica* **2002**, *126*, 89–97. [CrossRef]

211. Biddulph, T.; Mares, D.; Plummer, J.A.; Setter, T. Drought and high temperature increases preharvest sprouting tolerance in a genotype without grain dormancy. *Euphytica* **2005**, *143*, 277–283. [CrossRef]

212. Footitt, S.; Douterelo-Soler, I.; Clay, H.; Finch-Savage, W.E. Dormancy cycling in *Arabidopsis* seeds is controlled by seasonally distinct hormone-signaling pathways. *Proc. Natl. Acad. Sci. USA* **2011**, *108*, 20236–20241. [CrossRef] [PubMed]

213. Kulwal, P.L.; Singh, R.; Balyan, H.S.; Gupta, P.K. Genetic basis of pre-harvest sprouting tolerance using single-locus and two-locus QTL analyses in bread wheat. *Funct. Integr. Genom.* **2004**, *4*, 94–101. [CrossRef] [PubMed]

214. Imtiaz, M.; Ogbonnaya, F.C.; Oman, J.; van Ginkel, M. Characterization of quantitative trait loci controlling genetic variation for preharvest sprouting in synthetic backcross-derived wheat lines. *Genetics* **2008**, *178*, 1725–1736. [CrossRef] [PubMed]

215. Liu, S.; Bai, G. Dissection and fine mapping of a major QTL for preharvest sprouting resistance in white wheat Rio Blanco. *Theor. Appl. Genet.* **2010**, *121*, 1395–1404. [CrossRef] [PubMed]

216. Flintham, J. Different genetic components control coat-imposed and embryo-imposed dormancy in wheat. *Seed Sci. Res.* **2000**, *10*, 43–50. [CrossRef]

217. Lohwasser, U.; Röder, M.S.; Börner, A. QTL mapping of the domestication traits pre-harvest sprouting and dormancy in wheat (*Triticum aestivum* L.). *Euphytica* **2005**, *143*, 247–249. [CrossRef]

218. Chao, S.; Xu, S.; Elias, E.; Faris, J.; Sorrells, M. Identification of Chromosome Locations of Genes Affecting Preharvest Sprouting and Seed Dormancy Using Chromosome Substitution Lines in Tetraploid Wheat (*Triticum turgidum* L.). *Crop Sci.* **2010**, *50*, 1180–1187. [CrossRef]

219. Lohwasser, U.; Rehman, M.A.; Börner, A. Discovery of loci determining pre-harvest sprouting and dormancy in wheat and barley applying segregation and association mapping. *Biol. Plant* **2013**, *57*, 663–674. [CrossRef]

220. Borner, A.; Nagel, M.; Agacka-Moldoch, M.; Gierke, P.U.; Oberforster, M.; Albrecht, T.; Mohler, V. QTL analysis of falling number and seed longevity in wheat (*Triticum aestivum* L.). *J. Appl. Genet.* **2018**, *59*, 35–42. [CrossRef] [PubMed]

221. Liu, S.; Cai, S.; Graybosch, R.; Chen, C.; Bai, G. Quantitative trait loci for resistance to pre-harvest sprouting in US hard white winter wheat Rio Blanco. *Theor. Appl. Genet.* **2008**, *117*, 691–699. [CrossRef] [PubMed]

222. Singh, A.K.; Knox, R.E.; Clarke, J.M.; Clarke, F.R.; Singh, A.; DePauw, R.M.; Cuthbert, R.D. Genetics of pre-harvest sprouting resistance in a cross of Canadian adapted durum wheat genotypes. *Mol. Breed.* **2014**, *33*, 919–929. [CrossRef] [PubMed]

223. Lin, M.; Cai, S.; Wang, S.; Liu, S.; Zhang, G.; Bai, G. Genotyping-by-sequencing (GBS) identified SNP tightly linked to QTL for pre-harvest sprouting resistance. *Theor. Appl. Genet.* **2015**, *128*, 1385–1395. [CrossRef] [PubMed]

224. Cao, L.; Hayashi, K.; Tokui, M.; Mori, M.; Miura, H.; Onishi, K. Detection of QTLs for traits associated with pre-harvest sprouting resistance in bread wheat (*Triticum aestivum* L.). *Breed. Sci.* **2016**, *66*, 260–270. [CrossRef] [PubMed]

225. Dale, Z.; Jie, H.; Luyu, H.; Cancan, Z.; Yun, Z.; Yarui, S.; Suoping, L. An Advanced Backcross Population through Synthetic Octaploid Wheat as a "Bridge": Development and QTL Detection for Seed Dormancy. *Front. Plant Sci.* **2017**, *8*, 2123. [CrossRef] [PubMed]

226. Lin, Y.; Liu, S.; Liu, Y.; Liu, Y.; Chen, G.; Xu, J.; Deng, M.; Jiang, Q.; Wei, Y.; Lu, Y.; et al. Genome-wide association study of pre-harvest sprouting resistance in Chinese wheat founder parents. *Genet. Mol. Biol.* **2017**, *40*, 620–629. [CrossRef] [PubMed]

227. Torada, A.; Ikeguchi, S.; Koike, M. Mapping and validation of PCR-based markers associated with a major QTL for seed dormancy in wheat. *Euphytica* **2005**, *143*, 251–255. [CrossRef]

228. Lazo, G.R.; Chao, S.; Hummel, D.D.; Edwards, H.; Crossman, C.C.; Lui, N.; Matthews, D.E.; Carollo, V.L.; Hane, D.L.; You, F.M.; et al. Development of an expressed sequence tag (EST) resource for wheat (*Triticum aestivum* L.): EST generation, unigene analysis, probe selection and bioinformatics for a 16,000-locus bin-delineated map. *Genetics* **2004**, *168*, 585–593. [CrossRef] [PubMed]

229. Iyer-Pascuzzi, A.S.; McCouch, S.R. Functional markers for xa5-mediated resistance in rice (*Oryza sativa*, L.). *Mol. Breed.* **2007**, *19*, 291–296. [CrossRef]

230. Yang, Y.; Zhao, X.-L.; Zhang, Y.; Chen, X.-M.; He, Z.-H.; Yu, Z.; Xia, L.-Q. Evaluation and Validation of Four Molecular Markers Associated with Pre-harvest Sprouting Tolerance in Chinese Wheat. *Acta Agron. Sin.* **2008**, *34*, 17–24. [CrossRef]

231. Guo, F.; Liang, W.; Fan, Q.; Huang, C.; Gao, Q.; Li, G. Distribution and evolution of allelic variation of *Vp1B3* in Shandong wheat. *J. Triticeae Crops* **2009**, *29*, 575–578.

232. Xia, L.Q.; Yang, Y.; Ma, Y.Z.; Chen, X.M.; He, Z.H.; Röder, M.S.; Jones, H.D.; Shewry, P.R. What can the *Viviparous-1* gene tell us about wheat pre-harvest sprouting? *Euphytica* **2009**, *168*, 385–394. [CrossRef]

233. Zhao, B.; Wan, Y.X.; Wang, R. Screening of wheat cultivar resources with pre-harvest sprouting resistance. *J. Anhui Agric. Sci.* **2010**, *38*, 8900–8902.

234. Rasheed, A.; Wen, W.; Gao, F.; Zhai, S.; Jin, H.; Liu, J.; Guo, Q.; Zhang, Y.; Dreisigacker, S.; Xia, X.; et al. Development and validation of KASP assays for genes underpinning key economic traits in bread wheat. *Theor. Appl. Genet.* **2016**, *129*, 1843–1860. [CrossRef] [PubMed]

235. Wang, Y.; Wang, X.L.; Meng, J.Y.; Zhang, Y.J.; He, Z.H.; Yang, Y. Characterization of *Tamyb10* allelic variants and development of STS marker for pre-harvest sprouting resistance in Chinese bread wheat. *Mol. Breed.* **2016**, *36*, 148. [CrossRef] [PubMed]

236. Brenchley, R.; Spannagl, M.; Pfeifer, M.; Barker, G.L.A.; D'Amore, R.; Allen, A.M.; McKenzie, N.; Kramer, M.; Kerhornou, A.; Bolser, D.; et al. Analysis of the bread wheat genome using whole-genome shotgun sequencing. *Nature* **2012**, *491*, 705. [CrossRef] [PubMed]

6

Low Lignin Mutants and Reduction of Lignin Content in Grasses for Increased Utilisation of Lignocellulose

Cecilie S. L. Christensen and Søren K. Rasmussen *

Department of Plant and Environmental Sciences, University of Copenhagen, DK-1871 Frederiksberg C, Denmark; cslc@plen.ku.dk

* Correspondence: skr@plen.ku.dk

Abstract: Biomass rich in lignocellulose from grasses is a major source for biofuel production and animal feed. However, the presence of lignin in cell walls limits its efficient utilisation such as in its bioconversion to biofuel. Reduction of the lignin content or alteration of its structure in crop plants have been pursued, either by regulating genes encoding enzymes in the lignin biosynthetic pathway using biotechnological techniques or by breeding naturally-occurring low lignin mutant lines. The aim of this review is to provide a summary of these studies, focusing on lignin (monolignol) biosynthesis and composition in grasses and, where possible, the impact on recalcitrance to bioconversion. An overview of transgenic crops of the grass family with regulated gene expression in lignin biosynthesis is presented, including the effect on lignin content and changes in the ratio of *p*-hydroxyphenyl (H), guaiacyl (G) and syringyl (S) units. Furthermore, a survey is provided of low-lignin mutants in grasses, including cereals in particular, summarising their origin and phenotypic traits together with genetics and the molecular function of the various genes identified.

Keywords: brown midrib; cell wall; gold hull and internode; grass family; lignin; monolignol pathway; mutational breeding; orange lemma; transgenic cereals

---

1. Introduction

Cereals are a basic food supply for humans and animals worldwide and include rice, maize, wheat, barley and sorghum. They are mainly grown for their nutritional grains that provide dietary calories for human consumption, animal feed and alcoholic beverages. However, whole-crop silage is also a major product in agriculture and is used for animal fodder. Straw from grain production is often considered a by-product, but it is still essential for animal bedding and feed or can be returned to the soil to maintain soil fertility. Additionally, cereals are used in bioindustries for the production of biofuel, textiles, paper, and biochemicals (for a detailed list see [1,2]). The worldwide demand for cereals is growing, but a decrease in their production is starting to be seen [3]. It is therefore crucial to understand the barriers to efficient utility and breeding for new varieties with improved (utility) benefit as feedstuff for animals and bioproducts. In particular, the concept of the multi-purpose crop, in which the grains are used for food and feed and the straw for bioenergy seeks to overcome the food–feed–fuel dilemma by improving the ligno-cellulosic material from straw in second-generation bioethanol production [4].

Lignocellulose is the main component of plant cell walls and the most abundant organic material on earth. It is primarily composed of energy-rich polysaccharides in the form of cellulose, hemicellulose and pectin, rigid phenolic polymers forming lignin and structural (glyco) proteins. The structure is vital for plant growth and serves as a scaffold providing structural and mechanical strength to the plant and protection against external stresses; it encloses each cell individually and facilitates water and solute flux in the vascular systems [5,6]. Besides these properties, lignocellulose is also an essential source of animal feed and used in various bioindustries [2].

The composition of the lignocellulosic material differs depending on the biomass source, but it usually consists of 20–50% cellulose, 20–30% hemicellulose, 7–30% lignin and 5–35% pectin, with lower amounts of structural proteins that all depend on the plant species, as reviewed by [5,7,8]. Plant cells are made up by two types of cell walls, i.e., primary cell walls (PCW) and secondary cell walls (SCW) placed between the middle lamella and the plasma membrane (Figure 1). PCWs surround all plant cells and are continuously formed during cell growth. The structure is thin and flexible, suitable for elongating cells, but still sufficiently strong to withstand arising turgor pressure [9,10]. It consists primarily of cellulose and hemicellulose, with higher amounts of pectin and proteins in dicots compared to monocots [5,11]. SCWs are formed between the PCW and the plasma membrane in specialised cells such as sclerenchyma and xylem vessels after cell elongation has been completed. They are composed of a greater amount of cellulose and hemicellulose than PCW, and pectin is also partly replaced by lignin. These components form a thicker cross-linked matrix than in PCWs. As mentioned above, the function of lignocellulose is to provide mechanical strength to the cells and to facilitate fluid transport. Lignin is the fundamental component for forming that scaffolding structure and its occurrence has also been documented in PCW and the middle lamella [5,6].

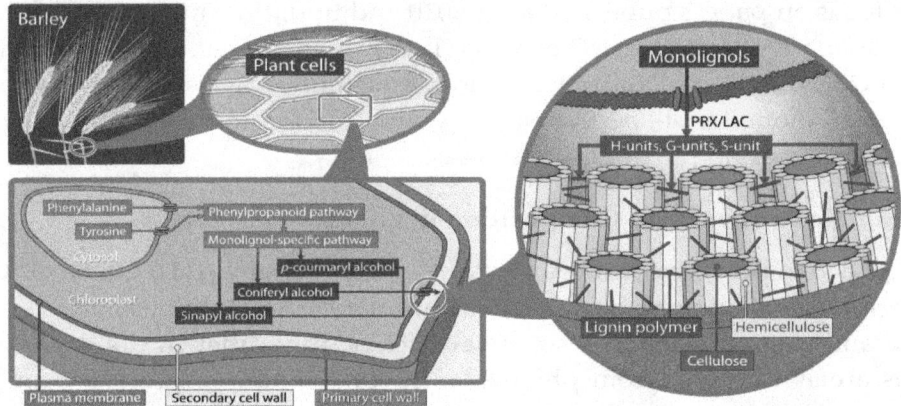

Figure 1. Schematic illustration of the lignocellulosic matrix in the secondary cell wall of the grass family. The main polymers shown are cellulose, hemicellulose, and lignin (shown simplified and not to scale: for microscopic pictures see [12]). They are organised in structures called microfibrils that give structural stability to the plant cell wall. Lignin is the component providing the recalcitrant structure embedding cellulose together with hemicellulose. Lignin is mainly composed of p-hydroxyphenyl (H), guaiacyl (G) and syringyl (S) units, which are derived from 4-hydroxycinnamyl alcohols also known as monolignol, p-coumaryl alcohol, coniferyl alcohol and sinapyl alcohol. The monolignols are synthesised in the cytosol from phenylalanine and tyrosine (grasses only) through the phenylpropanoid pathway and monolignol-specific pathway, then exported across the plasma membranes into the secondary cell wall and oxidized by cell wall-bound peroxidase (PRX) and laccase (LAC), before polymerization into the lignin polymer. (Illustration: Martin Mook).

The recalcitrant structure of lignin is the major limitation of utilising SCWs´ nutritional polysaccharides for animal feedstock and producing bioproducts. Lignin also serves as a mechanical defence barrier and is known to accumulate under pathogenic attacks [13–15]. It has also been demonstrated that genes in the monolignol pathway are directly affected by fungal infection [16–18]. For those reasons, lignin biosynthesis has received significant attention, making it one of the most studied pathways [19]. The expression of genes in the pathway has been modified in order to decrease lignin or alter its composition, thus making the pathway a perfect target for precise genome editing [19]. The involvement of transcription factors in lignin biosynthesis has recently been reviewed [20] and will not be discussed further here. Furthermore, both naturally spontaneous and chemically-induced mutants have been identified and commercialised for animal fodder, showing increased efficiencies for digestion, and are therefore used in breeding programmes. However, in terms of decreasing lignin recalcitrance to bioconversion, there is often a risk of disease infections and dwarfing, depending on the gene being modified [21]. Promising target genes for reduction of lignin recalcitrance without compromising biomass, yield and quality are final genes in the pathway such as *CAD* encoding cinnamyl alcohol dehydrogenase and *COMT* encoding caffeic acid *O*-methyltransferase [22,23]. CAD is responsible for reducing cinnamaldehydes to cinnamyl alcohols, the precursors of the building blocks of lignin, also known as monolignols, whereas COMT is a multifunctional enzyme, but with a preference for methylations of 5-hydroxyconiferaldehyde to sinapaldehydes and therefore primarily affecting the synthesis of syringyl monolignol [24,25] The genes responsible for the brown midrib phenotype in (*bm1, bm3*) maize and (*bmr6, bmr12*) sorghum, which are known for reduced lignin, have mutations within the *CAD* and *COMT* genes affecting their expression. These naturally-occurring low-lignin mutants are of interest for academia and the fodder industry as an alternative source for animal feed and bioproducts [26]. Promoting these well-described varieties avoids the issue of transgenic regulation in Europe, thus increasing the marketing area and also including the organic market. Therefore, downregulating these genes will resemble the naturally-occurring mutants with reduced lignin identified in several cereal crops in the early 20th century.

This review focuses on lignin reduction in important cereals for animal feed (and bioproducts), with a particular focus on papers published after 2010 and updating an earlier review paper, but still including references to primary papers. The aim is (1) to present the monolignol biosynthetic pathway, (2) to provide an overview of recent biotechnology/bioengineering studies targeting genes in the phenylpropanoid and monolignol-specific pathway, and (3) to introduce natural low-lignin mutants with regards to occurrence and phenotypic studies.

2. Lignin Biosynthetic Pathway and Composition in Grasses

Lignin is a phenolic polymer of three units: *p*-hydroxyphenyl (H), guaiacyl (G) and syringyl (S), which are derivatives of hydroxycinnamyl alcohol, also called monolignols, *p*-coumaryl alcohol, coniferyl alcohol and sinapyl alcohol, respectively. They only differ in the degree of methylation. The monolignols are synthesised from phenylalanine or tyrosine (exclusively for grasses) [27,28] through the general phenylpropanoid pathway, which is the precursor for numerous specialised metabolites, including flavonoids, tannins and coumarins, and monolignol-specific pathways in the cytosol, before polymerisation in the cell wall. The steps involved in the synthesis are well documented [29,30]. Briefly, phenylalanine and tyrosine are products of the shikimate pathway synthesised in the chloroplasts and exported to the cytosol, where the monolignols are synthesised via a series of enzymatic reactions, illustrated in Figure 2. Deposition of monolignols from the cytosol to the secondary cell wall is unclear, and it is being debated whether they are exported through passive diffusion or actively transported [31]. However, the monolignol-specific pathway is very plastic with numerous inter-specific variations and co-regulated genes. This is explicit with the complex constellation of the lignin polymer, varying in composition between plants and even between cell types. Lignin of grasses primarily consists of S- and G-units. Additionally, grasses also contain H-units and significantly larger amounts of ferulic acid (FA) and *p*-coumaric acid (*p*CA) [11,32]. The FA and *p*CA cross-link to the lignocellulosic matrix, providing structural integrity of the cell wall. They form covalent linkages or ether bonds between polysaccharide and lignin components [33]. Furthermore tricin, a member of the flavonoid family, has recently been discovered in the lignin polymer and designated an initiator of lignin chains [34,35]. Tricin is also thought to be found almost exclusively in grasses, with a little amount in other monocots and a few traces in alfalfa [36]. Importantly, the composition of the lignin polymer is relevant in terms of recalcitrance to bioconversion after the lignocellulosic material has undergone thermochemical pretreatment followed by enzymatic or acid/alkaline hydrolysis. The monolignols are coupled with recalcitrant C-C and C-O-C (ether) bonds, providing their recalcitrant structure. However, the coupling of monomers differs: H- and G-units can couple via β–5 (from monomer–monomer and monomer–oligomer reactions) and 5–5 (from oligomer–oligomer reactions) coupling modes with C-C linkages, whereas S-units are linked with β–O–4 which are more easily degraded [37]. A ratio between the monolignols (S/G ratio) is often used as a validation factor to draw conclusions about cell wall degradation ability. Shortly, a high ratio (above 1.0) indicates more S-units than G-units and a low ratio (below 1.0) indicates less S-unit than G-units. It is often stated that a high ratio favours digestibility, and the reason for that is discussed. One hypothesis is that more S-units compared to G-units increases the number of labile β–O–4 bonds and thereby affects enzymatic digestibility positively. On the contrary, increased S-units lead to a more linear structure with uncondensed (high β-O-4´) lignin, which provides higher coverage and interaction with the cellulose fibres and thereby lower enzymatic digestibility. Therefore, using the S/G ratio as a validation factor only partially contributes to biomass recalcitrance. Furthermore, *p*CA linkage with S-units via the ether bond and *p*CA is thought to inhibit fermentation due to toxic effect on yeast [38,39]. Similarly, changes in FA compounds using the monolignol ferulate transferase (*FMT*) gene also affect recalcitrance by introducing more easily broken ester bonds [40,41]. Lignin composition and content can be changed with regards to saccharification by regulating genes in the monolignol-specific pathway without compensating for biomass. Therefore lignin has been a target for genetic manipulation for several decades and remains of interest today.

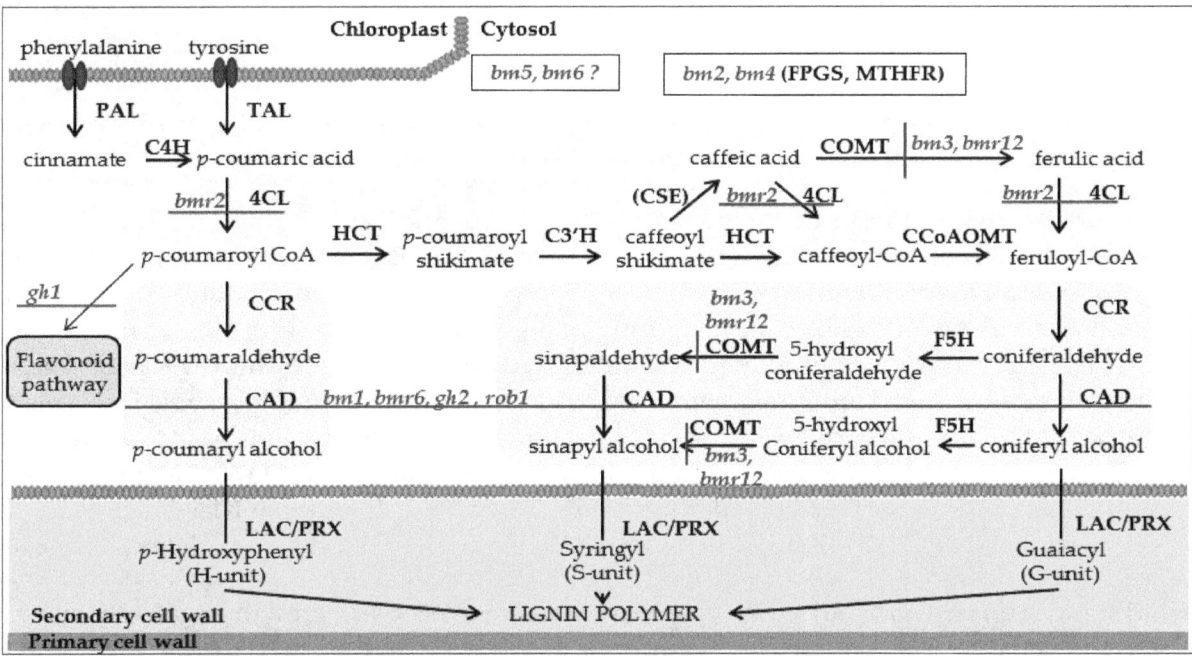

Figure 2. Monolignol biosynthetic pathway for grasses based on studies on *Brachypodium distachyon* [30,32] which is regarded model plant for grasses. The illustration was inspired by [42]. The green box represents the general phenylpropanoid pathway, the blue boxes represent the monolignol-specific pathway, and the light red box indicates *p*-coumarate-CoA as the precursor for the flavonoid pathway. Monolignols *p*-coumaryl alcohol, coniferyl alcohol, and sinapyl alcohol are synthesised in the cytosol and exported to the secondary cell wall where they undergo oxidation by cell wall-bound peroxidase (PRX) and laccase (LAC) prior to radical coupling in the lignin polymer. Red italic abbreviations for mutants: *brown midrib* maize (*bm*), *brown midrib* sorghum (*bmr*), *orange lemma* barley (*rob*) and *gold hull and internode* rice (*gh*), with identified mutations impairing respective gene enzyme activity indicated by red lines. Enzyme abbreviations: phenylalanine ammonia-lyase (PAL), tyrosine ammonia-lyase (TAL), cinnamate 4-hydroxylase (C4H), 4-coumarate coenzyme A ligase (4CL), *p*-hydroxycinnamoyl- CoA:quinate/shikimate hydroxycinnamoyl transferase (HCT), *p*-coumarate 3-hydroxylase (C3'H), caffeoyl shikimate esterase (CSE), caffeoyl-CoA *O*-methyltransferase (CCoAOMT), cinnamoyl CoA reductase (CCR), ferulate 5-hydroxylase (F5H), caffeic acid/5-hydroxyferulic acid *O*-methyltransferase (COMT) and cinnamyl alcohol dehydrogenase (CAD). Folylpolyglutamate synthase (FPGS) and methylenetetrahydrofolate reductase (MTHFR) are not structural genes in the phenylpropanoid pathway.

3. Biotechnology and Bioengineering of Monolignol Pathway in Grasses

The economic advantages of increasing cereals´ nutritional value and replacing fossil fuels with biofuels have driven scientists to investigate and regulate nine of the genes in the monolignol biosynthetic pathway (Table 1), making it an intensively studied pathway. Furthermore, the visual red/orange colouration appearing in stems after downregulating certain genes also makes it attractive as an easy target for new bioengineering methods. The most-used method for regulation and study of the function of genes is "downregulating expression" (of genes) using RNA interference (RNAi). This method introduces small regulatory RNAs (siRNA and miRNA) to the cell, which bind with the RNA-induced silencing complex, Argonaute and other effector proteins, that destroy messenger RNA (mRNA) and thereby prevent the formation of proteins [43]. However, the genes still function and the expression/formation of proteins varies greatly. Repression can be lost completely over a few generations. Furthermore, repression of gene expression does not give a complete picture of the function of a gene, although, it is still a very widely used method. In contrast, using CRISPR/Cas9 to directly knock out gene function by creating stable indel mutations is a more advantageous way of studying gene function [44]. However, in contrast to chemically-induced mutations, CRISPR/Cas9 site-directed mutagenesis requires that the nucleotide sequence of the candidate gene is known before the precise indel mutation can be designed, with stable inheritance over a few generations. This is a relatively new method that has only been used in the most recent studies. However, in July 2018 the EU officially declared that mutations created by CRISPR/Cas9 technology, in contrast to induced mutations, are not exempt from the GMO regulation [45].

It is mostly lignin biosynthetic genes in maize (8) and switchgrass (7) that have been studied by a transgenic approach, with a few in rice (4), Brachypodium (4) and barley (1) (Table 1). Generally, downregulating or knocking out genes leads to a reduced lignin content. However, the estimates of lignin concentration vary greatly depending on the method used for extraction. The most commonly used methods are the gravimetrically determined Klason lignin and the spectrophotometric acetyl bromide lignin method. Briefly, Klason lignin measures insoluble lignin after sulfuric acid hydrolysis of cell walls [46], whereas acetyl bromide lignin is based on the solubility of lignin and measures phenolic compounds´ UV absorbance at 280 nm [47]. Large studies have examined and compared quantification methods of lignin and concluded that Klason lignin estimates higher concentrations than acetyl bromide lignin, although both methods are widely used [48,49].

Modifying *PAL, 4CL* and *C3H* gene expression tends to affect plant growth negatively and induce sterility. However, downregulating genes later in the pathway (*F5H, CCoAMT, CCR, COMT,* and *CAD*) does not have any negative effect on growth (Table 1). This is in contrast with what has been reported for *bm3* mutants, which have mutations in the *COMT* gene [21]. It can be explained by RNAi only reducing gene expression, whereas a complete gene knock out of the candidate gene would have a more drastic effect. The amount of S- and G-units differs greatly between the studies and genes investigated, but there is a general tendency for an overall reduction in S-units. Most studies show that reducing *COMT* gene expression primarily affects the formation of S-units. One study [72] showed that downregulating the *CAD* gene in maize does not result in lignin reduction. This could be due to compensation by other *CAD* genes. Additionally, the expected pigmented phenotype does not appear in any of the grass species when *CAD* is downregulated; it was only observed in *COMT*-downregulated plants. This is in contrast to naturally-occurring low-lignin mutants where both *cad* and *comt* mutants exhibit the pigmented phenotype [26].

Table 1. Transgenic grasses with regulated gene expression in monolignol biosynthesis. The table summarises changes in Klason lignin (KL) or acetyl bromide lignin (ABL) content and changes in the composition of lignin polymer with regards to the amount of syringyl (S), guaiacyl (G) and p-hydroxyphenyl (H) units. Notes on other properties affected by gene expression are included, such as a change in growth, resistance, other compounds (mainly ferulic acid (FA) and p-coumaric acid (pCA)), saccharification based on sugar release, pigmented phenotype and other traits highlighted. The abbreviations for genes are the same as those listed in Figure 1; n.a.: data not available; ↑: increased, ↓: reduced, =: no change compared to wild type.

Gene	Species	Method	Lignin Content	S, G, H	Key Features	References
PAL	Brachypodium	RNAi	↓ 43% (KL)	↑S, ↓G, ↑H	↓growth, ↓pathogenic resistance, ↑saccharification, ↓FA, ↓pCA	[50]
PTAL1	Brachypodium	RNAi	↓ 43%	↓S, ↑G, ↑H	↓flavone and flavonol derivatives, ↑FA, ↓4CA	[32]
C4H-3	Maize	asRNA	↓ 14–17% (ABL)	n.a.	n.a.	[51]
4CL-1	Switchgrass	CRISPR/Cas9	↓ 8–30% (ABL)	↓S, ↓G, =H	Pigmented phenotype, ↑saccharification, ↑FA, ↑pCA, linkage bonds changed	[52]
4CL-1	Switchgrass	RNAi	↓22%	=S, ↓G, ↑H	=growth, pigmented phenotype, ↑saccharification	[53]
C3H-1	Maize	RNAi	↓22% (KL)	↓S, ↓G, ↑H	↓growth, sterility ↑saccharification, ↑anthocyanins, ↑FA, ↑tricin	[54]
C3H	Rice	RNAi	↓30% (KL)	↓S, ↓G, ↑H	=growth, ↑saccharification, ↓FA, ↑pCA, ↑tricin	[55]
C3H	Rice	CRISPR	n.a.	n.a.	↓growth, ↑death before maturity	[55]
F5H	Brachypodium	Overexpression	↓18% (KL)	↑S, ↓G, ↑H	↑saccharification	[56]
F5H	Rice	RNAi/overexpression	↑/ =	↓S, ↑G, =H/ ↑S, ↓G, =H	=growth, =FA, =pCA/↓growth, ↑sterility, =FA, ↓pCA	[57]
F5H	Rice	CRISPR	↑25%	↓S, ↑G, =H	=growth, =saccharification, ↑FA, =pCA	[58]
F5H	Sugarcane	RNAi	=	↓S, ↑G	=growth, ↑saccharification	[59]
CCoAOMT-2	Maize	Overexpression	↑	n.a.	↑pathogenic resistance	[60]
CCoAOMT	Maize	RNAi	↓22.4% (KL)	↑S, ↓G	=growth, ↑saccharification	[61]
CCoAOMT	Sugarcane	RNAi	=	n.a.	=growth, ↑saccharification	[59]
CCR-1	Maize	RNAi	↓7–8.7% (KL)	n.a.	bm phenotype, =growth, ↑saccharification	[62]
COMT6 *	Brachypodium	amiRNA	↓24–31.5% (ABL)	↓S, ↓G, =H	Earlier flowering time, ↑saccharification	[63]

Table 1. *Cont.*

Gene	Species	Method	Lignin Content	S, G, H	Key Features	References
COMT-1,2	Barley	RNAi	↓7–15% (KL)	↓S, ↑G, =H	↑saccharification, ↓*p*CA, =FA	[64]
COMT	Maize	Antisense downregulation	↓~17%	n.a.	*bm* phenotype, ↑saccharification	[65]
COMT	Maize	Antisense downregulation	↓25–30% (KL)	↓S, ↑G, ↓H	*bm* phenotype, ↑saccharification, ↓*p*CA, =FA	[66]
COMT	Sugarcane	RNAi	=	↓S, ↑G	=growth, ↑saccharification	[59]
COMT	Sugarcane	RNAi	↓4–14% (ABL)	↓S, =G	↓growth, pigmented phenotype, ↑saccharification	[67]
COMT	Sugarcane	RNAi	↓6–12% (ABL)	↓S, ↑G	↓growth, ↑saccharification, =FA, ↓*p*CA	[68]
COMT	Sugarcane	TALEN	↓29–32% (ABL)	↓S, =G	↓growth, pigmented phenotype, ↑hemicellulose	[69]
COMT	Switchgrass	RNAi	↓11–16%	↓	=growth, ↑saccharification, =pathogenic resistance	[70]
COMT	Switchgrass	RNAi	↓8–9% (ABL)	↓S, =G	=growth, *bmr* phenotype, ↑saccharification	[71]
COMT	Switchgrass	RNAi	↓11–13% (ABL)	↓S, ↓G	=growth, *bmr* phenotype, ↑saccharification	[22]
CAD1	Brachypodium	amiRNA	= (ABL)	↓S, ↑G, ↑H	↑growth, delayed flowering, pigmented phenotype, ↑saccharification	[63]
CAD	Maize	RNAi	= (KL)	↓S, ↑G, ↑H	=growth, ↑saccharification	[72]
CAD	Rice	RNAi	n.a.	n.a.	*gh* phenotype	[73]
CAD	Switchgrass	RNAi	↓14–22% (ABL)	↓S, ↓G	=growth, ↑saccharification, =*p*CA	[23]
CAD	Switchgrass	RNAi	↓23%	↓S, ↓G	↑saccharification	[74]

* BdCOMT6 (Bradi3g16530) was named BdCOMT4 in the paper [63]. However, based on the accession number and naming in other papers [75,76], BdCOMT6 was chosen. RNAi: RNA interference.

4. Mutants with Reduced Lignin

Naturally-occurring mutants with reduced lignin were identified in cereals such as barley and maize in the early 20th century [77–79]. The mutants are recognised by colour differences: an orange pigmentation occurs in node, lemma and rachis of barley (*rob*) mutants [80], in maize mutants a brown midrib is recognised in the leaves, hence the name 'brown midrib' (*bm*) [81], and rice mutants called 'gold hull and internode' (*gh*) exhibit a reddish brown pigmentation in the hull and internode [82]. Furthermore, induced mutants with a similar phenotype to *bm* maize have also been identified in sorghum *brown midrib* (*bmr*) mutants and the model plant *Brachypodium* [83,84]. Firstly, *brown midrib* mutants of maize and sorghum were investigated and marketed for ease of forage digestibility [85,86]. With the development of plant molecular biology, the genes responsible for the phenotype have been identified and several biochemical analyses performed [26]. Additionally, low lignin mutants are of great interest in bioethanol production as a replacement for fossil fuel [87]. The sections below give an overview for selected grasses.

4.1. Maize Brown Midrib (bm)

Maize (*Zea mays* L.) carrying *bm* mutations are by far the most studied species of all cereals identified with this phenotype. This is because maize silage is an important feed source for dairy cows and other animals. Improving feeding value can affect dairy production and is therefore of high agronomic interest. The first evidence of the positive effect of *bm* mutants on feeding value was obtained in 1971 [88], and since then analysis has expanded, mostly focusing on the *bm3* mutation. Data concerning feeding efficiency of *bm3* mutants from 1976–2017 have been combined and presented in a newly published review paper by [89]. They conclude that a diet based on *bm3* hybrids has an overall beneficial impact on milk production by dairy cows and reduces the need for energy concentrates. Additionally, knowledge of the impact of other *bm* mutants on cell wall digestibility is still of interest. In total six *bm* mutants have been identified [90,91] and listed in the MaizeGDB database (www.maizegdb.org). A literature search resulted in 191 studies on *bm* mutants, with 60 papers focusing only on *bm3* mutants and just a few on the other mutations. However, some studies investigate several mutants and include double mutants for comparison purposes [17,92,93]. With regards to review papers, previous publications have already discussed identified *bm* mutants and they can roughly be divided into three focus areas: (1) animal feed [89,94–96], (2) bioenergy [97] and (3) biochemical properties and molecular analysis [26,96], with some combining all three subjects [98]. The most recent review published by [89] describes the function of all six *bm* and provides an in-depth analysis of data in relation to animal fodder for *bm3*. However, a short overview of each *bm* is given below.

4.1.1. *bm1*

This *bm* was the first to be identified in maize. The phenotype/trait was discovered by the distinguishable orange/brown midrib in the leaves at three different events [77,99,100] and was described as a simple Mendelian recessive trait. With the discovery of other *bm* loci, it was renamed *bm1*. The *bm1* locus was mapped to chromosome 5 and co-segregates with the *CAD2* (Zm00001d015618) gene [101]. It has been argued that *bm1* only affects the expression of the *CAD2* gene and is not a null mutation. However, it is only recently that *bm1* has finally linked with the *CAD2* gene by sequencing and several different mutations (alleles) in the gene have been identified as being responsible for the phenotype [102]. Phenotypic properties of *bm1* mutants are reduced lignin content, reduced S- and G-units, reduced FA and *p*-CA, increased aldehydes, change in linkage bonds and normal growth as reviewed in [89,102], as well as agronomic properties of increased digestibility and bioethanol.

4.1.2. *bm2*

First described in 1932 by Burnham and Brink [103], *bm2* was mapped fairly recently to the methylenetetrahydrofolate reductase (*MTHFR*, GRMZM2G347056, EC 1.5.1.20) gene at chromosome 1 [81] localised in the cytoplasm [104]. Briefly, MTHFR affects methylation of S-adenosyl-L-methionine (SAM) in the methionine cycle, which acts as a methyl donor for CCoAOMT and COMT and thereby the formation of G- and S-units [81,105,106]. Regulating MTHFR thus affects the accumulation of both G- and S-units, described by [81]. The *bm2* mutant is caused by a miniature inverted-repeat transposable element (MITEs) insertion, thereby downregulating the function of *MTHFR* [107]. They observed an altered lignin composition in reduced G- (and C-) units, with little change in S-units, which did not affect the total amount of bromide acetyl lignin or growth. It also led to a significant improvement in cell wall saccharification efficiency. Other studies have also observed reduced lignin content and alteration with an increased S:G ratio caused by greatly reduced G-units, a slight increase or unchanged S units and unaffected H-units, reviewed in [26]. Moreover, it has been observed that the *bm2* mutant has the lowest susceptibility to fungus *Ustilago maydis* infection compared to *bm1*, *bm3* and *bm4* mutants [17].

4.1.3. *bm3*

Maize *bm3* was described in 1935 [78] and later linked to chromosome 4, affecting the *COMT* (Zm00001d049541) gene owing to two different mutation events [108,109]. The *bm3* is by far the most studied brown midrib mutant, probably because of its improved feeding values for cattle. It is closely associated with reduced lignin and improved digestion efficiency. The S:G ratio is greatly reduced with *p*-coumarates. Agronomic traits and chemical properties for this mutant have been reviewed very recently [89]. However, there have been no reports on any negative impact associated with *bm3*, except for one study which shows that the *bm3* mutant has the highest susceptibility for fungal infection when compared to *bm1, bm2* and *bm4* mutants [17].

4.1.4. *bm4*

Maize *bm4* was first described by [110] and has been mapped to a putative folylpolyglutamate synthase (*FPGS*, GRMZM2G393334, EC 6.3.2.17) gene at chromosome 9 and expressed in the cytoplasm [104]. FPGS catalyses the polyglutamylation of tetrahydrofolate (THF), which subsequently catalyses *bm2*-encoded MTHFR, thus affecting the formation of G- and S-units, similar to *bm2* mutants. The *bm4* mutant is caused by polymorphism in the form of deletions, resulting in a frameshift and premature stop codons. Furthermore, expression analysis indicates that the *bm4* allele is leaky [104]. The effects of *bm2* and *bm4* are correlated [111], however the review by Sattler, Funnell-Harris and Pedersen [26] concludes that they only have modest changes in lignin composition. With regards to biofuel production, a slight increase in glucose release with acid and base pretreatment has been observed for *bm4*, however, the amount is still lower compared to the *bm3* mutant [93]. Moreover, the *bm4* mutant has a reduced defence barrier for pathogenic infection [17].

4.1.5. *bm5*

This natural mutation *bm5* was identified by [112]. It has not yet been linked with a gene, only mapped to chromosome 5 close to *bm1*, but not allelic [113]. There have not been many studies on *bm5*. One study by Mechin, Laluc, Legee, Cezard, Denoue, Barriere and Lapierre [113] observed an increase in H- and S-units with a reduction of G- units, changing the lignin composition, and a reduction in Klason lignin was quantified. Additionally, reduced *p*CA but increased feruloyl esters were linked to the lignin polymer. Finally, it has been suggested that *bm5* is linked to the cinnamoyl CoA reductase gene, based on the incorporation of FA and thereby an increase in the weak *bis* 8-O-4 acetal linkage bonds [113], which can be associated with CCR deficiency [114].

4.1.6. *bm6*

This was first identified by [112] and later mapped to chromosome 2 near bin 2.02 [115]. Only a few analyses have been conducted on *bm6*, but it exhibits reduced height and increased cell wall digestibility [115].

4.1.7. Double Mutants

Several double mutants have been created. They often have adverse growth performances and decreased defence barriers compared to single mutants, however, the rate depends on mutant combination. The defence barrier for fungal infection is substantially reduced for *bm3-bm4*, compared to *bm2-bm3* and single mutants, however *bm2* has a similar infection rate to wild type [17]. In terms of growth performance, double mutants *bm2-bm4* show severely reduced growth and a significantly low maturity rate compared to other double and single mutants, including a reduced lignin content and a reduction in both S- and G-units. In addition, this double mutant also displays a darker brown midrib [111]. Investigations were conducted before *bm2* and *bm4* were linked to a specific gene.

4.2. Barley Orange Lemma (rob1)

Barley (*Hordeum vulgare* L.) mutants linked with reduced lignin exhibit an orange colouration in internode, lemma, palea and rachis (Figure 3), hence the locus name "Orange lemma 1" and locus symbol *rob1*. The mutants carrying this phenotype have been identified on several occasions, from both spontaneous and induced mutations (Barley Genetic Newsletter BGS254) [80]. Additionally, germplasm is stored and accessions can be obtained from the U.S. National Plant Germplasm System (https://www.ars-grin.gov/npgs/index.html). Even though the *rob1* mutants have been known for almost a century, only a few studies have investigated its utility with regards to animal feed or biofuel production [116–118]. This is in spite of barley being ranked fourth in cereal production and thus being a major lignocellulosic source. The greatest production is in Europe and Russia, but it is also grown worldwide. It is mainly produced for its nutritional grains for human consumption, animal feed or as malt, with the straw used for animal bedding in rural areas or mostly considered as a waste product [119].

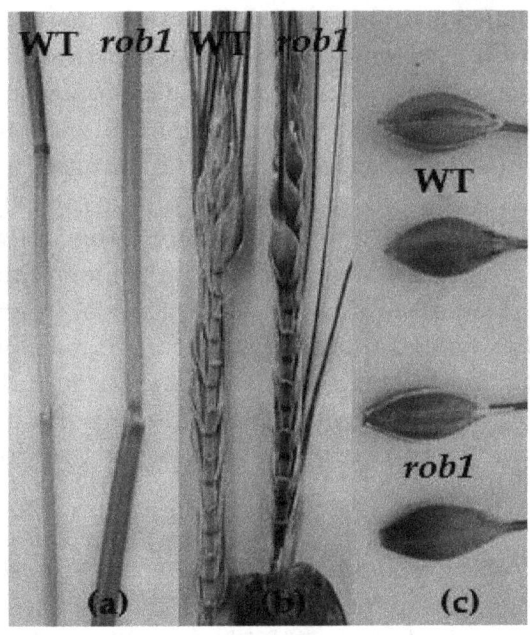

Figure 3. Picture of wild type (WT) barley cv. Golden Promise and barley *rob1* mutant (Rob 13/33) displaying the orange lemma phenotype. (**a**) Stem, *rob1* shows orange-coloured internodes, (**b**) spike, *rob1* shows brown rachis and (**c**) central spikelet, *rob1* show orange/brown palea and lemma close to rachis.

rob1

Rob1 was initially used in inheritance studies and considered to be monofactorial recessive following Mendel with a 3:1 ratio [79,120]. The mutation is located on chromosome 6 near the male-sterile 36 locus and the uniculm 2 locus [121] and used as a morphological markers [122–124]. With regards to chemical analysis, one published poster presents the results of *rob1* forage quality, however no differences have been identified between the mutant and the elite cultivars [117], despite measurement of lignin content being 10–15% lower in *rob1* mutants of different backgrounds, as well as altered lignin composition with decreased S:G ratio and increased saccharification efficiency compared to wild type [116,118]. The *rob1* is mapped to the *HvCAD2* gene, similar to *bm1* in maize [116]. However, the detected mutations responsible for the *rob1* mutant have not yet been published.

4.3. Rice Gold Hull and Internode (gh)

Rice (*Oryza sativa* L.) displaying the *gold hull and internode* (*gh*) phenotype has been identified in a number of mutants (*gh1, gh2, gh3* and *gh4*) listed in the Oryzabase (https://shigen.nig.ac.jp/rice/oryzabase/). They are recognised by their reddish-brown pigment in the internode and yellow coloration of the hull. Even though this phenotype was described as early as 1917 [82] and has been used as a marker for a long time [125], it is only recently that a few studies have investigated the genetics behind *gh1* and *gh2* and undertaken biochemical analysis with regards to lignin [82,126,127]. Rice is the second most produced cereal after maize, and it is estimated to be the staple food for one-fifth of the world's population [128]. It is mainly grown in Asia for its grain and its straw is generally used as a waste product. Furthermore, little is used for compost and only a small portion is used for animal feed, conceivably because the leaves are simply too sharp to be used as animal feed due to their high silicon content. This is a major lignocellulosic source with great potential for utilisation to make various products such as biofuel [129] and byproducts. Therefore, it is suggested that more research on *gh* mutants is needed.

4.3.1. *gh1*

Rice *gh1* is mapped to the chalcone isomerase (*CHI*, Os03g 0819600, EC 5.5.1.6) gene on chromosome 3 with a Dasheng retrotransposon inserted causing loss of function [127]. However, this gene is part of the flavonoid pathway, which is derived from the general phenylpropanoid pathway as well as the monolignol pathway [127]. Briefly, the CHI enzyme converts naringenin chalcone, a yellow pigment, into naringenin, and an accumulation of this product causes a yellow pigmentation [126,130]. Since both the flavonoid pathway and the monolignol pathway use the same precursors, one study investigated whether the *gh1* mutant has an effect on lignin formation [126]. Its results showed an increased saccharification efficiency and altered lignin composition with a reduced S:G ratio caused by significantly reduced S-units and increased H- and G-units (and FA). Lignin content differed depending on the extraction method, with reduced thioglycolic lignin content but no change in Klason lignin compared to wild type. Additionally, the *gh1* mutant shows no reduction in biomass or lodging resistance, however reduced grain yield has been reported. This indicates that regulation of genes in the flavonoid pathway affects monolignol formation and lignin composition.

4.3.2. *gh2*

Rice *gh2* phenotype is caused by mutations in the *CAD2* gene (Os02g0187800) on chromosome 2. The original spontaneous *gh2* mutant (Zhefu802) is caused by a point mutation in exon 4 which changes expression level and exhibits the *gh* phenotype [82], while the *gh2* mutant line created with *Tos17* insertion in exon 2 is a null mutant (https://tos.nias.affrc.go.jp/) and displays the *bm* phenotype [73]. Expression analysis of the original *gh2* shows reduced CAD and SAD activity differentiating between tissues, which indicates an additional function of CAD-isoenzymes. Klason lignin content is only slightly reduced, even though a dramatic reduction is shown for lignin monomers [82]. Additionally, the *Tos17*-generated *gh2* mutant shows less lignin and increased saccharification efficiency compared to both wild type and spontaneous *gh2* mutant. Furthermore, H- and S-units are also significantly reduced [73]. These two studies indicate the importance of the location of the mutation on the gene. For future research, biomass, grain yield and lodging resistance need to be investigated in order to evaluate the potential of *gh2* as a biofuel crop.

4.4. Sorghum Brown Midrib (bmr)

Sorghum (*Sorghum bicolor* (L.) Moench) *brown midrib (bmr)* mutants exhibit a similar phenotype to *bm* maize. As the name indicates, a brown coloration in the midrib of leaves is exhibited. The first identified *bmr* mutants were developed via chemical mutagenesis using diethyl sulfate in 1978. Nineteen *bmr* mutants were identified and six mutants (*bmr2, bmr6, bmr12, bmr14, bmr18* and *bmr19*) had a significantly reduced lignin compared to wild type [84]. Later, spontaneous *bmr* mutants were also identified by Dr. Gebisa Ejeta (Purdue University, unpublished results) and described in [131,132] and listed consecutively *bmr* 1-28 including the induced *bmr* mutants [87,133,134]. Additionally, a TILLING (Targeting Induced Local Lesions in Genomes) population was examined and even more *bmr* mutants identified [131,135,136]. Allelism tests have been performed and four allelic classes identified—*bmr2, bmr6, bmr12,* and *bmr19*—with *bmr6* and *bmr12* being the most widely used in breeding programmes [132,137]. *Bmr19* has been reported as having insignificantly reduced lignin and is therefore not of interest to the forage industry [132]. It will therefore not be discussed further in this review. Hence, many *bmr* mutants have been identified and linked to the same locus. In order to obtain a better overview, they have been organised by additional numbers (see [131]). Sorghum is ranked fifth in cereal production. It is mainly distributed in arid areas of Africa, Central America and South Asia, where it is grown for its grains utilised by humans or as silage for animal feed. Additionally, the stems are used for alcoholic beverages. Considerable research and biochemical analysis have been conducted on *bmr* mutants with regards to both silage and biofuel production. For farmers, the *bmr* phenotype is a visual marker that can be observed in the field to verify the quality trait. A literature search resulted in more than 200 papers published since 1978 when the first *bmr* mutants were developed [84]. Furthermore, many reviews have focused on digestion efficiency, lignin composition and improved saccharification [21,26,86,87,138,139]. Here *bmr2*, *bmr6* and *bmr12* are presented.

4.4.1. *bmr2* Group

Sorghum *bmr2* group, which includes *bmr2*, *bmr5* and *bmr14*, shows a reduction in both G- and S-units, which are all described in [132]. *Bmr2* is the most studied of the three mutants and described as two different alleles *bmr2-ref* [132] and *bmr2-2* [135]. The *bmr2* gene encodes 4CL located on chromosome 4, and sequencing reveals two point mutations within the coding sequence responsible for the phenotype. However, the gene *4CL* is part of a family with several isoforms varying in expression regulating different substrates. For a detailed description see [140].

4.4.2. *bmr6* Group

Sorghum *bmr6* group includes *bmr3, 4, 6, 20, 22–24, 27* and *28* [132]. The *bmr6* phenotype was mapped to the *CAD2* (Sb04g005950) gene on chromosome 4 [141], and different mutations responsible for the *bmr6* phenotype have been revealed by sequencing, resulting in premature STOP-codon or loss of important catalytic domains [141–143]. Reduced CAD2 activity resulted in decreased lignin content with low amount of G-units and increased level of cinnamaldehydes [144,145]. Another study observed a significant reduction in all lignin subunits, particularly S-units resulting in reduced S:G ratio [142] In-depth knowledge of the chemical composition, improved saccharification efficiency and decreased lignin content of *bmr6* and *bmr12* question whether the S:G ratio is a valid indicator for lignin recalcitrance and it has been concluded that more knowledge is needed [146]. In terms of agronomic values, lodging is not affected by *bmr6* in either forage [147] or grain sorghum [148], although negative effects on biomass have been reported for forage sorghum [147] and grain yield in grain sorghum [148]. Despite these negative effects, in terms of diets for dairy cows the *bmr6* forage sorghum performs better than wild type [149].

4.4.3. *bmr12* Group

The sorghum *bmr12* group includes *bmr7, 12, 15, 18, 25* and *26* [132] and are all mapped to the *COMT* gene with premature stop codons giving rise to the *bmr* mutants [150]. Other mutations have also been identified for *bmr12* mutants and characterised by [151]. Overall, the *bmr12* mutants in biomass sorghum all have reduced lignin and generally contribute positively to bioconversion and digestion efficiency [139]. However, negative impacts on agronomical traits have also been reported, such as reduced yield in grain sorghum [148] and biomass in forage sorghum [147], and thus do not affect susceptibility to disease [18]. However, a recent study concludes that weather conditions have a greater impact and in some cases free phenolic compounds even act as a defence mechanism, depending on the diseases reviewed [152].

4.5. Pearl Millet

Pearl millet (*Pennisetum glaucum* L.) is a highly drought-tolerant annual forage plant that is utilised for grain production or as silage for animal feed.

Three brown midrib mutants have been identified in pearl millet, which assembles the same colouration as *bm* maize and *bmr* sorghum. The mutations occurred spontaneously or were induced using dimethyl sulfate. However, only a few studies have investigated the properties of these mutant lines. The agronomic potential is reviewed by [26] and they conclude that a significant yield reduction is associated with *bmr* pearl millets and is therefore not of interest as breeding material.

4.6. Brachypodium

Brachypodium distachyon (Brachypodium) is a small grass with a relatively short growing season. It is diploid and the genome is fully sequenced and similar in size to rice. It is therefore used as a model plant for grasses [153].

A chemically-induced mutant population of Brachypodium was developed with a TILLING platform [75]. Several lines were identified with induced mutations in genes involved in the lignin biosynthesis such as *C4H, 4C,* and *COMT*. The same study analysed the effect of mutations in the *COMT6* (Bradi3g16530) gene on lignin content and composition in several lines. They discovered reduced Klason lignin and altered composition with a decreased S:G ratio, where S-units were significantly reduced and G-units increased. This corresponds with *bm3* and *bmr12* COMT-deficient plants [89,151]. Further studies have been performed on line *Bd*5139, which had a missense mutation in the *COMT6* gene, and revealed a reduction in *p*CA esterified to S-units. However, *p*CA linked to arabinoxylans was not affected, which substantiates *comt6* affinity for *p*CA ester-linkage to S-units [76]. Another study also used chemical mutagenesis to create mutations in Brachypodium plants and lignin-deficient mutants were identified by a brownish/red colouration in nodes, lemma and rachis [83]. An SNP mutation was identified in the *CAD1* gene (Bradi3g06480) causing the phenotype; interestingly it was identical to the sorghum *bmr6-3* [141]. Overall the mutant shows reduced lignin and altered composition, which is similar to what has been observed in other species. Furthermore, a coexpression database (www.gene2function.de) has been developed for important genes involved in the lignification of the cell wall in many organs at different developmental stages in Brachypodium [56].

5. Conclusions

Lignocellulosic material from grasses is an essential source for bioethanol production and/or animal fodder. However, the recalcitrant structure of lignin limits decomposition and hence utilisation of the embedded cellulose fibrils. Naturally-occurring low-lignin mutants have been identified in several species and investigations show the great potential in promoting these mutants. So far, however, only *bm* maize and *bmr* sorghum containing mutations have been commercialised. Promoting *gh* rice and *rob1* barley would extend the feedstock source for animals, bioenergy and the emerging circular bioeconomy. Based on existing knowledge about *bm* maize and *bmr* sorghum, it is predicted that there is great potential for improving and developing new commercial varieties of *rob1* barley and *gh* rice with improved utilisation. Furthermore, results from various genetic manipulations of genes in the lignin biosynthesis offers detailed information about the function and its potential for further modification in future research. However, down-regulating genes by antisense/RNAi only provides valid information about gene function and is not useful in breeding. Instead, chemical mutagenesis and CRISPR/Cas9 have the potential to create stable mutations with loss of function, which resembles the natural low-lignin mutants. It has been predicted that CRISPR/Cas9 will revolutionise precision breeding, however there has been a declaration that it now comes under GMO regulations in the EU [45], which complicates the use of this technology. Instead, the screening of existing germplasm is suggested with the use of TILLING to identify new mutations in order to overcome current regulatory difficulties with regard to crop improvements.

Author Contributions: S.K.R. conceived and outlined the manuscript and C.S.L.C. drafted the manuscript, prepared the figures and table. Both S.K.R. and C.S.L.C. finalised the manuscript.

References

1.   Anwar, Z.; Gulfraz, M.; Irshad, M. Agro-industrial lignocellulosic biomass a key to unlock the future bio-energy: A brief review. *J. Rad. Res. Appl. Sci.* **2014,** *7,* 163–173. [CrossRef]

2.   Guerriero, G.; Hausman, J.F.; Strauss, J.; Ertan, H.; Siddiqui, K.S. Lignocellulosic biomass: Biosynthesis, degradation, and industrial utilization. *Eng. Life Sci.* **2016,** *16,* 1–16. [CrossRef]

3.   FAO. Global Cereal Production and Inventories to Decline but Overall Supplies Remain Adequate. Available online: www.fao.org/worldfoodsituation/csdb/en/ (accessed on 7 December 2018).

4. Graham-Rowe, D. Agriculture: Beyond food versus fuel. *Nature* **2011**, *474*, 6–8. [CrossRef]

5. Vogel, J. Unique aspects of the grass cell wall. *Curr. Opin. Plant Biol.* **2008**, *11*, 301–307. [CrossRef]

6. Doblin, M.S.; Pettolino, F.; Bacic, A. Plant cell walls: The skeleton of the plant world. *Funct. Plant Biol.* **2010**, *37*, 357–381. [CrossRef]

7. Iqbal, H.M.N.; Kyazze, G.; Keshavarz, T. Advances in the valorization of lignocellulosic materials by biotechnology: An overview. *Bioresources* **2013**, *8*, 3157–3176. [CrossRef]

8. Jorgensen, H.; Kristensen, J.B.; Felby, C. Enzymatic conversion of lignocellulose into fermentable sugars: Challenges and opportunities. *Biofuels Bioprod.Biorefin.* **2007**, *1*, 119–134. [CrossRef]

9. Carpita, N.C.; Gibeaut, D.M. Structural models of primary cell walls in flowering plants: Consistency of molecular structure with the physical properties of the walls during growth. *Plant J.* **1993**, *3*, 1–30. [CrossRef] [PubMed]

10. Hamant, O.; Traas, J. The mechanics behind plant development. *New Phytol.* **2010**, *185*, 369–385. [CrossRef]

11. Harrington, M.J.; Mutwil, M.; Barriere, Y.; Sibout, R. Molecular biology of lignification in grasses. In *Lignins: Biosynthesis, Biodegradation and Bioengineering*; Jouann, L., Lapierre, C., Eds.; Academic Press—Elsevier: Cambridge, MA, USA, 2012; Volume 61, pp. 77–112.

12. Yu, H.; Liu, R.; Shen, D.; Wu, Z.; Huang, Y. Arrangement of cellulose microfibrils in the wheat straw cell wall. *Carbohydr. Polym.* **2008**, *72*, 122–127. [CrossRef]

13. Miedes, E.; Vanholme, R.; Boerjan, W.; Molina, A. The role of the secondary cell wall in plant resistance to pathogens. *Front. Plant Sci.* **2014**, *5*, 358. [CrossRef]

14. Vance, C.; Kirk, T.; Sherwood, R. Lignification as a mechanism of disease resistance. *Annu. Rev. Phytopathol.* **1980**, *18*, 259–288. [CrossRef]

15. Hammond-Kosack, K.E.; Jones, J. Resistance gene-dependent plant defense responses. *Plant Cell* **1996**, *8*, 1773–1791. [CrossRef] [PubMed]

16. Bhuiyan, N.H.; Selvaraj, G.; Wei, Y.; King, J. Gene expression profiling and silencing reveal that monolignol biosynthesis plays a critical role in penetration defence in wheat against powdery mildew invasion. *J. Exp. Bot.* **2008**, *60*, 509–521. [CrossRef]

17. Tanaka, S.; Brefort, T.; Neidig, N.; Djamei, A.; Kahnt, J.; Vermerris, W.; Koenig, S.; Feussner, K.; Feussner, I.; Kahmann, R. A secreted *Ustilago maydis* effector promotes virulence by targeting anthocyanin biosynthesis in maize. *Elife* **2014**, *3*, e01355. [CrossRef]

18. Sattler, S.; Funnell-Harris, D. Modifying lignin to improve bioenergy feedstocks: Strengthening the barrier against pathogens? *Front. Plant Sci.* **2013**, *4*, 70. [CrossRef]

19. Vanholme, R.; Morreel, K.; Ralph, J.; Boerjan, W. Lignin engineering. *Curr. Opin. Plant Biol.* **2008**, *11*, 278–285. [CrossRef]

20. Rao, X.; Dixon, R.A. Current models for transcriptional regulation of secondary cell wall biosynthesis in grasses. *Front. Plant Sci.* **2018**, *9*, 399. [CrossRef] [PubMed]

21. Pedersen, J.F.; Vogel, K.P.; Funnell, D.L. Impact of reduced lignin on plant fitness. *Crop Sci.* **2005**, *45*, 812–819. [CrossRef]

22. Fu, C.; Mielenz, J.R.; Xiao, X.; Ge, Y.; Hamilton, C.Y.; Rodriguez, M.; Chen, F.; Foston, M.; Ragauskas, A.; Bouton, J. Genetic manipulation of lignin reduces recalcitrance and improves ethanol production from switchgrass. *Proc. Natl. Acad. Sci. USA* **2011**, *108*, 3803–3808. [CrossRef]

23. Fu, C.; Xiao, X.; Xi, Y.; Ge, Y.; Chen, F.; Bouton, J.; Dixon, R.A.; Wang, Z.-Y. Downregulation of cinnamyl alcohol dehydrogenase (CAD) leads to improved saccharification efficiency in switchgrass. *Bioenergy Res.* **2011**, *4*, 153–164. [CrossRef]

24. Bukh, C.; Nord-Larsen, P.H.; Rasmussen, S.K. Phylogeny and structure of the cinnamyl alcohol dehydrogenase gene family in *Brachypodium distachyon*. *J. Exp. Bot.* **2012**, *63*, 6223–6236. [CrossRef]

25. Wu, X.; Wu, J.; Luo, Y.; Bragg, J.; Anderson, O.; Vogel, J.; Gu, Y.Q. Phylogenetic, molecular, and biochemical characterization of caffeic acid *O*-methyltransferase gene family in *Brachypodium distachyon*. *Int. J. Plant Genom.* **2013**, *2013*, 1–12. [CrossRef]

26. Sattler, S.E.; Funnell-Harris, D.L.; Pedersen, J.F. Brown midrib mutations and their importance to the utilization of maize, sorghum, and pearl millet lignocellulosic tissues. *Plant Sci.* **2010**, *178*, 229–238. [CrossRef]

27. Rösler, J.; Krekel, F.; Amrhein, N.; Schmid, J. Maize phenylalanine ammonia-lyase has tyrosine ammonia-lyase activity. *Plant Physiol.* **1997**, *113*, 175–179. [CrossRef]

28. Neish, A.C. Formation of *m*- and *p*-coumaric acids by enzymatic deamination of the corresponding isomers of tyrosine. *Phytochemistry* **1961**, *1*, 1–24. [CrossRef]
29. Boerjan, W.; Ralph, J.; Baucher, M. Lignin biosynthesis. *Annu. Rev. Plant Biol.* **2003**, *54*, 519–546. [CrossRef]
30. Faraji, M.; Fonseca, L.L.; Escamilla-Treviño, L.; Barros-Rios, J.; Engle, N.L.; Yang, Z.K.; Tschaplinski, T.J.; Dixon, R.A.; Voit, E.O. A dynamic model of lignin biosynthesis in *Brachypodium distachyon*. *Biotechnol. Biofuels* **2018**, *11*, 253. [CrossRef]
31. Mottiar, Y.; Vanholme, R.; Boerjan, W.; Ralph, J.; Mansfield, S.D. Designer lignins: Harnessing the plasticity of lignification. *Curr. Opin. Biotechnol.* **2016**, *37*, 190–200. [CrossRef]
32. Barros, J.; Serrani-Yarce, J.C.; Chen, F.; Baxter, D.; Venables, B.J.; Dixon, R.A. Role of bifunctional ammonia-lyase in grass cell wall biosynthesis. *Nat. Plants* **2016**, *2*, 16050. [CrossRef]
33. Grabber, J.H.; Ralph, J.; Hatfield, R.D. Cross-linking of maize walls by ferulate dimerization and incorporation into lignin. *J. Agric. Food Chem.* **2000**, *48*, 6106–6113. [CrossRef]
34. Del Río, J.C.; Rencoret, J.; Prinsen, P.; Martínez, A.T.; Ralph, J.; Gutiérrez, A. Structural characterization of wheat straw lignin as revealed by analytical pyrolysis, 2D-NMR, and reductive cleavage methods. *J. Agric. Food Chem.* **2012**, *60*, 5922–5935. [CrossRef] [PubMed]
35. Lan, W.; Lu, F.; Regner, M.; Zhu, Y.; Rencoret, J.; Ralph, S.A.; Zakai, U.I.; Morreel, K.; Boerjan, W.; Ralph, J. Tricin, a flavonoid monomer in monocot lignification. *Plant Physiol.* **2015**, *167*, 1284–1295. [CrossRef]
36. Lan, W.; Rencoret, J.; Lu, F.; Karlen, S.D.; Smith, B.G.; Harris, P.J.; del Río, J.C.; Ralph, J. Tricin-lignins: Occurrence and quantitation of tricin in relation to phylogeny. *Plant J.* **2016**, *88*, 1046–1057. [CrossRef]
37. Ralph, J.; Lundquist, K.; Brunow, G.; Lu, F.; Kim, H.; Schatz, P.F.; Marita, J.M.; Hatfield, R.D.; Ralph, S.A.; Christensen, J.H. Lignins: Natural polymers from oxidative coupling of 4-hydroxyphenyl-propanoids. *Phytochem. Rev.* **2004**, *3*, 29–60. [CrossRef]
38. Martinez, P.M.; Punt, A.M.; Kabel, M.A.; Gruppen, H. Deconstruction of lignin linked *p*-coumarates, ferulates and xylan by NaOH enhances the enzymatic conversion of glucan. *Bioresour. Technol.* **2016**, *216*, 44–51. [CrossRef]
39. Ralph, J.; Hatfield, R.D.; Quideau, S.; Helm, R.F.; Grabber, J.H.; Jung, H.J.G. Pathway of *p*-coumaric acid incorporation into maize lignin as revealed by NMR. *J. Am. Chem. Soc.* **1994**, *116*, 9448–9456. [CrossRef]
40. Karlen, S.D.; Zhang, C.; Peck, M.L.; Smith, R.A.; Padmakshan, D.; Helmich, K.E.; Free, H.C.; Lee, S.; Smith, B.G.; Lu, F. Monolignol ferulate conjugates are naturally incorporated into plant lignins. *Sci. Adv.* **2016**, *2*, e1600393. [CrossRef]
41. Grabber, J.H.; Hatfield, R.D.; Lu, F.; Ralph, J. Coniferyl ferulate incorporation into lignin enhances the alkaline delignification and enzymatic degradation of cell walls. *Biomacromolecules* **2008**, *9*, 2510–2516. [CrossRef] [PubMed]
42. Kumar, M.; Campbell, L.; Turner, S. Secondary cell walls: Biosynthesis and manipulation. *J. Exp. Bot.* **2016**, *67*, 515–531. [CrossRef]
43. Saurabh, S.; Vidyarthi, A.S.; Prasad, D. RNA interference: Concept to reality in crop improvement. *Planta* **2014**, *239*, 543–564. [CrossRef] [PubMed]
44. Belhaj, K.; Chaparro-Garcia, A.; Kamoun, S.; Patron, N.J.; Nekrasov, V. Editing plant genomes with CRISPR/Cas9. *Curr. Opin. Biotechnol.* **2015**, *32*, 76–84. [CrossRef]
45. Court of Justice of the European Union. Case C-528/16. Available online: http://curia.europa.eu/juris/documents.jsf?num=C-528/16 (accessed on 25 July 2018).
46. Theander, O.; Westerlund, E.A. Studies on dietary fiber. 3. Improved procedures for analysis of dietary fiber. *J. Agric. Food Chem.* **1986**, *34*, 330–336. [CrossRef]
47. Morrison, I.M. Improvements in the acetyl bromide technique to determine lignin and digestibility and its application to legumes. *J. Sci. Food Agric.* **1972**, *23*, 1463–1469. [CrossRef]
48. Fukushima, R.S.; Hatfield, R.D. Comparison of the acetyl bromide spectrophotometric method with other analytical lignin methods for determining lignin concentration in forage samples. *J. Agric. Food Chem.* **2004**, *52*, 3713–3720. [CrossRef]
49. Fukushima, R.S.; Kerley, M.S.; Ramos, M.H.; Porter, J.H.; Kallenbach, R.L. Comparison of acetyl bromide lignin with acid detergent lignin and Klason lignin and correlation with in vitro forage degradability. *Anim. Feed Sci. Technol.* **2015**, *201*, 25–37. [CrossRef]

50. Cass, C.L.; Peraldi, A.; Dowd, P.F.; Mottiar, Y.; Santoro, N.; Karlen, S.D.; Bukhman, Y.V.; Foster, C.E.; Thrower, N.; Bruno, L.C.; et al. Effects of *PHENYLALANINE AMMONIA LYSASE (PAL)* knockdown on cell wall composition, biomass digestibility, and biotic and abiotic stress responses in *Brachypodium. J. Exp. Bot.* **2015**, *66*, 4317–4335. [CrossRef]

51. Abdel-Rahman, M.M.; Mousa, I.E. Effects of down regulation of lignin content in maize (*Zea mays* L.) plants expressing *C4H3* gene in the antisense orientation. *Biofuels* **2016**, *7*, 289–294. [CrossRef]

52. Park, J.-J.; Yoo, C.G.; Flanagan, A.; Pu, Y.; Debnath, S.; Ge, Y.; Ragauskas, A.J.; Wang, Z.-Y. Defined tetra-allelic gene disruption of the 4-coumarate:coenzyme A ligase 1 (*Pv4CL1*) gene by CRISPR/Cas9 in switchgrass results in lignin reduction and improved sugar release. *Biotechnol. Biofuels* **2017**, *10*, 284. [CrossRef]

53. Xu, B.; Escamilla-Treviño, L.L.; Sathitsuksanoh, N.; Shen, Z.; Shen, H.; Percival Zhang, Y.H.; Dixon, R.A.; Zhao, B. Silencing of 4-coumarate:coenzyme A ligase in switchgrass leads to reduced lignin content and improved fermentable sugar yields for biofuel production. *New Phytol.* **2011**, *192*, 611–625. [CrossRef]

54. Fornale, S.; Rencoret, J.; Garcia-Calvo, L.; Capellades, M.; Encina, A.; Santiago, R.; Rigau, J.; Gutierrez, A.; del Rio, J.-C.; Caparros-Ruiz, D. Cell wall modifications triggered by the down-regulation of coumarate 3-hydroxylase-1 in maize. *Plant Sci.* **2015**, *236*, 272–282. [CrossRef]

55. Takeda, Y.; Tobimatsu, Y.; Karlen, S.D.; Koshiba, T.; Suzuki, S.; Yamamura, M.; Murakami, S.; Mukai, M.; Hattori, T.; Osakabe, K.; et al. Downregulation of *p*-coumaroyl ester 3-hydroxylase in rice leads to altered cell wall structures and improves biomass saccharification. *Plant J.* **2018**, *95*, 796–811. [CrossRef]

56. Sibout, R.; Proost, S.; Hansen, B.O.; Vaid, N.; Giorgi, F.M.; Ho-Yue-Kuang, S.; Legée, F.; Cézart, L.; Bouchabké-Coussa, O.; Soulhat, C. Expression atlas and comparative coexpression network analyses reveal important genes involved in the formation of lignified cell wall in *Brachypodium distachyon. New Phytol.* **2017**, *215*, 1009–1025. [CrossRef]

57. Takeda, Y.; Koshiba, T.; Tobimatsu, Y.; Suzuki, S.; Murakami, S.; Yamamura, M.; Rahman, M.M.; Takano, T.; Hattori, T.; Sakamoto, M. Regulation of coniferaldehyde 5-hydroxylase expression to modulate cell wall lignin structure in rice. *Planta* **2017**, *246*, 337–349. [CrossRef] [PubMed]

58. Takeda, Y.; Suzuki, S.; Tobimatsu, Y.; Osakabe, K.; Osakabe, Y.; Ragamustari, S.K.; Sakamoto, M.; Umezawa, T. Lignin characterization of rice *CONIFERALDEHYDE 5-HYDROXYLASE* loss-of-function mutants generated with the CRISPR/Cas9 system. *Plant J.* **2018**, *97*, 543–554. [CrossRef]

59. Bewg, W.P.; Poovaiah, C.; Lan, W.; Ralph, J.; Coleman, H.D. RNAi downregulation of three key lignin genes in sugarcane improves glucose release without reduction in sugar production. *Biotechnol. Biofuels* **2016**, *9*, 270. [CrossRef]

60. Yang, Q.; He, Y.; Kabahuma, M.; Chaya, T.; Kelly, A.; Borrego, E.; Bian, Y.; El Kasmi, F.; Yang, L.; Teixeira, P.; et al. A gene encoding maize caffeoyl-CoA *O*-methyltransferase confers quantitative resistance to multiple pathogens. *Nat. Genet.* **2017**, *49*, 1364. [CrossRef]

61. Li, X.; Chen, W.; Zhao, Y.; Xiang, Y.; Jiang, H.; Zhu, S.; Cheng, B. Downregulation of caffeoyl-CoA *O*-methyltransferase (CCoAOMT) by RNA interference leads to reduced lignin production in maize straw. *Genet. Mol. Biol.* **2013**, *36*, 540–546. [CrossRef] [PubMed]

62. Park, S.-H.; Mei, C.; Pauly, M.; Ong, R.G.; Dale, B.E.; Sabzikar, R.; Fotoh, H.; Thang, N.; Sticklen, M. Downregulation of maize cinnamoyl-coenzyme a reductase via RNA interference technology causes brown midrib and improves ammonia fiber expansion-pretreated conversion into fermentable sugars for biofuels. *Crop Sci.* **2012**, *52*, 2687–2701. [CrossRef]

63. Trabucco, G.M.; Matos, D.A.; Lee, S.J.; Saathoff, A.J.; Priest, H.D.; Mockler, T.C.; Sarath, G.; Hazen, S.P. Functional characterization of cinnamyl alcohol dehydrogenase and caffeic acid *O*-methyltransferase in *Brachypodium distachyon. BMC Biotechnol.* **2013**, *13*, 61. [CrossRef] [PubMed]

64. Daly, P.; McClellan, C.; Maluk, M.; Oakey, H.; Lapierre, C.; Waugh, R.; Stephens, J.; Marshall, D.; Barakate, A.; Tsuji, Y.; et al. RNAi-suppression of barley caffeic acid *O*-methyltransferase modifies lignin despite redundancy in the gene family. *Plant Biotechnol. J.* **2018**, *17*, 549–607. [CrossRef]

65. He, X.; Hall, M.B.; Gallo-Meagher, M.; Smith, R.L. Improvement of forage quality by downregulation of maize *O*-methyltransferase. *Crop Sci.* **2003**, *43*, 2240–2251. [CrossRef]

66. Piquemal, J.; Chamayou, S.; Nadaud, I.; Beckert, M.; Barriere, Y.; Mila, I.; Lapierre, C.; Rigau, J.; Puigdomenech, P.; Jauneau, A.; et al. Down-regulation of caffeic acid *O*-methyltransferase in maize revisited using a transgenic approach. *Plant Physiol.* **2002**, *130*, 1675–1685. [CrossRef]

67. Jung, J.H.; Fouad, W.M.; Vermerris, W.; Gallo, M.; Altpeter, F. RNAi suppression of lignin biosynthesis in sugarcane reduces recalcitrance for biofuel production from lignocellulosic biomass. *Plant Biotechnol. J.* **2012**, *10*, 1067–1076. [CrossRef]

68. Jung, J.H.; Vermerris, W.; Gallo, M.; Fedenko, J.R.; Erickson, J.E.; Altpeter, F. RNA interference suppression of lignin biosynthesis increases fermentable sugar yields for biofuel production from field-grown sugarcane. *Plant Biotechnol. J.* **2013**, *11*, 709–716. [CrossRef]

69. Jung, J.H.; Altpeter, F. TALEN mediated targeted mutagenesis of the *caffeic acid O-methyltransferase* in highly polyploid sugarcane improves cell wall composition for production of bioethanol. *Plant Mol. Biol.* **2016**, *92*, 131–142. [CrossRef]

70. Baxter, H.L.; Mazarei, M.; Fu, C.; Cheng, Q.; Turner, G.B.; Sykes, R.W.; Windham, M.T.; Davis, M.F.; Dixon, R.A.; Wang, Z.-Y.; et al. Time course field analysis of COMT-downregulated switchgrass: Lignification, recalcitrance, and rust susceptibility. *Bioenergy Res.* **2016**, *9*, 1087–1100. [CrossRef]

71. Liu, S.; Fu, C.; Gou, J.; Sun, L.; Huhman, D.; Zhang, Y.; Wang, Z.-Y. Simultaneous downregulation of MTHFR and COMT in switchgrass affects plant performance and induces lesion-mimic cell death. *Front. Plant Sci.* **2017**, *8*, 982. [CrossRef]

72. Fornale, S.; Capellades, M.; Encina, A.; Wang, K.; Irar, S.; Lapierre, C.; Ruel, K.; Joseleau, J.-P.; Berenguer, J.; Puigdomenech, P.; et al. Altered lignin biosynthesis improves cellulosic bioethanol production in transgenic maize plants down-regulated for cinnamyl alcohol dehydrogenase. *Mol. Plant* **2012**, *5*, 817–830. [CrossRef]

73. Koshiba, T.; Murakami, S.; Hattori, T.; Mukai, M.; Takahashi, A.; Miyao, A.; Hirochika, H.; Suzuki, S.; Sakamoto, M.; Umezawa, T. *CAD2* deficiency causes both brown midrib and gold hull and internode phenotypes in *Oryza sativa* L. cv. Nipponbare. *Plant Biotechnol.* **2013**, *30*, 365–373. [CrossRef]

74. Saathoff, A.J.; Sarath, G.; Chow, E.K.; Dien, B.S.; Tobias, C.M. Downregulation of cinnamyl-alcohol dehydrogenase in switchgrass by RNA silencing results in enhanced glucose release after cellulase treatment. *PLoS ONE* **2011**, *6*, e16416. [CrossRef] [PubMed]

75. Dalmais, M.; Antelme, S.; Ho-Yue-Kuang, S.; Wang, Y.; Darracq, O.; d'Yvoire, M.B.; Cézard, L.; Légée, F.; Blondet, E.; Oria, N. A TILLING platform for functional genomics in *Brachypodium distachyon*. *PLoS ONE* **2013**, *8*, e65503. [CrossRef] [PubMed]

76. Ho-Yue-Kuang, S.; Alvarado, C.; Antelme, S.; Bouchet, B.; Cezard, L.; Le Bris, P.; Legee, F.; Maia-Grondard, A.; Yoshinaga, A.; Saulnier, L.; et al. Mutation in *Brachypodium* caffeic acid *O*-methyltransferase 6 alters stem and grain lignins and improves straw saccharification without deteriorating grain quality. *J. Exp. Bot.* **2016**, *67*, 227–237. [CrossRef] [PubMed]

77. Jorgenson, L.R. *Brown midrib* in maize and its lineage relations. *ASA* **1931**, *23*, 549–577.

78. Emerson, R.; Beadle, G.W.; Fraser, A.C. A summary of linkage studies in maize. *Cornell Univ. Agric. Exp. Stn. Memoir* **1935**, *180*, 1–83.

79. Buckley, G.F.H. Inheritance in barley with special reference to the color of caryopsis and lemma. *Sci. Agric.* **1930**, *10*, 460–492. [CrossRef]

80. NordGen. BGS 254, *Orange Lemma 1, rob1*. Available online: www.nordgen.org/bgs/system/export_pdf.php?bgs=254 (accessed on 19 November 2018).

81. Tang, H.M.; Liu, S.; Hill-Skinner, S.; Wu, W.; Reed, D.; Yeh, C.-T.; Nettleton, D.; Schnable, P.S. The maize *brown midrib2* (*bm2*) gene encodes a methylenetetrahydrofolate reductase that contributes to lignin accumulation. *Plant J.* **2014**, *77*, 380–392. [CrossRef]

82. Zhang, K.; Qian, Q.; Huang, Z.; Wang, Y.; Li, M.; Hong, L.; Zeng, D.; Gu, M.; Chu, C.; Cheng, Z. *GOLD HULL AND INTERNODE2* encodes a primarily multifunctional cinnamyl-alcohol dehydrogenase in rice. *Plant Physiol.* **2006**, *140*, 972–983. [CrossRef] [PubMed]

83. D'Yvoire, M.B.; Bouchabke-Coussa, O.; Voorend, W.; Antelme, S.; Cezard, L.; Legee, F.; Lebris, P.; Legay, S.; Whitehead, C.; McQueen-Mason, S.J.; et al. Disrupting the *cinnamyl alcohol dehydrogenase 1* gene (*BdCAD1*) leads to altered lignification and improved saccharification in *Brachypodium distachyon*. *Plant J.* **2013**, *73*, 496–508. [CrossRef]

84. Porter, K.; Axtell, J.; Lechtenberg, V.; Colenbrander, V. Phenotype, fiber composition, and in vitro dry matter disappearance of chemically induced brown midrib (*bmr*) mutants of sorghum. *Crop Sci.* **1978**, *18*, 205–208. [CrossRef]

85. Barriere, Y.; Argillier, O.; Chabbert, B.; Tollier, M.T.; Monties, B. Breeding silage maize with brown-midrib genes—Feeding value and biochemical characteristics. *Agronomie* **1994**, *14*, 15–25. [CrossRef]

86. Cherney, J.; Cherney, D.; Akin, D.; Axtell, J. Potential of brown-midrib, low-lignin mutants for improving forage quality. In *Advances in Agronomy*; Sparks, D., Ed.; Elsevier: Amsterdam, The Netherlands, 1991; Volume 46, pp. 157–198.

87. Vermerris, W.; Saballos, A.; Ejeta, G.; Mosier, N.S.; Ladisch, M.R.; Carpita, N.C. Molecular breeding to enhance ethanol production from corn and sorghum stover. *Crop Sci.* **2007**, *47*, 142–153. [CrossRef]

88. Barnes, R.F.; Muller, L.D.; Bauman, L.F.; Colenbrander, V.F. In-vitro dry matter disappearance of brown midrib mutants of maize. *J. Anim. Sci.* **1971**, *31*, 881–884. [CrossRef]

89. Barriere, Y. Brown-midrib genes in maize and their efficiency in dairy cow feeding. Perspectives for breeding improved silage maize targeting gene modifications in the monolignol and *p*-hydroxycinnamate pathways. *Maydica* **2017**, *62*, 1–19.

90. Guillaumie, S.; Pichon, M.; Martinant, J.-P.; Bosio, M.; Goffner, D.; Barriere, Y. Differential expression of phenylpropanoid and related genes in brown-midrib *bm1, bm2, bm3*, and *bm4* young near-isogenic maize plants. *Planta* **2007**, *226*, 235–250. [CrossRef]

91. Kuc, J.; Nelson, O.E. The abnormal lignins produced by the *brown-midrib* mutants of maize: I. The *brown-midrib*-1 mutant. *Arch. Biochem. Biophys.* **1964**, *105*, 103–113. [CrossRef]

92. Provan, G.J.; Scobbie, L.; Chesson, A. Characterisation of lignin from CAD and OMT deficient *bm* mutants of maize. *J. Sci. Food Agric.* **1997**, *73*, 133–142. [CrossRef]

93. Santoro, N.; Cantu, S.L.; Tornqvist, C.-E.; Falbel, T.G.; Bolivar, J.L.; Patterson, S.E.; Pauly, M.; Walton, J.D. A high-throughput platform for screening milligram quantities of plant biomass for lignocellulose digestibility. *Bioenergy Res.* **2010**, *3*, 93–102. [CrossRef]

94. Barriere, Y.; Argillier, O. Brown-midrib genes of maize—A review. *Agronomie* **1993**, *13*, 865–876. [CrossRef]

95. Barriere, Y.; Emile, J.C.; Traineau, R.; Surault, F.; Briand, M.; Gallais, A. Genetic variation for organic matter and cell wall digestibility in silage maize. Lessons from a 34-year long experiment with sheep in digestibility crates. *Maydica* **2004**, *49*, 115–126.

96. Barriere, Y.; Guillet, C.; Goffner, D.; Pichon, M. Genetic variation and breeding strategies for improved cell wall digestibility in annual forage crops. A review. *Anim. Res.* **2003**, *52*, 193–228. [CrossRef]

97. Barriere, Y.; Mechin, V.; Riboulet, C.; Guillaumie, S.; Thomas, J.; Bosio, M.; Fabre, F.; Goffner, D.; Pichon, M.; Lapierre, C.; et al. Genetic and genomic approaches for improving biofuel production from maize. *Euphytica* **2009**, *170*, 183–202. [CrossRef]

98. Courtial, A.; Soler, M.; Chateigner-Boutin, A.-L.; Reymond, M.; Mechin, V.; Wang, H.; Grima-Pettenati, J.; Barriere, Y. Breeding grasses for capacity to biofuel production or silage feeding value: An updated list of genes involved in maize secondary cell wall biosynthesis and assembly. *Maydica* **2013**, *58*, 67–102.

99. Kiesselbach, T.A. *Corn Investigations*; University of Nebraska: Lincoln, NE, USA, 1922; Volume 20.

100. Eyster, W.H. Chromosome VIII in maize. *Science* **1926**, *64*, 22. [CrossRef] [PubMed]

101. Halpin, C.; Holt, K.; Chojecki, J.; Oliver, D.; Chabbert, B.; Monties, B.; Edwards, K.; Barakate, A.; Foxon, G.A. *Brown-midrib* maize (*bm1*)—A mutation affecting the cinnamyl alcohol dehydrogenase gene. *Plant J.* **1998**, *14*, 545–553. [CrossRef]

102. Barriere, Y.; Chavigneau, H.; Delaunay, S.; Courtial, A.; Bosio, M.; Lassagne, H.; Derory, J.; Lapierre, C.; Mechin, V.; Tatout, C. Different mutations in the *ZmCAD2* gene underlie the maize brown-midrib1 (*bm1*) phenotype with similar effects on lignin characteristics and have potential interest for bioenergy production. *Maydica* **2013**, *58*, 6–20.

103. Burnham, C.; Brink, R. Linkage relations of a second brown midrib gene (*bm2*) in maize. *Agron. J.* **1932**, *24*, 960–963. [CrossRef]

104. Li, L.; Hill-Skinner, S.; Liu, S.; Beuchle, D.; Tang, H.M.; Yeh, C.-T.; Nettleton, D.; Schnable, P.S. The maize *brown midrib4* (*bm4*) gene encodes a functional folylpolyglutamate synthase. *Plant J.* **2015**, *81*, 493–504. [CrossRef]

105. Green, A.R.; Lewis, K.M.; Barr, J.T.; Jones, J.P.; Lu, F.; Ralph, J.; Vermerris, W.; Sattler, S.E.; Kang, C. Determination of the structure and catalytic mechanism of *Sorghum bicolor* caffeic acid *O*-methyltransferase and the structural impact of three *brown midrib12* mutations. *Plant Physiol.* **2014**, *165*, 1440–1456. [CrossRef]

106. Ye, Z.-H.; Kneusel, R.E.; Matern, U.; Varner, J.E. An alternative methylation pathway in lignin biosynthesis in *Zinnia. Plant Cell* **1994**, *6*, 1427–1439. [CrossRef]

107. Wu, Z.; Ren, H.; Xiong, W.; Roje, S.; Liu, Y.; Su, K.; Fu, C. Methylenetetrahydrofolate reductase modulates methyl metabolism and lignin monomer methylation in maize. *J. Exp. Bot.* **2018**, *69*, 3963–3973. [CrossRef]

108. Vignols, F.; Rigau, J.; Torres, M.A.; Capellades, M.; Puigdomenech, P. The *brown midrib3* (*bm3*) mutation in maize ocurs in the gene encoding caffeic acid *O*-methyl transferase. *Plant Cell* **1995**, *7*, 407–416. [CrossRef] [PubMed]

109. Morrow, S.L.; Mascia, P.; Self, K.A.; Altschuler, M. Molecular characterization of a brown midrib3 deletion mutation in maize. *Mol. Breed.* **1997**, *3*, 351–357. [CrossRef]

110. Burnham, C.R. Cytogenetic studies of a translocation between chromosome-1 and chromosome-7 in maize. *Genetics* **1948**, *33*, 5–21.

111. Vermerris, W.; Sherman, D.M.; McIntyre, L.M. Phenotypic plasticity in cell walls of maize *brown midrib* mutants is limited by lignin composition. *J. Exp. Bot.* **2010**, *61*, 2479–2490. [CrossRef]

112. Ali, F.; Scott, P.; Bakht, J.; Chen, Y.; Luebberstedt, T. Identification of novel *brown midrib* genes in maize by tests of allelism. *Plant Breed.* **2010**, *129*, 724–726. [CrossRef]

113. Mechin, V.; Laluc, A.; Legee, F.; Cezard, L.; Denoue, D.; Barriere, Y.; Lapierre, C. Impact of the brown-midrib *bm5* mutation on maize lignins. *J. Agric. Food Chem.* **2014**, *62*, 5102–5107. [CrossRef]

114. Ralph, J.; Kim, H.; Lu, F.; Grabber, J.H.; Leplé, J.C.; Berrio-Sierra, J.; Derikvand, M.M.; Jouanin, L.; Boerjan, W.; Lapierre, C. Identification of the structure and origin of a thioacidolysis marker compound for ferulic acid incorporation into angiosperm lignins (and an indicator for cinnamoyl CoA reductase deficiency). *Plant J.* **2008**, *53*, 368–379. [CrossRef]

115. Chen, Y.; Liu, H.; Ali, F.; Scott, M.P.; Ji, Q.; Frei, U.K.; Luebberstedt, T. Genetic and physical fine mapping of the novel brown midrib gene *bm6* in maize (*Zea mays* L.) to a 180 kb region on chromosome 2. *Theor. Appl. Genet.* **2012**, *125*, 1223–1235. [CrossRef]

116. Stephens, J.; Halpin, C. Barley 'Orange Lemma' is a Mutant in the *CAD* Gene. Unpublished poster. 2008.

117. Meyer, D.W.; Franckowiak, J.D.; Nudell, R.D. Forage quality of barley hay. In *Agronomy Abstracts*; ASA: Madison, WI, USA, 1994.

118. Daly, P.; Stephens, J.; Halpin, C. Barley *'Orange Lemma'*—A Mutant in Lignin Biosynthesis? unpublished poster. 2007.

119. Newton, A.C.; Flavell, A.J.; George, T.S.; Leat, P.; Mullholland, B.; Ramsay, L.; Revoredo-Giha, C.; Russell, J.; Steffenson, B.J.; Swanston, J.S. Crops that feed the world 4. Barley: A resilient crop? Strengths and weaknesses in the context of food security. *Food Secur.* **2011**, *3*, 141–178. [CrossRef]

120. Myler, J.L.; Stanford, E.H. Color inheritance in barley. *Agron. J.* **1942**, *34*, 427–436. [CrossRef]

121. Falk, D.E. Linkage data with genes near the centromere of barley chromosome 6. *Barley Genet. Newsl.* **1980**, *10*, 13–16.

122. Kutcher, H.R.; Bailey, K.L.; Rossnagel, B.G.; Franckowiak, J.D. Linked morphological and molecular markers associated with common root rot reaction in barley. *Can. J. Plant Sci.* **1996**, *76*, 879–883. [CrossRef]

123. Falk, D.E. Presowing selection of male sterile barley plants for the production of outcrossed seeds. *Barley Genet. Newsl.* **1984**, *14*, 25–27.

124. Falk, D.E. Creation of a marked telo 6S trisomic for chromosome 6. *Barley Genet. Newsl.* **1994**, *23*, 33–35.

125. Zeng, D.-L.; Qian, Q.; Dong, G.-J.; Zhu, X.-D.; Dong, F.-G.; Teng, S.; Guo, L.-B.; Cao, L3.-Y.; Cheng, S.-H.; Xiong, Z.-M. Development of isogenic lines of morphological markers in indica rice. *Acta Bot. Sin.* **2003**, *45*, 1116–1120.

126. Hirano, K.; Masuda, R.; Takase, W.; Morinaka, Y.; Kawamura, M.; Takeuchi, Y.; Takagi, H.; Yaegashi, H.; Natsume, S.; Terauchi, R.; et al. Screening of rice mutants with improved saccharification efficiency results in the identification of *CONSTITUTIVE PHOTOMORPHOGENIC 1* and *GOLD HULL AND INTERNODE 1*. *Planta* **2017**, *246*, 61–74. [CrossRef]

127. Hong, L.; Qian, Q.; Tang, D.; Wang, K.; Li, M.; Cheng, Z. A mutation in the rice chalcone isomerase gene causes the *golden hull and internode 1* phenotype. *Planta* **2012**, *236*, 141–151. [CrossRef] [PubMed]

128. IRRI, A.C. *Global Rice Science Partnership (GRiSP)*; Council for Partnership on Rice Research in Asia: Metro Manila, Philippines, 2010.

129. Park, J.-y.; Kanda, E.; Fukushima, A.; Motobayashi, K.; Nagata, K.; Kondo, M.; Ohshita, Y.; Morita, S.; Tokuyasu, K. Contents of various sources of glucose and fructose in rice straw, a potential feedstock for ethanol production in Japan. *Biomass Bioenergy* **2011**, *35*, 3733–3735. [CrossRef]

130. Mouradov, A.; Spangenberg, G. Flavonoids: A metabolic network mediating plants adaptation to their real estate. *Front. Plant Sci.* **2014**, *5*, 620. [CrossRef]

131. Sattler, S.E.; Saballos, A.; Xin, Z.; Funnell-Harris, D.L.; Vermerris, W.; Pedersen, J.F. Characterization of novel sorghum *brown midrib* mutants from an EMS-mutagenized population. *G3* **2014**, *4*, 2115–2124. [CrossRef]

132. Saballos, A.; Vermerris, W.; Rivera, L.; Ejeta, G. Allelic association, chemical characterization and saccharification properties of *brown midrib* mutants of sorghum (*Sorghum bicolor* (L.) Moench). *Bioenergy Res.* **2008**, *1*, 193–204. [CrossRef]

133. Vogler, R.; Ejeta, G.; Johnson, K.; Axtell, J. Characterization of a new brown midrib sorghum line. In *Agronomy Abstracts*; ASA: Madison, WI, USA, 1994; p. 124.

134. Gupta, S.C. Allelic relationships and inheritance of brown midrib trait in sorghum. *J. Hered.* **1995**, *86*, 72–74. [CrossRef]

135. Xin, Z.; Wang, M.L.; Barkley, N.A.; Burow, G.; Franks, C.; Pederson, G.; Burke, J. Applying genotyping (TILLING) and phenotyping analyses to elucidate gene function in a chemically induced sorghum mutant population. *BMC Plant Biol.* **2008**, *8*, 103. [CrossRef]

136. Xin, Z.; Wang, M.L.; Burow, G.; Burke, J. An induced sorghum mutant population suitable for bioenergy research. *Bioenergy Res.* **2009**, *2*, 10–16. [CrossRef]

137. Bittinger, T.; Cantrell, R.; Axtell, J. Allelism tests of the brown-midrib mutants of sorghum. *J. Hered.* **1981**, *72*, 147–148. [CrossRef]

138. Li, X.; Weng, J.K.; Chapple, C. Improvement of biomass through lignin modification. *Plant J.* **2008**, *54*, 569–581. [CrossRef] [PubMed]

139. Da Silva, M.J.; Souza Carneiro, P.C.; de Souza Carneiro, J.E.; Borges Damasceno, C.M.; Lacerda Duraes Parrella, N.N.; Pastina, M.M.; Ferreira Simeone, M.L.; Schaffert, R.E.; da Costa Parrella, R.A. Evaluation of the potential of lines and hybrids of biomass sorghum. *Ind. Crops Prod.* **2018**, *125*, 379–385. [CrossRef]

140. Saballos, A.; Sattler, S.E.; Sanchez, E.; Foster, T.P.; Xin, Z.; Kang, C.; Pedersen, J.F.; Vermerris, W. *Brown midrib2* (*Bmr2*) encodes the major 4-coumarate: Coenzyme A ligase involved in lignin biosynthesis in sorghum (*Sorghum bicolor* (L.) Moench). *Plant J.* **2012**, *70*, 818–830. [CrossRef]

141. Saballos, A.; Ejeta, G.; Sanchez, E.; Kang, C.; Vermerris, W. A genomewide analysis of the cinnamyl alcohol dehydrogenase family in sorghum [*Sorghum bicolor* (L.) Moench] identifies *SbCAD2* as the *brown midrib6* gene. *Genetics* **2009**, *181*, 783–795. [CrossRef] [PubMed]

142. Sattler, S.E.; Saathoff, A.J.; Haas, E.J.; Palmer, N.A.; Funnell-Harris, D.L.; Sarath, G.; Pedersen, J.F. A nonsense mutation in a cinnamyl alcohol dehydrogenase gene is responsible for the sorghum *brown midrib6* phenotype. *Plant Physiol.* **2009**, *150*, 584–595. [CrossRef]

143. Scully, E.D.; Gries, T.; Funnell-Harris, D.L.; Xin, Z.; Kovacs, F.A.; Vermerris, W.; Sattler, S.E. Characterization of novel *Brown midrib 6* mutations affecting lignin biosynthesis in sorghum. *J. Integr. Plant Biol.* **2016**, *58*, 136–149. [CrossRef]

144. Pillonel, C.; Mulder, M.M.; Boon, J.J.; Forster, B.; Binder, A. Involvement of cinnamyl-alcohol dehydrogenase in the control of lignin formation in *Sorghum bicolor* L. Moench. *Planta* **1991**, *185*, 538–544. [CrossRef] [PubMed]

145. Bucholtz, D.L.; Cantrell, R.P.; Axtell, J.D.; Lechtenberg, V.L. Lignin biochemistry of normal and brown midrib mutant sorghum. *J. Agric. Food Chem.* **1980**, *28*, 1239–1241. [CrossRef]

146. Godin, B.; Nagle, N.; Sattler, S.; Agneessens, R.; Delcarte, J.; Wolfrum, E. Improved sugar yields from biomass sorghum feedstocks: Comparing low-lignin mutants and pretreatment chemistries. *Biotechnol. Biofuels* **2016**, *9*, 251. [CrossRef] [PubMed]

147. Oliver, A.L.; Pedersen, J.F.; Grant, R.J.; Klopfenstein, T.J. Comparative effects of the sorghum *bmr*-6 and *bmr*-12 genes: I. Forage sorghum yield and quality. *Crop Sci.* **2005**, *45*, 2234–2239. [CrossRef]

148. Oliver, A.L.; Pedersen, J.F.; Grant, R.J.; Klopfenstein, T.J.; Jose, H.D. Comparative effects of the sorghum *bmr*-6 and *bmr*-12 genes: II. Grain yield, stover yield, and stover quality in grain sorghum. *Crop Sci.* **2005**, *45*, 2240–2245. [CrossRef]

149. Oliver, A.L.; Grant, R.J.; Pedersen, J.F.; O'Rear, J. Comparison of brown midrib-6 and -18 forage sorghum with conventional sorghum and corn silage in diets of lactating dairy cows. *J. Dairy Sci.* **2004**, *87*, 637–644. [CrossRef]

150. Bout, S.; Vermerris, W. A candidate-gene approach to clone the sorghum *brown midrib* gene encoding caffeic acid *O*-methyltransferase. *Mol. Genet. Genom.* **2003**, *269*, 205–214. [CrossRef]

151. Sattler, S.E.; Palmer, N.A.; Saballos, A.; Greene, A.M.; Xin, Z.; Sarath, G.; Vermerris, W.; Pedersen, J.F. Identification and characterization of four missense mutations in *brown midrib 12* (*bmr12*), the caffeic *O*-methyltranferase (COMT) of sorghum. *Bioenergy Res.* **2012**, *5*, 855–865. [CrossRef]

152. Funnell-Harris, D.L.; O'Neill, P.M.; Sattler, S.E.; Gries, T.; Berhow, M.A.; Pedersen, J.F. Response of sorghum stalk pathogens to brown midrib plants and soluble phenolic extracts from near isogenic lines. *Eur. J. Plant Pathol.* **2017**, *148*, 941–953. [CrossRef]

153. Catalan, P.; Chalhoub, B.; Chochois, V.; Garvin, D.F.; Hasterok, R.; Manzaneda, A.J.; Mur, L.A.; Pecchioni, N.; Rasmussen, S.K.; Vogel, J.P. Update on the genomics and basic biology of *Brachypodium*: International *Brachypodium* Initiative (IBI). *Trends Plant Sci.* **2014**, *19*, 414–418. [CrossRef] [PubMed]

7

Optimizing ddRADseq in Non-Model Species

Natalia Cristina Aguirre [1,*,†], Carla Valeria Filippi [1,†], Giusi Zaina [2], Juan Gabriel Rivas [1], Cintia Vanesa Acuña [1], Pamela Victoria Villalba [1], Martín Nahuel García [1], Sergio González [1], Máximo Rivarola [1], María Carolina Martínez [1], Andrea Fabiana Puebla [1], Michele Morgante [2], Horacio Esteban Hopp [1,3], Norma Beatriz Paniego [1,†] and Susana Noemí Marcucci Poltri [1,†]

[1] Instituto de Agrobiotecnología y Biología Molecular—IABiMo—INTA-CONICET, Instituto de Biotecnología, Centro de Investigaciones en Ciencias Agronómicas y Veterinarias, Instituto Nacional de Tecnología Agropecuaria, Dr. Nicolás Repetto y de los Reseros S/N, Hurlingham B1686IGC, Argentina

[2] Department of Agricultural, Food, Environmental and Animal Sciences, University of Udine, 33100 Udine, Italy

[3] Laboratorio de Agrobiotecnología, FBMC, Facultad de Ciencias Exactas y Naturales, Universidad de Buenos Aires, Ciudad Universitaria, Buenos Aires C1428EHA, Argentina

* Correspondence: aguirre.natalia@inta.gob.ar

† These authors contributed equally to this work.

Abstract: Restriction site-associated DNA sequencing (RADseq) and its derived protocols, such as double digest RADseq (ddRADseq), offer a flexible and highly cost-effective strategy for efficient plant genome sampling. This has become one of the most popular genotyping approaches for breeding, conservation, and evolution studies in model and non-model plant species. However, universal protocols do not always adapt well to non-model species. Herein, this study reports the development of an optimized and detailed ddRADseq protocol in *Eucalyptus dunnii*, a non-model species, which combines different aspects of published methodologies. The initial protocol was established using only two samples by selecting the best combination of enzymes and through optimal size selection and simplifying lab procedures. Both single nucleotide polymorphisms (SNPs) and simple sequence repeats (SSRs) were determined with high accuracy after applying stringent bioinformatics settings and quality filters, with and without a reference genome. To scale it up to 24 samples, we added barcoded adapters. We also applied automatic size selection, and therefore obtained an optimal number of loci, the expected SNP locus density, and genome-wide distribution. Reliability and cross-sequencing platform compatibility were verified through dissimilarity coefficients of 0.05 between replicates. To our knowledge, this optimized ddRADseq protocol will allow users to go from the DNA sample to genotyping data in a highly accessible and reproducible way.

Keywords: SNP; SSR; next generation sequencing; genotyping by sequencing

1. Introduction

Efficient plant genome sampling, with sufficient and informative genetic markers, plays a key role in breeding, conservation, and evolution studies. In recent decades, researchers have developed different types of useful molecular markers, although nowadays SNPs have become the markers of choice. This selection is based on their high abundance in genomes, stability, co-dominance, and automation of the genotyping process [1].

SNP arrays are high-throughput and cost-effective tools, with the extra advantage of generating relatively reduced amount of missing data. These features make them one of the most popular genotyping tools for major crops and forest tree species. However, the development of a novel SNP array is costly, making them unaffordable for non-commercial plant species. Restriction site-associated DNA sequencing (RADseq) [2], genotyping by sequencing (GBS) [3], and their derived protocols (reviewed in [4] and [5]) are techniques that have recently emerged as promising genomic approaches for SNP discovery at a genome-wide scale. They are based on reduced representation sequencing of multiplexed samples, do not require a reference genome or previous polymorphism knowledge, and combine marker discovery and genotyping in a one-step process. Thus, they provide a rapid, high-throughput, and cost-effective strategy for carrying out multiple genome-wide analyses for several non-model species and germplasm sets.

These approaches involve digesting the DNA with restriction enzymes and then sequencing a specific size-selected range of generated fragments. Aiming to ameliorate some of the weaknesses of the original RADseq, specifically with regard to the dependence of the length of the generated fragments on random shearing effects [5,6], researchers have developed many derived methodologies, including 2b-RADseq [7], ezRADseq [8] and ddRADseq [9]. Double-digest restriction site-associated DNA sequencing, or ddRADseq, uses two different restriction enzymes to cut the DNA: one rare cutter (i.e., an enzyme with a large recognition site) and a frequent cutter. Only the fragments falling between both restriction sites and within a specific size range are sequenced [6]. This reduces the depth of sequencing needed to reach optimal coverage, as well as the percentage of missing data, in comparison with RADseq.

The original ddRADseq protocol was built and trained based on animal data, and it has been widely applied in SNP marker development and genotyping for several species in this kingdom. Applications of this technology to plants have been reported [10–14], especially in forest and fruit trees (reviewed in [15]), and were specifically improved [16,17]. However, most researchers still use universal default protocols that carry some limitations. Because of the diversity and complexity of plant genomes, the different steps of the ddRADseq protocol require revision to achieve better results in non-model plant species. The steps that would need revision include the selection of the pair of restriction enzymes, the determination of the optimum size range, the suitability and performance of the sequencing platform, the sequence depth, and the variant calling strategy. Moreover, because this could involve testing steps, the development of an optimized protocol for setting up the methodology in a small number of plant samples is mandatory, mainly for labs with low budgets.

A collateral application of this Next Generation Sequencing (NGS) technique in plants is the cost-effective discovery of other genetic variants like polymorphic SSR loci [18,19]. SSRs have numerous uses, including linkage map development, quantitative trait locus (QTL) mapping, marker-assisted selection, cultivar or clone fingerprinting, and population structure and genetic diversity studies, among others [20].

For *Eucalyptus*, several genotyping platforms, such as the recent SNP array EUChip60K [21], have been developed [22,23]. However, some species, such as the important forest species *Eucalyptus dunnii* Maiden (hereafter *E. dunnii*) are less represented in this case.

The present study involves the development of an optimized and lower-cost ddRADseq protocol in *E. dunnii* through the setting up of a small number of samples. This optimized protocol may be easily applied to any plant species. Additionally, this study presents the scaling up of the first protocol to a second one, which allows its application to a larger number of samples. To our knowledge, to date this is the most comprehensive and detailed ddRADseq report allowing users to optimize the protocol from the DNA sample to the molecular marker data in an easy and accessible way.

2. Materials and Methods

2.1. Plant Material and DNA Extraction

A ddRADseq derived protocol was optimized and applied on two samples of *E. dunnii* (A and B), and subsequently scaled up to another 24 samples (1 to 24). The samples belong to the INTA *Eucalyptus* breeding program (Supplementary Table S1). Fresh young leaves were collected, dried in a freeze dryer (Labconco Corporation, Kansas City, MO, USA) and conserved in silica gel until DNA extraction. Genomic DNA (gDNA) was extracted from the lyophilized leaves following the CTAB method described by Hoisington et al. [24] with modifications for *E. dunnii* species as described in Marcucci et al. [25]. Their quality was verified by Nanodrop (Thermo Fisher Scientific, Waltham, MA, USA) and agarose gel electrophoresis analysis. DNA was quantified using a Qubit 2.0 fluorometer (Thermo Fisher Scientific).

2.2. Evaluation of Enzymes and Size Selection Range

Several in silico digestions of *E. grandis* v2.0 reference genome (available on Phytozome https://phytozome.jgi.doe.gov/pz/portal.html, [26]) were performed to assess both the optimal set of restriction enzymes for the *E. dunnii* genome and the number of DNA fragments to be recovered by different size selections [9,15]. Simulations were performed with SimRAD package [27]. The evaluated restriction enzyme combinations PstI-MspI and SphI-MboI were selected based on the studies of Peterson et al. [16] and Scaglione et al. [28], respectively. In addition, different size selections were evaluated to achieve between $1e^4$ to $5e^4$ fragments in an optimal size selection window of 50 to 100 base pairs (bp), as suggested by Peterson et al. [9], or even of 150 bp. The average insert size was set from 295 to 420 bp, which led to a final library size range between 350 and 600 bp. This size range is suitable for bridge amplification in Illumina platforms and allows minimum overlapping of Paired End (PE) Run reads of 150 bp or longer.

The insilico.digest routine was applied for both enzyme combinations and the adapt.select routine was used to simulate the amplification of the fragments with both enzyme cutting site endings. Finally, size.select was used to select different subpopulations of fragments per digestion. For double digestion, the considered means were of 320, 370, 420 bp, with two window widths simulating manual (agarose gel electrophoresis, 100 or 150 bp) and automatic selection (by SAGE ELF, 70 or 140 bp, for one or two elution wells). Subsequently, in vitro *E. dunnii* gDNA digestions were run by using reaction conditions described elsewhere [28]. The profile of the obtained fragments was visualized in agarose gel (Figure 3 of the Supplementary File S1) and capillary electrophoresis in a 5200 Fragment Analyzer System (Advanced Analytical Technologies, Inc., Santa Clara, CA, USA) using the DNA high sensitivity kit (Agilent Technologies, Santa Clara, CA, USA). Supplementary File S2 contains the SimRAD command lines used for the simulation.

2.3. Protocol 1 (P1): Optimized ddRADseq

Digestion: For samples A and B, 150 ng of each gDNA was completely digested using SphI-HF and MboI (2.4 U per enzyme, New England Biolabs (NEB), Ipswich, MA, USA), and incubated at 37 °C for 90 min. The reaction was inactivated at 65 °C for 20 min and purified with 1.5 volumes (×) of Ampure XP bead (Beckman Coulter, Brea, CA, USA) [28]. At this point, the complete digestion of gDNA was assessed by electrophoresis in a Fragment Analyzer System (Advanced Analytical Technologies, Inc.). A homogeneous distributed fragment population shorter than about 3 kb was expected.

Ligation: The common adapters (double-stranded oligonucleotides) published by Peterson et al. [16] were used (Supplementary Table S2). Specifically, Adapter 2 (A2) had a "Y" form for the specific amplification of fragments with different cut site endings. Adapter 1 (A1) and A2 were modified by changing their sticky ends for SphI and MboI, respectively. The final ligation was done using 2 pmol and 5 pmol of A1 and A2, respectively, and 2.4 Weiss units of T4 DNA ligase (Invitrogen, Carlsbad, CA, USA). This final selection was based on the following tests: A1 and A2: 2 and 5 pmol similar to Scaglione et al.'s protocol [28], 2 pmol of both adapters, as reported by Elshire et al. [3], and 0.1 and 15 pmol, as reported by Peterson et al. [16]. The reaction was incubated for 1 h at 23 °C, followed by an additional incubation for 1 h at 20 °C; finally, the reaction was inactivated for 20 min at 65 °C [28]. A1 × Ampure XP bead purification per sample was done before performing PCR (Polymerase Chain Reaction).

PCR: The dual-indexed primers designed by Lange et al. [29] (Supplementary Table S2) were used for the reactions. The oligonucleotides have a portion for sequencing on the Illumina platforms plus an index (8 bp), which allows the identification of each library. NEB Phusion High-Fidelity DNA polymerase was used with the following cycling parameters [28]: 3 min of initial denaturation (95 °C), 10 cycles of amplification (30 s at 95 °C, 30 s at 60 °C, 45 s at 72 °C), and 2 min of a final extension (72 °C). A1.2× Ampure XP bead purification per PCR was subsequently performed.

Pooling: After adding the indexed primers by PCR, the obtained libraries were pooled based on concentration (according to Qubit 2.0 fluorometer analysis) and concentrated in a SpeedVac (Eppendorf, Hamburg, Germany).

Size Selection: A manual size selection was applied (a range between 450 and 550 bp, which corresponds to DNA fragment size of interest between 310 and 410 bp) in low-melting 1.5% agarose gel electrophoresis (Bio-Rad Laboratories, Hercules, CA, USA). Finally, the selected fragments were purified from the gel with QIAquick Gel Extraction kit (Qiagen N.V., Hilden, Germany) [28].

Sequencing: The final libraries were quantified by Qubit 2.0 fluorometer (HS dsDNA kit, Thermo Fisher Scientific) and their quality was checked on a Fragment Analyzer system (DNA High Sensitivity kit, Agilent). A PE sequencing run (2 × 151 bp) was performed on MiSeq (Illumina Inc., San Diego, CA, USA) for both samples (Figure 1).

Supplementary File S1 displays an extended version of P1.

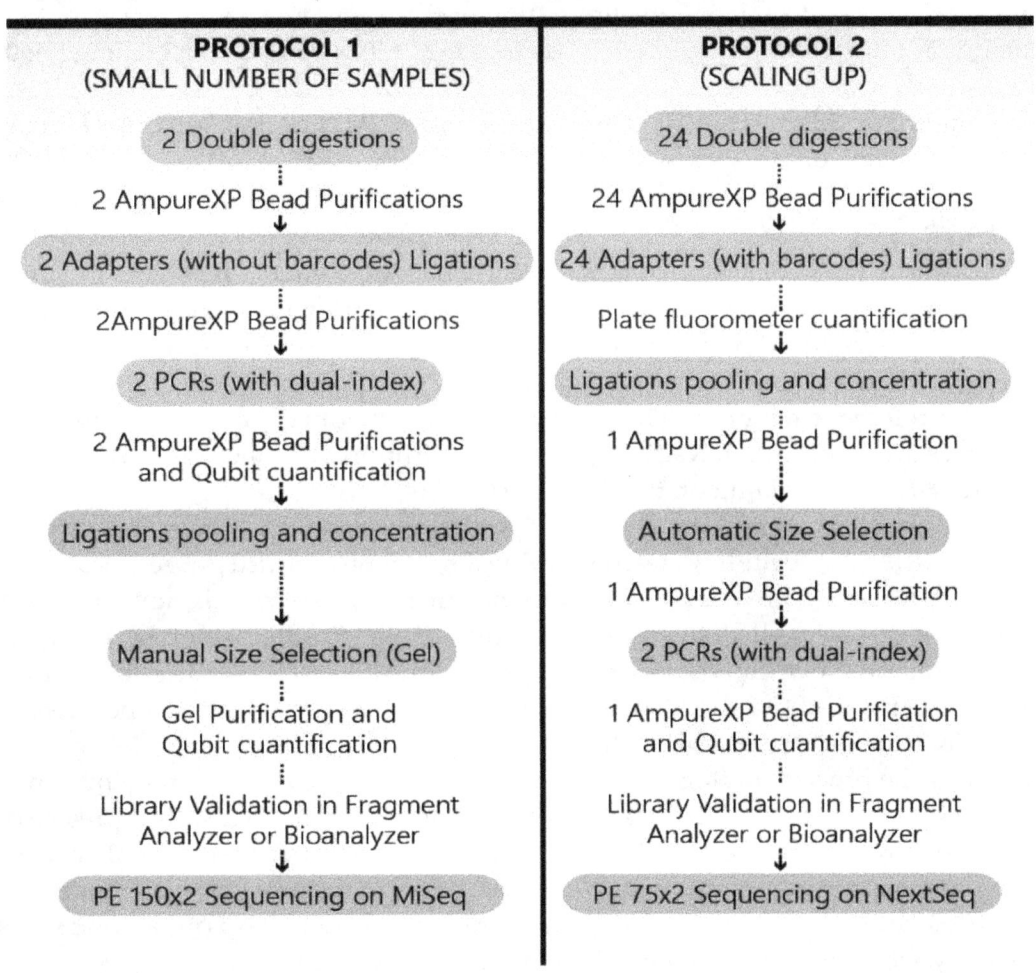

Figure 1. Workflow of the two optimized ddRADseq protocols.

2.4. Protocol 2 (P2): Optimized ddRADseq (Scaling Up to 24 Samples)

The Optimized ddRADseq P1 was subsequently scaled up by using the 24-plex P2 (samples 1 to 24, Supplementary Table S1). First at all, P2 was set up in 23 samples and consisted of P1 with some modifications as follows.

Ligation: 24 variable-length (4 to 9 bp) barcodes designed by Poland et al. [30] were added, in order to avoid low sequence quality of the first bases due to the restriction site [3,17] (Supplementary Table S2).

Pooling: Ligations were mixed by equal DNA quantity in a 23-plex pool, then concentrated and finally cleaned by one 1× Ampure XP bead purification per pool.

PCR: A PCR was performed per pool of libraries (a pair of indexes identifying each pool).

Sequencing: The libraries were sequenced on a very low depth PE (2× 250 bp) run of a MiSeq instrument (Illumina Inc.).

Finally, the P2 was customized to the definitive protocol (Figure 1). This led to the construction of new libraries from the gDNA of the same 23 samples plus an additional sample (24-plex), as detailed below.

Ligation: each reaction was done with 160 U of ligase (NEB Cohesive End Ligation).

Pooling: the ligations were 24-plex pooled, based on the concentration of each digestion quantified by Picogreen (Sigma-Aldrich) in a FluorStar Optima Fluorometer (BMG Labtech, Ortenberg, Germany).

Size selection: an automatic size selection run was performed in a 2% agarose cassette in the SAGE ELF (Sage Science, Inc., Beverly, MA, USA) and the fragments of 450 bp on average (between 415 and 485 bp) were collected from one well. Subsequently, an extra step of 0.8× Ampure XP bead purification was performed to ensure the elimination of the fragments below 300 bp.

Sequencing: as a final step, the pool was sequenced PE (2× 75 bp) on a NextSeq 500 sequencer (Illumina Inc.).

Supplementary File S3 presents an extended version of P2. Figure S1 of Supplementary Tables displays a schematic view of the library construction.

2.5. ddRADseq Data Analyses

The sequencing quality of each sample was checked using FastqC [31].

Although many bioinformatics software and R packages (R-3.5.2, R core team, Vienna, Austria) can handle this kind of reduced representation sequencing data, Stacks [32,33] (v1.48, University of Oregon, Eugene, OR, USA) is one of the software packages that performs equally well when working with or without a reference genome. This software was developed mainly for organisms without reference genomes and high-depth RAD sequencing. Additionally, Stacks is between the pipelines, with high accuracy for SNPs calling on this kind of data [34,35].

Herein, data obtained using both protocols were analyzed with different components of the software Stacks v1.48 [32,33], including cleaning raw reads, defining the ddRADseq loci and determining the SNPs. Samples A and B were used to compare the efficiency between de novo and with reference analyses for the species, as well as to assess the utility of P1. In the case of P2, on the other hand, samples 1 to 24 (run on NextSeq 500) were analyzed with reference. Additionally, libraries of the samples that were sequenced twice (MiSeq and NextSeq) were analyzed (both repetitions together) to evaluate the performance of the Illumina platforms.

First, by using the process_radtag.pl component, reads were removed if they presented uncalled bases, low Phred score (lower than 10), absence of enzyme recognition sites, and presence of adapter sequence. Additionally, A and B samples were trimmed to 145 bp, because of quality drop in the last bases according to mean average inspection using FastqC [31]. Otherwise, the raw data of samples 1 to 24 were demultiplexed and truncated to 66 bp after removing up to 9 bp of barcode sequences. For downstream analyses, paired and unpaired clean reads were considered.

Subsequently, the denovo_map.pl pipeline was used to search loci and SNPs de novo (only for P1), whereas ref_map.pl .pl was selected to assess SNPs after mapping cleaned reads to the *E. grandis* reference genome with Bowtie2 (default parameters) [36]. In all of the analyses, a minimum of three reads was used to define an allele (or stack) within an individual (-m 3). Particularly for de novo analysis, three mismatches between alleles were allowed to construct a locus within an individual (- M 3) and three mismatches were allowed between loci to build the catalog (-n 3).

After running each pipeline, the rxstacks program was applied to filter out putative sequencing errors of genotype and haplotype calls and, subsequently, the components cstacks and sstacks were rerun. Thus, the bounded SNP model was applied, and loci with log-likelihoods higher than (minus) -10 were kept. Furthermore, a proportion of individuals with confounded loci up to 0.05 were admitted, and excess haplotypes from individual loci were pruned according to their prevalence in the population.

Finally, the pipeline Populations was run with different filter combinations, resulting in three VCF (variant call format) files for the SNPs and ddRADseq polymorphic loci. For samples A and B, the basic data matrix (Total markers) was obtained. A second matrix without missing data was built (Shared markers). For the third matrix, allele data from samples 1 to 24 derived from P2 were filtered by a minimum allele frequency (MAF) of 0.05, and presentation of a locus by a minimum of 80% of individuals in order to be considered. Supplementary File S2 presents all the command lines used to run Stacks (with and without the reference genome).

2.6. SSR Identification

SSRs for samples A and B were identified using the software MIcroSAtellite (Institute of Plant Genetics and Crop Plant Research, Gatersleben, Germany) identification tool, also known as MISA [37], as in Qin et al.'s study [38]. The fasta_sample option of the Population module (Stacks v1.48) was used to obtain the sequences of the two haplotypes of each sample for each locus in FASTA format. Then, according to the same criterion used by Torales et al. [19], SSRs with a minimum of five repeats for dinucleotide, four repeats for trinucleotide, and three repeats for tetra, penta and hexanucleotide motives were searched. The polymorphic SSRs were also analyzed. Supplementary File S2 displays all the command lines used to run MISA.

2.7. Evaluation of Robustness—Sequencing Platform Comparison

The robustness of the protocol was evaluated by comparing MiSeq and NextSeq data sets from 46 libraries (23 MiSeq and 23 NextSeq). The VCF file was filtered by missing data and MAF lower than 20% and 0.05 respectively (Populations pipeline of Stacks v1.48). A dissimilarity matrix between all the samples was calculated directly from the filtered VCF using the R package SNPrelate [39]. The dendrogram was plotted using the R package ggplot2 [40].

3. Results

3.1. Evaluation of Enzymes and Size Selection Range

According to the in silico simulations of genome *E. grandis v2.0* digestion, the enzyme pair SphI-MboI generated 2,499,866 fragments (Figure 2a, grey area), of which 248,275 have both enzyme cutting site endings (type AB and BA fragments, data not shown). The enzyme pair PstI-MspI produced almost half (1,090,783) of the fragments of SphI-MboI (Figure 2b, grey area) and 174,771 of these fragments contain the expected pair of ends (AB+BA).

Table 1 displays the predicted AB+BA fragments generated for both enzyme combinations of different size selection ranges (270 to 420 bp, in manual size selection; 285 to 415 bp in automated size selection).

Table 1. Subpopulations of fragments obtained by in silico simulations of *E. grandis* genome v2.0.

Enzymes	Window	Manual Size Selection				Automated Size Selection	
		Insert Size 100 or 150 bp Window (mean)	Fragment Size in Gel (Protocol 1)	Hypothetical Fragment Size in Gel (Protocol 2; mean)	N° of Predicted Fragments (100 or 150 bp window)	Insert Size 70 bp (1 or 2 wells)	N° of Predicted Fragments (100 or 150 bp window)
SphI-MboI	100 or 70 bp	270–370 (320)	400–500	350–450 (400)	28,107	285–355 (1)	19,184
		320–420 (370)	450–550	400–500 (450)	24,508	335–405 (1)	17,317
		370–470 (420)	500–600	450–550	19,347	385–455 (1)	12,906
	150 bp	220–370 (295)	350–500	300–450	45,655	225–365 (2)	~ manual selection [a]
		270–420 (345)	400–550	350–500	39,122	265–415 (2)	~ manual selection [a]
PstI-MspI	100 or 70 bp	270–370 (320)	400–500	350–450	13,102	285–355 (1)	9137
		320–420 (370)	450–550	400–500 (450)	12,026	335–405 (1)	8359
		370–470 (420)	500–600	450–550	10,940	385–455 (1)	7595
	150 bp	220–370 (295)	350–500	300–450	20,749	225–365 (2)	~ manual selection [a]
		270–420 (345)	400–550	350–500	18,826	265–415 (2)	~ manual selection [a]

[a] The final number of predicted fragments will be almost the same for both manual an automated size selection. This is so because two contiguous elution wells of 70 bp range each are required to select a fraction of 150 bp using Automated Size Selection in SAGE ELF.

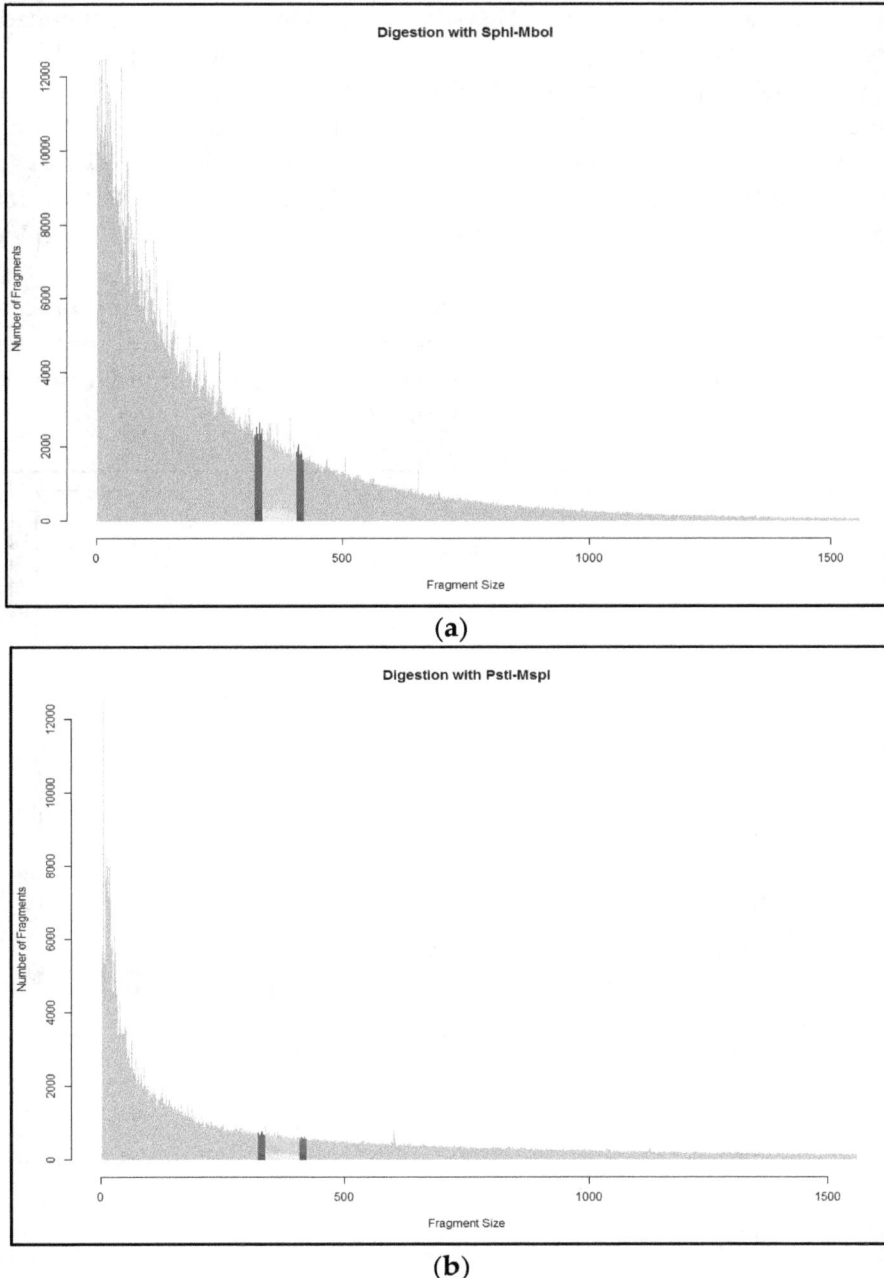

Figure 2. Histograms of in silico simulations (frequency versus fragment size). (**a**) In silico digestion with SphI-MboI. (**b**) In silico digestion with PstI-MspI. Total fragments obtained in digestion (grey); subpopulation of fragments obtained by manual size selection (the whole colored area); subpopulation of fragments obtained by automatic size selection (pink + light blue area); and subpopulation of AB+BA fragments selected, amplified and sequenced (manual: blue + light blue areas; automatic: light blue area).

We selected the SphI-MboI enzyme combination for *Eucalyptus*, because of the larger number of fragments in the thin window widths (100 and 70 bp). Specifically, we selected an average DNA fragment population size of 370 bp. This size gave the minimum overlapping between 150 bp PE reads (P1) in sequenced libraries. For this average fragment size, 24,508 AB+BA fragments fell within the range of 320 to 420 bp for the manual size selection in P1 (actually, the library fragment size was 450 to 550 bp, including adapters and primers, Figure 2a, blue + light blue areas). On the other hand, the automatic selection retrieved 17,317 AB+BA fragments in the range between 335 and 405 bp in P2 (Figure 2a, light blue area). For PE sequencing, this is 24,508 × 2 = 49,036 predicted sequenced ddRADseq *loci* for manual size selection and $17,317 \times 2 = 34,634$ predicted sequenced ddRADseq loci for automated size selection (70 bp range, due to the restriction of the equipment).

The enzyme pair PstI-MspI retrieved 12,026 AB+BA DNA fragments between a manual size selection window of 100 bp (between 320 and 420 bp; Figure 2b, blue + light blue areas), whereas it gave 8359 between an automatic size selection window of 70 bp (between 335 and 405 bp; Figure 2b, light blue area). Again, for PE sequencing, this was $12,026 \times 2 = 24,052$ predicted sequenced ddRADseq loci for manual size selection and $8359 \times 2 = 16,718$ predicted sequenced ddRADseq loci for automated size selection.

Moreover, in vitro digestion analyses showed that the SphI-MboI enzyme combination displays a more homogeneous pattern (Figure 3a). In accordance with the results from the in silico simulations, these enzymes gave higher frequencies of lower-sized fragments within the range of selection than those obtained with the PstI-MspI combination (Figure 3b).

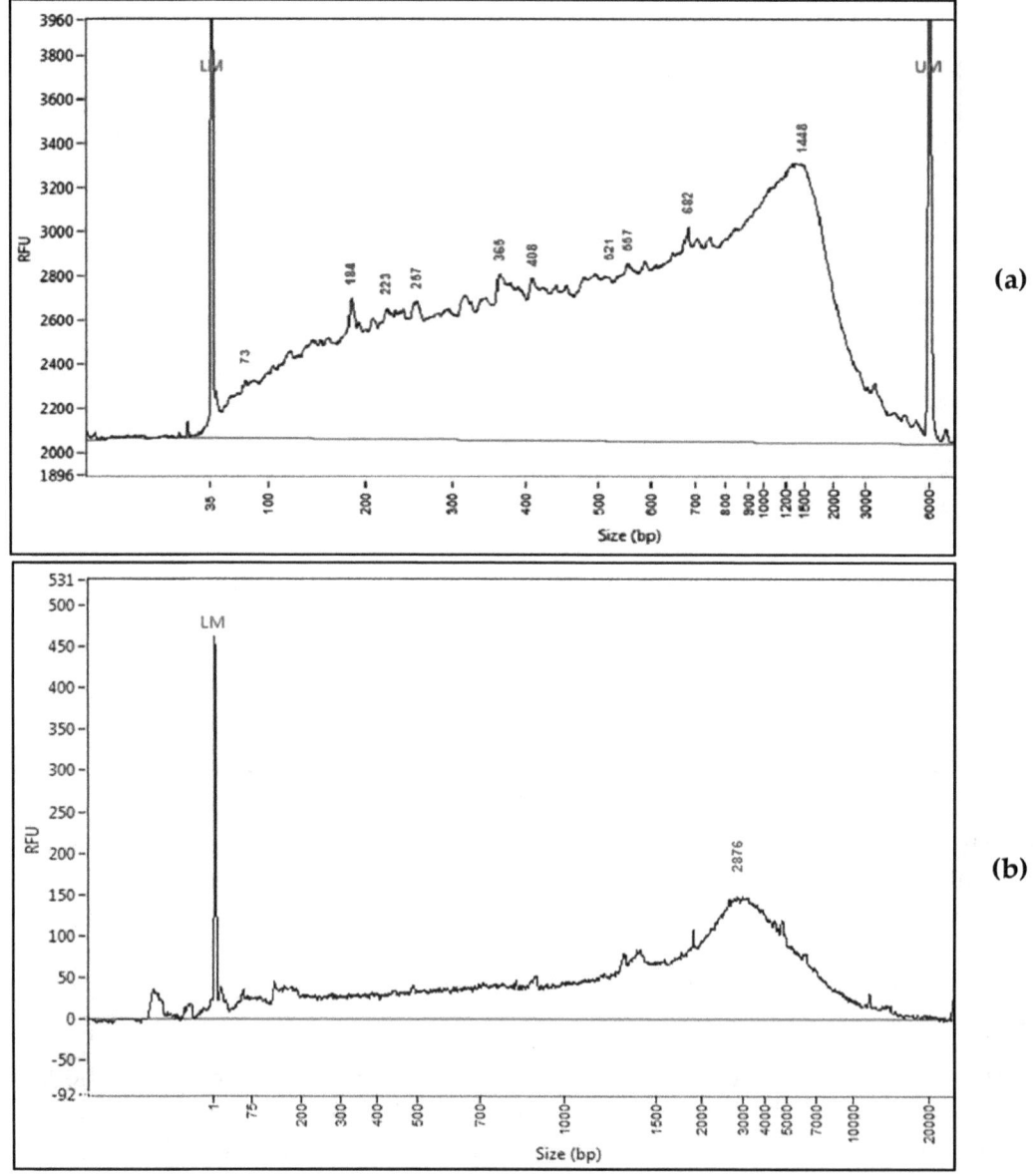

Figure 3. In vitro digestions of *E. dunnii* genomic DNA. Fragment Analyzer system runs. (**a**) SphI-MboI. (**b**) PstI-MspI.

Other enzymes and size selection combinations yielded a similar number of predicted fragments. For example, the PstI-MspI enzyme combination with an average fragment size selection of 345 bp and a window width of 150 bp retrieved 18,826 predicted fragments. However, our size selection equipment (SAGE ELF) at a window width of 70 bp retrieved several DNA fragment subpopulations of different ranges at the same time, and consequently a size selection with a width of 150 bp can only be done by collecting two elution wells or by manual selection.

3.2. Protocol 1: Analysis in Samples A and B

From the MiSeq sequencing, we obtained 1,984,145 and 2,294,900 PE reads of 151 bp for samples A and B, respectively. The overall read quality, according to FastqC visualization [31], was high enough for further analysis. Filtering by quality with the process_radtag.pl allowed us to obtain samples that retained more than 96% of the reads, with a mean of 2,066,064.5 reads.

The use of Bowtie2 [36] with the default parameters as the aligner allowed us to map approximately 82% of the reads against the *E. grandis* reference genome. The ref_map.pl pipeline of Stacks identified a total of 77,885 ddRADseq loci for sample A and 71,395 ddRADseq loci for sample B. These results showed a mean depth coverage of 24.16× and were used to build a catalogue. Then, a subsequent filtering by quality (with the rxstacks module) retained 41,834 ddRADseq loci. This result is at the expected order of magnitude according to in silico simulation (49,016 = 2 loci on 24,508 fragments using PE sequencing). Within these ddRADseq loci, 9299 were polymorphic (i.e., they had at least one SNP) and held 19,525 SNPs (a mean of 2.1 SNPs per locus) and 4246 SSRs. Moreover, both samples shared 7346 of these ddRADseq loci with 15,792 SNPs (Table 2). Additionally, sample A and B shared 420 SSRs with different motifs of repetition and 16 of these SSRs were polymorphic (Table 2, Supplementary Table S1).

An analysis using the denovo_map.pl routine implemented in Stacks [33] allowed us to obtain a higher number of ddRADseq loci (approximately the double: 156,013 and 135,501 for sample A and B, respectively) and polymorphic markers than the with reference analysis. In this case, the definitive catalog contained 125,432 loci. Within these de novo ddRADseq loci, 18,951 were polymorphic, and held 33,313 SNPs in all (a mean of 1.8 SNPs per haplotype), as well as 1366 SSRs. Finally, the samples shared 14,423 loci, 25,778 SNPs and 55 polymorphic SSRs (Table 2; Supplementary Table S3).

Table 2. Comparison of ddRADseq loci and polymorphic markers identified in two samples using with reference and de novo analyses. Number of SNPs and SSRs markers discovered by with reference and de novo analyses with Protocol 1. Total: total discovered markers; Shared: markers shared between both samples.

Analysis	Total			Shared			
	SNPs	loci	SSRs	SNPs	loci	SSRs	SSRs Polym.
with reference	19,525	9299	4246	15,792	7346	420	16
de novo	33,313	18,951	7717	25,778	14,423	1366	55

Dinucleotides (AG/GA> AT/TA> TC/CT) were the most frequent motifs observed in both cases (SSRs discovered by with reference (16 SSRs) or de novo (55 SSRs) analysis), followed by tetra and trinucleotides (approximate 15:5:1 respectively). At least 30 SSRs were polymorphic in a heterozygous state (20 without reference analysis). According to the with reference analysis, polymorphic SSRs were distributed in all chromosomes, except for chromosome 3 and 9 (Supplementary Table S3).

3.3. Protocol 2: Analysis in 24 Samples (Scaling-Up)

The demultiplexing of the 24-plex pool sequenced on NextSeq platform retrieved 27,400,302 good quality PE reads, with a mean of 1,141,679.25 PE reads per sample. This number varied from 404,702 for sample 1 to 2,280,731 for sample 18, with a standard deviation of 440,542.6 and a variant coefficient (VC) of 0.39 (Figure 4a; Supplementary Table S1). Of these reads, a mean of 82.39% was successfully mapped against the *E. grandis* genome. The mean ddRADseq loci number per sample was 68,622.

This result doubles the expected value according to our in silico prediction (34,634). This loci number also varied between 31,733 and 110,951 per sample, and six samples showed more than 80,000 loci (Figure 4). This loci number variation per sample shows a higher correlation with the number of reads per samples (r^2: 0.8742) than with the mean coverage per sample (r^2: 0.3654). The overall depth of coverage was 11.56 × (sd: 2.44, Supplementary Table S1). We identified 138,624 SNPs in 62,487 polymorphic loci. After applying filters of MAF 0.05 and 20% of missing data, we obtained 16,371 SNPs distributed in 9,466 ddRADseq loci, with a mean of 1.73 SNPs per *locus*. Of these SNPs, 15,950 were located through all the 11 chromosomes of the *E. grandis* genome (Figure 5), whereas the rest were located in the scaffolds, and thus were discarded from further analysis.

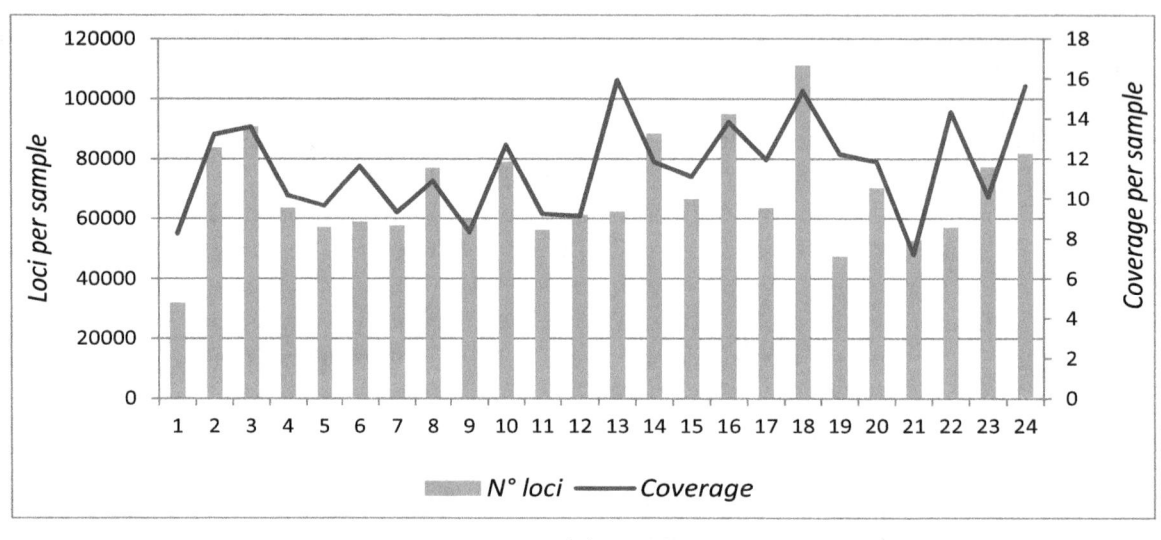

(a)

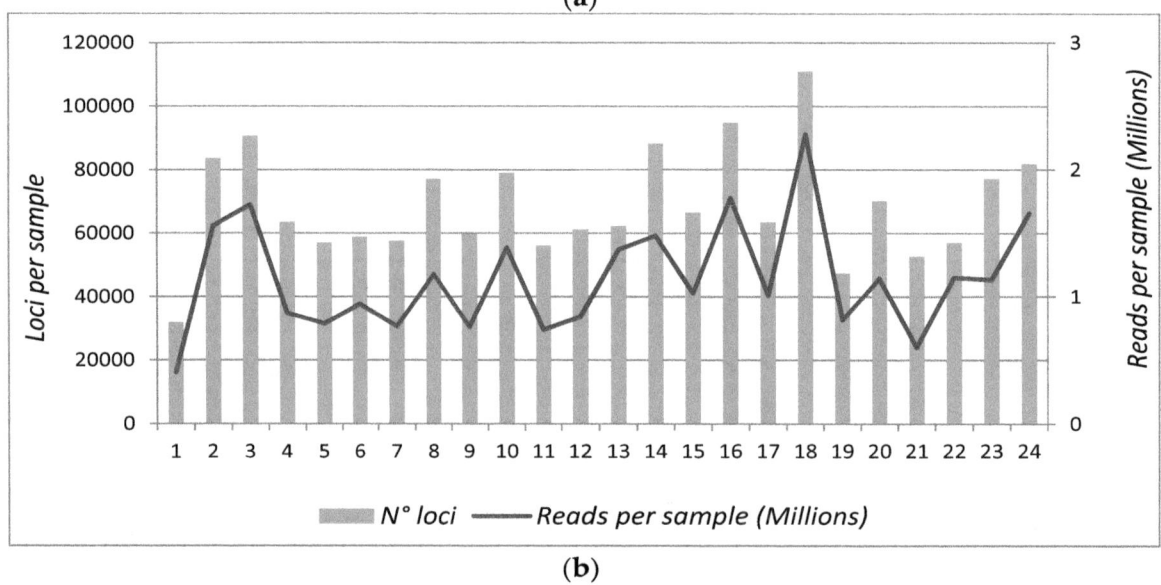

(b)

Figure 4. Data of 24 samples sequenced on NextSeq: Number of loci per sample compared with: (**a**) number of reads per sample and (**b**) mean depth of coverage (×) per sample.

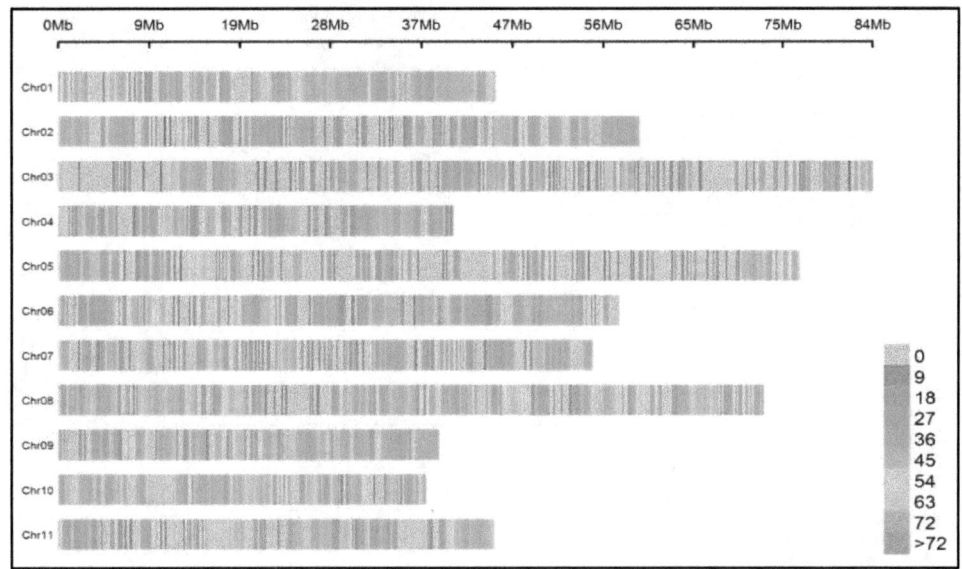

Figure 5. Distribution of 15,950 SNPs in the 11 chromosomes of *E. grandis* reference genome, NextSeq 24-plex run (1 Mb window).

3.4. Evaluation of Robustness—Sequencing Platform Comparison

The 23-plex pool of libraries was sequenced using MiSeq (Illumina Inc.) in low coverage (4.49×, with a range between 3.94 and 5.32×). From the overall number of 138,403 PE reads that were obtained per sample, 85% mapped successfully against the *E. grandis* reference genome. Subsequently, the 46 samples (23 replicates) of both pools sequenced in different platforms (NextSeq and MiSeq) gave 158,996 unfiltered SNPs in 294,212 loci with a mean of 16,807.63 loci per sample. However, after filtering them by quality with the rxstacks correction module, MAF lower than 0.05, and 20% missing data, a total of 1051 SNPs in 702 ddRADseq loci were kept. This final SNP matrix was used to construct a dendrogram (Figure 6). All replicates clustered together, with a dissimilarity coefficient lower than 0.05. This dissimilarity can be explained by the 20% missing data (mostly in MiSeq data, because of the low sequencing coverage), the expected error rate of sequencing and the differences between sequencing technologies. In addition, half-sibs (i.e., family samples 222, 247, 262) had the lowest dissimilarity coefficients (below 0.17), in accordance with the expected close relationships within families.

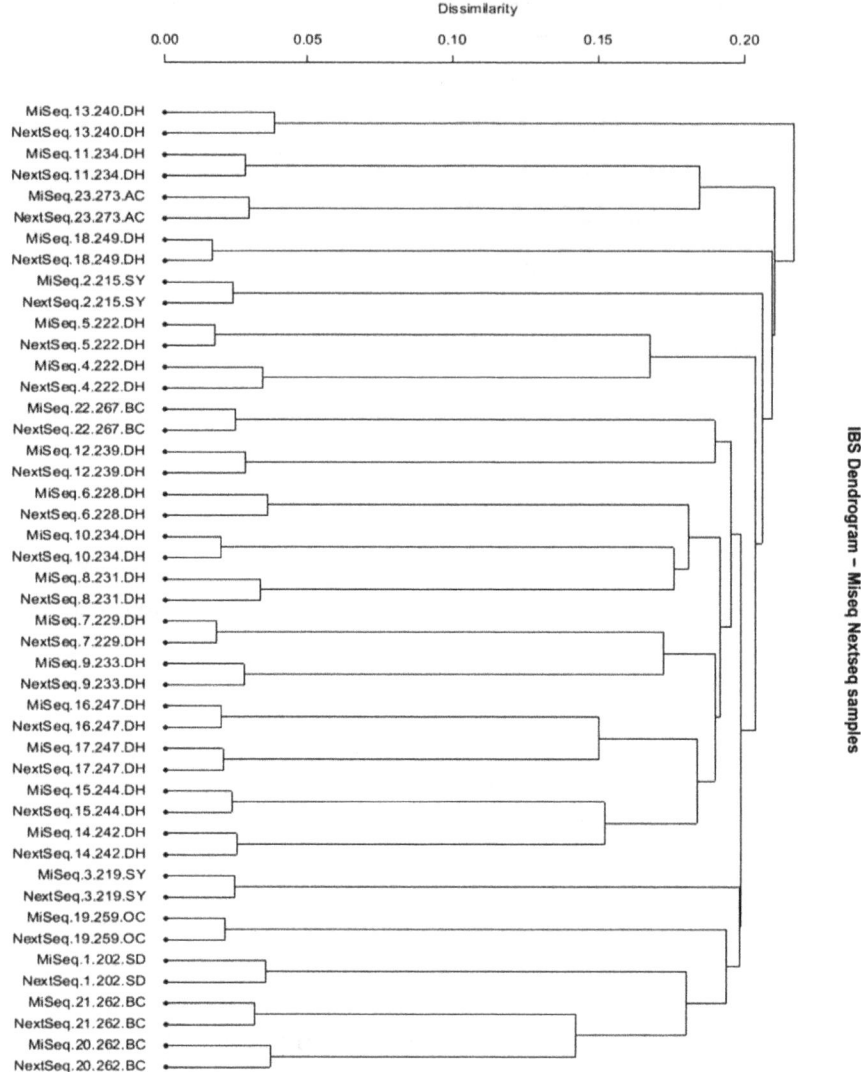

Figure 6. Dendrogram for the combined MiSeq and NextSeq dataset of 23 *E. dunnii* samples with a MAF of 0.05 and missing data below 20%. Each of 23 individuals has two sets of ddRADseq SNPs data: one set sequenced on a MiSeq and another sequenced on a NextSeq 500.

4. Discussion

Restriction-site associated DNA sequencing methodologies are becoming the most popular strategies in genomic data generation for a variety of applications related to crop and tree species breeding and genetics [15]. Nevertheless, for the *Eucalyptus* genus, the use of RADseq-derived methods is scarce. To date, the easy access to the commercial SNP array (EUChip60K) has led researchers to use it in the analysis of the genus [41–44], rather than RADseq-derived methods. Some species are poorly or less frequently represented in this chip than *E. grandis* (which is represented for its economic importance), and for this reason, the population allele frequencies and genetic relationships between individuals can be affected [45–47]. RADseq and GBS-based methodologies have the potential to avoid this type of bias [30], but the experience in *Eucalyptus* reported to date is not encouraging. Indeed, Duran et al. [41] applied GBS on 500 *E. globulus* individuals and only obtained 2597 polymorphic SNPs between them. The low number of markers suggests that the protocol should be improved for this genus in order to get enough whole genome coverage markers to perform population-level studies, such as genome-wide association mapping and genomic selection, among others. In addition, there are technical inconveniences associated with GBS and its derived protocols, since these can only enrich populations of sequenced DNA fragments that are below ~350 bp. Moreover, these protocols also result in high levels of missing data (reviewed in [48] and [49]). RADseq, on the other hand, involves more steps and equipment, as well as higher quantities of initial gDNA, and shows high read depth variation.

This study describes the optimization of a ddRADseq derived protocol for *E. dunnii* genotyping. Unlike these last methods, ddRADseq uses two enzymes and reduces the subset of sampled fragments, which allows higher reproducibility, greater loci coverage, larger fragment sizes and more effective SNPs [9,49]. Although allele dropout increases in comparison with RADseq [5], all the mentioned characteristics make ddRADseq genotyping a putatively more appropriate strategy.

With this in mind, we developed a modified protocol for achieving an optimized low-cost ddRADseq (P1) for plant species. This protocol was set up with a small sample (only two individuals), and thereafter was scaled up to perform on larger population analyses (P2).

Like in Peterson et al.'s protocol [16], the ligation step in P1 involves universal adapters, and PCR is performed before pooling the samples. Thus, the size selection turns out to be the last step of the protocol. The development of the P1 involved the analysis of different enzyme combinations and three concentrations and proportions of the adapters. Additionally, instead of the 6 bp-length Index Illumina primers, we used 16 forward primers and 96 reverse primers with an 8 bp-length index, which were designed by Lange et al. [29]. This change allowed higher multiplexing of samples, not only for library construction, but also for sequencing (up to 1536 samples). Moreover, we changed automatic size selection for manual size selection by agarose gel. All these modifications allowed P1 to be easily applicable on a very low number of *E. dunnii* samples and with minimal cost. Thus, this strategy may be extrapolated to other plant species, becoming an attractive tool for low-budget labs.

Despite its potential utility for setting up a ddRADseq protocol in any plant species, P1 involved the management of each sample independently until almost the end of the protocol, precluding its use with a large number of samples. In this regard, we propose P2 as a scale-up of P1. The addition of 24 adapters with barcodes in the ligation step allowed the pooling of samples and the application of size selection before PCR, as reported in ddRADseq [9]. Moreover, the use of different barcode lengths (4 to 9 bp, Poland et al. [27]) allowed us to avoid sequencing phasing error at the beginning of reads, as reported in GBS and MiddRADseq [3,17], but not considered in the original ddRADseq [9]. For this scaled-up P2, we also proposed the use of automatic size selection, which would decrease the possibility of cross-contamination and increase the precision and consistency when applying the protocol for more than one pool of samples [5], as reported in the original ddRADseq [9].

The first and maybe most critical step in every RADseq method is obtaining good quality, quantity and integrity of DNA material. Even ddRADseq has this high quality gDNA requirement [49]. Thus, gDNA extraction has to be done with a method that ensures gDNA integrity, and this integrity must be checked (e.g., by using a Nanodrop®-type spectrophotometer). gDNA needs to be quantified through a sensitive method such as Qubit® (Thermofisher). For instance, if gDNA is degraded, or if the quantity is insufficient, the results may retrieve higher VC between the read numbers obtained for each sample. With the CTAB DNA extraction protocol, we were able to reach the required gDNA integrity and quantity (See Supplementary File S1) [50]. However, high concentrations of good quality gDNA are not always easy to achieve in all species. In this regard, the initial amount of gDNA needed for the protocol is something to be considered. Whereas some ddRADseq-derived protocols rely on a high amount of starting material (e.g., 1000 ng [51]), our protocol requires minimal quantities (only 150 ng). The VC obtained for our samples (0.39) is lower, but at the same order of magnitude, than the ones reported for other ddRADseq approaches (e.g., 0.42, [28], 0.47, [51]). Moreover, the obtained VC is clearly influenced by the *E. dunnii*-1.202.SD sample, which has the lowest number of sequenced reads, and thus the lowest amount of genotyped markers.

Regarding the criteria for enzyme selection, some authors have proposed selecting enzymes for a specific species based only on in silico prediction, whereas others have suggested using universal enzymes (e.g., after doing an in silico evaluation of many enzymes). For example, Yang et al. [17] reported the use of the single universal pair AvaII-MspI for all angiosperms, which include *Eucalyptus*.

Based on our results, the evaluation of enzyme combinations (one frequent and one rare cutter) through both in silico and in vitro methods is an essential step in optimizing ddRADseq in new species (e.g., [52]). Owing to the absence of a reference genome for *E. dunnii*, we used the reference genome of a species of the same genus (*E. grandis*) for in silico prediction instead. Nevertheless, if the species under study lacks a reference genome (or a species that may be taken as reference because of its close proximity), the in silico prediction can also be done based on other information such as GC content and genome size [27].

In this study, the combination SphI-MboI showed a homogeneous digestion profile with a high number of fragments in the size selection range evaluated by both in silico and in vitro digestions, in comparison with PstI-MspI. Therefore, we selected SphI-MboI for the subsequent steps.

Another critical step to be adjusted is the size selection range window. First, according to previous studies (i.e., [28,51]), if the size selection for a RADseq-derived protocol is done in gel, and with a 100 bp ladder, the use of multiple ranges of 50 bp or 100 bp is advisable in order to minimize hand excision errors. That is why we evaluated (in silico) windows of 100 bp or 150 bp in a range between 220 and 470 bp for insert DNA fragments of interest within a range of 350 to 600 bp, when manual size selection was performed in P1. This inconvenience does not occur when using automatic size selection equipment. However, the amplitude of the range is delimited by the capacity of the equipment used in this study (i.e., in Sage ELF 2%, the range of each size selected correctly is around 70 bp, sagescience.com). This last window size is comparable to the "wide" size selection (72 bp) applied in the original ddRADseq protocol [9].

On the other hand, final library fragment sizes should not be too small (i.e., <200 bp) to avoid overlapping of the PE sequences, which should result in SNPs overestimation when doing de novo and with reference analyses. This is because Stacks v1.48 considers PE reads as independent loci (i.e., the software does not perform contig assembly in overlapping reads, [32]). Neither should the fragments be too long (i.e., >800 bp), because long fragments retrieve lower base quality in Illumina PE sequencing [53].

In comparison with MiddRADseq [17], we also used in silico prediction to evaluate the size selection. However, while those researchers used a window size of 300 bp (400–700 bp), we selected a narrower window of 70 or 100 bp (in the range of 320–420 bp in the manual selection and a mean of 370 bp for the automated selection). In a recent publication, Kess et al. [51] reported the use of a 300-bp window size. The use of thinner ranges avoids potential PCR amplification bias that would increase when using fragments with different lengths, while declining data quantity and quality [54,55]. Moreover, fewer reads per sample are needed to reach an optimal mean coverage per *locus*.

In terms of the number of ddRADseq loci generated with P1, we obtained 50% more loci per sample than the predicted loci (74,640 mean loci obtained per sample and 49,016 expected loci). Moreover, after filtering the catalog by rxstacks, we obtained a better correlation, with only 15% fewer loci than the predicted loci (41,834 ddRADseq loci). On the other hand, by using P2, we obtained a mean of 68,622.38 ddRADseq loci per sample. This result doubles the expected in silico prediction (34,634.00). According to Scaglione et al. [28], this phenomenon may be due to the stochastic possibility of each individual to yield loci that are out of the target [32]. In fact, ddRADseq loci present variability between samples, showing a higher correlation with the number of reads per samples (r^2: 0.8742) than with the mean coverage per sample (r^2: 0.3654). Moreover, the differences in genome sizes between the species and in the genome structures should also be considered. Indeed, only around 82% of the reads of *E. dunnii* were successfully mapped against the *E. grandis* reference genome.

With regard to the size selection methodology, we used both manual and automatic size selection. For P1, we applied the manual excision in agarose electrophoresis gels to reach a low-cost methodology, as in Scaglione et al. [28] and MiddRADseq [17], whereas for P2 we used the SAGE ELF device. In most publications, researchers use Pippin-prep as the automatic method of choice (e.g., [9]). However, we used SAGE ELF. We selected this method because it is easier to set up and gives tighter and higher DNA recovery in comparison with BluePippin [56]. As expected, the implementation of the manual size section (P1) method compared to the automatic one (P2) showed that the automatic method resulted in few recovered loci (74,640 vs. 68,622 loci). The lower number of loci is due to the restricted size range used in P2.

Another critical point for the set-up is the optimal concentration of adapters. In this work, we tested different concentrations (data not shown), and finally chose similar concentrations to those reported in Scaglione et al. [28]. Some protocols assess different adapter concentrations by titration [3,9]. Nevertheless, this procedure is not required for species with genomes below 20 Gb, such as *Eucalyptus* [17]. An excess of adapters can be used for proper ligation with DNA fragments, as in our protocols. Moreover, we used a Y-adapter form for the common restriction site. This generates ddRADseq libraries where Adapter 1 and Adapter 2 are on opposite ends of every amplified fragment. This type of library construction can reduce complexity [9,16,17,30]. Our P1 only requires a pair of adapters per set of enzymes, thus avoiding a substantial investment of funds at the beginning of the assay.

The P2 includes adapters with barcodes, as in the original ddRADseq and MiddRADseq protocols [9,17]. This addition simplifies downstream steps. For instance, many samples can be pooled in the same library, thus reducing the number of simultaneous reactions to just one. In addition, we specifically used 24 barcodes of different length designed for two-enzyme GBS protocol [30], as proposed in the original GBS protocol [3], and as performed in MiddRADseq [17], but not in the original ddRADseq [9]. The use of barcodes with different length avoids phasing error (low sequence quality). This error occurs when all the bases at the beginning of reads are the same in all clusters (Illumina sequencing) because of the restriction site. In P1, we solved this problem by using at least 5% of PhiX, as described in Peterson et al.'s protocol [16], or by mixing the ddRADseq libraries in the same sequencing run with other types of libraries with greater nucleotide variability in the first bases.

In the PCR step, an extra level of multiplexing can be achieved by using forward and reverse indexes that allow the inclusion of more libraries in the same sequencing lane. This is one of the main singularities of our protocols. In both protocols, we used the dual indexes developed by Lange et al. [29] within the PCR primers forward (16) and reverse (96). These combinatorial indexes allowed us to multiplex almost 1536 samples/libraries in the same lane. In this sense, our protocols are not limited by the number of Illumina Indexes, as in other ddRADseq methods [9,16,17]. Lower throughput sequencers (e.g., MiSeq, Illumina Inc.) may not support pooling such large numbers of libraries. By contrast, the use of higher throughput sequencers usually requires the capability of multiplexing to reduce budgets. With P2, we would be able to multiplex up to 36,864 samples (24 barcodes × 1536 primers). For example, we would be able to run them in low depth on a NovaSeq sequencer, which gives a maximum number of reads of 20 billion (for a dual S4 flow cell run on the NovaSeq 6000 System, Illumina Inc.).

Due to the absence of a reference genome for *E. dunnii*, we decided to work with Stacks [32] on both strategies, de novo and with reference analyses (called ref_map.pl and denovo_map.pl, respectively, in the software), to compare the obtained results in P1. Thus, we were able to apply both analyses to identify SNPs and SSRs with high accuracy after applying stringent bioinformatics settings and quality filters (Supplementary File S2). As expected, the de novo analysis retrieved more ddRADseq loci and markers, as all the reads were considered for marker identification, than with reference analysis, which only considered the reads that mapped against the reference genome (82% of the reads).

Both protocols achieved an optimal coverage (10–20× [5]), and consequently these can be efficiently used for a confident de novo loci calling. However, this strategy requires more stringent criteria and parameters when defining the loci, because of the larger number of false positives obtained using this method [57]. Thus, for further evaluation, we worked with the SNPs called using the *E. grandis* reference. Using the information of samples A and B generated with P1, we identified 7346 SNPs shared between the two samples. When applying P2 on 24 *E. dunnii* individuals, and after discarding the markers with high percentage of missing data, we identified and physically mapped 15,950 SNPs. These markers showed homogeneous distribution in the chromosomes. Even though a higher number of SNPs (138,624) was identified using P2 data. Based on our previous experience with imputation strategies for ddRADseq data [58], we applied a 20% missing data cut-off before performing further analysis.

We identified a mean of 1.95 SNPs within 145 bp between individuals A and B of *E. dunnii* with P1 (1 SNP each 74 bp) and 1.73 SNPs within 66 bp through 24 individuals with P2 (1 SNP each 38 bp). The difference between P1 and P2 also relies on the different number of samples tested between protocols (two individuals vs. 24, respectively). This causes P2 to yield higher polymorphism frequencies (or SNPs *locus* density). Even though there is no reported information for *E. dunnii*, these frequencies are in the same range than those observed for other species from the *Eucalyptus* genus. Indeed, Hendre et al. [59] reported 1 SNP per 65 pb in introns and per 108 pb in exons in *E. camaldulensis*, whereas Külheim et al. [60] detected 1 SNP in every 33 bp, 31 bp, 16 bp and 17 bp for *E. nitens*, *E. globulus*, *E. camaldulensis* and *E. loxophleba*, respectively.

The cross-platform compatibility of the obtained SNPs and the robustness in the ddRADseq derived SNPs calling are critical, but these issues have been studied less. Only one report [61] describes the assessment of the performance of Hiseq and NextSeq for ddRADseq-derived identification of SNPs in the butterfly genus. In our study, we sequenced the same 23 samples with both MiSeq and NextSeq sequencing platforms. As expected, the 23-sample NextSeq library (P2) recovered more loci than the 23-sample MiSeq library, with an overall higher mean read depth per *locus* and less missing data. This is attributed to the lower sequenced depth used in MiSeq data vs. NextSeq (4.49× vs. 11.56×, respectively). Both sequencing platforms achieve a high quality of data, according to FastqC report. More than 96% of the generated reads passed the Stacks quality filters and were kept for subsequent analysis. The low dissimilarity coefficient values between replicates (0.05) confirmed high reliability, despite the differences between the libraries' constructions and sequencing platforms.

P1 may be used in the first steps when applying a GBS/ddRADseq methodology in a laboratory. Its low cost relies mainly on the use of universal adapters for each enzyme, such as those used by Peterson et al. [16], the use of primers with 1536 combinatorial dual-index and the performing of a final size selection by agarose gel electrophoresis. Moreover, depending on the research focus, the generated sequences for a small number of samples (at least two) could be enough to obtain new marker information. It is interesting to notice that, whereas the cost per SNP genotyped in an array or an NGS derived technique falls when the number of interrogated SNPs rises, not all genomic studies relies on genotyping of a high number of markers. Because of the cost balance, many population studies, mainly related to conservation and evolution, give priority to raising the number of individuals sampled, rather than to adding more markers. A good example of this is the use of sequences for species-specific SSR identification, and even more so, for polymorphic SSRs and the heterozygous state of an individual. RADseq methods involve NGS, and the reads can consequently be used to design primers. These SSRs could then be used for population fingerprinting by using another genotyping strategy like fluorescent capillary electrophoresis. By using with reference genome analysis, we successfully identified SSRs (420 putative SSRs and 16 polymorphic in almost all chromosomes) using MISA [37] based on P1 data (sample A and B). A more comprehensive analysis of SSR identification using ddRADseq data can be found in a previous publication [38].

In summary, after setting an initial protocol P1 for the species of interest, P2 can be used for scaling up. The incorporation of adapters with custom-designed barcodes compatible with the enzyme restriction sites can make the method faster. This incorporation allowed us to pool 24 samples in the same library. This early barcoding simplified the following steps in the protocol. As with the original ddRADseq protocol, the approach described here can be used with a range of different restriction enzymes to produce a higher or lower complexity reduction of the genome being assayed.

5. Conclusions

The combined or individual use of our two protocols (P1 for setting up in a low number of samples and P2 for scaling up the number of samples) presented here show the pros of similar reported protocols but diminishes the drawbacks. Furthermore, the advantages of RADseq-derived methods, such as de novo marker discovery and removal of ascertainment bias in new germplasm, may make the ddRADseq technology one of the most promising genotyping approaches in the future.

Supplementary Materials: Supplementary Materials: Supplementary _ File_S1_Protocol1_ Aguirre _et _al_ 2019.docx: Protocol 1: Optimized ddRADseq (setting it up in newspecies with a low number of samples). Supplementary_File_S2_Command _lines_Aguirre_et_al_2019.docx: Command lines for performing in silico digestion of the reference genome and for ddRADseq data analysis. Supplementary_File_S3_Protocol2_Aguirre_ et_al_2019.docx: Protocol 2: Optimized ddRADseq (scaling up to a higher number of samples). Supplementary_ Tables_Aguirre_et_al_2019.xlsx: Table S1. General information of the two *Eucalyptus dunnii* individuals used in Protocol 1 and the 24 *E. dunnii* individuals used in Protocol 2. ddRADsequencing results (number of generated reads, % of reads mapping against the *E. grandis* reference genome), mean coverage per sample. Results of the analysis achieved using Stacks (Catchen et al. 2013). Table S2. List and sequences of the universal adapters, barcoded adapters and sequencing primers used in the optimized ddRADseq Protocols 1 and 2. Table S3. Polymorphic SSR identified in two *E. dunnii* ddRADseq data, considering with and without reference genome analysis. Figure S1. Schematic representation of the Final Library construction for both Protocol 1 and 2.

Author Contributions: Conceptualization, N.C.A., C.V.F., G.Z., M.C.M., H.E.H., N.B.P. and S.N.M.P.; methodology, N.C.A., C.V.F., G.Z., J.G.R., C.V.A., P.V.V. M.N.G. and M.C.M.; software and data analysis, N.C.A., C.V.F., G.Z., S.G. and M.R.; validation, N.C.A., C.V.F., G.Z., J.G.R., C.V.A., P.V.V., and M.N.G.; formal analysis, N.C.A., C.V.F., G.Z., A.F.P., and N.B.P.; investigation, N.C.A., C.V.F., G.Z., J.G.R., C.V.A., P.V.V., M.N.G., S.G., M.R., M.C.M., A.F.P., N.B.P. and S.N.M.P.; resources, M.M., H.E.H., N.B.P. and S.N.M.P.; data curation, N.C.A., C.V.F., G.Z., S.G., and M.R.; writing—original draft preparation, N.C.A., C.V.F., M.C.M., H.E.H., N.B.P. and S.N.M.P.; writing—review and editing, N.C.A., C.V.F., G.Z., M.C.M., M.M., H.E.H., N.B.P. and S.N.M.P.; visualization, N.C.A., C.V.F., G.Z., J.G.R., C.V.A., P.V.V., M.N.G., S.G., M.R., M.C.M., A.F.P., M.M., H.E.H., N.B.P. and S.N.M.P. supervision, M.C.M., A.F.P., G.Z., N.B.P. and S.N.M.P. project administration, A.F.P., N.B.P. and S.N.M.P.; funding acquisition, M.C.M., M.M., H.E.H., N.B.P. and S.N.M.P.

Acknowledgments: Special thanks to Juan A. Lópz, Martín A. Marcó, Javier Obershelp for selecting and collecting the plant material and Julieta M. Aguirre for designing Figure 1 and Julia Sabio y García for English corrections. Sequencing services using MiSeq (Illumina Inc.) were performed at INTA, Consorcio Argentino de Tecnología Genómica (CATG) funded by grant MinCyT PPL 2011 004 Genómica. This work used computational resources from BioCAD–Instituto de Biotecnología CICVyA, INTA, Consorcio Argentino de Tecnología Genómica, MinCyT PPL 2011 004; AECID PCI_ARG109, D/024562/09. We would like to thank the editor and reviewers for careful reading, and constructive suggestions.

References

1. Torkamaneh, D.; Boyle, B.; Belzile, F. Efficient Genome wide genotyping strategies and data integration in crop plants. *Theor. Appl. Genet.* **2018**, *131*, 499. [CrossRef] [PubMed]

2. Baird, N.A.; Etter, P.D.; Atwood, T.S.; Currey, M.C.; Shiver, A.L.; Lewis, Z.A.; Selker, E.U.; Cresko, W.A.; Johnson, E.A. Rapid SNP discovery and genetic mapping using sequenced RAD markers. *PLoS ONE* **2008**, *3*, e3376. [CrossRef] [PubMed]

3. Elshire, R.J.; Glaubitz, J.C.; Sun, Q.; Poland, J.A.; Kawamoto, K.; Buckler, E.S.; Mitchell, S.E. A robust, simple genotyping-by-sequencing (GBS) approach for high diversity species. *PLoS ONE* **2011**, *6*, e19379. [CrossRef] [PubMed]

4. Davey, J.W.; Hohenlohe, P.A.; Etter, P.D.; Boone, J.Q.; Catchen, J.M.; Blaxter, M.L. Genome-wide genetic marker discovery and genotyping using next-generation sequencing. *Nat. Rev. Genet.* **2011**, *12*, 499–510. [CrossRef] [PubMed]

5. Andrews, K.R.; Good, J.M.; Miller, M.R.; Luikart, G.; Hohenlohe, P.A. Harnessing the power of RADseq for ecological and evolutionary genomics. *Nat. Rev. Genet.* **2016**, *17*, 81–92. [CrossRef] [PubMed]

6. Timm, H.; Weigand, H.; Weiss, M.; Leese, F.; Rahmann, S. DDRAGE: A data set generator to evaluate ddRADseq analysis software. *Mol. Ecol. Resour.* **2018**, *18*, 681–690. [CrossRef]

7. Wang, S.; Meyer, E.; Mckay, J.K.; Matz, M.V. 2b-RAD: A simple and flexible method for genome-wide genotyping. *Nat. Methods* **2012**, *9*, 808–810. [CrossRef]

8. Toonen, R.J.; Puritz, J.B.; Forsman, Z.H.; Whitney, J.L.; Fernandez-Silva, I.; Andrews, K.R.; Bird, C.E. ezRAD: A simplified method for genomic genotyping in non-model organisms. *PeerJ* **2013**, *1*, e203. [CrossRef]

9.   Peterson, B.K.; Weber, J.N.; Kay, E.H.; Fisher, H.S.; Hoekstra, H.E. Double digest RADseq: An inexpensive method for *de novo* SNP discovery and genotyping in model and non-model species. *PLoS ONE* **2012**, *7*, e37135. [CrossRef]

10.   Nazareno, A.G.; Bemmels, J.B.; Dick, C.W.; Lohmann, L.G. Minimun sample sizes for population genomics: An empirical study from an Amazonian plant species. *Mol. Ecol. Resour.* **2017**, *17*, 1136–1147. [CrossRef] [PubMed]

11.   Pyne, R.; Honig, J.; Vaiciunas, J.; Koroch, A.; Wyenandt, C.; Bonos, S.; Simon, J. A first linkage map and downy mildew resistance QTL discovery for sweet basil (Ocimum basilicum) facilitated by double digestion restriction site associated DNA sequencing (ddRADseq). *PLoS ONE* **2017**, *12*, e0184319. [CrossRef] [PubMed]

12.   Roy, S.C.; Moitra, K.; De Sarker, D. Assessment of genetic diversity among four orchids based on ddRAD sequencing data for conservation purposes. *Physiol. Mol. Biol. Plants* **2017**, *23*, 169–183. [CrossRef] [PubMed]

13.   Vargas, O.M.; Ortiz, E.M.; Simpson, B.B. Conflicting phylogenomic signals reveal a pattern of reticulate evolution in a recent high-Andean diversification (Asteraceae: Astereae: Diplostephium). *New Phytol.* **2017**, *214*, 1736–1750. [CrossRef] [PubMed]

14.   Zhou, X.; Xia, Y.; Ren, X.; Chen, Y.; Huang, L.; Huang, S.; Liao, B.; Lei, Y.; Yan, L.; Jiang, H. Construction of SNP-based genetic linkage map in cultivated peanut basedon large scale marker development using next-generation double-digest restriction associated DNA sequencing (ddRADseq). *BMC Genom.* **2014**, *15*, 351. [CrossRef] [PubMed]

15.   Parchman, T.L.; Jahner, J.P.; Uckele, K.A.; Galland, L.M.; Eckert, A.J. RADseq approaches and applications for forest tree genetics. *Tree Genet. Genomes* **2018**, *14*. [CrossRef]

16.   Peterson, G.W.; Dong, Y.; Horbach, C.; Fu, Y.B. Genotyping-by-sequencing for plant genetic diversity analysis: A lab guide for SNP genotyping. *Diversity* **2014**, *6*, 665–680. [CrossRef]

17.   Yang, G.Q.; Chen, Y.M.; Wang, J.P.; Guo, C.; Zhao, L.; Wang, X.Y.; Guo, Y.; Li, L.; Li, D.Z.; Guo, Z.H. Development of a universal and simplified ddRAD library preparation approach for SNP discovery and genotyping in angiosperm plants. *Plant Methods* **2016**, *12*, 1–17. [CrossRef] [PubMed]

18.   Barchi, L.; Lanteri, S.; Portis, E.; Acquadro, A.; Valè, G.; Toppino, L.; Rotino, G.L. Identification of SNP and SSR markers in eggplant using RAD tag sequencing. *BMC Genom.* **2011**, *12*, 304. [CrossRef]

19.   Torales, S.L.; Rivarola, M.; Gonzalez, S.; Inza, M.V.; Pomponio, M.F.; Fernández, P.; Acuña, C.V.; Zelener, N.; Fornés, L.; Hopp, H.E.; et al. *De novo* transcriptome sequencing and SSR markers development for Cedrela balansae C.DC., a native tree species of northwest Argentina. *PLoS ONE* **2018**, *13*, e0203768. [CrossRef]

20.   Hodel, R.G.J.; Segovia-Salcedo, M.C.; Landis, J.B.; Crowl, A.A.; Sun, M.; Liu, X.; Gitzendanner, M.A.; Douglas, N.A.; Germain-Aubrey, C.C.; Chen, S.; et al. The report of my death was an exaggeration: A review for researchers using microsatellites in the 21st century. *Appl. Plant Sci.* **2016**, *4*, 1600025. [CrossRef]

21.   Silva-Junior, O.B.; Faria, D.A.; Grattapaglia, D. A flexible multi-species genome-wide 60K SNP chip developed from pooled resequencing of 240 *Eucalyptus* tree genomes across 12 species. *New Phytol.* **2015**, *206*, 1527–1540. [CrossRef] [PubMed]

22.   Sansaloni, C.P.; Petroli, C.D.; Carling, J.; Hudson, C.J.; Steane, D.A.; Myburg, A.A.; Grattapaglia, D.; Vaillancourt, R.E.; Kilian, A. A high-density Diversity Arrays Technology (DArT) microarray for genome-wide genotyping in Eucalyptus. *Plant Methods* **2010**, *6*, 1–11. [CrossRef] [PubMed]

23.   Grattapaglia, D.; de Alencar, S.; Pappas, G. Genome-wide genotyping and SNP discovery by ultra-deep Restriction-Associated DNA (RAD) tag sequencing of pooled samples of *E. grandis* and *E. globulus*. *BMC Proc.* **2011**, *5* (Suppl. 7), P45. [CrossRef]

24.   Hoisington, D.; Khairallah, M.; González-de-león, D. *Laboratory Protocols: CIMMYT Applied Molecular Genetics Laboratory*, 2nd ed.; CIMMYT: México, DF, Mexico, 1994.

25.   Marcucci Poltri, S.N.; Zelener, N.; Rodriguez Traverso, J.; Gelid, P.; Hopp, H.E. Selection of a seed orchard of *Eucalyptus dunnii* based on genetic diversity criteria calculated using molecular markers. *Tree Physiol.* **2003**, *23*, 625–632. [CrossRef] [PubMed]

26.   Myburg, A.A.; Grattapaglia, D.; Tuskan, G.A.; Hellsten, U.; Hayes, R.D.; Grimwood, J.; Jenkins, J.; Lindquist, E.; Tice, H.; Bauer, D. The genome of Eucalyptus grandis. *Nature* **2014**, *510*, 356–362. [CrossRef] [PubMed]

27.   Lepais, O.; Weir, J.T. SimRAD: An R package for simulation-based prediction of the number of *loci* expected in RADseq and similar genotyping by sequencing approach. *Mol. Ecol. Resour.* **2014**. [CrossRef] [PubMed]

28. Scaglione, D.; Fornasiero, A.; Pinto, C.; Cattonaro, F.; Spadotto, A.; Infante, R.; Meneses, C.; Messina, R.; Lain, O.; Cipriani, G.; et al. A RAD-based linkage map of kiwifruit (Actinidia chinensis Pl.) as a tool to improve the genome assembly and to scan the genomic region of the gender determinant for the marker-assisted breeding. *Tree Genet. Genomes* **2015**, *11*. [CrossRef]

29. Lange, V.; Böhme, I.; Hofmann, J.; Lang, K.; Sauter, J.; Schöne, B.; Paul, P.; Albrecht, V.; Andreas, J.M.; Baier, D.M. Cost-efficient high-throughput HLA typing by MiSeq amplicon sequencing. *BMC Genom.* **2014**, *15*. [CrossRef] [PubMed]

30. Poland, J.A.; Brown, P.J.; Sorrells, M.E.; Jannink, J.L. Development of high-density genetic maps for barley and wheat using a novel two-enzyme genotyping-by-sequencing approach. *PLoS ONE* **2012**, *7*, e32253. [CrossRef]

31. Andrews, S. FASTQC: A Quality Control Tool for High Throuput Sequencing Data. 2010. Available online: http://www.bioinformatics.babraham.ac.uk/projects/fastqc/ (accessed on 19 November 2015).

32. Catchen, J.; Hohenlohe, P.A.; Bassham, S.; Amores, A.; Cresko, W.A. Stacks: An analysis tool set for population genomics. *Mol. Ecol.* **2013**, *22*, 3124–3140. [CrossRef] [PubMed]

33. Catchen, J.M.; Amores, A.; Hohenlohe, P.; Cresko, W.; Postlethwait, J.H. Stacks: Building and Genotyping *Loci De Novo* From Short-Read Sequences. *G3 (Bethesda)* **2011**, *1*, 171–182. [CrossRef] [PubMed]

34. Torkamaneh, D.; Laroche, J.; Belzile, F. Genome-Wide SNP Calling from Genotyping by Sequencing (GBS) Data: A Comparison of Seven Pipelines and Two Sequencing Technologies. *PLoS ONE* **2016**, *11*, e0161333. [CrossRef] [PubMed]

35. Wickland, D.; Battu, G.; Hudson, K.A.; Diers, B.W.; Hudson, M.E. A comparison of genotyping-by-sequencing analysis methods on low-coverage crop datasets shows advantages of a new workflow, GB-eaSy. *BMC Bioinform.* **2017**, *18*, 586. [CrossRef] [PubMed]

36. Langmead, B.; Salzberg, S. Fast gapped-read alignment with Bowtie 2. *Nat. Methods* **2012**, *9*, 357–359. [CrossRef] [PubMed]

37. Thiel, T.; Michalek, W.; Varshney, R.; Graner, A. Exploiting EST databases for the development and characterization of gene-derived SSR-markers in barley (*Hordeum vulgare* L.). *Theor. Appl. Genet.* **2003**, *106*, 411–422. [CrossRef]

38. Qin, H.; Yang, G.; Provan, J.; Liu, J.; Gao, L. Using MiddRADseq data to develop polymorphic microsatellite markers for an endangered yew species. *Plant Divers.* **2017**, *39*, 294–299. [CrossRef]

39. Zheng, X.; Levine, D.; Shen, J.; Gogarten, S.M.; Laurie, C.; Weir, B.S. A High-performance Computing Toolset for Relatedness and Principal Component Analysis of SNP Data. *Bioinformatics* **2012**. [CrossRef]

40. Wickham, H. *ggplot2: Elegant Graphics for Data Analysis*; Springer: New York, NY, USA, 2016.

41. Durán, R.; Zapata-Valenzuela, J.; Balocchi, C.; Valenzuela, S. Efficiency of EUChip60K pipeline in fingerprinting clonal population of Eucalyptus globulus. *Trees* **2018**, *32*, 663. [CrossRef]

42. Cappa, E.P.; de Lima, B.M.; da Silva-Junior, O.B.; Garcia, C.C.; Mansfield, S.D.; Grattapaglia, D. Improving genomic prediction of growth and wood traits in Eucalyptus using phenotypes from non-genotyped trees by single-step GBLUP. *Plant Sci.* **2019**. [CrossRef]

43. Müller, B.S.F.; de Almeida Filho, J.E.; Lima, B.M.; Garcia, C.C.; Missiaggia, A.; Aguiar, A.M.; Takahashi, E.; Kirst, M.; Gezan, S.A.; Silva-Junior, O.B.; et al. Independent and Joint-GWAS for growth traits in Eucalyptus by assembling genome-wide data for 3373 individuals across four breeding populations. *New Phytol.* **2019**, *221*, 818–833. [CrossRef]

44. Suontama, M.; Klápště, J.; Telfer, E.; Graham, N.; Stovold, T.; Low, C.; McKinley, R.; Dungey, H. Efficiency of genomic prediction across two Eucalyptus nitens seed orchards with different selection histories. *Heredity* **2019**, *122*, 370–379. [CrossRef]

45. Albrechtsen, A.; Nielsen, F.C.; Nielsen, R. Ascertainment biases in SNP chips affect measures of population divergence. *Mol. Biol. Evol.* **2010**, *27*, 2534–2547. [CrossRef]

46. Bajgain, P.; Rouse, M.N.; Anderson, J.A. Comparing genotyping-by-sequencing and single nucleotide polymorphism chip genotyping for quantitative trait *loci* mapping in wheat. *Crop Sci.* **2016**, *56*, 232–248. [CrossRef]

47. Li, B.; Kimmel, M. Factors influencing ascertainment bias of microsatellite allele sizes: Impact on estimates of mutation rates. *Genetics* **2013**, *195*, 563–572. [CrossRef]

48. Poland, J.A.; Rife, T.W. Genotyping-by-Sequencing for Plant Breeding and Genetics. *Plant Genome* **2012**, *5*, 92–102. [CrossRef]

49. Scheben, A.; Batley, J.; Edwards, D. Genotyping-by-sequencing approaches to characterize crop genomes: choosing the right tool for the right application. *Plant Biotech. J.* **2017**, *15*, 149–161. [CrossRef]

50. Inglis, P.W.; Pappas, M.C.R.; Resende, L.V.; Grattapaglia, D. Fast and inexpensive protocols for consistent extraction of high quality DNA and RNA from challenging plant and fungal samples for high-throughput SNP genotyping and sequencing applications. *PLoS ONE* **2018**, *13*, e0206085. [CrossRef]

51. Kess, T.; Gross, J.; Harper, F.; Boulding, E.G. Low-cost ddRAD method of SNP discovery and genotyping applied to the periwinkleLittorina saxatilis. *J. Molluscan Stud.* **2016**, *82*, 104–109. [CrossRef]

52. Wang, Y.; Cao, X.; Zhao, Y.; Fei, J.; Hu, X.; Li, N. Optimized double-digest genotyping by sequencing (ddGBS) method with high-density SNP markers and high genotyping accuracy for chickens. *PLoS ONE* **2017**, *12*, e0179073. [CrossRef]

53. Tan, G.; Opitz, L.; Schlapbach, R.; Rehrauer, H. Long fragments achieve lower base quality in Illumina paired-end sequencing. *Sci. Rep.* **2019**, *9*, 2856. [CrossRef]

54. DaCosta, J.M.; Sorenson, M.D. Amplification biases and consistent recovery of loci in a double-digest RAD-seq protocol. *PLoS ONE* **2014**, *9*, e106713. [CrossRef]

55. Quail, M.A.; Kozarewa, I.; Smith, F.; Scally, A.; Stephens, P.J.; Durbin, R.; Swerdlow, H.; Turner, D.J. A large genome center's improvements to the Illumina sequencing system. *Nat. Methods* **2008**, *5*, 1005–1010. [CrossRef]

56. Heavens, D.; Garcia Accinelli, G.; Clavijo, B.; Clark, M.D. A method to simultaneously construct up to 12 differently sized Illumina Nextera long mate pair libraries with reduced DNA input, time, and cost. *BioTechniques* **2015**, *59*, 42–45. [CrossRef]

57. Rochette, N.C.; Catchen, J.M. Deriving genotypes from RAD-seq short-read data using Stacks. *Nat. Protoc.* **2017**, *12*, 2640–2659. [CrossRef]

58. Merino, G. Imputación de Genotipos Faltantes en Datos de Secuencación Masiva. Master's Thesis, Facultada de Ciencias Agrarias, Universidad Nacional de Córdoba, Córdoba, Argentina, 2018.

59. Hendre, P.S.; Kamalakannan, R.; Rajkumar, R.; Varghese, M. High-throughput targeted SNP discovery using Next Generation Sequencing (NGS) in few selected candidate genes in *Eucalyptus camaldulensis*. *BMC Proc.* **2011**, *5*, O17. [CrossRef]

60. Külheim, C.; Hui Yeoh, S.; Maintz, J.; Foley, W.; Moran, G. Comparative SNP diversity among four Eucalyptus species for genes from secondary metabolite biosynthetic pathways. *BMC Genom.* **2009**, *10*, 452. [CrossRef]

61. Campbell, E.O.; Davis, C.S.; Dupuis, J.R.; Muirhead, K.; Sperling, F.A.H. Cross-platform compatibility of *de novo*-aligned SNPs in a nonmodel butterfly genus. *Mol. Ecol. Resour.* **2017**, *17*, e84–e93. [CrossRef]

8

Biotechnological and Digital Revolution for Climate-Smart Plant Breeding

Francesca Taranto [1],*, Alessandro Nicolia [2], Stefano Pavan [3,4], Pasquale De Vita [1] and Nunzio D'Agostino [2],*

[1] CREA Research Centre for Cereal and Industrial Crops, 71121 Foggia, Italy; pasquale.devita@crea.gov.it

[2] CREA Research Centre for Vegetable and Ornamental Crops, 84098 Pontecagnano Faiano, Italy; alessandro.nicolia@crea.gov.it

[3] Department of Soil, Plant and Food Sciences, University of Bari Aldo Moro, 70126 Bari, Italy; stefano.pavan@uniba.it

[4] Institute of Biomedical Technologies, National Research Council (CNR), 70126 Bari, Italy

* Correspondence: francesca.taranto@crea.gov.it (F.T.); nunzio.dagostino@crea.gov.it (N.D.A.)

Abstract: Climate change, associated with global warming, extreme weather events, and increasing incidence of weeds, pests and pathogens, is strongly influencing major cropping systems. In this challenging scenario, miscellaneous strategies are needed to expedite the rate of genetic gains with the purpose of developing novel varieties. Large plant breeding populations, efficient high-throughput technologies, big data management tools, and downstream biotechnology and molecular techniques are the pillars on which next generation breeding is based. In this review, we describe the toolbox the breeder has to face the challenges imposed by climate change, remark on the key role bioinformatics plays in the analysis and interpretation of big "omics" data, and acknowledge all the benefits that have been introduced into breeding strategies with the biotechnological and digital revolution.

Keywords: climate change; mapping populations; genetic resources; mutation breeding; genome editing; new plant breeding techniques; "omics" data; bioinformatics

1. Climate Change is Increasing Pressure on Crop Breeding

Climate change is strongly influencing agricultural production and cultivation practices of all major crops with various and heterogeneous effects, which critically depend on geographical areas [1]. The climate variables that directly affect agricultural production are the rapid growth in mean temperatures and the increasing frequency and magnitude of extreme weather events [2].

Water deficit is a growth- and yield-limiting factor for crops worldwide [3]. It has been reported that water scarcity deeply influences flowering, pollination, and grain-filling of most grain crops; on the other hand, abundant rainfalls may have a positive impact on yield and end-use quality, but they may damage plants because of higher relative humidity, which predisposes plants to the outbreak of diseases [4]. Drought also has a major impact on crop yield; however, it has been demonstrated that the severity of the stress depends on the phenological status of the plant [5,6].

Impact of extreme heat waves has been analyzed in wheat [7,8], rice [9], maize [10], and soybean [11]. It has been noted that an increase of 1 °C of seasonal temperatures determines a decrease in yield ranging from 7.4% in maize to 3.1% in soybean [12].

The increase of atmospheric CO_2 has conflicting effects on crops: On one hand, it determines an increase in plant photosynthesis and growth; on the other hand, it negatively affects the nutritional quality of crops as well as their health status [13].

As an example, an increase in barley yellow dwarf virus infections has been observed in wheat under elevated CO_2 levels [14].

Breeding crop varieties for environmental stresses is a slow and challenging process, as the effects of stresses on crops are variable and complex especially when crops are exposed to multiple stresses [4,15,16]. Although various information is available on plant response to a single stress factor, much less is the knowledge on the response mechanisms of crops when exposed to a combination of biotic and abiotic stresses (i.e., simultaneous stresses). Clearly, plant response depends on the combination of specific stresses, on the intensity of each stress, and on the plant developmental stage [17]. Studies demonstrated that plant stress and defense responses are controlled by different, and sometimes conflicting, signaling pathways and that the plant activates specific signaling cascades and metabolic pathways, which differ depending on whether the plant is subjected to individual or multiple stresses [16,18].

Drought, heat stress, and their combination on growth-related traits have been widely investigated. Several studies demonstrated the negative effect of simultaneous high temperature and drought on the growth, development, and reproduction of cereals, thus affecting productivity [19–21]. The combination of drought and salt stress also decreased yield potential in barley [22]. Elevated temperatures combined with drought reduced the performance of grapevine in the Mediterranean basin, but elevated levels of CO_2 could mitigate such damaging effects [23]. Photosynthesis was shown to be sensitive to drought or heat stress. As reported by Feller (2016), the interaction between water scarcity and heat stress affects carbon assimilation in crops. Indeed, leaf temperature, stomatal opening, and water status are strongly interconnected, suggesting a complex regulatory network underlying plant adaptation processes and coordinating gene expression [24].

All these developing threats are leading to an increase in the incidence of weeds, pests, and pathogens, which generally were confined in particular geographical areas. According to predictive models, it is expected that between 2050 and 2100, *Fusarium oxysporum* spp. (Schltdl., 1824) will be the main cause of plant disease in European, Middle Eastern, and North African regions, posing risks to a number of cash crops [25]. At present, they are thriving worldwide because of the simultaneous occurrence of warming temperatures, increasing levels of humidity, CO_2, and ozone levels [26–28]. High temperature and moisture increases the production and germination of propagules and accelerates pathogen growth rates. Elevated temperatures and ozone levels favor infection by necrotropic pathogens. Otherwise, high levels of CO_2, temperature, and drought foster plant colonization by biotrophic pathogens. As an example, *Fusarium* head blight (FHB) and *Septoria tritici* (Desm.) Blotch (STB) diseases in wheat are increasing in China [29], United Kingdom [30], and in several countries of the European Union due to the altered weather patterns [31]. On the other hand, Rejeb et al. [17] reported several examples of cross-tolerance between abiotic and biotic stresses that may induce positive effects and enhanced resistance in plants with significant implications in plant breeding. For instance, drought stress induced an increase of abscisic acid levels with a significant increase of resistance response towards necrotophic fungus *Botrytis cinerea* (Pers., 1794) and *Oidium neolycopersici* (Kiss, 2001), while salt stress reduced *O. neolycopersici* infection [32].

In this challenging scenario, it is clear that we need miscellaneous strategies to develop climate-resilient cultivars and expedite the rate of genetic gains [33]. The understanding of the physiological, genetic, and molecular mechanisms that allow plants to adapt and respond to climate change and the identification of adaptation traits to variable environmental conditions triggered by climate change are among the main objectives of next generation breeding.

Next generation breeding relies on the availability of large plant breeding populations and germplasm collections, efficient high-throughput technologies, big data management tools, and downstream biotechnology and molecular breeding activities. It is allowing and will allow the scientific community to define, in a short time frame, one or more ideotypes suitable to satisfy the breeding demand and to discover superior alleles and haplotypes to be used in breeding programs.

Furthermore, recent advances in genomic knowledge and the increasing availability of information on genes as well as on in vitro regeneration technologies allow the development and use of second-generation biotechnologies, based on cisgenesis and genome editing [34–36], to produce a diverse array of novel value-added products that may be indispensable in addressing future challenges associated with sustainable agriculture.

Genome editing can breathe new life into plant breeding strategies. Indeed, genome editing is opening up novel opportunities for the precise and rapid modification of crops to boost yields and protect against pests, diseases, and abiotic stressors [37–39]. The great potential of the genome editing techniques relies on making crop breeding faster, more precise, and at lower production costs.

In this review, we provide a brief overview of the possibility of exploiting germplasm resources with diverse allelic combinations for genetics research and breeding. Then, we discuss the most recent strategies, cutting-edge technologies, methods, and tools for adapting crops to climate change, and remark on the key role bioinformatics plays in the analysis and interpretation of big "omics" data. Finally, we acknowledge the benefits that have been introduced into breeding strategies through the biotechnological and digital revolution, and we stress the concept that a "new figure" of breeder, with new specializations, is needed.

2. Browsing through the Literature: Trends of the Most Recent and Breakthrough Technologies to Advance Climate-Smart Breeding

In the 1995, the Intergovernmental Panel on Climate Change (IPCC) released the Second Assessment Report on the impact of climate change on the sustainable development of the society. This document has laid the foundations for achieving the international agreement linked to the United Nations Framework Convention on Climate Change, known as the Kyoto Protocol [40]. The report by IPCC describes the assessment of impact, adaptation, and mitigation of climate change with regard to environmental and socio-economic aspects. Following the dissemination of the ideas contained in the document, a growing interest by the scientific community has been observed in the study of the causes and effects of climate change.

The number of published academic papers is a powerful indicator for measuring the development tendencies of certain scientific researches. Literature related to "climate change" is vast and covers several branches of knowledge such as agronomy, molecular biology, physiology, and socio-economic disciplines [41,42]. Janssen et al. [41] and Wang et al. [42] performed a bibliometric analysis to determine qualitative and quantitative changes in the scientific research topics related to the resilience, vulnerability, and adaptation to climate change without taking into consideration the extent to which climate change impacts on plant breeding.

In this review, we analyzed, quite simply, the number of publications in which the most recent, popular, and breakthrough technologies applied to plant breeding were associated or not with climate change. The analysis was based on the information available in the Web of Science database (www.webofknowledge.com), category "Plant science", considering the time interval of 2000–2018. Different keywords (i.e., "plant breeding", "QTL (Quantitative Trait Loci)", "association mapping" and "GWAS (Genome Wide Association Studies)", "genomic selection (GS)" and "GS", "genome editing" and "mutagenesis") and Boolean operators were used to query the database (Figure 1).

The results showed that the largest number of publications was retrieved using "QTL*" as a keyword (Figure 1B), while the least number of publications affects those documents that included "genome editing" as a keyword (Figure 1E). This trend reflects the recent history of technological advances and methodological innovations in plant breeding. QTL mapping, in fact, is the oldest method used in plant breeding to identify genetic variants that influence the magnitude of measurable traits [43]. On the other hand, genome editing techniques have been introduced much more recently to support plant breeding and require the development of specific protocols that widely vary from species to species.

All the technologies taken into consideration showed an upward trend across the years, particularly after 2013. By contrast, mutagenesis with the exclusion of genome editing (Figure 1F) was the only method with a more stable trend across the years. Table 1 reports the top ten list of the most cited scientific articles retrieved by combining, in a single query, all the keywords mentioned above (Figure 1A).

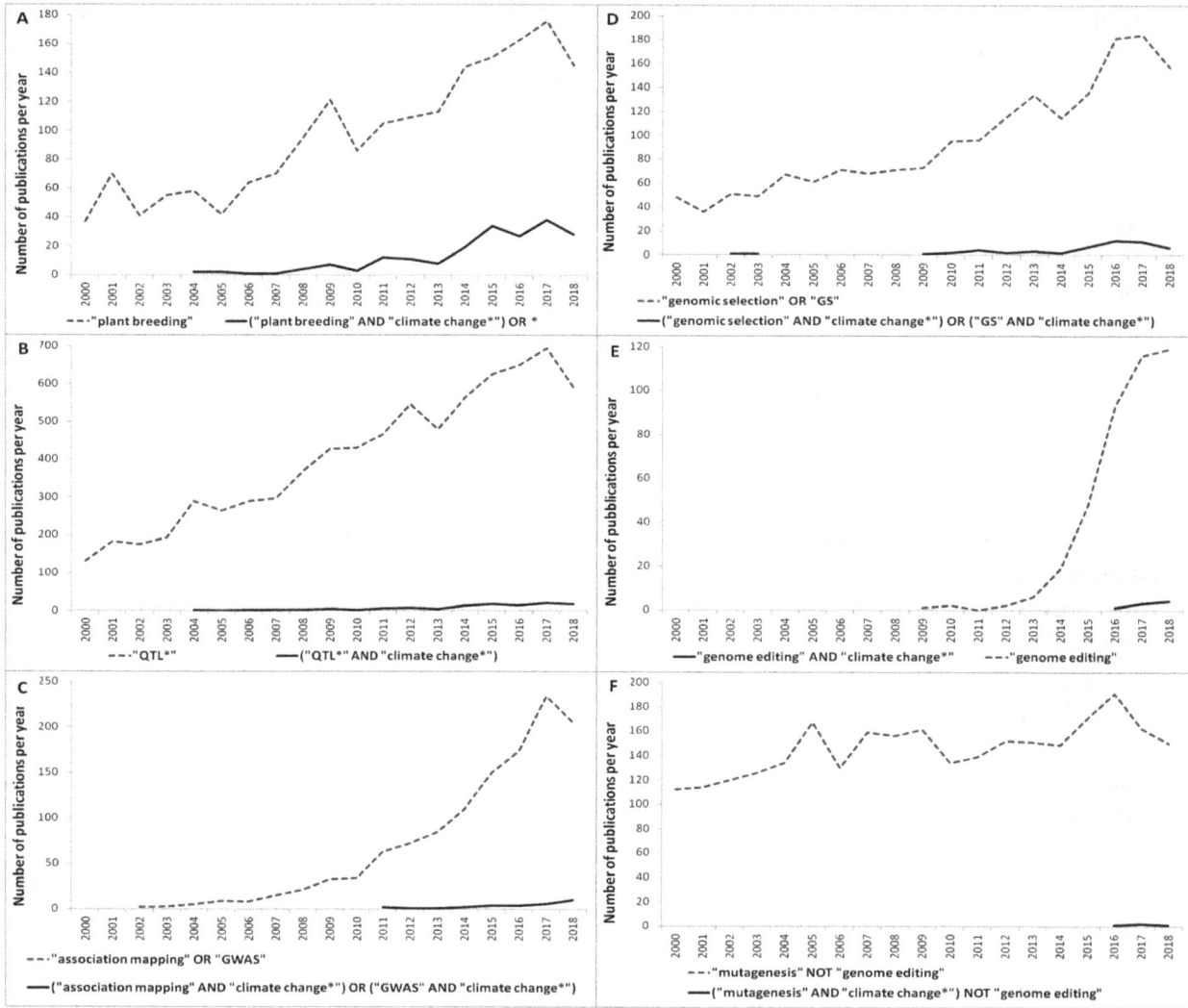

Figure 1. Number of publications in which the most recent and breakthrough technologies applied to plant breeding are associated (—) or not (---) with climate change. (**A**) Keyword: plant breeding; (**B**) keyword: QTL*; (**C**) keywords: association mapping or GWAS; (**D**) keywords: genomic selection or GS (**E**) keyword: genome editing; (**F**) keywords: mutagenesis NOT genome editing. * complete query = ("plant breeding" AND "climate change *") OR ("association mapping" AND "climate change*") OR ("gwas" AND "climate change*") OR ("genomic selection" AND "climate change*") OR ("mutagenesis" AND "climate change*") OR ("QTL*" AND "climate change*").

Table 1. Top ten list of the most cited scientific articles retrieved (up until 11/8/2018) from the Web of Science database, category "Plant science", using the following query: ([("plant breeding") AND ("climate change*")] OR [("association mapping") AND ("climate change*")] OR [("gwas") AND ("climate change*")] OR [("genomic selection") AND ("climate change*")] OR [("mutagenesis") AND ("climate change*")] OR [("QTL*") AND ("climate change*")]).

Reference Title	Journal	Publication Year	Total Citations
Genetic engineering for modern agriculture: challenges and perspectives [44]	Annual Review of Plant Biology	2010	356
Breeding for yield potential and stress adaptation in cereals [45]	Critical Reviewers in Plant Science	2008	278
Root system architecture: opportunities and constraints for genetic improvement of crops [46]	Trends in plant science	2007	255
The ozone component of global change: potential effects on agricultural and horticultural plant yield, product quality and interactions with invasive species [47]	Journal of Integrative Plant Biology	2009	156
Developments in breeding cereals for organic agriculture [48]	Euphytica	2008	147
Integrated genomics, physiology and breeding approaches for improving drought tolerance in crops [49]	Theoretical and Applied Genetics	2012	138
Genotyping-by-sequencing for plant breeding and genetics [50]	Plant Genome	2012	131
Climate change and diseases of food crops [51]	Plant Pathology	2011	99
The stay-green trait [52]	Journal of Experimental Botany	2014	82
Quantitative genetic analysis of biomass and wood chemistry of Populus under different nitrogen levels [53]	New Phytologist	2009	82

3. The Breeder's Toolbox for Facing the Challenges Imposed by Climate Change

3.1. Genetic Resources: A Cornerstone for Competitive Plant Breeding

A deep understanding of adaptive mechanisms to climate changes cannot be separated from detailed knowledge on the genetic background and phenotypic plasticity of crops [54].

Mapping populations are widely used to investigate the relationship between DNA polymorphisms and trait variation [55]. High-resolution trait mapping in crops implies the selection of adequate genetic material from which various germplasm resources can be developed in order to breed climate-resilient crops. The resolution and accuracy of mapping qualitative and quantitative trait loci (the latter referred to as QTLs) depends on the recombination rate and frequency, the effective population size (the larger the population, the higher the frequency of recombination and the higher the QTL resolution), and on trait heritability [56].

In order to dissect the genetic basis of complex traits in crops, geneticists generally use two different types of populations: namely, family-based mapping populations and association mapping populations. As it can be easily understood, the recombination rate and the linkage disequilibrium (LD) decay greatly differ between the two types of populations. Indeed, individuals in family-based mapping populations have accumulated a very low number of recombination events, leading to the presence of blocks of high LD [55].

Bi-parental and multi-parental mapping populations (MPPs) are both family-based mapping populations. Bi-parental mapping populations, classically used for QTL mapping, usually derive from the cross between two contrasting individuals differing for one or more target traits. Their main limitation is that QTL detection depends strongly on the phenotypic diversity of the two parents only and that a few recombination events occur during the development of the population.

Unlike bi-parental mapping populations, whose variation relies on a relatively narrow genetic base, MPPs have been proposed as suitable resources to define the genetic basis of complex traits as they are characterized by high levels of recombination events and larger phenotypic diversity [57–59].

Typically, multi-parent Advanced Generation Inter-Cross (MAGIC) mapping populations are developed by inter-crossing multiple (generally four, eight, or sixteen) parental lines so as to fully exploit their complex pedigree structure [60,61]. Developing a MAGIC population is not trivial, as it requires the identification of founder lines within worldwide germplasm collections, elite cultivars, landraces, and distant relatives with pronounced genetic and phenotypic differences. Generally, the mixing of multiple parents follows different crossing schemes depending on how many founders are taken into account [56]. Although the benefits of working with MAGIC populations are clear, it is also necessary to remark two constraints: (i) the alien introgressions that might occur in the population as a consequence of rearrangements; (ii) the time necessary to develop homozygous individuals derived by advanced inter-crossing. Indeed, it has been estimated that at least eight crop seasons are required to reach at least the S5 generation, which is associated with a residual heterozygosity below 3% [62].

Association mapping populations are developed by collecting hundreds of unrelated individuals among elite and old cultivars, landraces, and wild relatives, which represent an invaluable source of natural genetic variations. Many of these populations include individuals retrieved from different parts of the world and characterized by a wide diversity [63–65]. The great advantage of using association mapping populations relies on the higher allelic richness that is captured and that is essential for high-resolution QTL mapping.

Finally, Nested Association Mapping (NAM) populations have been developed by combining, in a single unified mapping population, the advantages of two different types of populations (i.e., bi-parental mapping populations and association mapping populations) with the purpose of further increasing the precision of QTL mapping [66,67]. Generally, NAM populations derive from the crossing of multiple lines (i.e., diversity donors) with a single "reference" inbreed line, possibly an elite cultivar improved for important agronomic traits and extensively used in breeding programs.

Crosses give rise to multiple bi-parental sub-populations, either as double haploid (DH) lines or as recombinant inbreed lines (RILs), each of which is subjected to self-fertilization for six generations before being genotyped. Finally, parental lines are, in turn, sequenced or genotyped, and the results are overlaid on the recombination blocks previously identified in each sub-population.

3.2. Cutting-Edge Technologies for Breeding Applications

3.2.1. QTL Mapping and Marker-Assisted Selection

The basic idea behind QTL mapping is the identification of DNA molecular markers (such as single nucleotide polymorphisms, SNPs) that correlate with a given trait in a segregant (mapping) population, thus allowing the positioning of QTLs within linkage maps. Quantitative traits can be controlled by a few loci with fairly large effects (i.e., major QTL), or by many loci, each with minute effects (i.e., minor QTLs). Different methods for QTL analysis have been developed so far, and over hundred QTL mapping software have been implemented (for an exhaustive review, see Sehgal et al. [68]).

The advent of new sequencing technologies greatly facilitated the study of genomic variation, as it led to the identification of a large number of DNA polymorphisms, especially SNP markers, at limited cost [69]. The development of dense and ultra-dense linkage maps [68] increased the accuracy of QTL mapping from a region of 10–30 centimorgan (cM) [70] to a region <1 cM on average [71]. As mentioned earlier, a broader genetic diversity (bi-parental vs MAGIC or NAM populations) gives high QTL resolution.

In addition, the availability of high-throughput plant phenomic tools is also of great importance for increasing the potential of QTL mapping [72]. The link between phenotypic traits and genotypic data is essential in explaining the genetic basis of complex traits.

QTLs affecting the phenotypes of interest can be also detected using LD mapping, which takes advantage of historical recombination events within the unobserved pedigree [73].

The resolution of QTL mapping can also be enhanced by combining linkage maps with LD maps [73]. Indeed, the existence of LD implies there are segments of a chromosome in the population which are descended from the same common ancestor. These identical-by-descent (IBD) chromosome segments carry both identical marker haplotypes and identical QTL alleles. This type of QTL mapping is referred to as LDLA (linkage disequilibrium linkage analysis) [73].

Recent studies report the molecular characterization of QTLs together with the identification of DNA polymorphisms underlying important traits, such as resistance to drought in barley and FHB resistance in wheat [74–76]. A large number of QTLs have been identified in cereals for agronomic and physiological traits under heat temperature and water stress conditions. As summarized in the review by Gupta et al. [77], several studies have already been conducted in wheat (*Triticum* L. spp.) using bi-parental interval mapping. Nine major and stable QTLs were detected for coleoptile length, root system, and grain yield, which represent the most relevant traits contributing to seedling emergence, grain yield, and adaption to drought environments [78]. Recently, high-density linkage maps were constructed using SNP markers in bread wheat RILs in order to detect QTLs for flag leaf-related traits, which play a key role in determining yield potential [79].

Once information on SNP-trait associations is available, it can be conveniently used to assist breeding programs. Marker assisted selection can be performed via medium- or high-throughput assays, such as KBioscience's Competitive Allele Specific-PCR SNP genotyping system (KASPar; http://www.lgcgenomics.com/) or high resolution melting (HRM) [80]. In case SNPs are associated with restriction endonuclease sites, they can be converted into cleaved amplified polymorphic sequence (CAPS), easily obtainable without the need of costly equipment [80,81].

3.2.2. Genome-Wide Association Studies and Genomic Selection

Genome-wide association studies investigate marker-trait associations based on the large nucleotide variability present within association mapping populations. The availability of a large number of SNPs is a necessary but not sufficient condition to improve the resolution of marker-trait association, which also strongly depends on the extent of LD decay over physical distance in a population [82]. The mating system of the species, recombination frequency, genetic drift, and the selection process of individuals are the most important factors affecting LD decay [83].

As clearly described by D'Agostino and Tripodi [69], once DNA variation has been captured, it is used to describe the genetic structure of the population under study. Assessment of population stratification (i.e., the presence of a systematic difference in allele frequency spectrum or in principal components between sub-populations) is essential to prevent false positive or negative SNP(s)–trait associations [69,84,85]. In addition, it is essential to have available robust phenotypic data for each individual in the population in such a way that significant genotype–phenotype associations can be scored. Association analysis can be performed with different tools (e.g., GAPIT [86] and GEMMA [87]) coupled with different model methods [88].

Based on GWAS, the genetic mechanisms underlying resistance and resilience traits to changing climate have been studied and their causative and predictive factors have been identified in several crops [89–91]. Specific SNPs or InDels have been used for functional marker-assistant selection in breeding programs.

Remarkable works have been conducted in cereals and leguminous to discover SNPs associated with a response to climate change and to develop new resilient crops. In sorghum, GWAS was used to identify SNPs associated with heat stress responses at the vegetative stage under field conditions [92]. SNPs associated with leaf firing and leaf blotching were located in candidate genes (transcription factors, heat-shock proteins, kinases, and phospholipases) that play a role in heat stress response or heat tolerance. A winter barley (*Hordeum vulgare* L.) collection was used to study the effect of CO_2 on biomass traits (aboveground biomass, ears, culms, and leaves) and detect SNPs located in genomic regions involved in the response to CO_2 and crop yield [93]. In chickpea, germplasm collections were used to evaluate drought tolerance, heat tolerance, and yield traits in order to identify significant marker-trait associations to be used for developing superior varieties with enhanced drought and heat tolerance [94]. In addition, Li et al. [95] found SNPs in auxin-related genes associated with yield-related traits under drought conditions.

Genomic selection (GS) may be considered a powerful tool to facilitate the selection of superior genotypes, accelerate the breeding cycle, and reduce the cost of breeding line development [96]. Firstly, a training population (TP) is assembled and is subjected to genotyping and phenotyping for the traits of interest. Then, data are integrated with pedigree information (i.e., a kinship square matrix quantifying pair-wise relationships among population individuals) to build a GS prediction model linking genome-wide marker data to phenotypes. Finally, the model is used on a different set of individuals, which have been previously genotyped but whose phenotype is undetermined (i.e., the breeding population, BP) to get information on their genomic estimated breeding value (GEBV). Clearly, knowing the GEBV of a breeding population allows hinging the selection on marker data without the need of time-consuming and costly phenotyping.

In wheat, GS models were largely developed to identify accessions that best adapt to the negative effects of climate change: FHB resistance [97], heading date as an important component of wheat adaptation [98] and water deficit stress [99]. Recently, Crain et al. [100] disclosed several GS methods in relation to the phenotypic information derived from high-throughput phenotyping platforms. Phenotypic data for drought and heat stresses were analyzed in two environments in more than one thousand advanced wheat lines for grain yield, available at the International Maize and Wheat Improvement Center (CIMMYT).

It was evident that GS, coupled with high-throughput genotyping and phenotyping approaches, increased prediction and selection accuracy in wheat breeding.

3.2.3. Mutation Breeding

Mutation breeding emerged in the middle of the last century with the purpose of artificially developing genetic variability. The use of chemical and physical agents to induce mutations has been successfully adopted worldwide since the 1930s to generate novel alleles, increase genetic diversity, and release mutant varieties in more than 170 different plant species [101]. However, this approach has been almost abandoned due to high costs and controversial opinions of the consequence of mutagenic agents on human health [102].

An alternative to chemical and physical mutagenesis is represented by techniques based on the use of biological agents. Indeed, site-directed mutagenesis and insertional mutagenesis represented alternative forward genetics methods to increase genetic diversity [103]. In the last two decades, mutation breeding has been recovered thanks also to advances in large-scale genome sequencing projects.

Targeting Induced Local Lesions in Genomes (TILLING) is a reverse genetic technique based on chemical induced mutagenesis coupled with a sensitive DNA screening-technique [104] which allows the discovery of rare mutations in populations. Traditionally, TILLING protocols were based on the use of enzymatic or physical methods to screen the population and select mutagenized lines. Loss-of-function, gain-of-function, and hypomorphic alleles can be identified and possibly associated with corresponding phenotypes [105].

By combining TILLING with the use of next generation sequencing coupled with multidimensional pooling, Tsai et al. [106] demonstrated that the identification of rare alleles in a population could be effectively expedited. TILLING by sequencing has been successfully applied to discover allelic variants underlying agronomic traits involved in the response to climate change [107,108]. In particular, TILLING was used to discover new allelic variants in the *Hsp26* gene family related to heat stress and thermal tolerance in wheat [109]. Barley mutants were generated by TILLING to study the nucleotide variations in the *era1* (*enhanced response to ABA1*) gene [110], which is differently regulated under drought tolerance in several species including wheat and soybean [111,112].

Modifications to the traditional TILLING or TILLING by sequencing methods have been subsequently proposed. De-TILLING (Deletion TILLING) is an alternative strategy that allows knock-out mutations to be exclusively detected [113]. EcoTILLING is a method developed by Comai et al. [114] to look for natural mutations in individuals. It could be an essential tool for discovering allelic variants responsible for crop adaptation to biotic and abiotic stresses derived by extreme agro-climate conditions [105].

3.2.4. Genome Editing

Genome editing technologies are listed under the larger group of the new plant breeding techniques (NPBT) [115] and can be classified into two categories: oligonucleotide-directed mutagenesis (ODM) and site-directed nucleases (SDNs). Both allow precise directed mutagenesis, gene transfer, and control of gene expression [116].

In the ODM, DNA fragments of 20 to 100 nucleotides in length are chemically synthesized and delivered into plant cells by common methods (e.g., PEG transfection, particle delivery) where they induce mutations in target sites with low efficiency (max. 0.05%) [115].

SDNs are enzymes that can specifically bind to short target DNA sequences ranging from 9 up to 40 nucleotides and exert different biochemical reaction *in situ* (introduction of double-strand breaks (DSBs), methylation, demethylation, acetylation, and deamination) to alter a biological activity (e.g., gene silencing, base editing, gene expression, etc.) [117]. Among all possible biochemical reactions mediated by SDNs, the introduction of DSBs is the most used so far.

In living cells, a DSB can be repaired either by non-homologous end joining (NHEJ) or by homologous recombination (HR); the former seems to be the most frequent in plants. The NHEJ pathway is error-prone, meaning random insertions/deletions (InDels) are usually introduced at the target site (SDN1); this can be exploited to knock-out or knock-down genes (e.g., to study gene function), alter gene expression, or remove domains (e.g., remove effector binding domain on susceptibility genes [118]). On the contrary, HR is an error-free template-based repair mechanism, which can be used to introduce non-random mutations (SDN2) or insert a large DNA fragment (SDN3) at a target site [115].

SDNs are classified in meganucleases (or homing endonucleases, HE), zinc finger nucleases (ZFNs), Transcription Activator-Like Effector Nucleases (TALENs), and clustered regularly interspaced short palindromic repeats (CRISPR)-associated protein (CRISPR/Cas). The scientific community and private companies have constantly subjected SDNs to investigation and optimization; however, only with the advent of CRISPR/Cas has genome editing become widely used [119].

Off-target activity is a common issue for all SDNs; however, in plants, the possibility of screening a large edited population and discarding "non-specific editing" makes this issue probably less important compared to the necessary development of transformation protocols, innovation in automation, and tissue-culture-free methods along with investment in transgene-free methods and genomic resources in crops [120]. Indeed, the knowledge on the target and off-target sequences, the availability of an efficient delivery system of SDNs into cells, and the ability to obtain edited homozygous plants are equally important steps in genome editing approaches (SND1-3) that cannot be easily pursued in all crops. For instance, in some important vegetables (pepper, artichoke, and pulses), the development of a reproducible transformation protocol is necessary. As a positive example, in cereals, after years of effort in developing transformation protocols, a recent major breakthrough by Lowe et al. [121] has allowed the boosting of transformation rates in a broad range of accessions.

Transgene-free methods rely on the possibility to transiently express SDNs in plant cells (e.g., protoplasts), and this can be achieved either by the transfer of a DNA-based expression cassette that does not undergo stable integration in the genome [122–124], or alternatively by the transfer of ribonucleoproteins (RNPs) [125,126]. The use of transgene-free methods can lead to genome-edited plants (SDN1 and SDN2 on case-by-case), which are indistinguishable by spontaneously mutated crops or mutants obtained by classical mutagenesis approaches (i.e., ethyl methanesulfonate, ionizing radiation) [115]. Therefore, in the European Union, a distinction in the legislation supporting the approval route toward commercialization of the edited plants deriving by SDN1, SDN2, and SDN3 methods was proposed [127,128]. However, the latest ruling by the European Court of Justice [129] requires that crops generated by using gene-editing techniques such as CRISPR must go through the same lengthy approval process as conventional genetically modified (GM) plants [130]. Surprisingly, no distinctions where made on SDN1, SDN2, or SDN3.

Editing of genes involved in responses to abiotic and biotic stresses has been reported, though only in a limited number of cases and exclusively using SDN1 [116,131]. One of the first successful applications has been the modification by TALEN of the promoter region of the rice bacterial blight susceptibility gene OsSWEET14. This change caused the removal of the effector binding element, thus giving resistance to major forms of bacterial blight [132]. Again, by TALEN, it was possible to simultaneously edit three (Mycoplasma Like Organism) homoalleles of the susceptibility gene MLO, resulting in powdery mildew resistance in bread wheat [133]. More recently, Nekrasov et al. [134] successfully applied the CRISPR/Cas9 technology in tomato to induce a loss-of-function mutation of the powdery mildew susceptibility gene, SlMLO1 [135].

Applications to abiotic stresses are still largely confined to model species (e.g., Arabidopsis), although some promising results have been recently reported in soybean (drought and salt tolerance by disrupting the Drb2a and Drb2b genes) [116,136] and more recently announced in cocoa [137]. Abiotic stresses are often controlled by complex genetic mechanisms, which may require simultaneous tuning of different genes (i.e., regulatory sequences, editing of SNPs); on the contrary, for biotic stresses, the knock-out of single genes (i.e., susceptibility genes [138]) is likely to produce the desired phenotype.

New target genes, further technical development allowing both SDN1 and SND2 approaches, and a harmonized legislation on edited crops are necessary to prompt the growth of a novel generation of breeders.

3.3. Bioinformatics and Data Mining: Next Generation Breeding is Going Digital

A large number of crop genomes have been released into the public domain due to major advances in DNA sequencing technologies and bioinformatics. If, on one hand, the availability of a reference genome sequence is of unquestionable value, then on the other hand, it does not represent the diversity within a particular species. As outlined in this work, information on DNA polymorphisms, available through whole genome re-sequencing [139], sequence capture, target-enrichment and re-sequencing methods [140], fractional genome sequencing strategies [141–143], and high-density genotyping arrays [144], is of paramount importance for crop breeding. Indeed, in the last few years, several works addressed the study of genetic diversity in major as well as in "orphan" crops [63,137,145–149].

Aiming to increase the effectiveness of QTL mapping, GWAS, and GS, it is becoming increasingly important to go over the "phenotyping bottleneck" [69] and choose automated technologies for high-throughput plant phenotyping in order to collect measurements of qualitative, agronomical, morphological, and physiological traits. The huge amount of phenotypic data points is challenging in its analysis, management, and accessibility to a greater extent than genotyping data.

As easily understood, bioinformatics is a rapidly expanding field of research as it is essential to extract knowledge from heterogeneous data (i.e., data mining). The analysis of a high number of SNPs and phenotypic data points is demanding and requires an adequate computational infrastructure as well as bioinformatic and shell scripting skills that are beyond the reach of a typical lab. In addition, it is becoming increasingly necessary to integrate various "omics" data (e.g., from genomics and phenomics) with mathematical and statistical models.

There is an urgent need for early training in bioinformatic skills in order to empower plant researchers and breeders to make use of their own data (i.e., for analysis and interpretation) [150]. However, to identify those who are adept at both bioinformatics and plant breeding is difficult and not trivial. A realistic approach is to build interdisciplinary working teams where researchers can share knowledge and expertise to impact on crop improvement.

It seems clear at this point that, similarly to biology, next generation breeding is going digital and that a new figure of breeder is required to cope with recent advances in genomics, transcriptomics, phenomics, and bioinformatics (Figure 2). With this, we do not mean that the next generation breeder will find the field and the computer indistinguishable, but rather that by combining expertise in complementary areas, they will have the greatest potential to be successful in breeding programs in a scenario of increasing climate variability.

Figure 2. A new figure of breeder is beginning to thrive in the niche created by biotechnological and digital revolution. By combining expertise in complementary areas (open-field trials, wet-lab techniques, big data analysis, and interpretation), they will have the greatest potential to be successful in developing climate-resilient crops.

4. Conclusions

Between 1950 and the late 1960s, the "Green Revolution" dramatically changed the field of agriculture with the aim of providing a solution for the world's food supply problem. Indeed, the global productivity increased drastically, especially in developing countries, thanks to the use of fertilizers, herbicides, pesticides, and high-yield varieties.

In the 21st century, agriculture will face new challenges, largely due to the need to increase global food supply under the declining availability of arable lands and increasing threats from climate change. With respect to this, a white paper was prepared in 2009 by the Food and Agriculture Organization in which the concept of Climate-Smart Agriculture, enabling the ability to cope with food security while facing the challenges of climate change, is emphasized [151]. A prerequisite for climate-smart breeding is the preservation and conservation of genetic resources. Indeed, climate change is altering the behavior of many species, thus affecting ecosystem dynamics. For these reasons, new strategies of germplasm characterization, selection, reproduction, and conservation should be played out so that suitable genetic resources are available to develop cultivars resilient to climate change.

In this review, the most recent popular and breakthrough technologies applied to plant breeding were described and several examples of their applications to breed climate resilient cultivars were provided.

Indeed, breeding for climate-smart agriculture is benefitting from a new revolution, which lays its foundation on the analysis and interpretation of big "omics" data and on NPBT, and which is expected to give fruitful results in the near future.

Nowadays, the breeder's skill set, although it continues to quickly evolve, is rich enough to allow us to start thinking of breeding with different tools than that in the past, as technological improvements in phenotypic and genotypic analysis, as well as the biotechnological and digital revolution, will reduce the breeding cycle in a cost-effective manner [152].

Author Contributions: Conceptualization, F.T. and N.D.A.; writing—original draft preparation, F.T, A.N., S.P. and N.D.A.; writing—review and editing, S.P and P.D.V; supervision, N.D.A and P.D.V.

Acknowledgments: We are really grateful to Annalisa Manfredi for her excellent contribution in designing and drawing Figure 2.

References

1.   Ackerly, D.D.; Loarie, S.R.; Cornwell, W.K.; Weiss, S.B.; Hamilton, H.; Branciforte, R.; Kraft, N.J.B. The geography of climate change: Implications for conservation biogeography. *Divers. Distrib.* **2010**, *16*, 476–487. [CrossRef]

2.   Gornall, J.; Betts, R.; Burke, E.; Clark, R.; Camp, J.; Willett, K.; Wiltshire, A. Implications of climate change for agricultural productivity in the early twenty-first century. *Phil. Trans. R. Soc. B-Biol. Sci.* **2010**, *365*, 2973–2989. [CrossRef] [PubMed]

3.   Rivero, R.M.; Kojima, M.; Gepstein, A.; Sakakibara, H.; Mittler, R.; Gepstein, S.; Blumwald, E. Delayed leaf senescence induces extreme drought tolerance in a flowering plant. *Proc. Natl. Acad. Sci. USA* **2007**, *104*, 19631–19636. [CrossRef] [PubMed]

4.   Rosenzweig, C.; Iglesias, A.; Yang, X.; Epstein, P.R.; Chivian, E. Climate change and extreme weather events; implications for food production, plant diseases, and pests. *Glob. Change Hum. Health* **2001**, *2*, 90–104. [CrossRef]

5.   Nuruddin, M.M.; Madramootoo, C.A.; Dodds, G.T. Effects of water stress at different growth stages on greenhouse tomato yield and quality. *HortScience* **2003**, *38*, 1389–1393.

6.   Greven, M.M.; Raw, V.; West, B.A. Effects of timing of water stress on yield and berry size. *Water Sci. Technol.* **2009**, *60*, 1249–1255. [CrossRef] [PubMed]

7.   Porter, J.R.; Gawith, M. Temperatures and the growth and development of wheat: A review. *European J. Agron.* **1999**, *10*, 23–36. [CrossRef]

8.   Ottman, M.; Kimball, B.A.; White, J.; Wall, G.W. Wheat growth response to increased temperature from varied planting dates and supplemental infrared heating. *Agron. J.* **2012**, *104*, 7. [CrossRef]

9.   Zhao, C.; Piao, S.; Wang, X.; Huang, Y.; Ciais, P.; Elliott, J.; Huang, M.; Janssens, I.A.; Li, T.; Lian, X.; et al. Plausible rice yield losses under future climate warming. *Nat. Plants* **2016**, *3*, 16202. [CrossRef] [PubMed]

10. Bassu, S.; Brisson, N.; Durand, J.-L.; Boote, K.; Lizaso, J.; Jones, J.W.; Rosenzweig, C.; Ruane, A.C.; Adam, M.; Baron, C.; et al. How do various maize crop models vary in their responses to climate change factors? *Glob. Change Biol.* **2014**, *20*, 2301–2320. [CrossRef] [PubMed]

11. Deryng, D.; Conway, D.; Ramankutty, N.; Price, J.; Warren, R. Global crop yield response to extreme heat stress under multiple climate change futures. *Environ. Res. Lett.* **2014**, *9*, 041001. [CrossRef]

12. Zhao, C.; Liu, B.; Piao, S.; Wang, X.; Lobell, D.B.; Huang, Y.; Huang, M.; Yao, Y.; Bassu, S.; Ciais, P.; et al. Temperature increase reduces global yields of major crops in four independent estimates. *Proc. Natl. Acad. Sci. USA* **2017**, *114*, 9326–9331. [CrossRef] [PubMed]

13. Irigoyen, J.J.; Goicoechea, N.; Antolín, M.C.; Pascual, I.; Sánchez-Díaz, M.; Aguirreolea, J.; Morales, F. Growth, photosynthetic acclimation and yield quality in legumes under climate change simulations: An updated survey. *Plant Sci.* **2014**, *226*, 22–29. [CrossRef] [PubMed]

14. Trębicki, P.; Nancarrow, N.; Cole, E.; Bosque-Pérez, N.A.; Constable, F.E.; Freeman, A.J.; Rodoni, B.; Yen, A.L.; Luck, J.E.; Fitzgerald, G.J. Virus disease in wheat predicted to increase with a changing climate. *Glob. Change Biol.* **2015**, *21*, 3511–3519. [CrossRef] [PubMed]

15. Atkinson, N.J.; Urwin, P.E. The interaction of plant biotic and abiotic stresses: From genes to the field. *J.f Exp. Bot.* **2012**, *63*, 3523–3543. [CrossRef] [PubMed]

16. Suzuki, N.; Rivero, R.M.; Shulaev, V.; Blumwald, E.; Mittler, R. Abiotic and biotic stress combinations. *New Phytol.* **2014**, *203*, 32–43. [CrossRef] [PubMed]

17. Rejeb, I.B.; Pastor, V.; Mauch-Mani, B. Plant responses to simultaneous biotic and abiotic stress: Molecular mechanisms. *Plants* **2014**, *3*, 458–475. [CrossRef] [PubMed]

18. Nguyen, D.; D'Agostino, N.; Tytgat, T.O.; Sun, P.; Lortzing, T.; Visser, E.J.; Cristescu, S.M.; Steppuhn, A.; Mariani, C.; van Dam, N.M. Drought and flooding have distinct effects on herbivore-induced responses and resistance in solanum dulcamara. *Plant Cell Environ.* **2016**, *39*, 1485–1499. [CrossRef] [PubMed]

19. Ihsan, M.Z.; El-Nakhlawy, F.S.; Ismail, S.M.; Fahad, S.; daur, I. Wheat phenological development and growth studies as affected by drought and late season high temperature stress under arid environment. *Front. Plant Sci.* **2016**, *7*. [CrossRef] [PubMed]

20. Zandalinas, S.I.; Mittler, R.; Balfagón, D.; Arbona, V.; Gómez-Cadenas, A. Plant adaptations to the combination of drought and high temperatures. *Physiol. Plant.* **2018**, *162*, 2–12. [CrossRef] [PubMed]

21. Hlaváčová, M.; Klem, K.; Rapantová, B.; Novotná, K.; Urban, O.; Hlavinka, P.; Smutná, P.; Horáková, V.; Škarpa, P.; Pohanková, E.; et al. Interactive effects of high temperature and drought stress during stem elongation, anthesis and early grain filling on the yield formation and photosynthesis of winter wheat. *Field Crops Res.* **2018**, *221*, 182–195. [CrossRef]

22. Ahmed, I.M.; Nadira, U.A.; Bibi, N.; Zhang, G.; Wu, F. Tolerance to combined stress of drought and salinity in barley. In *Combined Stresses in Plants*; Springer: Basel, Switzerland, 2015; pp. 93–121.

23. Kizildeniz, T.; Mekni, I.; Santesteban, H.; Pascual, I.; Morales, F.; Irigoyen, J.J. Effects of climate change including elevated co2 concentration, temperature and water deficit on growth, water status, and yield quality of grapevine (vitis vinifera l.) cultivars. *Agric. Water Manag.* **2015**, *159*, 155–164. [CrossRef]

24. Feller, U. Drought stress and carbon assimilation in a warming climate: Reversible and irreversible impacts. *J. Plant Physiol.* **2016**, *203*, 84–94. [CrossRef] [PubMed]

25. Shabani, F.; Kumar, L.; Esmaeili, A. Future distributions of fusarium oxysporum f. Spp. In european, middle eastern and north african agricultural regions under climate change. *Agric. Ecosyst. Environ.* **2014**, *197*, 96–105. [CrossRef]

26. Garrett, K.A.; Nita, M.; De Wolf, E.; Esker, P.D.; Gomez-Montano, L.; Sparks, A.H. Plant pathogens as indicators of climate change. In *Climate Change*, 2nd ed.; Elsevier: Amsterdam, The Netherlands, 2015; pp. 325–338.

27. Prasch, C.M.; Sonnewald, U. Simultaneous application of heat, drought and virus to arabidopsis thaliana plants reveals significant shifts in signaling networks. *Plant Physiol.* **2013**, *162*, 1849–1866. [CrossRef] [PubMed]

28. Elad, Y.; Pertot, I. Climate change impacts on plant pathogens and plant diseases. *J. Crop Improv.* **2014**, *28*, 99–139. [CrossRef]

29. Zhang, X.; Halder, J.; White, R.P.; Hughes, D.; Ye, Z.; Wang, C.; Xu, R.; Gan, B.; Fitt, B.D. Climate change increases risk of fusarium ear blight on wheat in central china. *Ann. Appl. Biol.* **2014**, *164*, 384–395. [CrossRef]

30. West, J.S.; Holdgate, S.; Townsend, J.A.; Edwards, S.G.; Jennings, P.; Fitt, B.D. Impacts of changing climate and agronomic factors on fusarium ear blight of wheat in the uk. *Fungal Ecol.* **2012**, *5*, 53–61. [CrossRef]

31. Fones, H.; Gurr, S. The impact of septoria tritici blotch disease on wheat: An eu perspective. *Fungal Genet. Biol.* **2015**, *79*, 3–7. [CrossRef] [PubMed]

32. Achuo, E.; Prinsen, E.; Höfte, M. Influence of drought, salt stress and abscisic acid on the resistance of tomato to botrytis cinerea and oidium neolycopersici. *Plant Pathol.* **2006**, *55*, 178–186. [CrossRef]

33. Varshney, R.K.; Bansal, K.C.; Aggarwal, P.K.; Datta, S.K.; Craufurd, P.Q. Agricultural biotechnology for crop improvement in a variable climate: Hope or hype? *Trends Plant Sci.* **2011**, *16*, 363–371. [CrossRef] [PubMed]

34. Cardi, T. Cisgenesis and genome editing: Combining concepts and efforts for a smarter use of genetic resources in crop breeding. *Plant Breed.* **2016**, *135*, 139–147. [CrossRef]

35. Cardi, T.; Neal Stewart, C., Jr. Progress of targeted genome modification approaches in higher plants. *Plant Cell Rep.* **2016**, *35*, 1401–1416. [CrossRef] [PubMed]

36. Rinaldo, A.R.; Ayliffe, M. Gene targeting and editing in crop plants: A new era of precision opportunities. *Mol. Breed.* **2015**, *35*, 40. [CrossRef]

37. Osakabe, Y.; Watanabe, T.; Sugano, S.S.; Ueta, R.; Ishihara, R.; Shinozaki, K.; Osakabe, K. Optimization of crispr/cas9 genome editing to modify abiotic stress responses in plants. *Sci. Rep.* **2016**, *6*, 26685. [CrossRef] [PubMed]

38. Courtier-Orgogozo, V.; Morizot, B.; Boëte, C. Agricultural pest control with crispr-based gene drive: Time for public debate: Should we use gene drive for pest control? *EMBO Rep.* **2017**, *18*, 878–880. [CrossRef] [PubMed]

39. Appiano, M.; Catalano, D.; Santillán Martínez, M.; Lotti, C.; Zheng, Z.; Visser, R.G.F.; Ricciardi, L.; Bai, Y.; Pavan, S. Monocot and dicot mlo powdery mildew susceptibility factors are functionally conserved in spite of the evolution of class-specific molecular features. *BMC Plant Biol.* **2015**, *15*, 257. [CrossRef] [PubMed]

40. Protocol, K. United nations framework convention on climate change. *Kyoto Protoc. Kyoto* **1997**, *19*.

41. Janssen, M.A.; Schoon, M.L.; Ke, W.; Börner, K. Scholarly networks on resilience, vulnerability and adaptation within the human dimensions of global environmental change. *Glob. Environ. Change* **2006**, *16*, 240–252. [CrossRef]

42. Wang, B.; Pan, S.-Y.; Ke, R.-Y.; Wang, K.; Wei, Y.-M. An overview of climate change vulnerability: A bibliometric analysis based on web of science database. *Nat. Hazards* **2014**, *74*, 1649–1666. [CrossRef]

43. Dhingani, R.M.; Umrania, V.V.; Tomar, R.S.; Parakhia, M.V.; Golakiya, B. Introduction to qtl mapping in plants. *Ann. Plant Sci.* **2015**, *4*, 1072–1079.

44. Mittler, R.; Blumwald, E. Genetic engineering for modern agriculture: Challenges and perspectives. *Ann. Rev. Plant Biol.* **2010**, *61*, 443–462. [CrossRef] [PubMed]

45. Araus, J.L.; Slafer, G.A.; Royo, C.; Serret, M.D. Breeding for yield potential and stress adaptation in cereals. *Crit. Rev. Plant Sci.* **2008**, *27*, 377–412. [CrossRef]

46. De Dorlodot, S.; Forster, B.; Pagès, L.; Price, A.; Tuberosa, R.; Draye, X. Root system architecture: Opportunities and constraints for genetic improvement of crops. *Trends Plant Sci.* **2007**, *12*, 474–481. [CrossRef] [PubMed]

47. Booker, F.; Muntifering, R.; McGrath, M.; Burkey, K.; Decoteau, D.; Fiscus, E.; Manning, W.; Krupa, S.; Chappelka, A.; Grantz, D. The ozone component of global change: Potential effects on agricultural and horticultural plant yield, product quality and interactions with invasive species. *J. Integr. Plant Biol.* **2009**, *51*, 337–351. [CrossRef] [PubMed]

48. Wolfe, M.S.; Baresel, J.P.; Desclaux, D.; Goldringer, I.; Hoad, S.; Kovacs, G.; Löschenberger, F.; Miedaner, T.; Østergård, H.; Lammerts van Bueren, E.T. Developments in breeding cereals for organic agriculture. *Euphytica* **2008**, *163*. [CrossRef]

49. Mir, R.R.; Zaman-Allah, M.; Sreenivasulu, N.; Trethowan, R.; Varshney, R.K. Integrated genomics, physiology and breeding approaches for improving drought tolerance in crops. *Theor. Appl. Genet.* **2012**, *125*, 625–645. [CrossRef] [PubMed]

50. Poland, J.A.; Rife, T.W. Genotyping-by-sequencing for plant breeding and genetics. *Plant Genome* **2012**, *5*, 92–102. [CrossRef]

51. Luck, J.; Spackman, M.; Freeman, A.; Trębicki, P.; Griffiths, W.; Finlay, K.; Chakraborty, S. Climate change and diseases of food crops. *Plant Pathol.* **2011**, *60*, 113–121. [CrossRef]

52. Thomas, H.; Ougham, H. The stay-green trait. *J. Exp. Bot.* **2014**, *65*, 3889–3900. [CrossRef] [PubMed]

53. Novaes, E.; Osorio, L.; Drost, D.R.; Miles, B.L.; Boaventura-Novaes, C.R.D.; Benedict, C.; Dervinis, C.; Yu, Q.; Sykes, R.; Davis, M.; et al. Quantitative genetic analysis of biomass and wood chemistry of populus under different nitrogen levels. *New Phytol.* **2009**, *182*, 878–890. [CrossRef] [PubMed]

54. Gao, S.-B.; Mo, L.-D.; Zhang, L.-H.; Zhang, J.-L.; Wu, J.-B.; Wang, J.-L.; Zhao, N.-X.; Gao, Y.-B. Phenotypic plasticity vs. Local adaptation in quantitative traits differences of stipa grandis in semi-arid steppe, China. *Sci. Rep.* **2018**, *8*. [CrossRef] [PubMed]

55. Xu, Y.; Li, P.; Yang, Z.; Xu, C. Genetic mapping of quantitative trait loci in crops. *Crop J.* **2017**, *5*, 175–184. [CrossRef]

56. Cockram, J.; Mackay, I. Genetic mapping populations for conducting high-resolution trait mapping in plants. In *Plant Genetics and Molecular Biology*; Varshney, R.K., Pandey, M.K., Chitikineni, A., Eds.; Springer International Publishing: Cham, Switzerland, 2018; pp. 109–138.

57. Ladejobi, O.; Elderfield, J.; Gardner, K.A.; Gaynor, R.C.; Hickey, J.; Hibberd, J.M.; Mackay, I.J.; Bentley, A.R. Maximizing the potential of multi-parental crop populations. *Appl. Trans. Genom.* **2016**, *11*, 9–17. [CrossRef] [PubMed]

58. Cavanagh, C.; Morell, M.; Mackay, I.; Powell, W. From mutations to magic: Resources for gene discovery, validation and delivery in crop plants. *Curr. Opin. Plant Biol.* **2008**, *11*, 215–221. [CrossRef] [PubMed]

59. Mackay, I.; Powell, W. Methods for linkage disequilibrium mapping in crops. *Trends Plant Sci.* **2007**, *12*, 57–63. [CrossRef] [PubMed]

60. Dell'Acqua, M.; Gatti, D.M.; Pea, G.; Cattonaro, F.; Coppens, F.; Magris, G.; Hlaing, A.L.; Aung, H.H.; Nelissen, H.; Baute, J.; et al. Genetic properties of the magic maize population: A new platform for high definition qtl mapping in zea mays. *Genome Biol.* **2015**, *16*. [CrossRef] [PubMed]

61. Kover, P.X.; Valdar, W.; Trakalo, J.; Scarcelli, N.; Ehrenreich, I.M.; Purugganan, M.D.; Durrant, C.; Mott, R. A multiparent advanced generation inter-cross to fine-map quantitative traits in arabidopsis thaliana. *PLOS Genetics* **2009**, *5*, e1000551. [CrossRef] [PubMed]

62. Huang, B.E.; Verbyla, K.L.; Verbyla, A.P.; Raghavan, C.; Singh, V.K.; Gaur, P.; Leung, H.; Varshney, R.K.; Cavanagh, C.R. Magic populations in crops: Current status and future prospects. *Theor. Appl. Genet.* **2015**, *128*, 999–1017. [CrossRef] [PubMed]

63. Taranto, F.; D'Agostino, N.; Greco, B.; Cardi, T.; Tripodi, P. Genome-wide snp discovery and population structure analysis in pepper (capsicum annuum) using genotyping by sequencing. *BMC Genom.* **2016**, *17*, 943. [CrossRef] [PubMed]

64. Rodriguez, M.; Rau, D.; Bitocchi, E.; Bellucci, E.; Biagetti, E.; Carboni, A.; Gepts, P.; Nanni, L.; Papa, R.; Attene, G. Landscape genetics, adaptive diversity and population structure in phaseolus vulgaris. *New Phytol.* **2016**, *209*, 1781–1794. [CrossRef] [PubMed]

65. Sacco, A.; Ruggieri, V.; Parisi, M.; Festa, G.; Rigano, M.M.; Picarella, M.E.; Mazzucato, A.; Barone, A. Exploring a tomato landraces collection for fruit-related traits by the aid of a high-throughput genomic platform. *PLoS ONE* **2015**, *10*, e0137139. [CrossRef] [PubMed]

66. McMullen, M.D.; Kresovich, S.; Villeda, H.S.; Bradbury, P.; Li, H.; Sun, Q.; Flint-Garcia, S.; Thornsberry, J.; Acharya, C.; Bottoms, C.; et al. Genetic properties of the maize nested association mapping population. *Science* **2009**, *325*, 737–740. [CrossRef] [PubMed]

67. Yu, J.; Holland, J.B.; McMullen, M.D.; Buckler, E.S. Genetic design and statistical power of nested association mapping in maize. *Genetics* **2008**, *178*, 539–551. [CrossRef] [PubMed]

68. Sehgal, D.; Singh, R.; Rajpal, V.R. Quantitative trait loci mapping in plants: Concepts and approaches. In *Molecular Breeding for Sustainable Crop Improvement*; Rajpal, V.R., Rao, S.R., Raina, S.N., Eds.; Springer International Publishing: Cham, Switzerland, 2016; Volume 2, pp. 31–59.

69. D'Agostino, N.; Tripodi, P. Ngs-based genotyping, high-throughput phenotyping and genome-wide association studies laid the foundations for next-generation breeding in horticultural crops. *Diversity* **2017**, *9*, 38. [CrossRef]

70. Kearsey, M.J.; Farquhar, A.G.L. QTL analysis in plants; where are we now? *Heredity* **1998**, *80*, 137. [CrossRef] [PubMed]

71. Yu, H.; Xie, W.; Wang, J.; Xing, Y.; Xu, C.; Li, X.; Xiao, J.; Zhang, Q. Gains in qtl detection using an ultra-high density snp map based on population sequencing relative to traditional rflp/ssr markers. *PLoS ONE* **2011**, *6*, e17595.

72.   Araus, J.L.; Cairns, J.E. Field high-throughput phenotyping: The new crop breeding frontier. *Trends Plant Sci.* **2014**, *19*, 52–61. [CrossRef] [PubMed]

73.   Pikkuhookana, P.; Sillanpää, M.J. Combined linkage disequilibrium and linkage mapping: Bayesian multilocus approach. *Heredity* **2014**, *112*, 351–360. [CrossRef] [PubMed]

74.   Gudys, K.; Guzy-Wrobelska, J.; Janiak, A.; Dziurka, M.A.; Ostrowska, A.; Hura, K.; Jurczyk, B.; Żmuda, K.; Grzybkowska, D.; Śróbka, J.; et al. Prioritization of candidate genes in qtl regions for physiological and biochemical traits underlying drought response in barley (*Hordeum vulgare* L.). *Front. Plant Sci.* **2018**, *9*, 769. [CrossRef] [PubMed]

75.   Sari, E.; Berraies, S.; Knox, R.E.; Singh, A.K.; Ruan, Y.; Cuthbert, R.D.; Pozniak, C.J.; Henriquez, M.A.; Kumar, S.; Burt, A.J.; et al. High density genetic mapping of fusarium head blight resistance qtl in tetraploid wheat. *PLoS ONE* **2018**, *13*, e0204362. [CrossRef] [PubMed]

76.   Buerstmayr, M.; Steiner, B.; Wagner, C.; Schwarz, P.; Brugger, K.; Barabaschi, D.; Volante, A.; Valè, G.; Cattivelli, L.; Buerstmayr, H. High-resolution mapping of the pericentromeric region on wheat chromosome arm 5as harbouring the fusarium head blight resistance qtl qfhs.Ifa-5a. *Plant Biotechnol. J.* **2018**, *16*, 1046–1056. [CrossRef] [PubMed]

77.   Gupta, P.; Balyan, H.; Gahlaut, V. Qtl analysis for drought tolerance in wheat: Present status and future possibilities. *Agronomy* **2017**, *7*, 5. [CrossRef]

78.   Djanaguiraman, M.; Prasad, P.; Kumari, J.; Rengel, Z. Root length and root lipid composition contribute to drought tolerance of winter and spring wheat. *Plant Soil* **2018**, 1–17. [CrossRef]

79.   Liu, K.; Xu, H.; Liu, G.; Guan, P.; Zhou, X.; Peng, H.; Yao, Y.; Ni, Z.; Sun, Q.; Du, J. Qtl mapping of flag leaf-related traits in wheat (*Triticum aestivum* L.). *Theor. Appl. Genet.* **2018**, *131*, 839–849. [CrossRef] [PubMed]

80.   Pavan, S.; Schiavulli, A.; Appiano, M.; Miacola, C.; Visser, R.G.F.; Bai, Y.; Lotti, C.; Ricciardi, L. Identification of a complete set of functional markers for the selection of er1 powdery mildew resistance in *pisum sativum* L. *Mol. Breed.* **2013**, *31*, 247–253. [CrossRef]

81.   Pavan, S.; Schiavulli, A.; Lotti, C.; Ricciardi, L. *Caps Technology as a Tool for the Development of Genic and Functional Markers: Study in Peas*; NOVA Publisher: New York, NY, USA, 2014.

82.   Wang, Y.; Rannala, B. In silico analysis of disease-association mapping strategies using the coalescent process and incorporating ascertainment and selection. *Am. J. Hum. Genet.* **2005**, *76*, 1066–1073. [CrossRef] [PubMed]

83.   Rafalski, A.; Morgante, M. Corn and humans: Recombination and linkage disequilibrium in two genomes of similar size. *Trends Genet.* **2004**, *20*, 103–111. [CrossRef] [PubMed]

84.   Wright, S.I.; Gaut, B.S. Molecular population genetics and the search for adaptive evolution in plants. *Mol. Biol. Evol.* **2005**, *22*, 506–519. [CrossRef] [PubMed]

85.   Ersoz, E.S.; Yu, J.; Buckler, E.S. Applications of linkage disequilibrium and association mapping in crop plants. In *Genomics-Assisted Crop Improvement: Vol. 1: Genomics Approaches and Platforms*; Varshney, R.K., Tuberosa, R., Eds.; Springer Netherlands: Dordrecht, The Nrtherlands, 2007; pp. 97–119.

86.   Tang, Y.; Liu, X.; Wang, J.; Li, M.; Wang, Q.; Tian, F.; Su, Z.; Pan, Y.; Liu, D.; Lipka, A.E. Gapit version 2: An enhanced integrated tool for genomic association and prediction. *Plant Genome* **2016**, *9*. [CrossRef] [PubMed]

87.   Zhou, X.; Stephens, M. Genome-wide efficient mixed model analysis for association studies. *Nat. Genet.* **2012**, *44*, 821–824. [CrossRef] [PubMed]

88.   Hayes, B. Overview of statistical methods for genome-wide association studies (GWAS). In *Genome-Wide Association Studies and Genomic Prediction*; Springer: Basel, Switzerland, 2013; pp. 149–169.

89.   Dawson, I.K.; Russell, J.; Powell, W.; Steffenson, B.; Thomas, W.T.; Waugh, R. Barley: A translational model for adaptation to climate change. *New Phytol.* **2015**, *206*, 913–931. [CrossRef] [PubMed]

90.   Shea, D.J.; Itabashi, E.; Takada, S.; Fukai, E.; Kakizaki, T.; Fujimoto, R.; Okazaki, K. The role of flowering locus c in vernalization of brassica: The importance of vernalization research in the face of climate change. *Crop Pasture Sci.* **2018**, *69*, 30–39. [CrossRef]

91.   Mousavi-Derazmahalleh, M.; Bayer, P.E.; Hane, J.K.; Babu, V.; Nguyen, H.T.; Nelson, M.N.; Erskine, W.; Varshney, R.K.; Papa, R.; Edwards, D. Adapting legume crops to climate change using genomic approaches. *Plant Cell Environ.* **2018**. [CrossRef] [PubMed]

92.   Chen, J.; Chopra, R.; Hayes, C.; Morris, G.; Marla, S.; Burke, J.; Xin, Z.; Burow, G. Genome-wide association study of developing leaves' heat tolerance during vegetative growth stages in a sorghum association panel. *Plant Genome* **2017**, *10*. [CrossRef] [PubMed]

93. Mitterbauer, E.; Enders, M.; Bender, J.; Erbs, M.; Habekuss, A.; Kilian, B.; Ordon, F.; Weigel, H.J. Growth response of 98 barley (*Hordeum vulgare* L.) genotypes to elevated co 2 and identification of related quantitative trait loci using genome-wide association studies. *Plant Breed.* **2017**, *136*, 483–497. [CrossRef]

94. Thudi, M.; Upadhyaya, H.D.; Rathore, A.; Gaur, P.M.; Krishnamurthy, L.; Roorkiwal, M.; Nayak, S.N.; Chaturvedi, S.K.; Basu, P.S.; Gangarao, N. Genetic dissection of drought and heat tolerance in chickpea through genome-wide and candidate gene-based association mapping approaches. *PLoS ONE* **2014**, *9*, e96758. [CrossRef] [PubMed]

95. Li, Y.; Ruperao, P.; Batley, J.; Edwards, D.; Khan, T.; Colmer, T.D.; Pang, J.; Siddique, K.H.; Sutton, T. Investigating drought tolerance in chickpea using genome-wide association mapping and genomic selection based on whole-genome resequencing data. *Front. Plant Sci.* **2018**, *9*. [CrossRef] [PubMed]

96. Crossa, J.; Pérez-Rodríguez, P.; Cuevas, J.; Montesinos-López, O.; Jarquín, D.; de los Campos, G.; Burgueño, J.; González-Camacho, J.M.; Pérez-Elizalde, S.; Beyene, Y.; et al. Genomic selection in plant breeding: Methods, models, and perspectives. *Trends Plant Sci.* **2017**, *22*, 961–975. [CrossRef] [PubMed]

97. Dong, H.; Wang, R.; Yuan, Y.; Anderson, J.; Pumphrey, M.; Zhang, Z.; Chen, J. Evaluation of the potential for genomic selection to improve spring wheat resistance to fusarium head blight in the pacific northwest. *Front. Plant Sci.* **2018**, *9*. [CrossRef] [PubMed]

98. Huang, M.; Mheni, N.; Brown-Guedira, G.; McKendry, A.; Griffey, C.; Van Sanford, D.; Costa, J.; Sneller, C. Genetic analysis of heading date in winter and spring wheat. *Euphytica* **2018**, *214*. [CrossRef]

99. Ly, D.; Huet, S.; Gauffreteau, A.; Rincent, R.; Touzy, G.; Mini, A.; Jannink, J.-L.; Cormier, F.; Paux, E.; Lafarge, S. Whole-genome prediction of reaction norms to environmental stress in bread wheat (*Triticum aestivum* L.) by genomic random regression. *Field Crops Res.* **2018**, *216*, 32–41. [CrossRef]

100. Crain, J.; Mondal, S.; Rutkoski, J.; Singh, R.P.; Poland, J. Combining high-throughput phenotyping and genomic information to increase prediction and selection accuracy in wheat breeding. *Plant Genome* **2018**. [CrossRef] [PubMed]

101. Fao/Iaea Mutant Variety Database (mvd). Available online: https://mvd.iaea.org/ (accessed on 8 October 2018).

102. Scarascia-Mugnozza, G.; D'amato, F.; Avanzi, S.; Bagnara, D.; Belli, M.L.; Bozzini, A.; Cervigni, T.; Devreux, M.; Donini, B.; Giorgi, B.; et al. Mutation breeding for durum wheat (*Triticum turgidum* ssp. Durum desf.) improvement in italy. In Proceedings of the International Symposium on the Contribution of Plant Mutation Breeding to Crop Improvement, Vienna, Austria, 8–22 June 1990; pp. 1–28.

103. Jankowicz-Cieslak, J.; Till, B.J. Forward and reverse genetics in crop breeding. In *Advances in Plant Breeding Strategies: Breeding, Biotechnology and Molecular Tools*; Springer: Cham, Switzerland, 2015; p. 215.

104. Slade, A.J.; Knauf, V.C. Tilling moves beyond functional genomics into crop improvement. *Transgenic Res.* **2005**, *14*, 109–115. [CrossRef] [PubMed]

105. Kurowska, M.; Daszkowska-Golec, A.; Gruszka, D.; Marzec, M.; Szurman, M.; Szarejko, I.; Maluszynski, M. Tilling—A shortcut in functional genomics. *J. Appl. Genet.* **2011**, *52*, 371–390. [CrossRef] [PubMed]

106. Tsai, H.; Howell, T.; Nitcher, R.; Missirian, V.; Watson, B.; Ngo, K.J.; Lieberman, M.; Fass, J.; Uauy, C.; Tran, R.K.; et al. Discovery of rare mutations in populations: Tilling by sequencing. *Plant Physiol.* **2011**, *156*, 1257. [CrossRef] [PubMed]

107. Tadele, Z.; Mba, C.; Till, B.J. Tilling for mutations in model plants and crops. In *Molecular Techniques in Crop Improvement, 2nd Edition*; Jain, S.M., Brar, D.S., Eds.; Springer Netherlands: Dordrecht, The Netherlands, 2009; pp. 307–332.

108. Thudi, M.; Gaur, P.M.; Krishnamurthy, L.; Mir, R.R.; Kudapa, H.; Fikre, A.; Kimurto, P.; Tripathi, S.; Soren, K.R.; Mulwa, R.; et al. Genomics-assisted breeding for drought tolerance in chickpea. *Funct. Plant Biol.* **2014**, *41*, 1178–1190. [CrossRef]

109. Comastri, A.; Janni, M.; Simmonds, J.; Uauy, C.; Pignone, D.; Nguyen, H.T.; Marmiroli, N. Heat in wheat: Exploit reverse genetic techniques to discover new alleles within the triticum durum shsp26 family. *Front. Plant Sci.* **2018**, *9*. [CrossRef] [PubMed]

110. Daszkowska-Golec, A.; Skubacz, A.; Sitko, K.; Słota, M.; Kurowska, M.; Szarejko, I. Mutation in barley era1 (enhanced response to aba1) gene confers better photosynthesis efficiency in response to drought as revealed by transcriptomic and physiological analysis. *Environ. Exp. Bot.* **2018**, *148*, 12–26. [CrossRef]

111. Manmathan, H.; Shaner, D.; Snelling, J.; Tisserat, N.; Lapitan, N. Virus-induced gene silencing of arabidopsis thaliana gene homologues in wheat identifies genes conferring improved drought tolerance. *J. Exp. Bot.* **2013**, *64*, 1381–1392. [CrossRef] [PubMed]

112. Ogata, T.; Nagatoshi, Y.; Yamagishi, N.; Yoshikawa, N.; Fujita, Y. Virus-induced down-regulation of gmera1a and gmera1b genes enhances the stomatal response to abscisic acid and drought resistance in soybean. *PLoS ONE* **2017**, *12*, e0175650. [CrossRef] [PubMed]

113. Rogers, C.; Wen, J.; Chen, R.; Oldroyd, G. Deletion-based reverse genetics in *Medicago truncatula*. *Plant Physiol.* **2009**, *151*, 1077. [CrossRef] [PubMed]

114. Comai, L.; Young, K.; Till, B.J.; Reynolds, S.H.; Greene, E.A.; Codomo, C.A.; Enns, L.C.; Johnson, J.E.; Burtner, C.; Odden, A.R.; et al. Efficient discovery of DNA polymorphisms in natural populations by ecotilling. *Plant J.* **2004**, *37*, 778–786. [CrossRef] [PubMed]

115. HLG-SAM. New Techniques in Agricultural Biotechnology. Explanatory note 02, 2017. Available online: https://ec.europa.eu/research/sam/pdf/topics/explanatory_note_new_techniques_agricultural_ biotechnology.pdf#view=fit&pagemode=none (accessed on 23 November 2018).

116. Cardi, T.; Batelli, G.; Nicolia, A. Opportunities for genome editing in vegetable crops. *Emerg. Topics Life Sci.* **2017**, *1*, 193. [CrossRef]

117. Puchta, H. Applying crispr/cas for genome engineering in plants: The best is yet to come. *Curr. Opin. Plant Biol.* **2017**, *36*, 1–8. [CrossRef] [PubMed]

118. Pavan, S.; Jacobsen, E.; Visser, R.G.; Bai, Y. Loss of susceptibility as a novel breeding strategy for durable and broad-spectrum resistance. *Mol. Breed.* **2010**, *25*. [CrossRef] [PubMed]

119. Adli, M. The crispr tool kit for genome editing and beyond. *Nat. Commun.* **2018**, *9*, 1911. [CrossRef] [PubMed]

120. Cardi, T.; D'Agostino, N.; Tripodi, P. Genetic transformation and genomic resources for next-generation precise genome engineering in vegetable crops. *Front. Plant Sci.* **2017**, *8*, 241. [CrossRef] [PubMed]

121. Lowe, K.; Wu, E.; Wang, N.; Hoerster, G.; Hastings, C.; Cho, M.-J.; Scelonge, C.; Lenderts, B.; Chamberlin, M.; Cushatt, J.; et al. Morphogenic regulators *Baby boom* and *Wuschel* Improve Monocot Transformation. *Plant Cell* **2016**, *28*, 1998. [CrossRef] [PubMed]

122. Andersson, M.; Turesson, H.; Nicolia, A.; Fält, A.-S.; Samuelsson, M.; Hofvander, P. Efficient targeted multiallelic mutagenesis in tetraploid potato (*Solanum tuberosum*) by transient crispr-cas9 expression in protoplasts. *Plant Cell Rep.* **2017**, *36*, 117–128. [CrossRef] [PubMed]

123. Clasen, B.M.; Stoddard, T.J.; Luo, S.; Demorest, Z.L.; Li, J.; Cedrone, F.; Tibebu, R.; Davison, S.; Ray, E.E.; Daulhac, A.; et al. Improving cold storage and processing traits in potato through targeted gene knockout. *Plant Biotechnol. J.* **2016**, *14*, 169–176. [CrossRef] [PubMed]

124. Nicolia, A.; Proux-Wéra, E.; Åhman, I.; Onkokesung, N.; Andersson, M.; Andreasson, E.; Zhu, L.-H. Targeted gene mutation in tetraploid potato through transient talen expression in protoplasts. *J. Biotechnol.* **2015**, *204*, 17–24. [CrossRef] [PubMed]

125. Woo, J.W.; Kim, J.; Kwon, S.I.; Corvalán, C.; Cho, S.W.; Kim, H.; Kim, S.-G.; Kim, S.-T.; Choe, S.; Kim, J.-S. DNA-free genome editing in plants with preassembled crispr-cas9 ribonucleoproteins. *Nat. Biotechnol.* **2015**, *33*, 1162. [CrossRef] [PubMed]

126. Andersson, M.; Turesson, H.; Olsson, N.; Fält, A.-S.; Ohlsson, P.; Gonzalez, M.N.; Samuelsson, M.; Hofvander, P. Genome editing in potato via crispr-cas9 ribonucleoprotein delivery. *Physiol. Plant.* **2018**. [CrossRef] [PubMed]

127. Wolt, J.D.; Wang, K.; Yang, B. The regulatory status of genome-edited crops. *Plant Biotechnol. J.* **2016**, *14*, 510–518. [CrossRef] [PubMed]

128. Scientific opinion addressing the safety assessment of plants developed using zinc finger nuclease 3 and other site-directed nucleases with similar function. *EFSA J.* **2012**, *10*, 2943. [CrossRef]

129. Callaway, E. Crispr plants now subject to tough gm laws in european union. *Nature* **2018**, *560*, 16. [CrossRef] [PubMed]

130. Casacuberta, J.M.; Puigdomènech, P. European politicians must put greater trust in plant scientists. *Nature* **2018**, *561*, 33. [CrossRef] [PubMed]

131. Arora, L.; Narula, A. Gene editing and crop improvement using crispr-cas9 system. *Front. Plant Sci.* **2017**, *8*. [CrossRef] [PubMed]

132. Li, T.; Liu, B.; Spalding, M.H.; Weeks, D.P.; Yang, B. High-efficiency talen-based gene editing produces disease-resistant rice. *Nat. Biotechnol.* **2012**, *30*, 390. [CrossRef] [PubMed]

133. Wang, Y.; Cheng, X.; Shan, Q.; Zhang, Y.; Liu, J.; Gao, C.; Qiu, J.-L. Simultaneous editing of three homoeoalleles in hexaploid bread wheat confers heritable resistance to powdery mildew. *Nat. Biotechnol.* **2014**, *32*, 947. [CrossRef] [PubMed]

134. Nekrasov, V.; Wang, C.; Win, J.; Lanz, C.; Weigel, D.; Kamoun, S. Rapid generation of a transgene-free powdery mildew resistant tomato by genome deletion. *Sci. Rep.* **2017**, *7*, 482. [CrossRef] [PubMed]

135. Zheng, Z.; Appiano, M.; Pavan, S.; Bracuto, V.; Ricciardi, L.; Visser, R.G.; Wolters, A.-M.A.; Bai, Y. Genome-wide study of the tomato slmlo gene family and its functional characterization in response to the powdery mildew fungus oidium neolycopersici. *Front. Plant Sci.* **2016**, *7*, 380. [CrossRef] [PubMed]

136. Curtin, S.J.; Xiong, Y.; Michno, J.M.; Campbell, B.W.; Stec, A.O.; Čermák, T.; Starker, C.; Voytas, D.F.; Eamens, A.L.; Stupar, R.M. Crispr/cas9 and talen s generate heritable mutations for genes involved in small rna processing of glycine max and medicago truncatula. *Plant Biotechnol. J.* **2018**, *16*, 1125–1137. [CrossRef] [PubMed]

137. Farrell, A.D.; Rhiney, K.; Eitzinger, A.; Umaharan, P. Climate adaptation in a minor crop species: Is the cocoa breeding network prepared for climate change? *Agroecol. Sustain. Food Syst.* **2018**, 1–22. [CrossRef]

138. Zaidi, S.S.-e.-A.; Mukhtar, M.S.; Mansoor, S. Genome editing: Targeting susceptibility genes for plant disease resistance. *Trends Biotechnol.* **2018**, *36*. [CrossRef] [PubMed]

139. Huang, X.; Feng, Q.; Qian, Q.; Zhao, Q.; Wang, L.; Wang, A.; Guan, J.; Fan, D.; Weng, Q.; Huang, T.; et al. High-throughput genotyping by whole-genome resequencing. *Genome Res.* **2009**, *19*, 1068–1076. [CrossRef] [PubMed]

140. Terracciano, I.; Cantarella, C.; Fasano, C.; Cardi, T.; Mennella, G.; D'Agostino, N. Liquid-phase sequence capture and targeted re-sequencing revealed novel polymorphisms in tomato genes belonging to the mep carotenoid pathway. *Sci. Rep.* **2017**, *7*. [CrossRef] [PubMed]

141. Elshire, R.J.; Glaubitz, J.C.; Sun, Q.; Poland, J.A.; Kawamoto, K.; Buckler, E.S.; Mitchell, S.E. A robust, simple genotyping-by-sequencing (gbs) approach for high diversity species. *PLoS ONE* **2011**, *6*, e19379. [CrossRef] [PubMed]

142. Miller, M.R.; Dunham, J.P.; Amores, A.; Cresko, W.A.; Johnson, E.A. Rapid and cost-effective polymorphism identification and genotyping using restriction site associated DNA (rad) markers. *Genome Res.* **2007**, *17*, 240–248. [CrossRef] [PubMed]

143. Davey, J.W.; Blaxter, M.L. Radseq: Next-generation population genetics. *Brief. Funct.Genom.* **2010**, *9*, 416–423. [CrossRef] [PubMed]

144. Wang, S.; Wong, D.; Forrest, K.; Allen, A.; Chao, S.; Huang, B.E.; Maccaferri, M.; Salvi, S.; Milner, S.G.; Cattivelli, L.; et al. Characterization of polyploid wheat genomic diversity using a high-density 90,000 single nucleotide polymorphism array. *Plant Biotechnol. J.* **2014**, *12*, 787–796. [CrossRef] [PubMed]

145. Pavan, S.; Lotti, C.; Marcotrigiano, A.R.; Mazzeo, R.; Bardaro, N.; Bracuto, V.; Ricciardi, F.; Taranto, F.; D'Agostino, N.; Schiavulli, A.; et al. A distinct genetic cluster in cultivated chickpea as revealed by genome-wide marker discovery and genotyping. *Plant Genome* **2017**, *10*. [CrossRef] [PubMed]

146. Pavan, S.; Marcotrigiano, A.R.; Ciani, E.; Mazzeo, R.; Zonno, V.; Ruggieri, V.; Lotti, C.; Ricciardi, L. Genotyping-by-sequencing of a melon (*Cucumis melo L.*) germplasm collection from a secondary center of diversity highlights patterns of genetic variation and genomic features of different gene pools. *BMC Genom.* **2017**, *18*. [CrossRef] [PubMed]

147. Pavan, S.; Curci, P.L.; Zuluaga, D.L.; Blanco, E.; Sonnante, G. Genotyping-by-sequencing highlights patterns of genetic structure and domestication in artichoke and cardoon. *PLoS ONE* **2018**, *13*, e0205988. [CrossRef] [PubMed]

148. D'Agostino, N.; Taranto, F.; Camposeo, S.; Mangini, G.; Fanelli, V.; Gadaleta, S.; Miazzi, M.M.; Pavan, S.; di Rienzo, V.; Sabetta, W. Gbs-derived snp catalogue unveiled wide genetic variability and geographical relationships of italian olive cultivars. *Sci. Rep.* **2018**, *8*. [CrossRef] [PubMed]

149. Varshney, R.K.; Ribaut, J.-M.; Buckler, E.S.; Tuberosa, R.; Rafalski, J.A.; Langridge, P. Can genomics boost productivity of orphan crops? *Nat. Biotechnol.* **2012**, *30*, 1172–1176. [CrossRef] [PubMed]

150. Brazas, M.D.; Blackford, S.; Attwood, T.K. Plug gap in essential bioinformatics skills. *Nature* **2017**, *544*, 161. [CrossRef] [PubMed]

151. Lipper, L.; Thornton, P.; Campbell, B.M.; Baedeker, T.; Braimoh, A.; Bwalya, M.; Caron, P.; Cattaneo, A.; Garrity, D.; Henry, K. Climate-smart agriculture for food security. *Nat. Climate Change* **2014**, *4*. [CrossRef]

152. Varshney, R.K.; Singh, V.K.; Kumar, A.; Powell, W.; Sorrells, M.E. Can genomics deliver climate-change ready crops? *Curr.Opin. Plant Biol.* **2018**, *45*, 205–211. [CrossRef] [PubMed]

9

Characterization and Gene Mapping of *non-open hull 1 (noh1)* Mutant in Rice (*Oryza sativa* L.)

Jun Zhang [†], Hao Zheng [†], Xiaoqin Zeng, Hui Zhuang, Honglei Wang, Jun Tang, Huan Chen, Yinghua Ling and Yunfeng Li *

Rice Research Institute, Key Laboratory of Application and Safety Control of Genetically Modified Crops, Academy of Agricultural Sciences, Southwest University, Chongqing 400715, China; JZ563156@163.com (J.Z.); hotzheng@126.com (H.Z.); swuzxq1991@163.com (X.Z.); amberzzh@163.com (H.Z.); WHL7991@163.com (H.W.); 18202398740@163.com (J.T.); chappyj@163.com (H.C.); lingyh65@swu.edu.cn (Y.L.)
* Correspondence: liyf1980@swu.edu.cn
† These authors contributed equally to this work.

Abstract: Hull opening is a key physiological process during reproductive development, strongly affecting the subsequent fertilization and seed development in rice. In this study, we characterized a rice mutant, *non-open hull 1 (noh1)*, which was derived from ethylmethane-sulfonate (EMS)-treated Xinong 1B (*Oryza sativa* L.). All the spikelets of *noh1* developed elongated and thin lodicules, which caused the failure of hull opening and the cleistogamy. In some spikelets of the *noh1*, sterile lemmas transformed into hull-like organs. qPCR analysis indicated that the expression of A- and E-function genes was significantly upregulated, while the expression of some B-function genes was downregulated in the lodicules of *noh1*. In addition, the expression of A-function genes was significantly upregulated, while the expression of some sterile-lemma maker genes was downregulated in the sterile lemma of *noh1*. These data suggested that the lodicule and sterile lemma in *noh1* mutant gained glume-like and lemma-like identity, respectively. Genetic analysis showed that the *noh1* trait was controlled by a single recessive gene. The *NOH1* gene was mapped between the molecular markers ZJ-9 and ZJ-25 on chromosome 1 with a physical region of 60 kb, which contained nine annotated genes. These results provide a foundation for the cloning and functional research of *NOH1* gene.

Keywords: gene mapping; lodicule; *non-open hull 1(noh1)*; rice

1. Introduction

Rice floret contains four whorls of floral organs, which are made up of one lemma, one palea, two lodicules, six stamens, and one carpel from outside to inside, respectively [1]. After fertilization, the lodicules and stamens gradually degenerate, while the carpel develops into seed. The lemma and palea hook together most of the time to protect the internal organs and seeds. Only during the flowering date, the lemma and palea in each floret open once from 09:00 to 11:00 under normal circumstances. This process generally lasts 40 to 90 min and then the lemma and palea close and never reopen. These two actions are called hull opening (the lemma and the palea together are called a hull) and hull closing, respectively. Hull opening is a necessary condition for plant fertilization, and hull closing protects the development of seed from outside interference after fertilization.

Lodicules, two small, fleshy, and lung-like floral organs, asymmetrically formed inside the lemma, play a crucial role in promoting hull opening/closing by their expansion and shrinkage. During flowering, the water potential of lodicule cells decreases, and then lodicules absorb water and cause the lodicules to swell. The swollen lodicules then push lemma outwards and simultaneously squeeze the palea inwards, which causes the hook of the lemma and palea to release, and thus the lemma and the palea are separated from each other. At this time, the filaments are rapidly elongated, the anthers stick out of the hulls and crack, and then the pollination begins [2]. It is also believed that the excessive water absorption in lodicule cells finally leads the cells to rupture and shrink [2]. With the lodicules losing the supports to the lemma, the lemma and palea lock together again and the floret gets closed [3].

The identity of lodicules is mainly determined by B-class genes *OsMADS2*, *OsMADS4*, and *OsMADS16* in the ABCDE model, all of which encode transcription factors containing MADS-box domain [4–6]. Another two MADS-box genes, *OsMADS6* and *OsMADS32*, are also responsible for regulating lodicule identity [7,8]. In addition to the MADS-box genes, the *STAMENLESS1 (SL1)* gene encoding a C2H2 zinc finger protein is also involved in the regulation of lodicule development [9]. These genes all regulate the development of the morphological characteristics of the lodicules. Their mutations usually cause the transformation of lodicules into other floral organs, which generally lose their normal function and fail to mediate hull opening/closing. In fact, cleistogamy may be a favorable trait for self-pollination crops (such as rice, wheat) to avoid failure of pollinating caused by adverse weather conditions, while it may prevent transgenes from spreading into the environment by reducing outcrossing rates.

In this study, a rice mutant with failure of hull opening, named *non-open hull 1 (noh1)*, is reported. The lodicules in *noh1* mutant transformed into glume-like organs, and lost the water swelling function, which caused the rice floret to remain in a closed state and complete cleistogamy during the whole development process. The *NOH1* gene has been located on chromosome 1, which provides a foundation for the cloning and further function research of *NOH1*.

2. Materials and Methods

2.1. Plant Materials

The *noh1* mutant derived from an EMS mutagenesis population which used *xian*-type *(indica)* maintainer line Xinong 1B cultivated by Rice Research Institute of Southwest University as donor. The *noh1* mutation was stably inherited through successive generations of self-crossing. F_1 population was generated by the cross of *xian*-type *(indica)* sterile line 56S with *noh1*. The parent plants, F_1 population and F_2 population, were all grown in Chongqing for genetic analysis, and then plants with mutant phenotype in F_2 population were used to map the *NOH1* gene.

2.2. Morphological and Histological Analysis of noh1

During flowering stage, the phenotypic characteristics of the mutants and wild-type spikelets were investigated using a Nikon SMZ1500 stereoscope (Nikon Instruments Shanghai Inc., China) and a Hitachi SU3500 (Hitachi High-Technologies Corporation, Tokyo, Japan) scanning electron microscope with a −20 °C cooling stage under a low-vacuum environment.

For paraffin section, the mutant and wild-type spikelets at heading stage were fixed in FAA (50% (*v/v*) ethanol, 0.9 N glacial acetic, and 3.7% (*v/v*) formaldehyde) and placed at 4 °C for at least 16 h after pumping the air in the tissue. Next, those spikelets were dehydrated by a gradient ethanol series, infiltrated by xylene, embedded into paraffin (Sigma-Aldrich Inc., Shanghai, China), cut into 8-μm-thick slices, and then pasted on the microscope slides (RM2245; Leica, Hamburg, Germany). These slices were dyed sequentially with 1% (*w/v*) safranin (Amresco Inc., Framingham, MA, USA) and 1% (*w/v*) Fast Green (Amresco Inc., Framingham, MA, USA) and then dehydrated through an ethanol series, infiltrated with xylene, and finally mounted beneath a coverslip. Light microscopy was performed using a Nikon E600 microscope.

2.3. Molecular Mapping of NOH1

Locating the target gene was performed according to the BSA (bulked segregant analysis) method [10]. DNA of the parents, F_2 population, wild-type, and mutant gene pools were extracted from similar sized fresh leaves following the previously reported CTAB (hexadecyl trimethyl ammonium bromide) method [11]. The quality and quantity of DNA were estimated using a NanoDrop 2000 spectrophotometer (Thermo Fisher Scientific Inc., Wilmington, DE, USA) and 1% (w/v) agarose gel electrophoresis. SSR (simple sequence repeats) markers that were distributed evenly on the 12 chromosomes were employed for gene mapping from the Web (http://www.gramene.org/ microsat/) (see Table S1 in the Supplementary Materials). All the primers were synthesized by the Shanghai Invitrogen Company. The total volume of the PCR amplifications was 15 μL, which contained 1.5 μL 10× PCR buffer, 1 μL 50 ng μL^{-1} DNA, 0.75 μL 2.5 mmol L^{-1} dNTPs, 9.5 μL ddH$_2$O, 1 μL 10 mmol L^{-1} forward and reverse primer, and 0.25 μL 5 U μL^{-1} rTaq DNA polymerase (Takara Bio Inc., Dalian, China). Amplification was performed with a MyCycler Thermal Cycler (Bio-Rad, Foster City, CA, USA) under the following conditions: 5 min at 94 °C for DNA strand separation, followed by 35 cycles of denaturing at 94 °C for 30 s, annealing at 56 °C for 30 s, extension at 72 °C for 30s, and finally an extension at 72 °C for 7 min. Amplified products were separated by electrophoresis on 10.0% polyacrylamide gels, and then silver staining was used to observe the color of the band patterns [12].

2.4. Linkage Map Construction

Bands for molecular markers were identified in both parents, 56S and *noh1*, and labeled as A and B, respectively, and the heterozygote that contains two parent bands was labeled as H. The linkage relationship was analyzed by MAPMAKER3.0 [13], and the recombination rate was represented as the number of recombinants. Meanwhile, we constructed the physical map based on the rice genome sequence information offered by the Gramene website (http://www.gramene.org/).

2.5. qPCR Analysis

After appearance of the non-open hull phenotype at the flowering stage, we collected about 60 pairs of lodicules from several plants with the corresponding phenotype using a Nikon SMA1500 stereoscope and extracted total RNA from the lodicules of wild-type and *noh1* florets using the RNAprep Pure Plant Kit (Tiangen Biochemical Technology Co., Ltd., Beijing, China). The first-strand cDNA was synthesized from 1 μg total RNA with oligo(dT)$_{18}$ primers in a 20 μL reaction volume using the PrimeScript® Reagent Kit With gDNA Eraser (Takara Bio Inc., Dalian, China). Several representative floral specific genes were selected by consulting the literature and their expression levels (*OsMADS1, OsMADS2, OsMADS4, OsMADS6, OsMADS16, OsMADS14, OsMADS15, OsMADS34, G1, ASP1* and *DL*) were detected by qPCR. The qPCR analysis was performed with three replicates using SYBR premix Ex Taq II Kit (Takara Bio Inc., Dalian, China) in an ABI 7500 Sequence Detection System (Applied Biosystems, Carlsbad, CA, USA). *ACTIN* was used as an endogenous control (see Table S4 in the Supplementary Materials).

3. Results

3.1. Morphological Analysis of noh1 Mutant

While no significant defect was observed during the vegetative stage, some significant abnormalities were exhibited in the *noh1* spikelets at the reproductive stage in comparison with the wild-type spikelets. Generally, the wild-type spikelet has two pairs of glumes (rudimentary glume and sterile lemma) and one fertile top floret, consisting of a lemma and a palea in whorl 1, two lodicules in whorl 2, six stamens in whorl 3, and one carpel in whorl 4 (Figure 1a,b). The lodicules of the wild-type look fleshy and semitransparent with smooth surfaces (Figure 1c–e). However, in the *noh1* floret, the lodicules were elongated and became much thinner and narrower than those in the wild-type, while the identities of other floral organs were normal (Figure 1h–l). Furthermore, it was observed that these slim lodicules developed linear cells in parallel arrays in their smooth upper epidermis, which was similar to the epidermis of wild-type sterile lemma (Figure 1e,g,l). In addition, most of the sterile lemmas in *noh1* mutants were also obviously elongated and showed abundant trichomes and protrusions in their upper epidermis, which was almost the same as that in the hull (lemma and palea) (Figure 1f,m).

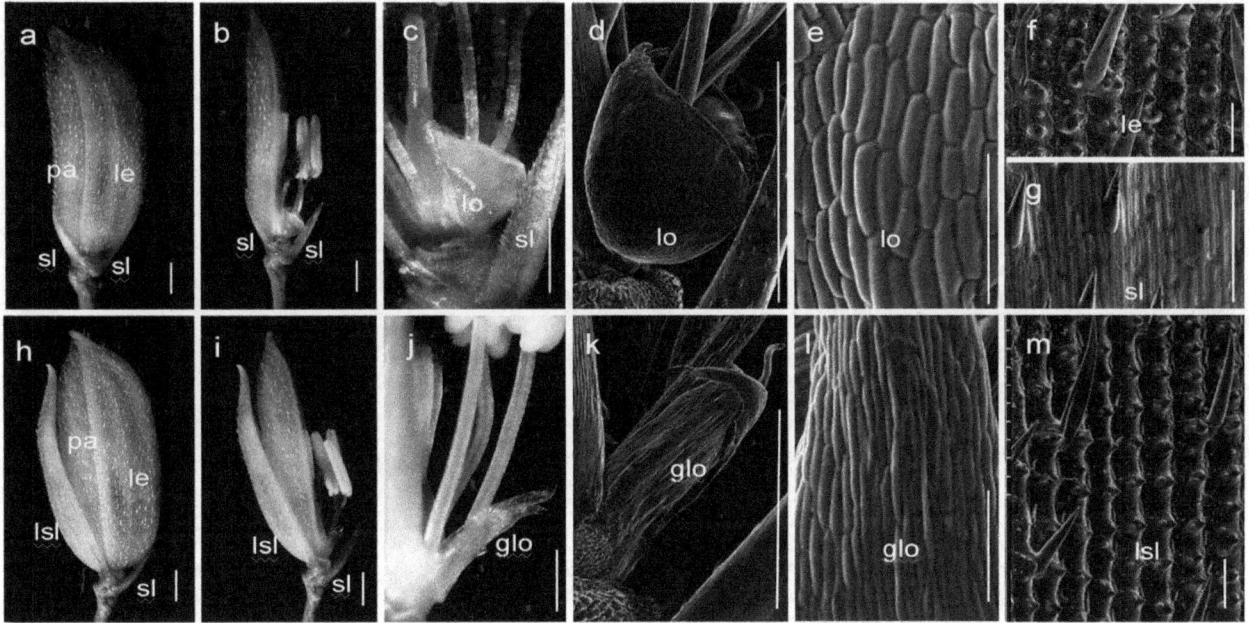

Figure 1. Morphological analysis of spikelet of the wild-type (Xinong 1B) and *noh1* mutant. (**a–c**) the wild-type spikelet; (**d–e**) the wild-type lodicule; (**f**) the wild-type lemma; (**g**) the wild-type sterile lemma; (**h–j**) the *noh1* spikelet; (**k–l**) the glume-like lodicule in *noh1*; (**m**) the lemma-like sterile lemma of *noh1*. glo: glume-like lodicule; lsl: lemma-like sterile lemma; le: lemma; lo: lodicule; pa: palea; sl: sterile lemma; bars represent 1000 μm (**a–d,h–k**) and 100 μm (**e–g,l,m**).

3.2. Histological Analysis of noh1 Mutant

To further clarify the identities of floral organs in *noh1* mutants, a histological analysis was performed. In the wild-type florets, the lodicules were composed of lots of large parenchymatous cells and radial small vascular bundles (Figure 2a,b), while the sterile lemma consisted of three layers including upper and lower epidermal cells and middle parenchymatous cells (Figure 2a,c). However, in the thinner lodicules of *noh1* florets, the number of parenchymatous cell layers decreased significantly, and even the number of vascular bundles was reduced to only one (Figure 2d,e), which caused it to look like the cell structure of the wild-type sterile lemma. Furthermore, the wild-type lemma consisted of four cell layers including the silicified upper epidermis, fibrous sclerenchyma, spongy parenchyma cells, and lower epidermis, with five vascular bundles. The wild-type glume contained three kinds of cell types, smooth upper epidermal cells, parenchymatous cells, and lower epidermal cells, with only one vascular bundle. It was found that the cell structure of the elongated sterile lemma in *noh1* mutants was not similar to the wild-type glume but to the wild-type lemma (Figure 2c,f).

3.3. Morphological Analysis of noh1 Lodicules during Flowering

In the wild-type floret of rice, about one hour before flowering, the lemma and the palea were still interlocked tightly and the lodicules were not plump enough (Figure 3a,e,i). At about 20 min after floret opening, the lodicules absorbed water and notably expanded, while the filaments were elongated and the anthers were cracked, which allowed pollination to occur (Figure 3b,f,j). About 1.5 h after blossoming, the lodicules gradually shrank and the hull began to close (Figure 3c,g,k). About 48 h after flowering, the hull had closed completely with full shrinkage of the lodicules, and the ovary began to develop (Figure 3d,h,l). Therefore, the process from hull opening to closing is highly relevant to the expansion and atrophy of the lodicules.

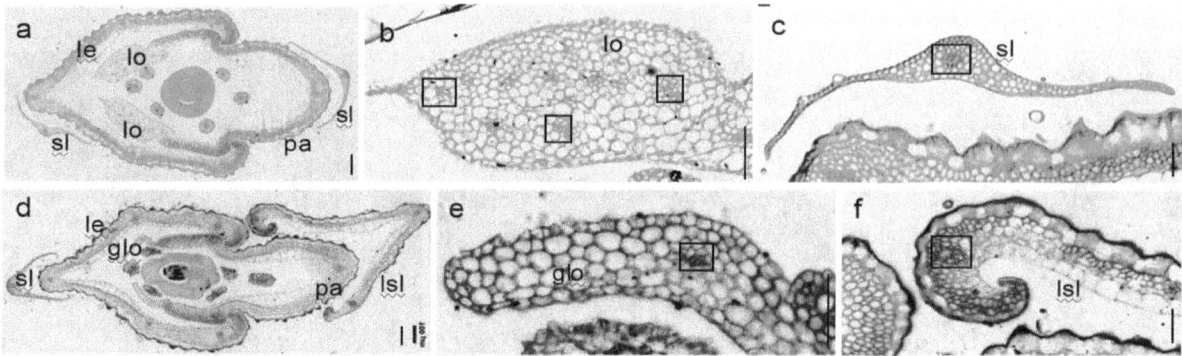

Figure 2. Histological analysis of spikelet at flowering stage in the wild-type and *noh1* plants; (**a–c**) the cross-sections of the wild-type spikelet; (**d–f**) the cross-sections of the *noh1* spikelet. glo: glume-like lodicule; lsl: lemma-like sterile lemma; le: lemma; lo: lodicule; pa: palea; sl: sterile lemma; □: vascular bundle; bars represent 100 μm (**a,d**) and 50 μm (**b,c,e,f**).

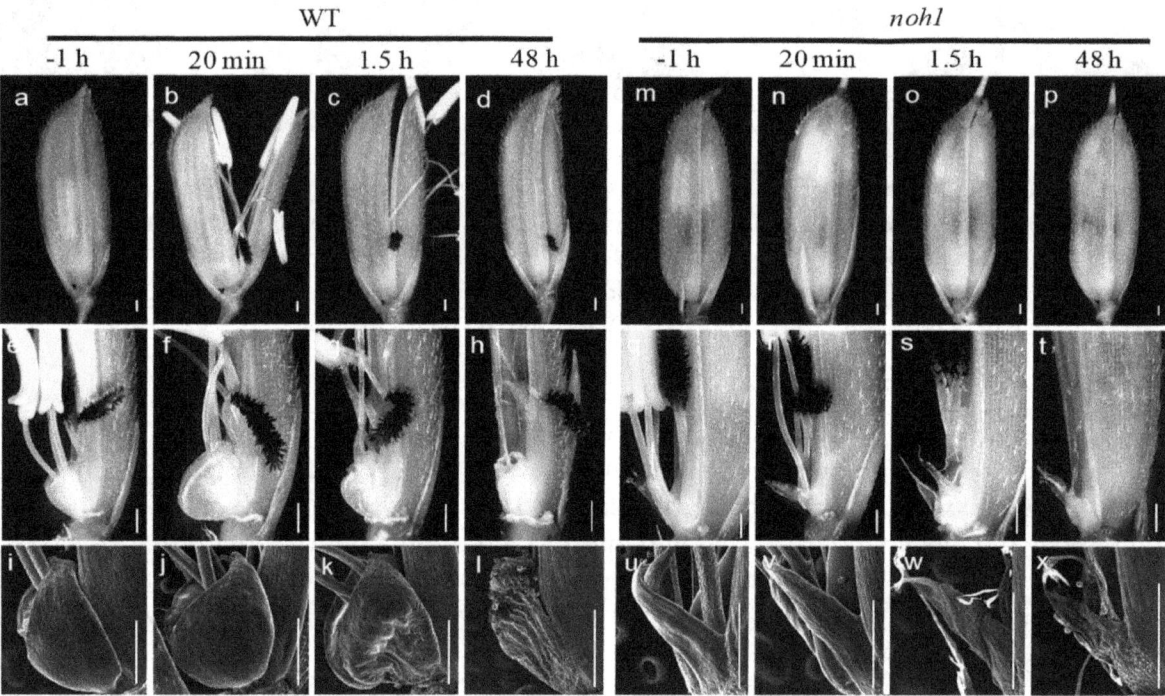

Figure 3. Morphological observation of wild-type (WT) and *noh1* mutant florets during flowering; (**a–d**) show successively the WT florets at one hour before opening, 20 min after opening, 1.5 h after opening. and 48 h after opening; (**m–p**) show successively the *noh1* florets at one hour before opening, 20 min after opening, 1.5 h after opening, and 48 h after opening; (**e–h**) show successively the morphological features of the WT lodicule at one hour before opening, 20 min after opening, 1.5 h after opening, and 48 h after opening; (**q–t**) show successively the morphological features of the *noh1* lodicule at one hour before opening, 20 min after opening, 1.5 h after opening, and 48 h after opening; (**i–l**) show successively the microscopic structure of the WT lodicule at one hour before opening, 20 min after opening, 1.5 h after opening, and 48 h after opening; (**u–x**) show successively the microscopic structure of the *noh1* lodicule at one hour before opening, 20 min after opening, 1.5 h after opening, and 48 h after opening; bars represent 500 μm.

Compared with the wild-type, obvious differences were observed during the *noh1* floret opening. Firstly, the hull in *noh1* florets never opened during the whole flowering stage (Figure 3m–p). Secondly, comparing the florets with elongated filaments (which means that the florets have been opened or pollinated) with the florets with non-elongated filaments (which means that the florets have not been opened or pollinated), no obvious swelling was observed in the lodicules during the *noh1* floret opening or pollination (Figure 3u,v). Fortunately, the other processes related to floret opening, such as filament elongation, anther dehiscence, and pollination, were almost normal so that the *noh1* florets were still fertile.

3.4. Expression Analysis of Floral Organ Identity Genes in Lodicule of noh1 and Wild-Type

Given that the *noh1* lodicules and sterile lemma showed glume/hull-like identity in morphological features, the expression of some spikelet/floral organ identity genes were detected, which includes *OsMADS6* (which was expressed in the lodicule and palea) [7], *DL* (which was expressed in the lemma and pistil) [14,15], *G1* (which was expressed in the sterile lemma and palea) [16], *ASP1* (which was mainly expressed in the sterile lemma) [17], A class genes *OsMADS14* and *OsMADS15* (which were expressed in the lemma and palea) [18,19], B class genes *OsMADS2*, *OsMADS4*, and *OsMADS16/SPW1* (which were expressed in the lodicule and stamen) [4–6], E class gene *OsMADS1* (which was mainly expressed in the lemma and palea), and *OsMADS34* (which was mainly expressed in the sterile lemma, lemma, and palea) [20,21]. Compared with that of the wild-type plants, it was not surprising that the relative expression level of three B class genes *OsMADS2*, *OsMADS4*, and *OsMADS16*, which were all involved in the development of lodicules, were downregulated in the lodicules of *noh1* mutants. On the contrary, the genes related to the identities of the sterile lemma and hull were significantly upregulated, such as *OsMADS1*, *OsMADS14* and *DL*, or even extremely upregulated, such as *OsMADS15* and *G1* (Figure 4a,b). In the elongated sterile lemma of *noh1*, the expression of *OsMADS34*, *G1*, and *ASP1* was significantly downregulated, while the expression levels of *OsMADS1*, *OsMADS14*, *OsMADS15*, and *DL* were upregulated with varying degrees (Figure 4c,d). In addition, the expression of *OsMADS6*, which was related to palea but not lemma, was also significantly downregulated in the elongated sterile lemma of *noh1* mutants.

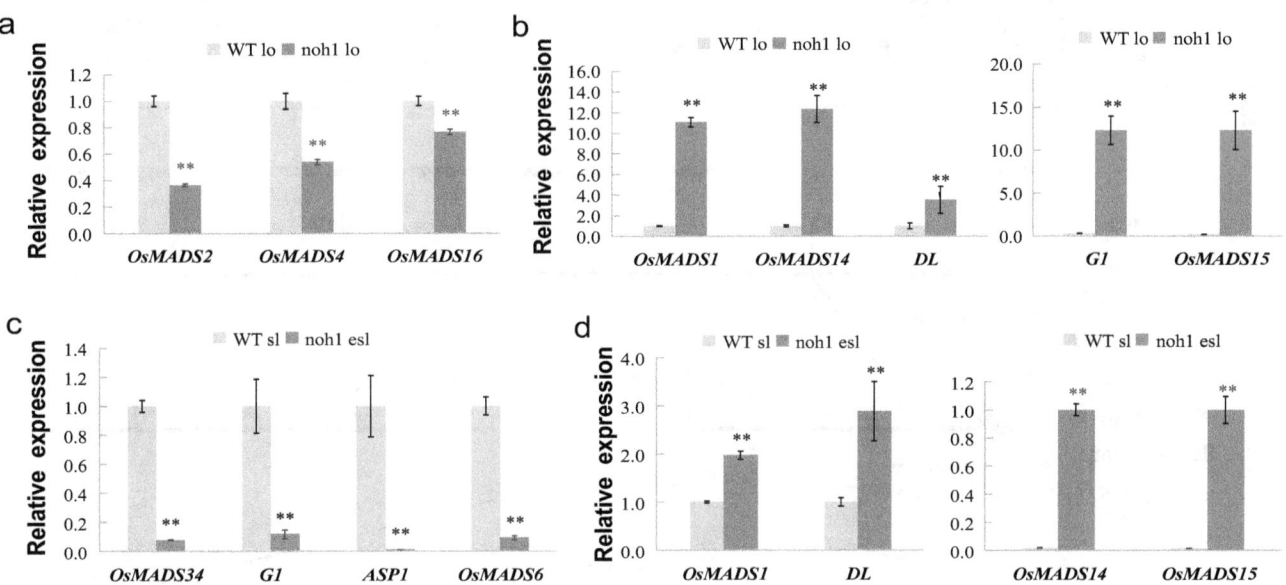

Figure 4. Relative expression levels of floral organ identity genes in the wild type (WT) and *noh1* floral organs; (**a**) B class genes downregulated in the *noh1* lodicule; (**b**) genes partially upregulated in the *noh1* lodicule; (**c**) genes partially downregulated in the *noh1* elongated sterile lemma (about 15 pairs of sterile lemmas with corresponding phenotype were collected by using a pointy tweezer to extract the total RNA); (**d**) genes partially upregulated in the *noh1* elongated sterile lemma. lo: lodicule; sl: sterile lemma; esl: elongated sterile lemma; error bars indicate SD. ** indicates $p < 0.01$.

3.5. Genetic Analysis

The *noh1* mutant was crossed with the sterile line 56S. All F_1 plants exhibited normal phenotype, while F_2 plants showed the normal phenotype or *noh1* mutant phenotype. Genetic analysis of the F_2 progeny showed that the segregation ratio of WT plants and mutant individuals was a good fit to 3:1 (536 of 2242 were mutant individuals; $\chi^2 = 1.37 < \chi^2_{0.05} = 3.84$), indicating that the *noh1* trait was controlled by a single recessive gene.

3.6. Gene Mapping of NOH1

Recessive individuals in the F_2 population were used as a mapping population to localize the *NOH1* gene. Ten wild-type plants and ten mutant plants in the F_2 segregating population were randomly selected to construct wild-type and mutant DNA pools, respectively. About 430 pairs of primers for SSR and IN/DEL (insertion-deletion) markers, which were uniformly distributed on the 12 chromosomes of the parents 56S and *noh1* (see Table S1 in the Supplementary Materials), were used. A total of 112 pairs of primers for molecular markers showed polymorphism between the parents and were employed to screen the wild-type gene pool and the mutant gene pool (see Table S2 in the Supplementary Materials). By the linkage analysis, the target gene was linked to the polymorphic markers M33, M69, M76, and M86 on chromosome 1 (Table 1). Therefore, we used the four markers to survey the 300 mutant individuals, and then the *NOH1* gene was localized between SSR markers M33 and M76 on the long arm of chromosome 1 with recombinants of 47 and 3, respectively (Figure 5a).

Table 1. Polymorphic markers for gene mapping.

Primer	Forward Sequence (5′–3′)	Revers Sequence (5′–3′)
M33	CTTGAGTTCGAAGCGAGAAGACG	CACTTGAGCTCGAGACGTAGCC
ZJ-9	CAGATGGAGTACATGAAGTGCCAATG	GCATTGTGTCAACAACTCAGGTCC
ZJ-25	CACGGTAATGTGCTAAAGCTCCTG	GTGGGTTGTGGAGAGACAACCTG
ZJ-30	CAAGAAGCTCAACCAGGACGGCTTC	GAGAGTAGAGTTGAGGCACCGAATCG
M69	CTCTACAGCTTGAGTTTGGTACATCC	GTGTTGGTGAGCTAGCTGTTGC
M76	GTCGACGGCTTCCTCAAGATTGG	TGAGACCTCTGTGAAGGCACTCG
M86	CTCACTCACTGACCCACAACTCC	TTAAGATGATGGCTCCTCTCTGC

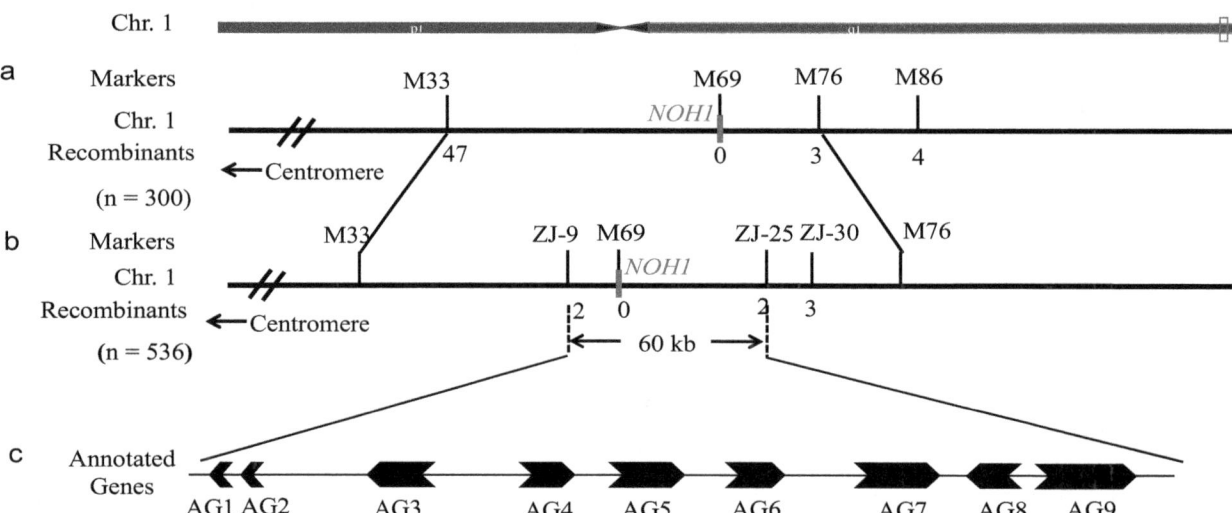

Figure 5. Localization of the target gene *NOH1* on Chromosome 1 (Chr.1) of rice. (**a**) Primary mapping of *NOH1* on chromosome 1 based on 300 individuals; (**b**) *NOH1* was fine-mapped to an interval of 60 kb by 536 individuals; (**c**) Nine genes were annotated on the 60 kb region. *AG1: miR172b*; *AG2: miR806a*; *AG3: LOC_Os01g74130*; *AG4: LOC_Os01g74140*; *AG5: LOC_Os01g74146*; *AG6: LOC_Os01g74152*; *AG7: LOC_Os01g74160*; *AG8: LOC_Os01g74170*; *AG9: LOC_Os01g74180*.

In order to further locate *NOH1*, 36 pairs of primers for new markers between M33 and M76 were synthesized (see Table S3 in the Supplementary Materials). Of those primers, ZJ-9, ZJ-25, and ZJ-30 exhibited polymorphism (Table 1). Then we used these primers to further analyze all of the 536 mutant individuals. Ultimately, the *NOH1* gene was localized between SSR markers ZJ-9 and ZJ-25 with recombinants of 2 and 2, respectively (Figure 5b). The physical distance was approximate 60 kb with nine annotated genes (seven Open Reading Frames and two non-coding RNA genes), referring to the BAC (Bacterial Artificial Chromosome) (AP003277) of the sequencing variety *Nipponbare* (Figure 5c).

4. Discussion and Conclusions

The lodicule is a grass-specific organ that is generally regarded as a homolog of dicot petal [22,23], which plays a vital role in floret opening in most grass species. The expansion and atrophy of lodicules can promote the opening and closing of the hull, respectively. It was reported earlier that the rice B-class genes *OsMADS2*, *OsMADS4*, and *OsMADS16* specified the identity of lodicules [4–6,15,24]. In the mutants that had non-functional copies of these genes, the lodicules were elongated and/or transformed into glume-like structures. Moreover, recent studies revealed that *OsMADS6* and *OsMADS32* were also responsible for the regulation of lodicule identity, and that mutations of these genes resulted in the transformation of the lodicule into a glume-like structure [25,26]. In the present study, compared with the WT plants, the lodicule identity genes *OsMADS2*, *OsMADS4*, and *OsMADS16* were significantly downregulated and the lemma identity-related genes *OsMADS1*, *OsMADS14*, *DL*, *OsMADS15*, and *G1* were largely upregulated in lodicules of *noh1* mutants, which indicated that those elongated lodicules in the *noh1* mutants lost the identity of the lodicule and might acquire the glume-like (both lemma and sterile lemma) identity to some extent. Considering morphological and histological characteristics, there was strong support for those elongated lodicules being partially transformed into sterile lemma-like organs. Therefore, this change of the *noh1* lodicule made it lose the water swelling function, which eventually led to the failure of the hulls to be pushed open, similar to the mutants of B-class genes, *OsMADS6*, *OsMADS16*, and *OsMADS32*. However, most mutants of these genes were either completely or partially sterile because of the defects of other floral organs, although the florets in *noh1* and *spw1-cls* (a allele mutation in the *OsMADS16* gene, which only led to transformation of lodicules, but not the stamens [27]) plants were perfectly fertile due to the normal function of the other three floral organs. Therefore, cleistogamy might be an efficient strategy to consider for preventing gene flow from genetically modified crops, and even has positive effect on the seeds setting rate for the self-pollination crops especially during the rainy season; thus, the *noh1* mutant could provide a favorable trait for conventional breeding [27].

The sterile lemma is a unique organ structure of Gramineae and is always considered a degenerate lemma [28,29]. In this study, some *noh1* sterile lemmas were elongated, and displayed rough upper epidermis with abundant trichomes and protrusions, which were similar to the wild-type lemma. Previous studies revealed that *G1*, *OsMADS34*, and *ASP1* genes determined the sterile lemma identity, and inhibited the transformation to the lemma [16,17,21,30–33]. In the elongated sterile lemma of *noh1* plants, the expression levels of *OsMADS1*, *OsMADS14*, *OsMADS15*, and *DL* (the lemma identity-related genes) were upregulated, while *OsMADS34*, *G1*, *ASP1*, and *OsMADS6* expressions were downregulated significantly. These results indicated that the elongated sterile lemma had lost the sterile lemma identity and simultaneously gained the identity of the lemma instead of palea in *noh1* mutants. Therefore, *NOH1* might take effect in the sterile lemma development by regulating one or several of these genes.

In this study, the *NOH1* gene was mapped on the long arm of chromosome 1 with a physical distance of 60 kb between the molecular markers ZJ-9 and ZJ-25. According to the gene annotation information provided by the Gramene website (http://www.gramene.org/), seven annotated genes and two non-coding RNA are included within this interval: a retrotransposon protein (*LOC_Os01g74130*), a transcription factor (*LOC_Os01g74140*), a cell cycle switch protein (*LOC_Os01g74146*), two enzymes (*LOC_Os01g74152* and *LOC_Os01g74160*), a expressed protein

(*LOC_Os01g74170*), an adaptin (*LOC_Os01g74180*), and non-coding RNA genes *miR172b* and *miR806a*. Among these genes, *LOC_Os01g74140* encodes a WRKY transcription factor, which regulates many metabolic processes and participates in the process of disease resistance, injury, senescence, growth, and gibberellin signal transduction in plants [34]. *LOC_Os01g74146* is a B-type cell cycle switch gene in rice that is highly expressed in floral organs and chaff. Particularly, *Cly1*, one of the *miR172* target genes in barley, was reportedly responsible for lodicule development, a single nucleotide substitution of which at the *miR172* target sequence led to smaller lodicules and cleistogamy [35]. In this study, we could not detect any sequence differences in all of the seven protein-coding genes between the *noh1* mutant and the wild-type. As there are a lot of repetitive regions near the *miR172b* and *miR806a*, it still remains unclear whether there is any mutation located there. Next, resequencing and/or epigenetic analysis will be used to determine which kind of mutation results in the *noh1* trait.

The *noh1* mutant displayed glume-like lodicules and lemma-like sterile lemma. The glume-like lodicules caused the *noh1* floret to lose the capability of hull opening and cleistogamy, which might be a favorable trait for conventional breeding. Genetic analysis revealed that the *noh1* trait was controlled by a single recessive gene and finally mapped between ZJ-9 and ZJ-25 on chromosome 1 with a 60 kb region. Results of our research lay a foundation for map-based cloning and function analysis of *NOH1* gene.

Supplementary Materials: Table S1: Polymorphic markers for polymorphic screening, Table S2: Polymorphic markers for gene pool screening, Table S3: Primers synthesized for fine mapping on chromosome 1, Table S4: Primers used for qPCR analysis.

Author Contributions: Data curation: J.Z. and H.Z.; Formal analysis: X.Z. and H.Z.; Funding acquisition: Y.L. (Yinghua Ling) and Y.L. (Yunfeng Li); Methodology: J.Z.; Software: H.Z.; Validation: H.W., J.T., and H.C.; Writing—original draft: J.Z.; Writing—review and editing: Y.L. (Yunfeng Li).

References

1. Yoshida, H.; Nagato, Y. Flower development in rice. *J. Exp. Bot.* **2011**, *62*, 4719–4730. [CrossRef] [PubMed]
2. Zeng, X.; Zhou, X.; Wu, X. Advances in study of opening mechanism in rice florets. *Sci. Agric. Sin.* **2004**, *37*, 188–195. [CrossRef]
3. Wang, Z.; Gu, Y.; Gao, Y. Studies on the mechanism of the anthesis of rice III. Structure of the lodicule and changes of its contents during flowering. *Acta Agron. Sin.* **1991**, *17*, 96–101. (In Chinese)
4. Xiao, H.; Wang, Y.; Liu, D.; Wang, W.; Li, X.; Zhao, X.; Xu, J.; Zhai, W.; Zhu, L. Functional analysis of the rice *AP3* homologue *OsMADS16* by RNA interference. *Plant Mol. Biol.* **2003**, *52*, 957–966. [CrossRef] [PubMed]
5. Yadav, S.R.; Prasad, K.; Vijayraghavan, U. Divergent regulatory *OsMADS2* functions control size, shape and differentiation of the highly derived rice floret second-whorl organ. *Genetics* **2007**, *176*, 283–294. [CrossRef]
6. Yao, S.G.; Ohmori, S.; Kimizu, M.; Yoshida, H. Unequal genetic redundancy of rice *PISTILLATA* orthologs, *OsMADS2* and *OsMADS4*, in lodicule and stamen development. *Plant Cell Physiol.* **2008**, *49*, 853–857. [CrossRef]
7. Li, H.; Liang, W.; Jia, R.; Yin, C.; Zong, J.; Kong, H.; Zhang, D. The *AGL6*-like gene *OsMADS6* regulates floral organ and meristem identities in rice. *Cell Res.* **2010**, *20*, 299–313. [CrossRef]
8. Wang, H.; Zhang, L.; Cai, Q.; Hu, Y.; Jin, Z.; Zhao, X.; Fan, W.; Huang, Q.; Luo, Z.; Chen, M.; et al. *OsMADS32* interacts with PI-like proteins and regulates rice flower development. *J. Integr. Plant Biol.* **2015**, *57*, 504–513. [CrossRef]
9. Xiao, H.; Tang, J.; Li, Y.; Wang, W.; Li, X.; Jin, L.; Xie, R.; Luo, H.; Zhao, X.; Meng, Z. *STAMENLESS 1*, encoding a single C2H2 zinc finger protein, regulates floral organ identity in rice. *Plant J.* **2009**, *59*, 789–801. [CrossRef]
10. Michelmore, R.W.; Paran, I.; Kesseli, R.V. Identification of markers linked to disease-resistance genes by bulked segregant analysis: A rapid method to detect markers in specific genomic regions by using segregating populations. *Proc. Natl. Acad. Sci. USA* **1991**, *88*, 9828–9832. [CrossRef]
11. Murray, M.G.; Thompson, W.F. Rapid isolation of high molecular weight plant DNA. *Nucleic Acids Res.* **1980**, *8*, 4321–4325. [CrossRef] [PubMed]
12. Luo, Z.; Yang, Z.; Zhong, B.; Li, Y.; Xie, R.; Zhao, F.; Ling, Y.; He, G. Genetic analysis and fine mapping of a dynamic rolled leaf gene, *RL10(t)*, in rice (*Oryza sativa* L.). *Genome* **2007**, *50*, 811–817. [CrossRef] [PubMed]

13. Lander, E.S.; Green, P.; Abrahamson, J.; Barlow, A.; Daly, M.J.; Lincoln, S.E.; Newberg, L.A. MAPMAKER: An interactive computer package for constructing primary genetic linkage maps of experimental and natural populations. *Genomics* **1987**, *1*, 174–181. [CrossRef]

14. Yamaguchi, T.; Nagasawa, N.; Kawasaki, S.; Matsuoka, M.; Nagato, Y.; Hirano, H.Y. The YABBY gene *DROOPING LEAF* regulates carpel specification and midrib development in *Oryza sativa*. *Plant Cell* **2004**, *16*, 500–509. [CrossRef] [PubMed]

15. Nagasawa, N.; Miyoshi, M.; Sano, Y.; Satoh, H.; Hirano, H.; Sakai, H.; Nagato, Y. *SUPERWOMAN1* and *DROOPING LEAF* genes control floral organ identity in rice. *Development* **2003**, *130*, 705–718. [CrossRef] [PubMed]

16. Yoshida, A.; Suzaki, T.; Tanaka, W.; Hirano, H.Y. The homeotic gene long sterile lemma (*G1*) specifies sterile lemma identity in the rice spikelet. *Proc. Natl. Acad. Sci. USA* **2009**, *106*, 20103–20108. [CrossRef]

17. Yoshida, A.; Ohmori, Y.; Kitano, H.; Taguchi-Shiobara, F.; Hirano, H.Y. *ABERRANT SPIKELET AND PANICLE1*, encoding a TOPLESS-related transcriptional co-repressor, is involved in the regulation of meristem fate in rice. *Plant J.* **2012**, *70*, 327–339. [CrossRef]

18. Pelucchi, N.; Fornara, F.; Favalli, C.; Masiero, S.; Lago, C.; Pè, M.E.; Colombo, L.; Kater, M.M. Comparative analysis of rice MADS-box genes expressed during flower development. *Sex. Plant Reprod.* **2002**, *15*, 113–122. [CrossRef]

19. Wang, K.J.; Tang, D.; Hong, L. L.; Xu, W.Y.; Huang, J.; Li, M.; Gu, M.H.; Xue, Y.B.; Cheng, Z.K. *DEP* and *AFO* Regulate Reproductive Habit in Rice. *PLoS Genet.* **2010**, *6*, e1000818. [CrossRef]

20. Prasad, K.; Parameswaran, S.; Vijayraghavan, U. *OsMADS1*, a rice MADS-box factor, controls differentiation of specific cell types in the lemma and palea and is an early-acting regulator of inner floral organs. *Plant J.* **2005**, *43*, 915–928. [CrossRef]

21. Gao, X.; Liang, W.; Yin, C.; Ji, S.; Wang, H.; Su, X.; Guo, C.; Kong, H.; Xue, H.; Zhang, D. The *SEPALLATA*-like gene *OsMADS34* is required for rice inflorescence and spikelet development. *Plant Physiol.* **2010**, *153*, 728–740. [CrossRef] [PubMed]

22. Bommert, P.; Satohnagasawa, N.; Jackson, D.; Hirano, H.Y. Genetics and evolution of inflorescence and flower development in grasses. *Plant Cell Physiol.* **2005**, *46*, 69–78. [CrossRef] [PubMed]

23. Whipple, C.J.; Zanis, M.J.; Kellogg, E.A.; Schmidt, R.J. Conservation of B class gene expression in the second whorl of a basal grass and outgroups links the origin of lodicules and petals. *Proc. Natl. Acad. Sci. USA* **2007**, *104*, 1081–1086. [CrossRef] [PubMed]

24. Prasad, K.; Vijayraghavan, U. Double-stranded RNA interference of a rice *PI/GLO* paralog, *OsMADS2*, uncovers its second-whorl-specific function in floral organ patterning. *Genetics* **2003**, *165*, 2301–2305. [CrossRef] [PubMed]

25. Ohmori, S.; Kimizu, M.; Sugita, M.; Miyao, A.; Hirochika, H.; Uchida, E.; Nagato, Y.; Yoshida, H. *MOSAIC FLORAL ORGANS1, an AGL6*-like MADS box gene, regulates floral organ identity and meristem fate in rice. *Plant Cell* **2009**, *21*, 3008–3025. [CrossRef] [PubMed]

26. Sang, X.; Li, Y.; Luo, Z.; Ren, D.; Fang, L.; Wang, N.; Zhao, F.; Ling, Y.; Yang, Z.; Liu, Y.; He, G. *CHIMERIC FLORAL ORGANS 1*, encoding a Monocot-specific MADS-box Protein, Regulates Floral Organ Identity in Rice. *Plant Physiol.* **2012**, *160*, 788–807. [CrossRef] [PubMed]

27. Yoshida, H.; Itoh, J.I.; Ohmori, S.; Miyoshi, K.; Horigome, A.; Uchida1, E.; Kimizu, M.; Matsumura, Y.; Kusaba, M.; Satoh, H.; Nagato, Y. *superwoman1-cleistogamy*, a hopeful allele for gene containment in GM rice. *Plant Biotechnol J.* **2007**, *5*, 835–846. [CrossRef] [PubMed]

28. Kellogg, E.A. The evolutionary history of *Ehrhartoideae, Oryzeae*, and *Oryza*. *Rice* **2009**, *2*, 1–14. [CrossRef]

29. Ren, D.; Li, Y.; Zhao, F.; Sang, X.; Shi, J.; Wang, N.; Guo, S.; Ling, Y.; Zhang, C.; Yang, Z.; He, G. *MULTI-FLORET SPIKELET1*, which encodes an *AP2/ERF* protein, determines spikelet meristem fate and sterile lemma identity in rice. *Plant Physiol.* **2013**, *162*, 872–884. [CrossRef] [PubMed]

30. Hong, L.; Qian, Q.; Zhu, K.; Tang, D.; Huang, Z.; Gao, L.; Li, M.; Gu, M.; Cheng, Z. ELE restrains empty glumes from developing into lemmas. *J. Genet. Genom.* **2010**, *37*, 101–115. [CrossRef]

31. Kobayashi, K.; Maekawa, M.; Miyao, A.; Hirohiko, H.; Kyozuka, J. *PANICLE PHYTOMER2 (PAP2)*, encoding a SEPALLATA subfamily MADS-box protein, positively controls spikelet meristem identity in rice. *Plant Cell Physiol.* **2010**, *51*, 47–57. [CrossRef] [PubMed]

32.   Lin, X.; Wu, F.; Du, X.; Shi, X.; Liu, Y.; Liu, S.; Hu, Y.; Theißen, G.; Meng, Z. The pleiotropic *SEPALLATA*-like gene *OsMADS34* reveals that the 'empty glumes'of rice (*Oryza sativa*) spikelets are in fact rudimentary lemmas. *New Phytol.* **2014**, *202*, 689–702. [CrossRef] [PubMed]

33.   Li, W.; Yoshida, A.; Takahashi, M.; Maekawa, M.; Kojima, M.; Sakakibara, H.; Kyozuka, J. *SAD1*, an RNA polymerase I subunit A34.5 of rice, interacts with Mediator and controls various aspects of plant development. *Plant J.* **2015**, *81*, 282–291. [CrossRef] [PubMed]

34.   Yu, Y.; Qiao, M.; Liu, Z.; Xiang, F. Diversification function of WRKY transcription factor. *Chin. Bull. Life Sci.* **2010**, *22*, 345–351. [CrossRef]

35.   Nair, S.K.; Wang, N.; Turuspekov, Y.; Pourkheirandish, M.; Sinsuwongwat, S.; Chen, G.; Sameri, M.; Tagiri, A.; Honda, I.; Watanabe, Y.; et al. Cleistogamous Flowering in Barley Arises from the Suppression of MicroRNA-Guided *HvAP2* mRNA Cleavage. *Proc. Natl. Acad. Sci. USA* **2010**, *107*, 490–495. [CrossRef] [PubMed]

10

Metabolic Profiling of Phloem Exudates as a Tool to Improve Bread-Wheat Cultivars

S. Marisol L. Basile [1,*], **Mike M. Burrell** [2], **Heather J. Walker** [2], **Jorge A. Cardozo** [1], **Chloe Steels** [2], **Felix Kallenberg** [2], **Jorge A. Tognetti** [3], **Horacio R. DallaValle** [1] and **W. John Rogers** [1]

[1] Laboratorio de Biología Funcional y Biotecnología (CICPBA-BIOLAB AZUL) CIISAS, Facultad de Agronomía, Universidad Nacional del Centro de la Provincia de Buenos Aires (UNCPBA), CONICET-INBIOTEC, Av. Rep. Italia 780, CC 47, 7300 Azul, Province of Buenos Aires, Argentina; jorge_alejos_cardozo@yahoo.com.ar (J.A.C.); dallavalle@speedy.com.ar (H.R.D.); rogers@faa.unicen.edu.ar (W.J.R.)

[2] Biomics Facility, Department of Animal and Plant Sciences, Alfred Denny Building, University of Sheffield, Western Bank, Sheffield S10 2TN, UK; m.burrell@sheffield.ac.uk (M.M.B.); h.j.walker@sheffield.ac.uk (H.J.W.); chloe-steels@hotmail.co.uk (C.S.); felix.kallenberg@gmx.de (F.K.)

[3] Laboratorio de Fisiología Vegetal, Facultad de Ciencias Agrarias, Universidad Nacional de Mar del Plata, Ruta 226 Km 73,5, Balcarce, Provincia de Buenos Aires, y Comisión de Investigaciones Científicas de la Provincia de Buenos Aires, 7620 Balcarce, Province of Buenos Aires, Argentina; jtognetti2001@yahoo.com.ar

* Correspondence: marisol_basile@yahoo.com.ar

Abstract: In a proof of concept study aimed at showing that metabolites in bread wheat (*Triticum aestivum* L. ssp. *aestivum*), phloem exudates have potential as biochemical markers for cultivar discrimination, Argentinean cultivars from three quality groups (groups 1, 2, and 3 of high, intermediate, and low quality, respectively) were grown under two nitrogen (N) availabilities and analysed for metabolic profile by electrospray ionisation mass spectrometry. Data as signal strengths of mass/charge (m/z) values binned to a resolution of 0.2 Daltons were subjected to principal component analysis and orthogonal projections to latent structures discriminant analysis. Certain bins were influential in discriminating groups taken in pairs and some were involved in separating all three groups. In high N availability, group 3 cultivars clustered away from the other cultivars, while group 1 cultivars clustered tightly together; group 2 cultivars were more scattered between group 1 and group 3 cultivars. In low N availability, the cultivars were not clustered as tightly; nonetheless, group 1 cultivars tended to cluster together and mainly separated from those of group 2. m/z values also showed potential for discrimination between N availability. In conclusion, phloem exudate metabolic profiles could provide biochemical markers for selection during breeding and for discerning the effects of N fertiliser application.

Keywords: phloem metabolites; electrospray ionisation; mass spectrometry; cultivar; quality groups; nitrogen

1. Introduction

The aim of the work reported in this paper was to examine whether metabolic profiling of wheat phloem exudates could be used to discriminate between cultivars with different bread-making qualities and to discern the effects of nitrogen (N) fertiliser application. Since the development of the wheat grain depends on the supply of nutrients in the phloem [1], such measurements could potentially also provide insights into aspects such as N use efficiency. The identification of biochemical markers useful for discriminatory purposes and for selecting combinations of good quality, high yield, and good tolerance to biotic and abiotic stresses appears feasible [2]. Metabolic profiling has been used to distinguish between leaf and fruit extracts of species and cultivars of tomato and examine the effects of N on metabolism in these organs [3–5].

In bread wheat (*Triticum aestivum* L. ssp. *aestivum*), one important target trait for improvement is grain protein content (GPC), which is one of the determinants of international market price [1]. Argentinean wheat cultivars are currently classified into three quality groups for bread-making based upon a range of tests, including grain protein content, wet gluten content, properties in the alveograph and farinograph, and loaf volume. Group 1 is the best quality (including wheats that can be blended with others to correct their visco-elastic properties and that are suitable for industrial bread-making), and group 3, the poorest quality, which tends to include high-yielding cultivars of deficient quality for bread-making (suitable for only short fermentation times, of less than eight hours). Group 2 is of intermediate quality, including cultivars that are not correctors and that are suitable for traditional bread-making and for fermentation of more than eight hours. Groups 1, 2, and 3 are expected to rank from high to low for protein content and other quality traits [6], and phloem composition would be expected to be related to such differences.

Soil or foliar applications of N, applied at different rates and stages of growth, are commonly used to improve GPC. One of the principal contributors to GPC is the process of remobilisation from leaves during organ senescence, but N runoff from soils creates environmental issues. Barneix [1] demonstrated that more than 50–70% of final grain N is accumulated by the plant before anthesis and is remobilised to the grain later. The relationship between the N supplied and that absorbed by the plant is not linear; rather, there is a limit to the potential GPC, which lowers fertiliser use efficiency when high N doses are supplied. The concentration of free amino acids in the phloem acts as a signal to the roots that indicates the N status of the plant, which activates or inhibits NO_3 uptake by the plant. As a consequence, N metabolism in the shoot dictates the rate of NO_3 uptake [1]. The concentration of the majority of the amino acids in phloem exudates is proportional to the concentration in the leaves [7], if no pathogens are involved [8].

The concentration of the majority of the amino acids in phloem exudates is proportional to the concentration in the leaves [7], if no pathogens are involved [8]. Grain filling is mainly dependent on remobilisation from the flag leaf and the adjacent leaf. Furthermore, the final GPC has been correlated with the amount of free amino acids in the flag leaf during grain filling [9,10]. Therefore, the analysis of the metabolites present in the phloem exudates provides a valid indicator of the compounds that will be present in the future grains and of the effect of N fertilisation, given that secondary metabolites serve as a N reserve [11].

These relationships, showing the potential importance for the quality of phloem exudate composition, gave rise to the current work. Since direct injection mass spectrometry (DIMS) through electrospray ionisation time-of-flight (ESI-TOF-MS) provides a rapid method to obtain an initial metabolic profile of samples [3,12], it was chosen as an approach for profile analysis of phloem exudate samples in this proof of concept study.

2. Materials and Methods

2.1. Experimental Design

The field trial was carried out at the Experimental Field of the Faculty of Agronomy, UNCPBA (Lat.: 36°45′ S; Long.: 59°50′ W; Height above sea level: 132 m) situated on the Ruta Nacional No. 3 Km 307, in Azul, Province of Buenos Aires, Argentina.

The trial was a randomised complete block design with three replicate blocks, with a split plot design where the main plot was N fertiliser level and the sub-plot was cultivar. Eight cultivars of contrasting quality were sown: 'ACA 304', 'Klein Proteo', and 'Buck Meteoro' of group 1 (high bread-making quality); 'Bio 3003', 'Bio 1000', and 'Buck Malevo' of group 2 (medium bread-making quality); and 'Klein Gavilán' and 'Klein Guerrero' of group 3 (low bread-making quality). These cultivars have been previously evaluated for quality traits [13–19], amongst others.

Soil N content before fertilisation was approximately 35 kg N per hectare and the field area was fertilised with triple superphosphate (100 ppm) prior to sowing. Sowing date was adjusted according to the cultivar cycle (short or long) with the aim of ensuring that the cultivars arrived simultaneously at the required physiological stage for extracting phloem exudates. Plots measured 9.5 m × 2.8 m and sowing density was 350 plants/m^2. Half of each plot, randomly designated, was fertilised to achieve 100 ppm of N availability as urea by hand broadcasting. The herbicide DICAMBA-SPA was applied for weed control. At exudate sampling the plants showed good general health status, with differences in colour intensity between the well-fertilised treatments compared to the rest.

2.2. Extraction Method

Phloem exudate extraction was done according to Urquhart and Joy [20] as modified in [21]. Samples were taken when the flag leaf was fully expanded, which occurred within a lapse of eight days covering all cultivars, meaning the different sowing dates for short and long cycled cultivars resulted in the desired limited range of dates for the expanded leaf physiological stage, minimising the effect of the cycle.

The exudates should mainly reveal the contents of the phloem; while we cannot rule out the possibility of xylem contents and cellular leakage also occurring in the samples, such contamination would be expected to be slight [20,22]. The flag leaf and the adjacent leaf were harvested according to [21]. The extremes were cut and complete leaves kept for 15 min in a 20 mm pH 6 ethylenediaminetetraacetic acid (EDTA) solution, which was subsequently discarded to minimise such contamination. Leaves were washed in new EDTA solution for one minute and then left for three hours in a new aliquot of the EDTA solution to allow the phloem to exude. The whole procedure was carried out in the dark to minimise leaf transpiration and absorption of the solution by the xylem. The exudates were stored at −4 °C and lyophilised to −40 °C and 76 mm Hg for 24 h.

2.3. Sample Processing

Samples were resuspended in 1 mL of H_2O and an aliquot of 20 μL was diluted in 100 μL of methanol, 0.1 μL of formic acid, and distilled water to a final volume of 200 μL. The samples were analysed by direct injection mass spectrometry (DIMS) through electrospray ionisation time-of-flight (ESI-TOF-MS) using an Applied Systems/MDS Sciex (Foster City, CA, USA) hybrid quadrupole time-of-flight Q-Star Pulsar-I mass spectrometer. For all samples three technical replicates were analysed.

Of the 48 exudate samples (8 cultivars × 3 blocks × 2 N levels), those from ACA 304 without added N (−N) block II, Bio 3003 −N block I, Bio 1000 with added N (+N) block III and Meteoro +N block III went missing during processing.

2.4. Data Analysis

The mass spectra were processed according to Overy et al. [3]. Only mass/charge (m/z) values occurring in all three technical replications were included and the data was binned to a resolution of 0.2 Daltons. The metabolic profiles thus obtained were then analysed by unsupervised statistics using principal component analysis (PCA) with SIMCA14 (Umetrics, Umeå, Sweden). When clustering was observed the data was processed by orthogonal projections to latent structures discriminant analysis (OPLS-DA) to resolve the bins contributing towards the clustering of samples and putative metabolites were assigned to these bins from empirical formulae.

Analysis of variance (ANOVA) was carried out using the software Infogen, Córdoba, Argentina [23], for individual bins between quality groups and between N availability.

3. Results

3.1. Resolution of Cultivars

An overall PCA of all data (all samples including all detected masses) showed that the proportion of the total variation accounted for by principal components 1 and 2 was 56.5%. Since the analytical approach used was completely untargeted, the data were examined to reveal any clustering, by using an unsupervised statistical approach and then overlaying the known variables on the plot. Initially the data for the plots with and without added N were analysed separately. From the analysis of plots with added N (Figure 1a), it can be seen that the group 3 cultivars Gavilán and Guerrero cluster away from the other cultivars, while the group 1 cultivars ACA 304, Proteo, and Meteoro cluster tightly together. The group 2 cultivars are more scattered between the group 1 and group 3 cultivars, as might be expected from their intermediate characteristics. For the plots without added N (Figure 1b), the cultivars are not clustered as tightly. The group 3 variety replicates are more scattered, but the group 1 cultivars tend to cluster in the right hand two quartiles and, apart from one Proteo replicate, are separated from the group 2 cultivars. If fewer cultivars are included in the analysis to reduce the total variance, the groups are more clearly separated (data not shown).

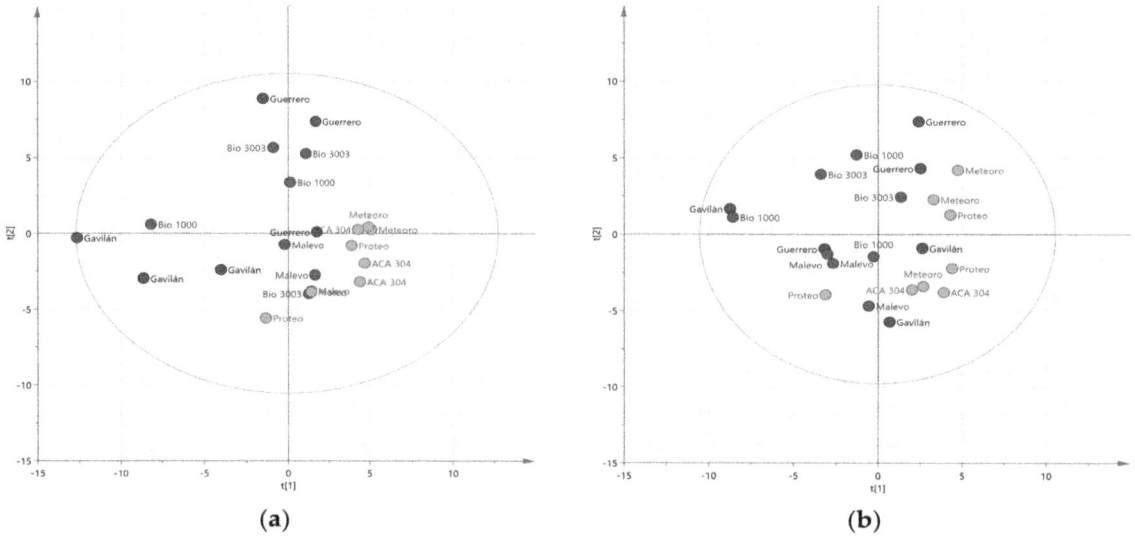

Figure 1. Principal components analysis: PC1 vs. PC2 of the cultivars under study (group 1—green; group 2—blue; group 3—red). (**a**) With, and (**b**) without added N.

Since there is evidence of clustering of the data it was examined with an OPLS-DA analysis to identify those bins that contribute most towards the separation. These values are obtained from the loadings plot of the first predictive component of the model and correspond to the covariance of the X-variables and the predictive score vectors. It is interesting to note that many of the bins that separate group 3 from group 2 also separate group 3 from group 1 (Table 1). The cut-off point of the top ten values was an arbitrary decision to illustrate that some bins contribute to the separation of all three groups, such as bin 383; putative metabolites were assigned to these influential bins (Supplementary Table S1).

ANOVAs carried out for individual bins of high influence in the original overall analysis also showed differences between quality groups (Table 2). For example, mean values for bin 203, given in Table 1 as being influential in separating group 3 from the other groups, ranked group 1 < group 2 < group 3 in the ANOVAs, and so may be associated with poor quality. In contrast, bin 370.6 ranked group 1 > group 2 > group 3, and so may favour quality (Table 2). Bin 301.2, influential in separating group 3 from group 1 and group 2 from group 1 (Table 1), gave a ranking consistent with this, of 3 = 2 > 1, and bin 329, influential in separating group 2 from group 1 (Table 1), gave a ranking consistent with this, of 2 > 1 = 3.

Table 1. m/z bins that separate groups. The top 10 bins showing the strongest contribution in order of their p and q values.

Group 3 from Group 2	Group 3 from Group 1	Group 2 from Group 1
203	203	301.2
140	140	471
239	301.2	361
118	383	157
383	239	227
701.2	261	173
204	539	449
539	441.2	329
156	245	195
261	156	383

Table 2. Influential mass bins showing significant differences between quality groups in ANOVA.

Ranking Pattern	Mass Bin	Means		
		Group 1	Group 2	Group 3
$3^c > 2^b > 1^a$	203	3.205	6.589	10.256
$1^c > 2^b > 3^a$	370.6	0.141	0.096	0.046
$2^c > 1^b > 3^a$	143	0.326	0.425	0.27
	165	0.12	0.148	0.093
$1^b > 2^a = 3^a$	131	0.124	0.094	0.094
	315	10.939	5.972	5.517
	651	0.202	0.099	0.061
	673	1.276	0.717	0.598
	695	3.607	2.311	1.804
	696	0.581	0.27	0.242
	717	1.481	1.077	0.835
	316	1.05	0.444	0.508
	375	1.144	0.661	0.668
	376	0.039	0.015	0.023
$2^b = 1^b > 3^a$	172	0.112	0.131	0.064
	371	0.018	0.02	0.008
	382	0.404	0.415	0.278
$1^b = 2^b > 3^a$	144	0.011	0.009	0.004
	255	0.223	0.208	0.089
	269	0.128	0.108	0.062
	287.2	0.189	0.142	0.075
	370	1.097	0.894	0.402
	761	0.335	0.297	0.148
$3^b > 1^a = 2^a$	135	0.025	0.013	0.11
	156	0.193	0.178	0.303
	204	0.228	0.111	0.405
	205	0.074	0.053	0.107
	219	0.405	0.232	0.881
	261	0.3	0.188	0.517
	383	0.314	0.16	0.766
$2^b = 3^b > 1^a$	167	0.056	0.092	0.09
	181	0.06	0.094	0.088
	307.2	0.083	0.114	0.11

Table 2. *Cont.*

Ranking Pattern	Mass Bin	Means		
		Group 1	Group 2	Group 3
$3^b = 2^b > 1^a$	140	5.687	6.743	7.397
	183	0.041	0.07	0.075
	301.2	3.458	5.521	5.763
	302.2	0.428	0.572	0.583
	561	0.068	0.165	0.17
$2^b > 1^a = 3^a$	222	0.013	0.02	0.014
	249.2	0.11	0.202	0.119
	295.2	0.056	0.07	0.059
	299.2	0.142	0.42	0.252
	309.2	0.149	0.272	0.193
	403	0.657	1.218	0.663
	180	0.01	0.014	0.008
	195	0.125	0.181	0.123
	199	0.377	0.568	0.367
	300.2	0.008	0.02	0.008
	329	0.111	0.202	0.085
	392	0.199	0.345	0.1
	449	0.069	0.148	0.058
	560	0.162	0.272	0.073
$1^b = 3^b > 2^a$	353	0.479	0.34	0.443
$2^b > 1^{ab} > 3^a$	99	0.011	0.012	0.006
	102	0.021	0.028	0.013
	189	0.112	0.127	0.093
	263.2	0.064	0.074	0.053
	269.2	0.064	0.073	0.054
	279	0.445	0.491	0.362
$3^b > 2^{ab} > 1^a$	118	1.887	2.135	2.553
	539	0.233	0.383	0.529
$2^b > 3^{ab} > 1^a$	193.2	0.159	0.222	0.19
	251.2	0.187	0.254	0.23
	267.2	0.207	0.28	0.23
	303.2	0.031	0.043	0.036
	321.2	0.071	0.106	0.088
$1^b > 2^{ab} > 3^a$	257	0.933	0.76	0.556
	359	14.285	12.586	10.842
	739	1.107	0.9	0.6

Group numbers followed by different letters differed significantly (LSD, $p < 0.05$). No mass bins showed patterns $2^c > 3^b > 1^a$; $3^c > 1^b > 2^a$; $1^c > 3^b > 2^a$; $3^b = 1^b > 2^a$; $3^b > 1^{ab} > 2^a$ and $1^b > 3^{ab} > 1^a$.

3.2. Effect of Nitrogen

The effect of adding N was examined by cultivar (Figure 2a (Malevo) and Figure 2b (Guerrero)). Adding N clearly alters the metabolic profiles of the exudates, but an OPLS-DA analysis of these two cultivars do not reveal common masses in the top ten discriminant bins. We hypothesised that the variation in field N could be creating additional variation and, therefore, examined the effect of revealing the blocks. Clearly there is both an effect of block and addition of N on the metabolic profiles of the exudates (Figure 3a,b). Block 3 clusters in the bottom left quartile. This field effect is creating much variation in the samples and, with the small number of replicates used, limits the resolution of the difference in metabolic profiles caused by N.

Effects of N could also be discerned with ANOVA of individual bins, where, in an overall analysis including all cultivars in the three quality groups, nine mass bins were found to show significant differences for N availability (Table 3). Some of these mass bins were also influential in the original overall analysis separating quality groups (bins 301.2 and 329 in Table 1).

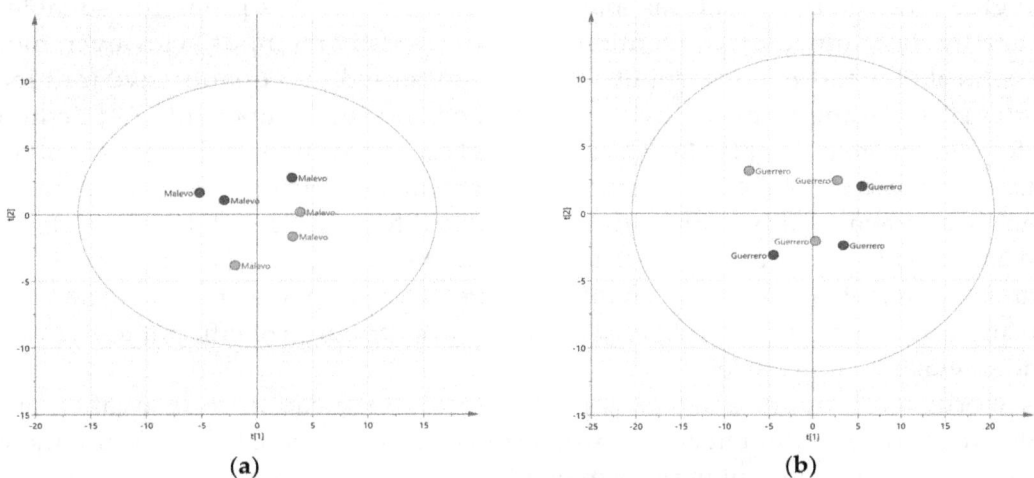

Figure 2. Principal component analysis: PC 1 vs. PC 2 of (**a**) Malevo; and (**b**) Guerrero. With N—blue; without N—green.

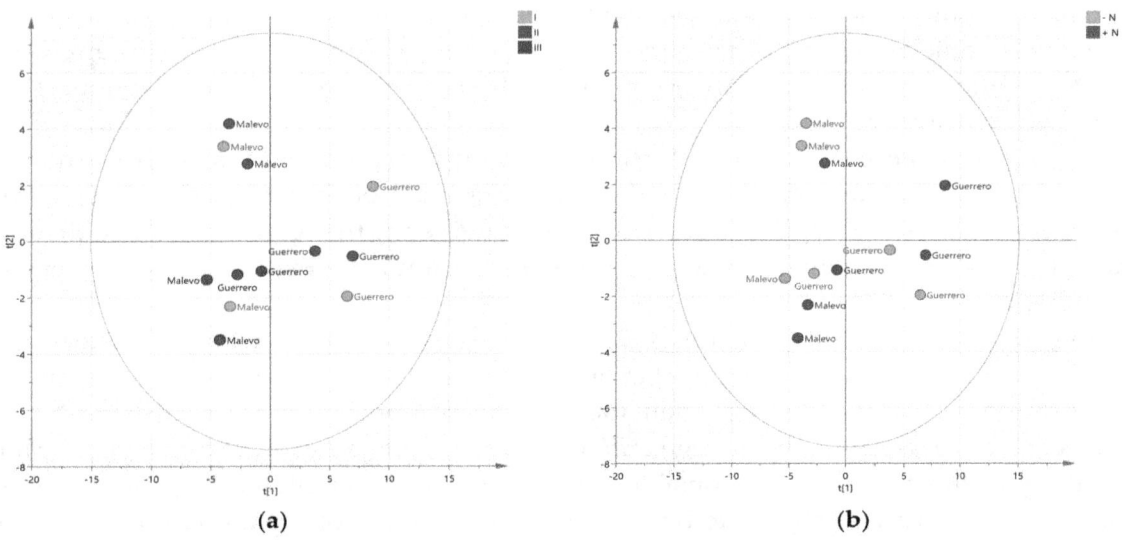

Figure 3. Principal component analysis: PC 1 vs. PC 2 of Malevo and Guerrero. (**a**) Block effect; and (**b**) N effect.

Table 3. Mass bins responding to fertiliser levels in ANOVAs applied to quality groups.

Mass Bin	Means	
	−N	+N
91	0.053[a]	0.111[b]
149	0.087[a]	0.114[b]
183	0.059[a]	0.065[b]
301.2	4.482[a]	5.346[b]
339	0.146[a]	0.211[b]
442.2	0.404[a]	0.435[b]
211	0.200[a]	0.252[b]
329	0.117[a]	0.148[b]
255	0.148[a]	0.198[b]

Values for a certain mass followed by a different letter were significantly different (LSD, $p < 0.05$).

4. Discussion

Bread-making quality is determined by genetic and environmental factors. The balance between gliadins and glutenins determines, at least partially, that the dough has properties suitable for baking. Quality is usually determined by conventional and time-costly methods. However, more recently, the genetic control of bread-making quality has been dissected into quantitative trait loci (QTL) in numerous studies, for example, micro-alveograph testing and sub-components [24], sedimentation in sodium dodecyl sulphate (SDS) [24–26], GPC [24,26–34], and hardness [24], amongst others. As well as genetic factors, these traits are influenced by environmental factors and management practices, such as N and water availability, temperature and light intensity [35–37]; for example, increases in GPC can be achieved by N addition, but after incremental additions of N fertilizer, GPC reaches a maximum and then remains constant, without any increase in N uptake or remobilization by the crop, thus decreasing the efficiency of N fertilizer [1], hence, the importance of efficient tools to assist genetic selection and management practices.

In this work, we propose a novel approach to study bread-making quality by metabolic profiling. Electrospray ionisation mass spectrometry is highly suitable for analysing the wide range of non-volatile compounds present in phloem exudates: it is a soft ionisation method at atmospheric pressure; it is practical for molecules the size of metabolites; it provides good sensitivity; it is adaptable to a wide range of aqueous and organic solvents and, therefore, can be used directly with metabolite extracts prepared from plants; and it can analyse a wide range of different types of molecules, including highly polar molecules, such as peptides, oligonucleotides, and oligosaccharides, as well as small polar molecules, ionic metal complexes, and other soluble inorganic analytes. The analysis of ions by MS-TOF is simple and allows analyses of a virtually unlimited mass range with a resolution of 0.0001 and high sensitivity.

The approach taken in this demonstration of principle experiment was to use a non-targeted analysis of phloem exudates. Non-targeted approaches have been previously used in studies on tomatoes and *Arabidopsis* [3,12]. PCA provides strong evidence that cultivars can be distinguished from each other, as well as between quality groups. For example, in Figure 1, Gavilán germplasm is clearly different to ACA 304.

The overall PCA identified PC 1 and PC 2, accounting for 56.5% of the total variation, which was relatively high compared to values obtained in other work, such as Rogers et al. [38]; presumably these cultivars differed in many aspects of their metabolism responsible for their different agronomic performance. Some mass bins showed large differences between quality groups and could be important for accounting for the differences responsible for belonging to different quality groups. For example, the mass bin 203, which was shown to be the first in the ranking of the top ten bins for separating group 3 from 2 and group 3 from 1 (Table 1), showed a ranking 1< 2 < 3 in the ANOVAs, meaning it may be associated with poor quality. Another mass bin showing large differences between quality groups in the ANOVAs was 370.6, that showed a ranking of 1 > 2 > 3 (Table 2). These mass bins could be markers for the rapid selection of cultivars for quality.

Other mass bins showing different rankings for quality groups in the ANOVAs (Table 2) give similar patterns to those observed in the top ten of Table 1. For example, ranking 3 > 1:301.2, 140, 283, 539, and 156; ranking 3 > 2:204, 156, and 383; ranking 2 > 1:301.2, 449, and 329.

Differences between N fertiliser levels identified by the PCA were to a certain extent masked by the differences between cultivars, but could still be extracted from the data. For example, mass bins 203 and 305 identified responses to N treatment in the cultivars.

When only three replicates are used, as in this study, the variation between replicates and the large amount of variation between cultivars clouds the separation. When the experimental blocks are overlaid on the data (Figure 3) it is clear that block three samples cluster tightly, whereas blocks one and two are more scattered. Thus, there is a large field effect in this experiment, meaning it may be beneficial in future experiments to increase the number of blocks in order to better take into account such heterogeneity [39]. In spite of this, clear effects on the addition of N on the metabolic profiles of the exudates were observed in PCA. Future work will also de aimed at widening the number of cultivars under study and to relate their detailed quality characteristics with the masses analysed here; as mentioned in the introduction, these cultivars have been studied for quality characteristics and we propose to add our own quality data obtained from controlled field trials to these in future studies.

The period over which exudates were collected was short and the environmental conditions changed little over this time; for example, mean maximum, minimum, and mean temperatures were 22.9 ± 3.01, 6.73 ± 2.67, and 15 ± 1.7 °C, respectively, and rainfall was minimal (0.3 and 0.2 mm). Hence, we would not expect phloem exudate composition to be significantly affected. Moreover, Overy et al. [3] and Tetyuk et al. [40] collected exudates from plants differing in age (three days in the former and up to fourteen in the latter) and made no observations on this. The difference in the latter study resulted in collection from different phenological stages; we, in our study, were interested in analysing exudates from plants of the same stage, rather than the same age.

The results presented show that metabolic profiling may be used to extract biochemical markers that may be of potential use in selection, in the discrimination of cultivars of differing quality, and in elucidating the effect of N fertilisation. The challenge will be to identify which metabolites are those associated with that performance and to determine the definitive identification of the metabolites that goes beyond the putative metabolites included in the current study, in order to allow possible reasons to be postulated for those differences, and to generate the potential for biochemical marker selection for the important traits; future work will be directed towards this. Some of the mass bins identified as differing significantly between the cultivars may be involved in grain protein composition and quality characteristics. While these are possible candidates for explaining differences in agronomic performance, further analysis will be needed in order to establish these relationships and, as previously mentioned, the evaluation of a larger number of cultivars will be required. As data becomes available from field trials over several years, designed to explore the consistency of responses for the mass bins, this detailed analysis will be a focus for our resources.

5. Conclusions

In summary, we believe metabolic profiling has the potential to be developed into a breeding tool to refine plant breeding efficiency. In particular, it would appear to be useful to identify when field effects could be clouding the results of other screening methods.

Supplementary Materials: Supplementary materials can be found at Table S1: Putative metabolites for the top ten m/z bins that separate quality groups in Table 1.

Acknowledgments: We wish to thank Carla Caputo for her generous guidance with sample lyophilisation. We are indebted to the CIC-PBA, CONICET, and CIISAS, Facultad de Agronomía, UNCPBA, for funding this work. S.M.L.B. has been in receipt of a doctoral fellowship from the ANPCyT and from CONICET. This work is part of the doctoral thesis of S.M.L.B.

Author Contributions: S.M.L.B., J.A.C., H.D.V., M.M.B., and W.J.R. conceived and designed the experiments; S.M.L.B., M.M.B., H.J.W., J.A.C., C.S., F.K., J.A.T., H.R.V., and W.J.R. performed the experiments; S.M.L.B., M.M.B., J.A.T., and W.J.R. analysed the data; M.M.B. contributed reagents, materials, and analytical tools; and S.M.L.B., M.M.B., J.A.T., and W.J.R. wrote the paper.

References

1.   Barneix, A.J. Physiology and biochemistry of source-regulated protein accumulation in the wheat grain. *J. Plant Physiol.* **2007**, *164*, 581–590. [CrossRef] [PubMed]

2.   Hall, R.; Brouwer, I.; Fitzgerald, M. Plant metabolomics and its potential application for human nutrition. *Physiol. Plant.* **2008**, *132*, 162–175. [CrossRef] [PubMed]

3.   Overy, S.A.; Walker, H.J.; Malone, S.; Howard, T.P.; Baxter, C.J.; Sweetlove, L.J.; Hill, S.A.; Quick, W.P. Application of metabolite profiling to the identification of traits in a population of tomato introgression lines. *J. Exp. Bot.* **2005**, *56*, 287–296. [CrossRef] [PubMed]

4.   Schauer, N.; Zamir, D.; Fernie, A.R. Metabolic profiling of leaves and fruit of wild species tomato: A survey of the Solanumlycopersicum complex. *J. Exp. Bot.* **2005**, *56*, 297–307. [CrossRef] [PubMed]

5.   Urbanczyk-Wochniak, E.; Fernie, A.R. Metabolic profiling reveals altered nitrogen nutrient regimes have diverse effects on the metabolism of hydroponically-grown tomato (*Solanum lycopersicum*) plants. *J. Exp. Bot.* **2005**, *56*, 309–321. [CrossRef] [PubMed]

6.   Cuniberti, M. Propuesta de clasificación del trigo argentino. *IDIA XXI* **2004**, *6*, 21–25.

7.   Caputo, C.; Criado, M.V; Roberts, I.N.; Gelso, M.A.; Barneix, A.J. Regulation of glutamine synthetase 1 and amino acids transport in the phloem of young wheat plants. *Plant Physiol. Biochem.* **2009**, *47*, 335–342. [CrossRef] [PubMed]

8.   Gai, Y.P.; Han, X.J.; Li, Y.Q.; Yuan, C.Z.; Mo, Y.Y.; Guo, F.Y.; Liu, Q.X.; Ji, X.L. Metabolomic analysis reveals the potential metabolites and pathogenesis involved in mulberry yellow dwarf disease. *Plant Cell Environ.* **2014**, *37*, 1474–1490. [CrossRef] [PubMed]

9.   Millet, E.; Zaccai, M.; Feldman, M. Paternal and maternal effects on grain wheat and protein percentages in crosses between hexaploid and tetraploid high protein and low protein wheat genotypes. *Genome* **1992**, *35*, 257–260. [CrossRef]

10.   Barneix, A.J.; Guitman, M.R. Leaf regulation of the nitrogen concentration in the grain of wheat plants. *J. Exp. Bot.* **1993**, *44*, 1607–1612. [CrossRef]

11.   Poulton, J.E. Cyanogenesis in plants. *Plant Physiol.* **1990**, *94*, 401. [CrossRef] [PubMed]

12.   Walker, T.S.; Pal Bais, H.; Halligan, K.M.; Stermitz, F.R.; Vivanco, J.M. Metabolic profiling of root exudates of Arabidopsis thaliana. *J. Agric. Food Chem.* **2003**, *51*, 2548–2554. [CrossRef] [PubMed]

13.   Villar, J.; Cencig, G. *Evaluación De Cultivares De Trigo 2004/2005 Y Recomendaciones Para La Próxima Campaña*; INTA—Estación Experimental Agropecuaria Rafaela, Información técnica de trigo campaña 2005, N° 103; Publicación Miscelánea, INTA EEA: Rafaela, Argentina, 2005.

14.   Villar, J.; Cencig, G. *Evaluación De Cultivares De Trigo 2008 Y Recomendaciones Para La Próxima Campaña*; Información técnica de trigo y otros cultivos de invierno campaña 2009, N° 113; Publicación Miscelánea, INTA EEA: Rafaela, Argentina, 2009.

15.   Villar, J.; Cencig, G. *Evaluación De Cultivares De Trigo 2010 Y Recomendaciones Para La Próxima Campaña*; Información técnica de trigo y otros cultivos de invierno campaña 2011, N° 119; Publicación Miscelánea, INTA EEA: Rafaela, Argentina, 2011.

16.   Molfese, E.R.; Astiz, V. *Calidad Del Trigo Pan En El Sur Bonaerense*; Cosecha 2014–2015, Colección divulgación; Ediciones, INTA EEA: Barrow, Argentina, 2015.

17.   Donaire, G.; Masiero, B.; Gutierrez, C.; Conde, B.; Salines, J.; Chiacchera, S.; Bertram, N.; Amigone, M.; Fraschina, J.; Gómez, D.; et al. *Caracterización De Cultivares De Trigo Doble Propósito*; Planteos ganaderos; Aapresid: Rosario, Argentina, 2010.

18.   Albrecht, R.; Quaino, O.; Perez, D.; Martins, L. *Evaluación De Cultivares De Trigo*; Campaña 2008/2009 San Fabián—Departamento San Jerónimo—Santa Fe, INTA—Estación Experimental Agropecuaria Rafaela, Información técnica de trigo y otros cultivos de invierno, Campaña, N° 113; Publicación Miscelánea, INTA EEA: Rafaela, Argentina, 2009.

19.   Bainotti, C.T.; Fraschina, J.; Salines, J.; Alberione, E.; Gómez, D.; Donaire, G.; Nisi, J.; Masiero, B.; Conde, B.; Cuniberti, M.; et al. *Evaluación De Cultivares De Trigo En La Eea Marcos Juárez-Campaña 2007/2008*; Trigo Actualización 2008, Informe de Actualización Técnica N° 8; INTA EEA: Marcos Juárez, Argentina, 2008.

20.   Urquhart, A.A.; Joy, K.W. Use of phloem exudate technique in the study of amino acid transport in pea plants. *Plant Physiol.* **1981**, *68*, 750–754. [CrossRef] [PubMed]

21.   Caputo, C.; Barneix, A.J. The Relationship between Sugar and Amino Acid Export to the Phloem in Young Wheat Plants. *Ann. Bot.* **1999**, *84*, 33–38. [CrossRef]

22.   Cao, T.; Lahiri, I.; Singh, V.; Louis, J.; Shah, J.; Ayre, B.G. Metabolic engineering of raffinose-family oligosaccharides in the phloem reveals alterations in carbon partitioning and enhances resistance to green peach aphid. *Front. Plant Sci.* **2013**, *4*, 263. [CrossRef] [PubMed]

23.   Balzarini, M.; Di Renzo, J. *Infogen: Software Para Análisis Estadísticos De Marcadores Genéticos*; Facultad de Ciencias Agropecuarias, Universidad Nacional de Córdoba: Córdoba, Argentina, 2003.

24.   Perretant, M.R.; Cadalen, T.; Charmet, G.; Sourdille, P; Nicolas, P.; Boeuf, C.; Tixier, M.H.; Branlard, G.; Bernard, S.; Bernard, M. QTL analysis of bread-making quality in wheat using a doubled haploid population. *Theor. Appl. Genet.* **2000**, *100*, 1167–1175. [CrossRef]

25.   Blanco, A.; Bellomo, M.P.; Lotti, C.; Maniglio, T.; Pasqualone, A.; Simeone, R.; Troccoli, A.; Di Fonzo, N. Genetic mapping of sedimentation volume across environments using recombinant inbred lines of durum wheat. *Plant Breed.* **1998**, *117*, 413–417. [CrossRef]

26. Rousset, M.; Brabant, P.; Kota, R.S.; Dubcovsky, J.; Dvorak, J. Use of recombinant substitution lines for gene mapping and QTL analysis of bread making quality in wheat. *Euphytica* **2001**, *119*, 81–87. [CrossRef]

27. Groos, C.; Robert, N.; Bervas, E.; Charmet, G. Genetic analysis of grain protein-content, grain yield and thousand-kernel weight in bread wheat. *Theor. Appl. Genet.* **2003**, *106*, 1032–1040. [CrossRef] [PubMed]

28. Wang, L.; Wang, J.; Cui, F.; Jun, L.; Ding, A.; Zhao, C.; Li, X.; Feng, D.; Gao, J.; Wang, H. Conditional QTL mapping of protein content in wheat with respect to grain yield and its components. *J. Genet.* **2012**, *91*, 303–312. [CrossRef] [PubMed]

29. Cormier, F.; Le Gouis, J.; Dubreuil, P.; Lafarge, S.; Praud, S. A genome-wide identification of chromosomal regions determining nitrogen use efficiency components in wheat (*Triticum aestivum* L.). *Theor. Appl. Genet.* **2014**, *127*, 2679–2693. [CrossRef] [PubMed]

30. Li, C.; Bai, G.; Chao, S.; Carver, B.; Wang, Z. Single nucleotide polymorphisms linked to quantitative trait loci for grain quality traits in wheat. *Crop J.* **2016**, *4*. [CrossRef]

31. Krishnappa, G.; Singh, A.M.; Chaudhary, S.; Ahlawat, A.K.; Singh, S.K.; Shukla, R.B.; Jaiswal, J.P.; Singh, G.P.; Singh Solanki, I. Molecular mapping of the grain iron and zinc concentration, protein content and thousand kernel weight in wheat (*Triticum aestivum* L.). *PLoS ONE* **2017**, *12*, e0174972. [CrossRef] [PubMed]

32. Nedelkou, I.P.; Maurer, A.; Schubert, A.; Léon, J.; Pillen, K. Exotic QTL improve grain quality in the tri-parental wheat population SW84. *PLoS ONE* **2017**, *12*, e0179851. [CrossRef] [PubMed]

33. Monostori, I.; Szira, F.; Tondelli, A.; Árendás, T.; Gierczik, K.; Cattivelli, L.; Galiba, G.; Vágújfalvi, A. Genome-wide association study and genetic diversity analysis on nitrogen use efficiency in a Central European winter wheat (*Triticum aestivum* L.) collection. *PLoS ONE* **2017**, *12*, e0189265. [CrossRef] [PubMed]

34. Liu, J.; Feng, B.; Xu, Z.; Fan, X.; Jiang, F.; Jin, X.; Cao, J.; Wang, F.; Liu, Q.; Yang, L.; Wang, T. A genome-wide association study of wheat yield and quality-related traits in southwest China. *Mol. Breed.* **2018**, *38*. [CrossRef]

35. Carrillo, J.M.; Rousset, M.; Qualset, C.O.; Kasarda, D.D. Use of recombinant inbred lines of wheat for study of associations of high-molecular-weight glutenin subunit alleles to quantitative traits. 1. Grain yield and quality prediction tests. *Theor. Appl. Genet.* **1990**, *79*, 321–330. [CrossRef] [PubMed]

36. Silvela, L.; Ayuso, M.G.; Gil-Delgado, L.G.; Solaices, L. Genetic and environmental contributions to bread-wheat flour quality using the SDS sedimentation test as an index. *Theor. Appl. Genet.* **1993**, *86*, 889–894. [CrossRef] [PubMed]

37. Blanco, A.; Mangini, G.; Giancaspro, A.; Giove, S.; Colasuonno, P.; Simeone, R.; Signorile, A.; De Vita, P.; Mastrangelo, A.M.; Cattivelli, L.; et al. Relationships between grain protein content and grain yield components through quantitative trait locus analyses in a recombinant inbred line population derived from two elite durum wheat cultivars. *Mol. Breed.* **2012**, *30*, 79–92. [CrossRef]

38. Rogers, W.J.; Cogliatti, M.; Burrell, M.M.; Steels, C.; Kallenberg, F.; Bongiorno, F.; DallaValle, H. Mejoramiento y metabolómica del cultivo de alpiste (*Phalaris canariensis*). In *Cereales de Invierno. Investigación Científico-Técnica*; Steinglein, S.A., Moreno, M.V., Cogliatti, M., Rogers, W.J., Carmona, M.A., Lavado, R.S., Eds.; Universidad Nacional del Centro de la Provincia de Buenos Aires: Tandil/Buenos Aires, Argentina, 2012; pp. 147–154. ISBN 978-950-658-301-9.

39. Montgomery, D.C.; Runger, G.C. Design and analysis of single-factors experiments: The analysis of variance. In *Applied Statistics and Probability for Engineers*, 3rd ed.; John Wiley & Sons, Inc.: New York, NY, USA, 2003; Chapter 13; pp. 468–504. ISBN 0-471-20454-4.

40. Tetyuk, O.; Benning, U.F.; Hoffmann-Benning, S. Collection and Analysis of Arabidopsis Phloem Exudates Using the EDTA-facilitated Method. *J. Vis. Exp.* **2013**, *80*, e51111. [CrossRef] [PubMed]

11

Characterization of Genetic Diversity in Accessions of *Prunus salicina* Lindl: Keeping Fruit Flesh Color Ideotype while Adapting to Water Stressed Environments

Cintia V. Acuña [1,*], Juan G. Rivas [1], Silvina M. Brambilla [1], Teresa Cerrillo [2], Enrique A. Frusso [3], Martín N. García [1], Pamela V. Villalba [1], Natalia C. Aguirre [1], Julia V. Sabio y García [1], María C. Martínez [1], Esteban H. Hopp [1,4] and Susana N. Marcucci Poltri [1]

[1] Instituto de Agrobiotecnología y Biología Molecular—IABiMo—INTA-CONICET, Instituto de Biotecnología, Centro de Investigaciones en Ciencias Agronómicas y Veterinarias, Instituto Nacional de Tecnología Agropecuaria, Dr. Nicolás Repetto y de los Reseros S/N, Hurlingham, Buenos Aires B1686IGC, Argentina
[2] EEA Delta del Paraná, Río Paraná y Canal Laurentino Comas, Campana (2804), Buenos Aires B1686IGC, Argentina
[3] Instituto de Recursos Biológicos, Centro de Investigación de Recursos Naturales, Instituto Nacional de Tecnología Agropecuaria, Dr. Nicolás Repetto y de los Reseros S/N, Hurlingham, Buenos Aires B1686IGC, Argentina
[4] Lab. Agrobiotecnología, FBMC, Facultad de Ciencias Exactas y Naturales, Universidad de Buenos Aires, Intendente Güiraldes 2160, Ciudad Universitaria, CABA C1428EGA, Argentina
* Correspondence: acuna.cintia@inta.gob.ar

Abstract: The genetic diversity of 14 Japanese plum (*Prunus salicina* Lindl) landraces adapted to an ecosystem of alternating flooding and dry conditions was characterized using neutral simple sequence repeat (SSR) markers. Twelve SSRs located in six chromosomes of the *Prunus persica* reference genome resulted to be polymorphic, thus allowing identification of all the evaluated landraces. Differentiation between individuals was moderate to high (average shared allele distance (DAS) = 0.64), whereas the genetic diversity was high (average indices polymorphism information content (PIC) = 0.62, observed heterozygosity (Ho) = 0.51, unbiased expected heterozygosity (uHe) = 0.70). Clustering and genetic structure approaches grouped all individuals into two major groups that correlated with flesh color. This finding suggests that the intuitive breeding practices of growers tended to select plum trees according to specific phenotypic traits. These neutral markers were adequate for population genetic studies and cultivar identification. Furthermore, we assessed the SSR flanking genome regions (25 kb) in silico to search for candidate genes related to stress resistance or associated with other agronomic traits of interest. Interestingly, at least 26 of the 118 detected genes seem to be related to fruit quality, plant development, and stress resistance. This study suggests that the molecular characterization of specific landraces of Japanese plum that have been adapted to extreme agroecosystems is a useful approach to localize candidate genes which are potentially interesting for breeding.

Keywords: Japanese plum; SSR; diversity; genetic structure; candidate genes

1. Introduction

The genus *Prunus* consists of more than 400 species, of which plum, peach, almond, and cherry trees stand out for their commercial importance worldwide. The Japanese plum (*Prunus salicina* Lindl) is one of the most commercially important plum species. This diploid ($2x = 2n = 16$) species has been cultivated in different environments and was introduced to North America from China in the 19th century [1]. Today, most Japanese plum cultivars grown worldwide come from the early selections originated in California at the end of the 19th century [2]. Besides, most planted plum orchards in other extensive production regions of Argentina derive from a few introduced cultivars of global distribution.

By contrast, European immigrant growers early in the last century who settled in the Paraná River Delta (PRD) introduced old traditional cultivars from their European countries. They selected wild plants emerged by the spontaneous sowing of seeds of the introduced varieties that they cultivated based on good performance and high-quality traits. Growers from the PRD had to face a particularly harsh environmental ecosystem.

The PRD is a unique ecosystem dominated by floods because of water discharges mainly from the Paraná River followed by discharges from the Uruguay and Gualeguay rivers. Tidal and storm surges from the Río de la Plata estuary as well as local rainfalls also contribute to generate these wetlands [3]. Precipitations in this region are influenced by the El Niño Southern Oscillation (ENSO) [4]. Indeed, three important floods associated with El Niño took place in PRD in 1982, 1998, and 2007 [3,5]. Furthermore, the alternation of wet and dry periods in the PRD influences the variability of the ecosystem. All these features give place to at least 15 landscape units [3] with a high ecological diversity distinctly adapted to annual hydrological cycles [6].

In this climate context, growers from the PRD generated local plum landraces that were cultivated after multiplication by grafting or from seeds, and therefore obtained the best wetland-adapted plants. In this way, they intuitively created a specific fruit tree germplasm for fruit production in an ecosystem with harsh environmental conditions that alternates river flooding periods with extreme dry season. The generated germplasm, which presents different harvest times (November to February, Table 1), is highly variable regarding pulp and skin coloration, as well as in its organoleptic characteristics.

Table 1. Characteristics of the landraces collected from the Paraná River Delta (PRD).

Landrace	Origin	Flesh Color	Harvest
Ratto	Seedling Delta	red	Very late season—End of January
Severiana	Seedling Delta	red	Very late season—End of January to mid-February
Remolacha de Berisso	Seedling Delta	red	Very late season—End of January to early February
Fragata	Seedling Delta	red	Very late season—End of January
Remolacha de Leber	Seedling Delta	red	Middle season—Early/middle December
Gigaglia	Seedling Delta	red	Middle season—Mid-November to early December
Capri	Seedling Delta	red	Late season—Late December
Giordano	Seedling Delta	red	Middle season—Late November to early December
Ciervita	Seedling Delta	pink	Early season—Mid-to late November
Reina de oro	Seedling Delta	yellow	Late season—Early January
Juanita	Seedling Delta	yellow	Early season—Mid-November to early December
Gascón	Seedling Delta	yellow	Middle season—Late November to early December
Tricerri	Seedling Delta	yellow	Mid-November
"X"	Seedling Delta	yellow	Mid-November

Thus, Japanese plum landraces selected in the PRD are a unique genetic resource in the world, and one of the few adapted to delta edaphic conditions and to humid temperate climate [7]. These landraces could constitute the closest reservoirs of "useful" alleles for future genetic improvement in the context of climate change. They contain the genetic variants naturally or artificially selected by growers because of their adaptation, productivity, or resistance to different stresses in the territory [4,5]. The knowledge of this interesting germplasm collection's diversity is essential for its characterization, conservation, and maintenance, as proposed for sweet cherry [8].

Genomics has triggered a revolution in the study of diversity [9,10] and conservation [11] by providing methods for the genetic characterization of individuals, populations, and species. The availability of *Prunus* genomic resources such as peach whole-genome sequences [12] allows the search for sequences of interest for breeding purposes (see for review [13]).

Mnejja et al. [2] isolated 27 microsatellites or simple sequence repeats (SSRs) in *P. salicina*, whereas Carrasco et al. [14] analyzed genetic diversity and correlation among 29 Japanese plum cultivars using a combination of inter simple sequence repeat (ISSR) and SSR markers. On the other hand, Klabunde et al. [15] genotyped 47 cultivars with eight microsatellite markers. A unique genetic map based on Restriction Fragment Length Polymorphism (AFLP) has been generated for this species [16]. Recently, González et al. [17] reported a set of EST-SSR markers for *P. salicina* developed from specific genes, which determine different organoleptic properties of the fruit, and analyzed 29 cultivars with these markers.

Microsatellite markers generate genotype profiles that can allow the identification and traceability of different cultivars in the diverse stages of their breeding. Therefore, the use of microsatellite markers could be a valuable tool to describe and protect Japanese plum germplasm. In addition, the genome position of SSR on a reference genome, together with potential candidate genes for fruit properties and self-incompatibility systems, could also be a useful tool for breeding purposes [13].

This unique plum germplasm was studied through the use of SSR markers to assess the diversity, genetic structure, and relationships between 14 landraces of Japanese plums from the PRD. Therefore, we also analyzed the flanking regions of the polymorphic SSR with bioinformatics tools to search for candidate genes. These candidate genes were in linkage disequilibrium with polymorphic SSR markers in an adaptive evolution to water stress conditions and fruit quality.

2. Materials and Methods

2.1. Plant Material

Fourteen landraces of *P. salicina* and two introduced commercial cultivars were assessed to evaluate amplification and polymorphic status of the 12 selected pairs of primers from Mnejja et al. [2]. The analyzed landraces were "Ratto", "Severiana", "Remolacha de Berisso", "Fragata", "Remolacha de Leber", "Gigaglia", "Capri", "Ciervita", "Giordano", "Reina de oro", "Juanita", "Tricerri", "X" (unknown origin), "Gascón". The introduced cultivars consisted of one Japanese plum of American origin ("Santa Rosa") and one European plum cultivar (*P. domestica*; "Reina Claudia"). Two peach cultivars (*P. persica*) "Sol de Mayo" and "Zelanda" were included as outgroups (Table 1). All samples were collected from one representative tree of each clonally reproduced genotype located in the Pacífico River, San Fernando Island of the PRD, Province of Buenos Aires, Argentina (34°12′01.62″ S, 58°40′07.66″ W).

2.2. DNA Extraction, PCR, and Gel Electrophoresis Conditions

Total genomic DNA from 20 mg of leaf samples was extracted with a Nucleo SpinR Plant II kit (Macherey-Nagel, Düren, Germany).

The samples were screened for polymorphisms with 12 SSR markers [2]. PCR was performed according to Mnejja et al. [2]. The amplification products were denatured for 5 min in denaturing loading buffer at 95 °C and separated by a 6% polyacrylamide gel electrophoresis (6% acrylamide/bisacrylamide 20:1, 7.5 M urea, 0.5 × TBE) along with a 25 bp DNA ladder standard (Invitrogen, Waltham, USA). The DNA silver-staining procedure of Promega (Madison, WI, USA) was used for visualization.

2.3. Genetic Analyses

The number (Na), effective number (Ne) and frequency of alleles, as well as the observed heterozygosity (Ho), unbiased expected heterozygosity (uHe), and private alleles by population were determined for the 14 local Japanese plums samples using the GenAlEx 6.5 program [18]. The polymorphism information content (PIC) of each marker was calculated according to Botstein et al. [19].

Genetic diversity analysis was performed on plum and peach genotypes. Shared allele distance (DAS) [20] was implemented between individuals using POPULATIONS 1.2.28 [21]. The unweighted pair group method using arithmetic averages (UPGMA) algorithm was used for cluster analysis and the development of the corresponding dendrogram2.4. Population Structure.

The population structure pattern was assessed by performing a Bayesian analysis with the software STRUCTURE v.2.3.3 [22]. Assignment of individuals to a group was evaluated according to the membership coefficient Q criteria ($Q \geq 0.8$). The model used was an admixture model with ten replicates for each number of genetic groups ($K = 1$–10) and 100,000 iterations of burn-in followed by 250,000 Markov chain Monte Carlo (MCMC) iterations. The outputs of the genetic group analysis were extracted in Structure Harvester [23]. The optimal K-value was determined using the delta K method, as described by Evanno et al. [24].

Because the assumptions underlying the population genetic model in STRUCTURE (e.g., Hardy–Weinberg or linkage equilibrium (LD)) may limit its use, this analysis was complemented with a discriminant analysis of principal components (DAPC) [25]. The value of K was determined using the Bayesian Information criterion (BIC) value given by the software [26]. The critical membership value was set at 0.8. The DAPC was implemented in the R package adegenet 2.0 [27,28].

2.4. Mapping of SSR Markers on the Prunus persica Genome and Identification of Their Flanking Genes

The obtained amplification sequences for the 12 SSR were mapped to the *P. persica* genome [29] (Phytozome–*Prunus persica* v2.1 (phytozome.jgi.doe.gov)). Mapping was performed using the Bowtie2 Alignment tool with default settings [30]. A custom Perl script (Higgins J., personal communication) was used to determine the annotated genes of the *P. persica* genome within a flanking region of 50 kb ($\pm$ 25 kb adjacent to each SSR locus). This window size was selected based on the high macrosynteny found between *Prunus* species [31] and studies in sweet cherry (*P. avium* L), which presents the lowest LD in *Prunus* genus. Indeed, in sweet cherry, the intra-chromosomal LD is lower than peach, therefore declining among 0.05 and 0.1 Mb [8].

3. Results

3.1. Diversity Study

All SSRs generated polymorphic bands in the 14 *P. salicina* landraces (Table 2). The alleles were clearly differentiated, with no discrepancies in the banding pattern.

The 12 polymorphic SSRs in the 14 evaluated local landraces of Japanese plums generated 66 different alleles (Electronic Supplementary Material, Table S1). The Na per locus ranged from four to seven, with a mean of 5.5 (SD 1.0), and the Ne ranged from 2.2 to 4.2, with a mean of 3.1 (SD 0.6). Additionally, PIC, Ho, and uHe ranged from 0.51 to 0.73, 0.15 to 0.86, and 0.56 to 0.79, with global means of 0.62 (SD 0.06), 0.51 (SD 0.23), and 0.70 (SD 0.06), respectively (Table 2). Thus, the genetic diversity levels were moderate to high, and kept 37 private alleles in 12 SSRs (20 with a frequency below 0.08) with respect to 4 private alleles in 4 SSRs (all of them with a frequency of 0.25) found in the analyzed outgroups.

3.2. Cluster and Genetic Structure Analyses

These 12 SSRs allowed the unambiguous genotype differentiation of the 14 landraces studied. DAS values between plum individuals fluctuated from 0.14 (between "Ratto" and "Severiana", which shared 19 alleles out of 24) to 1 (between "Reina de oro" and "Remolacha de Berisso"; with no shared alleles), with an average DAS of 0.64 among all the landraces (Electronic Supplementary Material, Table S2).

In order to easily represent the genetic relationship among all samples, a DAS-based dendrogram was developed. Therefore, the genotypes were grouped in three main groups (Figure 1). One group contained nine accessions with red flesh fruits (Group 1: "Ratto", "Severiana", "Remolacha de Berisso",

"Fragata", "Remolacha de Leber", "Gigaglia", "Capri", "Ciervita", and "Giordano") and within this group "Ratto" and "Severiana" were the closest genotypes (DAS = 0.14). A second group consisted of seven accessions with yellow flesh fruits (Group 2: "Reina de oro", "Juanita", "Tricerri", "X", "Gascón", and the two introduced commercial genotypes "Santa Rosa" and "Reina Claudia"). The sample from the "X" landrace showed higher genetic similarity with "Tricerri" (DAS = 0.31) but differed from all the other samples. Finally, the third group contained the outgroup varieties, that is, the two peach samples (Group 3: "Sol de Mayo" and "Zelanda").

Table 2. Information of the 12 analyzed polymorphic simple sequence repeat (SSR) markers in 14 local landraces of Japanese plums.

SSR Marker	Location	AR (pb)	N	Na	Ne	Ho	uHe	PIC
CPSCT011 (AY426199.1)	C 5 (4163117–4164052)	171–189	13	6	3.04	0.38	0.70	0.63
CPSCT018 (AY426204.1)	C 8 (123199–124139) 3e-	151–172	13	6	2.18	0.23	0.56	0.51
CPSCT021 (AY426206.1)	C 2 (27308766–27309699)	132–157	11	6	3.66	0.36	0.76	0.69
CPSCT022 (AY426207.1)	C 5 (16620819–16621314)	159–179	13	7	2.91	0.77	0.68	0.62
CPSCT024 (AY426209.1)	C 1 (28058204–28058903)	157–179	13	4	2.89	0.61	0.68	0.60
CPSCT025 (AY426210.1)	C 3 (6709070–6709740)	181–223	12	5	2.51	0.66	0.64	0.53
CPSCT026 (AY426211.1)	C 7 (11365276–11365753)	176–195	14	4	2.78	0.43	0.66	0.60
CPSCT027 (AY426212.1)	C 1 (23010058–23010528)	137–156	13	5	2.66	0.31	0.65	0.57
CPSCT030 (AY426215.1)	C 5 (15121388–15122031)	179–200	14	7	4.21	0.86	0.79	0.73
CPSCT034 (AY426219.1)	C 2 (29946149–29946699)	177–224	14	6	3.70	0.64	0.76	0.68
CPSCT042 (AY426226.1)	C 7 (16682143–16682742)	167–185	14	5	3.53	0.71	0.74	0.67
CPSCT044 (AY426228.1)	C 2 (20793617–20794126)	218–241	13	5	3.35	0.15	0.73	0.65
Mean				5.50	3.12	0.51	0.70	0.62
SD				1.00	0.58	0.23	0.06	0.06

Location on the reference genome of *P. persica*, C: chromosome, AR: allele size ranges (pb), N: sample size, Na: number of alleles, Ne: effective number of alleles, Ho: observed heterozygosity, uHe: unbiased expected heterozygosity, PIC: polymorphism information content, SD: standard deviation.

Subsequently, the Bayesian analysis [22] supported the existence of a genetic structure among the studied genotypes, with a most probable value of $K = 3$ subpopulations. All the accessions except for "Ciervita" ($Q = 0.77$) were assigned to a group with Q values above 0.85. The low membership value (Figure 2a) for "Ciervita" suggests a possible genetic mixed origin of this variety.

Similarly, the DAPC analysis also retrieved $K = 3$ genetic groups according to cultivar flesh color (Figure 2b). This analysis supported the results obtained by the Bayesian model and the genetic distance results.

These three main groups had several private alleles, and half of these alleles were at a very low frequency in plums. For instance, the red pulp group (9 genotypes) presented 13 private alleles (9 SSRs) and 5 alleles with a frequency below 0.06. The yellow pulp group (7 genotypes) contained 26 private alleles (11 SSRs), of which 10 had a frequency below 0.08. Both peaches had 10 private alleles (8 SSRs), and all these alleles showed a frequency above 0.25.

Comparison of the genotypic and phenotypic data enabled the identification of a marker/allele related to flesh color phenotype. Marker CPSCT25 revealed an allele of 193 bp that was exclusively present in all the red flesh plums. Allele "193-bp" was present in a heterozygous or homozygous state (Electronic Supplementary Material, Table S1). BLASTx search using the sequences harboring CPSCT25 against the GenBank non-redundant protein database revealed significant matches with *P. persica* protein FAR1-Related sequence 6.

Another SSR marker, CPSCT11, presented a private allele, allele "171-bp". This allele was present in all the yellow flesh plums, except for the "Reina de Oro" cultivar, but absent from the red flesh plums (Electronic Supplementary Material, Table S1). BLASTx search using the sequences harboring CPSCT11 revealed significant matches with *P. mume* abscisic acid receptor PYR1 isoform X1. Both FAR1 and PYR1 are related to the abscisic-acid-mediated signaling pathway and to seed germination regulation [32].

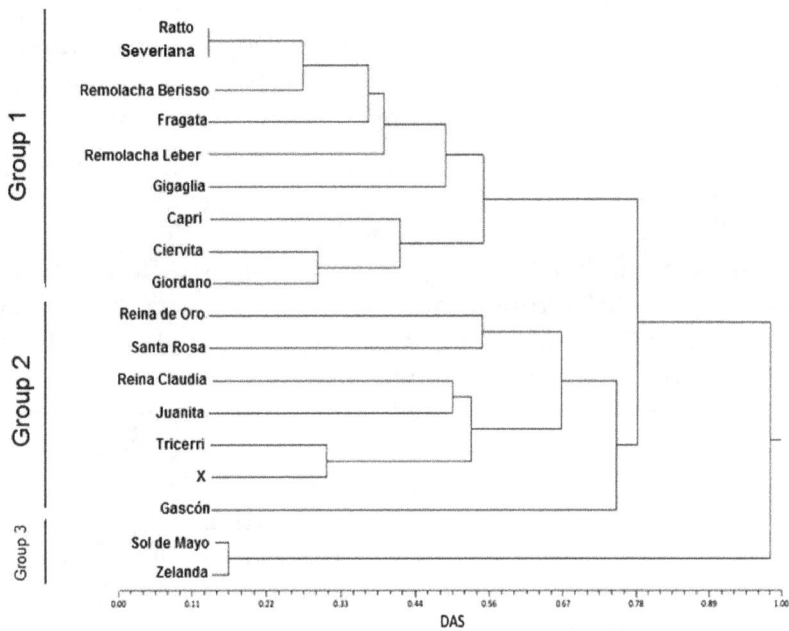

Figure 1. Cluster analysis. Dendrogram obtained by unweighted pair group method using arithmetic averages (UPGMA) calculated by shared allele distance (DAS) based on 12 SSRs. Group 1: red flesh plums, Group 2: yellow flesh plums, Group 3: outgroups (peach samples).

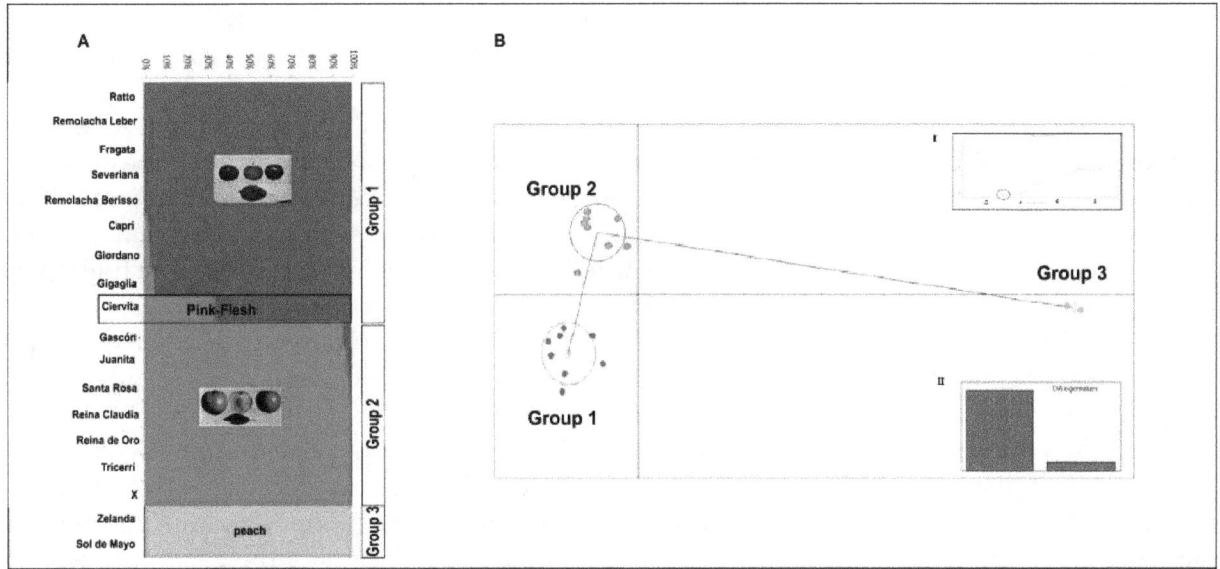

Figure 2. Estimated population structure. Genetic structure of 16 plum genotypes (14 landraces and two introduced cultivars) analyzed as estimated by (**A**) model-based Bayesian cluster analysis implemented in the program Structure; and (**B**) discriminant analysis of principal components (DAPC) implemented in the adegenet package of R software. Inset I: Inference of the number of clusters (Clusters vs. Bayesian information criterion (BIC)). Inset II: Relative importance of the two most important axes.

3.3. Mapping of SSR Markers on the Prunus persica *Genome and Identification of Their Flanking Genes*

The sequences containing polymorphic SSR were mapped and annotated on the reference genome of *P. persica*. The loci were physically mapped to a unique position on six out of eight chromosomes (1, 2, 3, 5, 7, and 8) (Table 2) with sequence similarities above 85%.

By an in-silico analysis, we identified 118 genes located within the 50 kb windows flanking the SSRs.

Interestingly, within these genes, some candidates were related to stress resistance (NAC transcription factor and glutathione S-transferases (GSTs)), fruit quality (Squamosa, AGAMOUS-like MADS box, Transducin/WD40 repeat-like superfamily protein and possibly NAC), plant growth and development (VQ motif-containing proteins and microspore-specific promoter proteins) (Electronic Supplementary Material, Table S3).

4. Discussion

In this study, we detected high levels of diversity within 14 different plum landraces from the PRD region that have been selected through the years by local growers. The diversity indexes were lower than those described by Carrasco et al. [14] (He = 0.80 and Ho = 0.90) and Ferrero Klabunde et al. [15] (PIC = 0.80 and Ho = 0.77) with eight SSR markers. However, this was expected due to a higher number of *P. salicina* cultivars analyzed (29 and 47, respectively). On the other hand, the PIC value detected in the present study was closer to the value obtained in a study of 24 cultivars from different areas in China evaluated with 16 SSR markers (PIC = 0.7) [33]. Likewise, the levels of diversity in our study were similar to or higher than those of other *Prunus* species, such as peach (Ho = 0.35; He = 0.55 [34]), apricot (Na = 3.5; Ho = 0.58 [35]), and cherry (He = 0.66; [36]).

Although the number of genotypes analyzed in this study was small, the degree of diversity was high. This could be due to the mating system of *P. salicina*. Indeed, *P. salicina* requires cross-pollination because of its strong gametophytic self-incompatibility system, and this characteristic makes it more diverse [37] and heterozygous. Cross-pollination may have played an important role in the evolution of landraces in the PRD region because early 20th century producers very likely multiplied their trees by seed planting instead of vegetative cloning.

The 12 SSRs allowed the unambiguous identification of the genotypes studied (DAS values between plum individuals ≥ 0.14). Thus, the 14 evaluated plum landraces were genetically unique (i.e., with SSR genotypes different from the rest). Cases of synonyms among samples with different names but identical SSR profiles at the 12 analyzed loci were discarded.

The cluster analysis (UPGMA) grouped the plum genotypes into two groups: plums with red and yellow flesh, respectively. Half of the private alleles of both groups occurred in a low frequency. Furthermore, the genetic relationship analysis supported the existence of genetic structure within the studied landraces. The two different methods of accessing genetic structure (Bayesian and DAPC analyses) coincided in differentiating the landraces based on the color of fruit flesh. The high Q values in the Bayesian study could reflect the selection history in response to flesh color.

Particularly, the membership of "Ciervita" to a group varied depending on the software applied. This characteristic suggests a genetic mixed origin in this landrace. Interestingly, "Ciervita" has fruits with a pink flesh phenotype.

Another interesting result is that "Santa Rosa" and "Reina Claudia" were grouped within Group 2, that is, based on a phenotypic characteristic of fruits (yellow flesh fruits). Because both are introduced commercial cultivars and have some private alleles, we would have expected that these cultivars were within another group, or at least that they would differ more from the rest of the genotypes within this group. With the use of more (or different) SSRs, we may obtain different results, and these independent cultivars would be separated from these groups.

By contrast, González et al. [17] found that the cultivars they evaluated in their study were grouped based on other characteristics. They used SNPs and EST–SSR markers developed from the putative flavonoid pathway transcription factors to study the genetic structure. Only when using the specific markers EST-SSR (PsMYB10, PsMYB1, and PsbHLH35) did they obtain three clusters of cultivars related to the skin color: two cultivars with red skin fruits and a third cluster that grouped all the cultivars with yellow skin fruits.

Although nutritional traits like the anthocyanin content of plum fruit flesh are interesting, the reason for characterizing PRD landraces was because of their unique adaptation to a peculiar water-stressed ecosystem. Many genes may contribute to this adaptation, and for this reason the first goal was to represent-as much as possible-a plum genome with this rather limited number of molecular markers. The genetic mapping of SSR loci to the *P. persica* v2.1 genome enabled us to establish the position of the 12 polymorphic SSR markers used. The next step was to evaluate if these polymorphic markers were flanked, probably linked, to genes that could be associated to water stress conditions. The information of annotated genes nearby polymorphic SSRs (< 25 kb) allowed us to describe the context of the markers and provided a first step to future association studies by identifying candidate genes for the expression of important agronomical characteristics.

We identified 118 genes located within the 50-kb windows. Of these genes, 106 corresponded to homologous genes described in *Arabidopsis*. Anatomical and physiological comparisons between *Arabidopsis* and *Prunus* species indicated marked similarities between them [35]. Cautiously taking into account the phylogenetic distance between *Prunus* and *Arabidopsis*, as well as their enormous biologic differences, we found some similarities in predicted gene functions that were also found in various plant taxonomic groups. Therefore, we detected candidate genes related to stress resistance, fruit quality, and other interesting candidate genes (Electronic Supplementary Material, Table S3).

For instance, genes encoding Acyl-CoA N-acyltransferases (NAT) superfamily proteins found close to marker CPSCT021 are associated with flavor generation during ripening in some fruit species, such as apricot [38] and peach [39]. Marker CPSCT022 was also close to other genes associated with fruit ripening and quality: SQUAMOSA MADS Box [40] and AGAMOUS-like MADS box genes [41,42]. Interestingly, these two genes have been also associated with the metabolic pathway of anthocyanins responsible for fruit color. Indeed, they were associated with the regulation of anthocyanin accumulation or synthesis in pear and *Arabidopsis* [43,44], as well as in bilberry [45]. Anthocyanin biosynthesis is cooperatively regulated by transcriptional regulators, including WD40 proteins (adjacent to CPSCT042). These regulators form a complex that binds to promoters and activates the transcription of structural genes of the anthocyanin biosynthetic pathway [46]. In addition, an alpha/beta-hydrolases superfamily protein involved in delaying fruit senescence under low temperature in strawberry fruit [47] was close to marker CPSCT025 (Electronic Supplementary Material, Table S3).

Stress-related genes corresponding to the NAC (No Apical Meristem) domain transcriptional regulator superfamily protein and tau-type glutathione S-transferases (GSTs) [48,49] were located close to CPSCT024. Additionally, stress-related genes encoding reversibly glycosylated polypeptide (RGP family) [50] and the bZIP transcriptional regulator gene [51] were adjacent to CPSCT022 and CPSCT042, respectively.

The NAC protein consists of a large and complex family of transcription factors that are involved in multiple biological processes in plants, including perception of biotic/abiotic stress, signal transduction, transcription control, and gene activation. The NAC transcription factors regulate the differentiation of cells specialized for water conduction in vascular plants, and their conserved genetic basis suggests roles for NAC proteins in the adaptation of plants to land [52].

Interestingly, NAC (close to CPSCT024) was also related to cold-stress responses in *P. mume* [53], and to the metabolic pathway of anthocyanins responsible for fruit color in different plants such as peach, plum, and *Arabidopsis* [54–56].

On the other hand, the bZIP families described by Janiak et al. [51] have a role in gene expression regulation in roots and may have an impact on root development under drought stress conditions.

Furthermore, genes involved in plant growth and development encoding VQ motif-containing proteins and microspore-specific promoter proteins [57,58] were close to CPSCT011.

Although the evaluated sample size was too small to define associated markers by association mapping analysis through haploblocking definition (genetic blocks where alleles are grouped in linkage disequilibrium without recombination), pedigree history could also explain this result [13].

Further analyses are needed to assess the potential contribution of these genes to the specific adaptation of the local landraces to water-stressed environments. Some, or even all, of the allele associations described above, could still be attributed to demographic or management reasons that have nothing to do with adaptation to the environment.

In peach, an analysis of 53 SSRs distributed through the genome was carried out in 104 landraces from six Chinese geographical regions, determining an LD decay across all populations of 2500 kb [59]. However, lower LD decay was found in grape (~1300 kb) [60] and in sweet cherry (~100 kb) [8], independently of the number and type of molecular markers used. In our work, the evaluated sequences for searching interesting candidate genes were up to 25 kb distance from the SSR markers, so we expect that linkage between them is high enough to be considered in future breeding studies with association markers.

The findings described here could provide an interesting working hypothesis for future research in plum molecular breeding. The incorporation of the markers linked to the annotated genes in this study could be of high value in marker-assisted selection breeding programs and in future genome-wide association studies (GWAS).

5. Conclusions

This is the first study characterizing a representative sample of different plum landraces adapted to water-stressed environments, particularly in the PRD region of Argentina.

Twelve SSRs allowed us to estimate the genetic diversity and structure of PRD plums. These landraces showed a high degree of diversity and a differentiation between accessions with red and yellow flesh.

The genetic fingerprinting profiles could allow the identification and traceability of different genotypes in the various stages of breeding. Furthermore, these profiles could contribute to registering the assessed genotypes in the National Register of Cultivars. The results highlight the utility of bioinformatics to identify genes involved in complex characters to contribute to the understanding of the genetics behind the phenotypic variation. Potentially, the markers close to candidate genes are suitable for comparative QTL mapping, molecular-marker-assisted breeding, and for population genetic studies across different species within the genus *Prunus*.

Supplementary Materials: Table S1 SSR Genotyping Matrix. Table S2 Shared allele distance (DAS) Matrix. Table S3 Mapping and annotation of genes nearby (< 25 kb window) to polymorphic SSR on the *Prunus persica* genome.

Author Contributions: Conceptualization, E.H.H., S.N.M.P., T.C. and E.A.F.; methodology, C.V.A., S.M.B.; formal analysis, C.V.A., J.G.R., P.V.V., N.C.A. and M.N.G.; investigation, C.V.A.; resources, T.C. and E.A.F.; writing—original draft preparation, C.V.A.; writing—review and editing, J.V.S.y.G., M.C.M, S.N.M.P. and E.H.H.; project administration, S.N.M.P.; funding acquisition, S.N.M.P.

Acknowledgments: Special thanks to David Gomez for selecting and collecting the plant material.

References

1.   Das, B. Prunus diversity—Early and present development: A review. *Int. J. Biodivers. Conserv.* **2012**, *3*, 14.
2.   Mnejja, M.; Garcia-Mas, J.; Howad, W.; Badenes, M.L.; Arús, P. Simple-sequence repeat (SSR) markers of Japanese plum (*Prunus salicina* Lindl.) are highly polymorphic and transferable to peach and almond. *Mol. Ecol. Notes* **2004**, *4*, 163–166. [CrossRef]
3.   Kandus, P.; Quintana, R.D. The Paraná River Delta. In *The Wetland Book*; Elsevier: Amsterdam, The Netherlands, 2016; pp. 1–9.
4.   Caffera, R.; Berbery, H. Climatologia de la Cuenca del Plata. *Clim. La Cuenca Del Plata* **2006**, *20*, 19–38.
5.   Zagare, V.M.E. Natural territory, urban growth and climate change in the Parana River Delta and Rio de la Plata estuarine system. *An Overv.* **2016**, 1–16.

6. Kandus, P.; Malvárez, A.I. Vegetation patterns and change analysis in the lower delta islands of the Parana River (Argentina). *Wetlands* **2006**, *24*, 620–632. [CrossRef]

7. Meis, M.; Llano, M.P. Modelado estadístico del caudal mensual en la baja Cuenca del Plata. *Meteorologica* **2018**, *43*, 63–77.

8. Campoy, J.A.; Lerigoleur-Balsemin, E.; Christmann, H.; Beauvieux, R.; Girollet, N.; Quero-García, J.; Dirlewanger, E.; Barreneche, T. Genetic diversity, linkage disequilibrium, population structure and construction of a core collection of *Prunus avium* L. landraces and bred cultivars. *BMC Plant Biol.* **2016**, *16*, 49. [CrossRef]

9. Acuña, C.V.; Fernandez, P.; Villalba, P.V.; García, M.N.; Hopp, H.E.; Marcucci Poltri, S.N. Discovery, validation, and in silico functional characterization of EST-SSR markers in *Eucalyptus globulus*. *Tree Genet. Genom.* **2012**, *8*, 289–301. [CrossRef]

10. Wünsch, A.; Carrera, M.; Hormaza, J.I. Molecular Characterization of Local Spanish Peach [*Prunus persica* (L.) Batsch] Germplasm. *Genet. Resour. Crop Evol.* **2006**, *53*, 925. [CrossRef]

11. Arif, I.A.; Khan, H.A.; Bahkali, A.H.; Al Homaidan, A.A.; Al Farhan, A.H.; Al Sadoon, M.; Shobrak, M. DNA marker technology for wildlife conservation. *Saudi J. Biol. Sci.* **2011**, *18*, 219–225. [CrossRef]

12. Verde, I.; Abbott, A.G.; Scalabrin, S.; Jung, S.; Shu, S.; Marroni, F.; Zhebentyayeva, T.; Dettori, M.T.; Grimwood, J.; Cattonaro, F.; et al. The high-quality draft genome of peach (*Prunus persica*) identifies unique patterns of genetic diversity, domestication and genome evolution. *Nat. Genet.* **2013**, *45*, 487–494. [CrossRef] [PubMed]

13. Aranzana, M.J.; Decroocq, V.; Dirlewanger, E.; Eduardo, I.; Gao, Z.S.; Gasic, K.; Iezzoni, A.; Jung, S.; Peace, C.; Prieto, H.; et al. Prunus genetics and applications after de novo genome sequencing: Achievements and prospects. *Hortic. Res.* **2019**, *6*, 58. [CrossRef]

14. Carrasco, B.; Díaz, C.; Moya, M.; Gebauer, M.; García-González, R. Genetic characterization of Japanese plum cultivars (*Prunus salicina*) using SSR and ISSR molecular markers. *Cienc. E Investig. Agrar.* **2012**, *39*, 533–543. [CrossRef]

15. Klabunde, G.H.F.; Dalbó, M.A.; Nodari, R.O. DNA fingerprinting of Japanese plum (*Prunus salicina*) cultivars based on microsatellite markers. *Crop. Breed. Appl. Biotechnol.* **2014**, *14*, 139–145. [CrossRef]

16. Vieira, E.A.; Onofre Nodari, R.; Cibele De Mesquita Dantas, A.; Henri, J.-P.; Ducroquet, J.; Dalbó, M.; Borges, C.V. Genetic Mapping of Japanese Plum. Crop Breeding and Applied Biotechnology: Vicosa, Brazil, 2005; p. 5.

17. Gonz#xE1;lez, M.; Salazar, E.; Castillo, J.; Morales, P.; Mura-Jornet, I.; Maldonado, J.; Silva, H.; Carrasco, B. Genetic structure based on EST–SSR: a putative tool for fruit color selection in Japanese plum (*Prunus salicina* L.) breeding programs. *Mol. Breed.* **2016**, *36*.

18. Peakall, R.; Smouse, P.E. Genalex 6: Genetic analysis in Excel. Population genetic software for teaching and research. *Mol. Ecol. Notes* **2006**, *6*, 288–295. [CrossRef]

19. Botstein, D.; White, R.L.; Skolnick, M.; Davis, R.W. Construction of a genetic linkage map in man using restriction fragment length polymorphisms. *Am. J. Hum. Genet.* **1980**, *32*, 314.

20. Chakraborty, R.; Jin, L. A unified approach to study hypervariable polymorphisms: Statistical considerations of determining relatedness and population distances. In *DNA Fingerprinting: State of the Science*; Birkhäuser Basel: Basel, Switzerland, 1993; pp. 153–175.

21. Langela, O. *Populations: A free Population Genetic Software*; UCLA: Los Angeles, CA, USA, 2002.

22. Pritchard, J.K.; Stephens, M.; Donnelly, P. Inference of Population Structure Using Multilocus Genotype Data. *Genetics* **2000**, *155*, 945–959.

23. Earl, D.A.; vonHoldt, B.M. STRUCTURE HARVESTER: A website and program for visualizing STRUCTURE output and implementing the Evanno method. *Conserv. Genet. Resour.* **2012**, *4*, 359–361. [CrossRef]

24. Evanno, G.; Regnaut, S.; Goudet, J. Detecting the number of clusters of individuals using the software structure: A simulation study. *Mol. Ecol.* **2005**, *14*, 2611–2620. [CrossRef]

25. Jombart, T.; Devillard, S.; Balloux, F. Discriminant analysis of principal components: A new method for the analysis of genetically structured populations. *BMC Genet.* **2010**, *11*, 94. [CrossRef]

26. Liu, N.; Zhao, H. A non-parametric approach to population structure inference using multilocus genotypes. *Hum. Genom.* **2006**, *2*, 353. [CrossRef]

27. Jombart, T. Adegenet: A R package for the multivariate analysis of genetic markers. *Bioinformatics* **2008**, *24*, 1403–1405. [CrossRef]

28. Jombart, T.; Ahmed, I. Adegenet 1.3-1: New tools for the analysis of genome-wide SNP data. *Bioinformatics* **2011**, *27*, 3070–3071. [CrossRef]

29. Arús, P.; Verde, I.; Sosinski, B.; Zhebentyayeva, T.; Abbott, A.G. The peach genome. *Tree Genet. Genomes* **2012**, *8*, 531–547. [CrossRef]

30. Langmead, B.; Salzberg, S.L. Fast gapped-read alignment with Bowtie 2. *Nat. Methods* **2012**, *9*, 357. [CrossRef]

31. Olmstead, J.W.; Sebolt, A.M.; Cabrera, A.; Sooriyapathirana, S.S.; Hammar, S.; Iriarte, G.; Wang, D.; Chen, C.Y.; Van Der Knaap, E.; Iezzoni, A.F. Construction of an intra-specific sweet cherry (*Prunus avium* L.) genetic linkage map and synteny analysis with the Prunus reference map. *Tree Genet. Genomes* **2008**, *4*, 897–910. [CrossRef]

32. Yan, A.; Chen, Z. The pivotal role of abscisic acid signaling during transition from seed maturation to germination. *Plant Cell Rep.* **2017**, *36*, 689–703. [CrossRef]

33. Li-hui, Z.; Zhi-xiao, H.; Hai-yong, L.; YANG Min-sheng, A. Analysis of Genetic Diversity of *Prunus salicina* from Different Producing Areas by SSR Markers. *Acta Hortic. Sin.* **2015**, *42*, 11–118.

34. Rojas, G.; Méndez, M.A.; Muñoz, C.; Lemus, G.; Hinrichsen, P. Identification of a minimal microsatellite marker panel for the fingerprinting of peach and nectarine cultivars. *Electron. J. Biotechnol.* **2008**, *11*, 4–5. [CrossRef]

35. Tani, E.; Polidoros, A.N.; Tsaftaris, A.S. Characterization and expression analysis of FRUITFULL- and SHATTERPROOF-like genes from peach (*Prunus persica*) and their role in split-pit formation. *Tree Physiol.* **2007**, *27*, 649–659. [CrossRef]

36. Schueler, S.; Tusch, A.; Schuster, M.; Ziegenhagen, B. Characterization of microsatellites in wild and sweet cherry *Prunus avium*—Markers for individual identification and reproductive processes. *Genome* **2003**, *46*, 95–102. [CrossRef]

37. Mnejja, M.; Garcia-Mas, J.; Audergon, J.M.; Arús, P. *Prunus* microsatellite marker transferability across rosaceous crops. *Tree Genet. Genomes* **2010**, *6*, 689–700. [CrossRef]

38. González-Agüero, M.; Troncoso, S.; Gudenschwager, O.; Campos-Vargas, R.; Moya-León, M.A.; Defilippi, B.G. Differential expression levels of aroma-related genes during ripening of apricot (*Prunus armeniaca* L.). *Plant Physiol. Biochem.* **2009**, *47*, 435–440.

39. Zhang, B.; Shen, J.; Wei, W.; Xi, W.; Xu, C.-J.; Ferguson, I.; Chen, K. Expression of Genes Associated with Aroma Formation Derived from the Fatty Acid Pathway during Peach Fruit Ripening. *J. Agric. Food Chem.* **2010**, *58*, 6157–6165. [CrossRef]

40. Xu, Z.; Sun, L.; Zhou, Y.; Yang, W.; Cheng, T.; Wang, J.; Zhang, Q. Identification and expression analysis of the SQUAMOSA promoter-binding protein (SBP)-box gene family in *Prunus mume*. *Mol. Genet. Genom.* **2015**, *290*, 1701–1715. [CrossRef]

41. Tani, E.; Polidoros, A.N.; Flemetakis, E.; Stedel, C.; Kalloniati, C.; Demetriou, K.; Katinakis, P.; Tsaftaris, A.S. Characterization and expression analysis of AGAMOUS-like, SEEDSTICK-like, and SEPALLATA-like MADS-box genes in peach *Prunus persica* fruit. *Plant Physiol. Biochem.* **2009**, *47*, 690–700. [CrossRef]

42. Roy Choudhury, S.; Roy, S.; Nag, A.; Singh, S.K.; Sengupta, D.N. Characterization of an AGAMOUS-like MADS Box Protein, a Probable Constituent of Flowering and Fruit Ripening Regulatory System in Banana. *PLoS ONE* **2012**, *7*, e44361. [CrossRef]

43. Wu, J.; Zhao, G.; Yang, Y.-N.; Le, W.-Q.; Khan, M.A.; Zhang, S.-L.; Gu, C.; Huang, W.-J. Identification of differentially expressed genes related to coloration in red/green mutant pear *Pyrus communis*. *Tree Genet. Genom.* **2013**, *9*, 75–83. [CrossRef]

44. Gou, J.-Y.; Felippes, F.F.; Liu, C.-J.; Weigel, D.; Wang, J.-W. Negative Regulation of Anthocyanin Biosynthesis in Arabidopsis by a miR156-Targeted SPL Transcription Factor. *Plant Cell* **2011**, *23*, 1512–1522. [CrossRef]

45. Jaakola, L.; Poole, M.; Jones, M.O.; Kämäräinen-Karppinen, T.; Koskimäki, J.J.; Hohtola, A.; Häggman, H.; Fraser, P.D.; Manning, K.; King, G.J.; et al. A SQUAMOSA MADS box gene involved in the regulation of anthocyanin accumulation in bilberry fruits. *Plant Physiol.* **2010**, *153*, 1619–1629. [CrossRef]

46. González, M.; Salazar, E.; Cabrera, S.; Olea, P.; Carrasco, B. Analysis of anthocyanin biosynthesis genes expression profiles in contrasting cultivars of Japanese plum *Prunus salicina* during fruit development. *Gene Expr. Patterns* **2016**, *21*, 54–62. [CrossRef]

47. Xu, X.; Ma, X.; Lei, H.; Yin, L.; Shi, X.; Song, H. MicroRNAs play an important role in the regulation of strawberry fruit senescence in low temperature. *Postharvest Biol. Technol.* **2015**, *108*, 39–47. [CrossRef]

48. Puranik, S.; Sahu, P.P.; Srivastava, P.S.; Prasad, M. NAC proteins: Regulation and role in stress tolerance. *Trends Plant Sci.* **2012**, *17*, 369–381. [CrossRef]

49. Lan, T.; Yang, Z.L.; Yang, X.; Liu, Y.J.; Wang, X.R.; Zeng, Q.Y. Extensive functional diversification of the *Populus* glutathione S-transferase supergene family. *Plant Cell Online* **2009**, *21*, 3749. [CrossRef]

50. Zavaliev, R.; Sagi, G.; Gera, A.; Epel, B.L. The constitutive expression of Arabidopsis plasmodesmal-associated class 1 reversibly glycosylated polypeptide impairs plant development and virus spread. *J. Exp. Bot.* **2010**, *61*, 131–142. [CrossRef]

51. Janiak, A.; Mirosłw., K.; Iwonas, S. Gene expression regulation in roots under drought. *J. Exp. Botany.* **2016**, *67*, 1003–1014. [CrossRef]

52. Xu, B.; Ohtani, M.; Yamaguchi, M.; Toyooka, K.; Wakazaki, M.; Sato, M.; Kubo, M.; Nakano, Y.; Sano, R.; Hiwatashi, Y.; et al. Contribution of NAC Transcription Factors to Plant Adaptation to Land. *Science* **2014**, *343*, 1505–1508. [CrossRef]

53. Zhuo, X.; Zheng, T.; Zhang, Z.; Zhang, Y.; Jiang, L.; Ahmad, S.; Sun, L.; Wang, J.; Cheng, T.; Zhang, Q. Genome-Wide Analysis of the NAC Transcription Factor Gene Family Reveals Differential Expression Patterns and Cold-Stress Responses in the Woody Plant *Prunus mume*. *Genes* **2018**, *9*, 494. [CrossRef]

54. Zhou, H.; Lin-Wang, K.; Wang, H.; Gu, C.; Dare, A.P.; Espley, R.V.; He, H.; Allan, A.C.; Han, Y. Molecular genetics of blood-fleshed peach reveals activation of anthocyanin biosynthesis by NAC transcription factors. *Plant J.* **2015**, *82*, 105–121. [CrossRef]

55. Fang, Z.-Z.; Zhou, D.-R.; Ye, X.-F.; Jiang, C.-C.; Pan, S.-L. Identification of Candidate Anthocyanin-Related Genes by Transcriptomic Analysis of 'Furongli' Plum *Prunus salicina* Lindl. during Fruit Ripening Using RNA-Seq. *Front. Plant Sci.* **2016**, *7*, 1–15. [CrossRef] [PubMed]

56. Morishita, T.; Kojima, Y.; Maruta, T.; Nishizawa-Yokoi, A.; Yabuta, Y.; Shigeoka, S. Arabidopsis NAC transcription factor, ANAC078, regulates flavonoid biosynthesis under high-light. *Plant Cell Physiol.* **2009**, *50*, 2210–2222. [CrossRef] [PubMed]

57. Cheng, Y.; Zhou, Y.-H.; Yang, Y.; Chi, Y.-J.; Zhou, J.; Chen, J.-Y.; Wang, F.; Fan, B.; Shi, K.; Zhou, Y.-H.; et al. Structural and Functional Analysis of VQ Motif-Containing Proteins in Arabidopsis as Interacting Proteins of WRKY Transcription Factors. *Plant Physiol.* **2012**, *159*, 810–825. [CrossRef] [PubMed]

58. Custers, J.B.M.; Oldenhof, M.T.; Schrauwen, J.A.M.; Cordewener, J.H.G.; Wullems, G.J.; van Lookeren Campagne, M.M. Analysis of microspore-specific promoters in transgenic tobacco. *Plant Mol. Biol.* **1997**, *35*, 689–699. [CrossRef]

59. Cao, K.; Wang, L.; Zhu, G.; Fang, W.; Chen, C.; Luo, J. Genetic diversity, linkage disequilibrium, and association mapping analyses of peach *Prunus persica* landraces in China. *Tree Genet. Genomes* **2012**, *8*, 975–990. [CrossRef]

60. Barnaud, A.; Lacombe, T.; Doligez, A. Linkage disequilibrium in cultivated grapevine, *Vitis vinifera.* L. *Theor. Appl. Genet.* **2006**, *112*, 708–716. [CrossRef] [PubMed]

12

Association Mapping of Fertility Restorer Gene for CMS PET1 in Sunflower

Denis V. Goryunov [1,2,*], **Irina N. Anisimova** [3], **Vera A. Gavrilova** [3], **Alina I. Chernova** [1],
Evgeniia A. Sotnikova [4], **Elena U. Martynova** [1], **Stepan V. Boldyrev** [1,5], **Asiya F. Ayupova** [1],
Rim F. Gubaev [1], **Pavel V. Mazin** [1], **Elena A. Gurchenko** [1], **Artemy A. Shumskiy** [1],
Daria A. Petrova [1], **Sergey V. Garkusha** [6], **Zhanna M. Mukhina** [6], **Nikolai I. Benko** [7],
Yakov N. Demurin [8], **Philipp E. Khaitovich** [1] and **Svetlana V. Goryunova** [1,5,*]

[1] Skolkovo Institute of Science and Technology, Moscow 121205, Russia; alin.chernova@gmail.com (A.I.C.); elenamartynovaster@gmail.com (E.U.M.); beibaraban34@gmail.com (S.V.B.); A.Ayupova@skoltech.ru (A.F.A.); rimgubaev@gmail.com (R.F.G.); iaa.aka@gmail.com (P.V.M.); Elena.Gurchenko@skoltech.ru (E.A.G.); artemy.shumskiy@gmail.com (A.A.S.); D.Petrova@skoltech.ru (D.A.P.); P.Khaitovich@skoltech.ru (P.E.K.)

[2] Belozersky Institute of Physico-Chemical Biology, Lomonosov Moscow State University, Moscow 119992, Russia

[3] N.I. Vavilov All-Russian Research Institute of Plant Genetic Resources, Saint Petersburg (ex Leningrad) 190000, Russia; irina_anisimova@inbox.ru (I.N.A.); v.gavrilova@vir.nw.ru (V.A.G.)

[4] Department of Computer Science and Control Systems, Bauman Moscow State Technical University, Moscow 105005, Russia; sotnikova.evgeniya@gmail.com

[5] Institute of General Genetics, Russian Academy of Science, Moscow 119333, Russia

[6] All-Russia Rice Research Institute, Krasnodar 350921, Russia; arrri_kub@mail.ru (S.V.G.); agroplazma@gmail.com (Z.M.M.)

[7] Breeding and Seed Production Company "Agroplazma", Krasnodar 350012, Russia; agroplasma@rambler.ru

[8] Pustovoit All-Russia Research Institute of Oil Crops, Krasnodar 350038, Russia; yakdemurin@yandex.ru

[*] Correspondence: D.Goryunov@skoltech.ru (D.V.G.); S.Goryunova@skoltech.ru (S.V.G.)

Abstract: The phenomenon of cytoplasmic male sterility (CMS), consisting in the inability to produce functional pollen due to mutations in mitochondrial genome, has been described in more than 150 plant species. With the discovery of nuclear fertility restorer (*Rf*) genes capable of suppressing the CMS phenotype, it became possible to use the CMS-*Rf* genetic systems as the basis for practical utilization of heterosis effect in various crops. Seed production of sunflower hybrids all over the world is based on the extensive use of the PET1 CMS combined with the *Rf1* gene. At the same time, data on *Rf1* localization, sequence, and molecular basis for the CMS PET1 type restoration of fertility remain unknown. Searching for candidate genes of the *Rf1* gene has great fundamental and practical value. Therefore, in this study, association mapping of fertility restorer gene for CMS PET1 in sunflower was performed. The genome-wide association study (GWAS) results made it possible to isolate a segment 7.72 Mb in length on chromosome 13, in which 21 candidates for *Rf1* fertility restorer gene were identified, including 20 pentatricopeptide repeat (PPR)family genes and one Probable aldehyde dehydrogenase gene. The results will serve as a basis for further study of the genetic nature and molecular mechanisms of pollen fertility restoration in sunflower, as well as for further intensification of sunflower breeding.

Keywords: cytoplasmic male sterility; fertility restoration; sunflower; *Rf1* gene; GWAS; Pentatricopeptide Repeats; *PPR* genes; association mapping; candidate genes

1. Introduction

The phenomenon of cytoplasmic male sterility, consisting in the inability to produce functional (viable) pollen due to mutations in mitochondrial genome, has been described in more than 150 plant species [1–3]. With the discovery of nuclear *Rf* genes capable of suppressing the cytoplasmic male sterility (CMS) phenotype, it became possible to use the CMS-*Rf* genetic systems as the basis for the practical utilization of heterosis effect in various crops (maize, sunflower, rice, sorghum, sugar beet, rapeseed, and others), and also as models to study the interaction mechanisms between the nuclear and mitochondrial genomes. The use of CMS lines as the female parent eliminates manual labor during crossing, and the use of fertile paternal lines carrying the gene (genes) for the restoration of pollen fertility (*Rf*) allows the production of highly fertile offspring that exhibit heterosis effect. When crossing the lines of annual cultivated sunflower (*Helianthus annuus* L.), the heterosis effect ranges from 28 to 40% according to different authors [4]. Sunflower has more than 70 sources of CMS, but in commercial hybrids breeding, the PET1 CMS obtained by P. Leclercq from the interspecific hybrid *H. petiolaris* Nutt. × *H. annuus* is used predominantly [5]. The PET1 mtDNA differs from the mtDNA of fertile forms by the presence of an inversion of 11 kb and insertion of 5 kb that leads to the appearance of a new open reading frame *orfH522*, which is co-transcribed with the atpA gene and encodes a 16 kDa protein [6,7]. Literature describes various types of inheritance of pollen fertility restoration in sunflower with CMS PET1-type and reports on the different number of genes that determine this character. Based on hybridological analysis, various authors distinguish from one to five genes responsible for the restoration of pollen fertility in CMS PET1 [8,9]. Seed production of sunflower hybrids is based on the extensive use of the *Rf1* and *Rf2* genes, which by interacting with each other give the effect of restoring pollen fertility. It is believed that the main gene is *Rf1*, which is responsible for the fertility restoration and is present in the vast majority of CMS PET1 fertility restoring lines [10]. Sometimes, as a result of hybridological analysis in the second generation of hybrids, monogenic segregation occurs, in such cases it is assumed that the second gene is present in both crossed forms.

The *Rf1* gene was originally assigned to the sixth linkage group [11] and was subsequently reassigned to the second linkage group [12]. On the integrated genetic map of sunflower, the genetic factor *Rf1* responsible for the restoration of pollen fertility is localized in the linkage group 13 [13, 14]. To identify the *Rf1* gene alleles in the genotype, the closely linked molecular markers were developed [15], the diagnostic value of which was confirmed by several researchers [4,16]. However, approximately 10% of clones from the N.I. Vavilov All-Russian Research Institute of Plant Genetic Resources (VIR) collection lacked markers, although the presence of the *Rf1* gene dominant allele in their genotypes was confirmed using other methods [17].

Markers designed based on the information on the primary nucleotide sequence of the *Rf* genes themselves are considered to be the most effective for the selection of the functional *Rf* alleles carriers [18]. The absence of such markers in sunflower is explained by the lack of information about the nature of the *Rf* gene (genes). To identify the *Rf1* gene, positional cloning method was used [19], and molecular markers based on polymorphic fragments of *PPR* genes have also been developed [20,21]. However, to this day, the sequence of the *Rf1* gene remains unknown.

Most of the *Rf* genes described so far encode PPR proteins that contain repetitive degenerate motifs of 35 amino acid residues (Pentatricopeptide Repeats, PPR), with a few exceptions—for example, the *Rf2* gene, which encodes maize aldehyde dehydrogenase [22] and the rice *Rf2* gene, the product of which is a protein containing a glycine-rich domain [23]. RF-PPR proteins have from 15 to 20 PPR motifs in their sequences [24]. In angiosperm plants, *PPR* genes belong to two main types, P and PLS, which differ in the structure of PPR domains [25,26]. In flowering plants, the family of *PPR* genes, containing up to 600 members, is involved in anterograde/retrograde regulation, which ensures the coordinated work of nuclear and organelle genomes. Genes the products of which have the function of restoring fertility are included into the *RFL-PPR* (Restoration of Fertility Like-PPR) subfamily. Unlike the numerous families of conservative *PPR* genes that regulate processing, as well as participate in the splicing, editing, stabilization and translation of organelle RNAs, *RFL-PPR* genes are organized into clusters and are characterized by an exceptionally high level of variability [27]. It is believed that allele-specific markers of *RFL-PPR* genes can be used for positional cloning of fertility restorer genes, as well as for efficient selection of the carriers of functional alleles of *Rf* loci [18,28]. In addition, an exceptionally high rate of evolution of the subfamily of *RFL-PPR* genes [29,30], as well as the important role of CMS and restoration of fertility in the formation of new species [31] allow us to consider them as a source of molecular markers for phylogenetic research.

Thus, the *Rf1* gene is a key element in obtaining heterotic sunflower hybrids based on CMS. At the same time, data on its localization in the genome remain controversial. In addition, the gene sequence and molecular basis for the CMS PET1 type restoration of fertility remains unknown. Therefore, searching for candidate genes and mapping of the *Rf1* gene may have a great fundamental and practical value. Therefore, in this study, association mapping of fertility restorer gene for CMS PET1 in sunflower was performed.

2. Materials and Methods

134 *Helianthus annuus* L. elite lines from "Agroplazma" breeding and seed production company (Krasnodar, Russia) were taken into the study (Table S1, Table S2). They included 74 restorer lines carrying dominant allele of the *Rf1* gene and 60 sterility maintainer lines with the recessive allele of the gene.

Genomic DNA was extracted from the etiolated seedlings after one week of germination in the dark. 100 mg of tissue for each sample was grounded using the FastPrep-96™ Automated Homogenizer (MP Biomedicals, Santa Ana, CA, USA). DNA was extracted using the NucleoSpin® Plant II plant DNA extraction kit (Macherey-Nagel, Düren, Germany) according to the manufacturer's recommendations and stored at −20 °C. The quality of the purified DNA samples and DNA concentration were assessed by gel electrophoresis and Qubit 3.0 Fluorometer (ThermoFisher Scientific, Waltham, MA, USA). Restriction site associated DNA sequencing (RAD) libraries were prepared using HindIII and NlaIII endonucleases as previously described [32] with minor modifications and sequenced in Illumina HiSeq4000 (San Diego, CA, USA). Raw sequence data are available on NCBI SRA under project number PRJNA515598.

Preprocessing of raw reads was performed with the aid of the Trimmomatic software (version 0.30) [33]. Genome variants were called in Tassel 5 GBS v2 pipeline [34] with the following command line arguments: -kmerLength 65, -minMAPQ 20, and -mnQS 20. Bowtie2 [35] was used to map tags to the HanXRQr1.0 reference genome [36] with —very-sensitive—very-sensitive preset. Principal component analysis (PCA) and linkage disequilibrium (LD) analyses were accomplished in Tassel 5 software and visualized by means of ggplot2 R library (version 3.1.0) [37]. Statistical analysis using the mixed linear models (MLMs) [38] implemented in the Tassel 5 software was performed for association mapping with PCA and kinship matrixes as covariates. Multiallelic variants and those with the high missing call rates, MAF below 0.01 as well as the samples with many missing calls were filtered out in PLINK 1.9 [39,40] before genome-wide association study (GWAS) analysis. Significant loci were identified based on Bonferroni and FDR adjusted q-values with 0.01 alpha significance level.

Genome-wide association study results were visualized with the aid of the qqman R package (version 0.1.4) [41].

3. Results and Discussion

3.1. Genotyping and GWAS Analysis

Sequencing of RAD-libraries and subsequent analysis has identified 28,153 SNP (Single nucleotide polymorphisms) in 134 sunflower accessions. Overall transitions to transversions ratio was 1.83.

PCA analysis revealed significant population structure. Restorer lines and sterility maintainers form separate groups on the scatterplot (Figure 1).

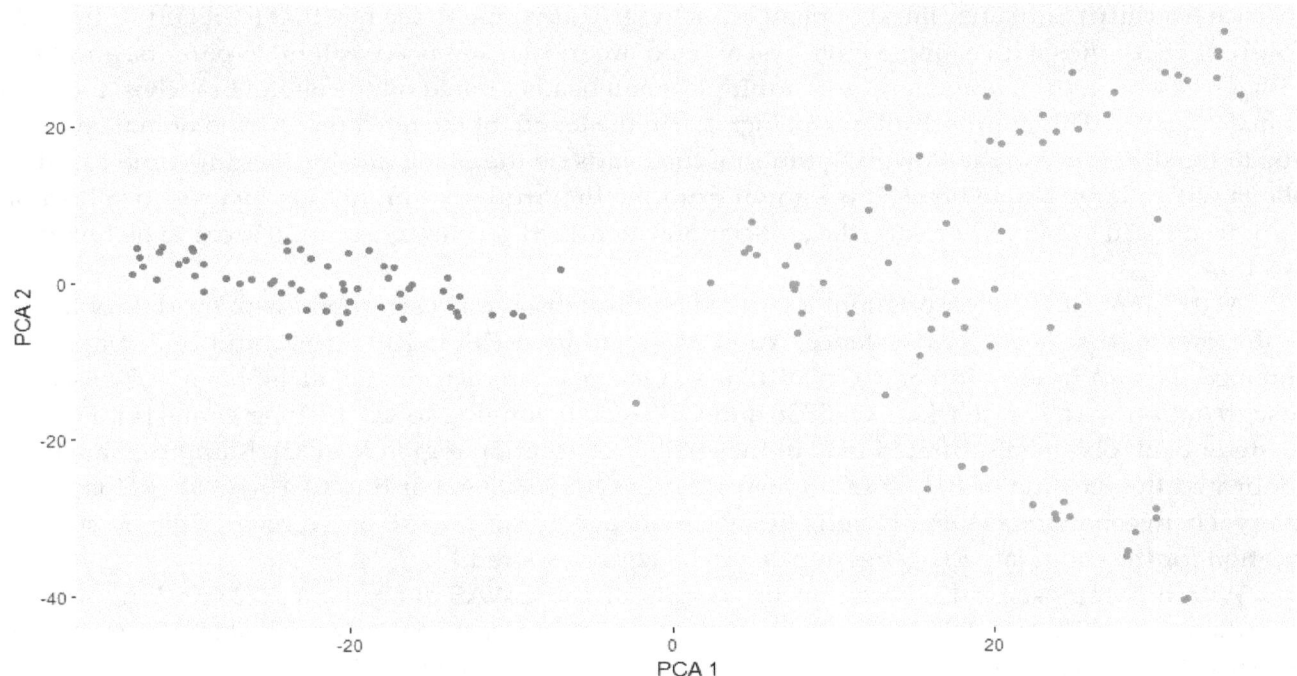

Figure 1. Principal component analysis (PCA) plot of sunflower lines based on Restriction site associated DNA (RAD) sequencing data. Pink dots–sterility maintainers, blue dots–restorer lines.

GWAS analysis revealed four loci associated with the ability to suppress CMS phenotype. Single significant marker was revealed at both 8 and 17 linkage groups. Most of the markers significantly associated with the trait under study, as well as the markers with the highest *p*-values, were located at 10 and 13 LG (Figure 2, Figure S1, Table S3).

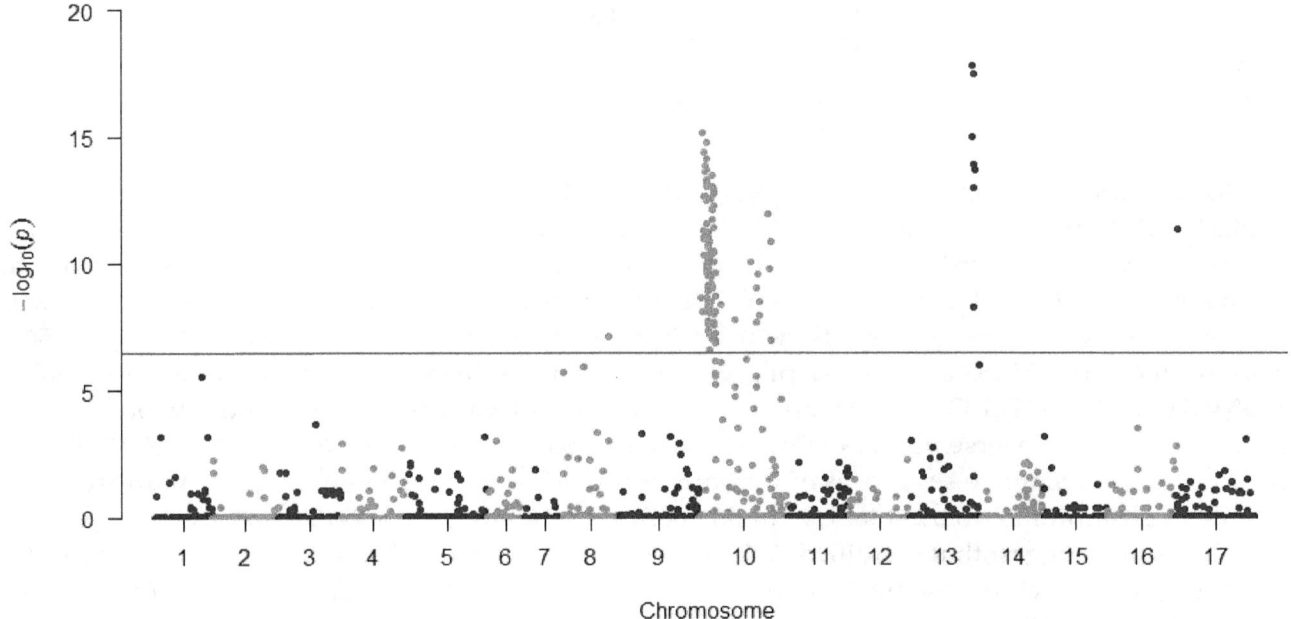

Figure 2. Manhattan plot of associations with the ability to suppress CMS (cytoplasmic male sterility) phenotype. Red line indicates the significance threshold based on the Bonferroni multiple testing correction (alpha = 0.01).

In addition to the difference in the ability to restore pollen fertility in the crosses with sterile lines with PET1-type cytoplasm, the analyzed sunflower lines differed by the presence (restorer lines) or absence (sterility maintainer lines) of plant branching. This is due to the fact that to obtain F1 hybrids, non-branched lines with a single large apical head are most often used as female parents, and lines with a recessive type of branching, with multiple small heads located on the lateral branches, are used as male parents. This approach allows an increase in the length of the flowering period of male parents due to the difference in the flowering times of the heads on the plant, and at the same time to get F1 plants with a large single head. It is known from the literature that branching locus is localized on chromosome 10 [42,43]. Therefore, the associations identified on chromosome 10 seem to be linked to this trait.

At the same time, the associations identified on chromosome 13 correspond with the data obtained in the previous studies. For instance, Yu et al. combined RFLP, RFLP-SSR, and SSR maps and obtained data for localization of *Rf1* in LG13 [44]. One year later Kusterer et al. [45] map *Rf1* based on cosegregation with SSR markers ORS388 and ORS1030 belonging to LG 13 Tang et al. [14]. Further Kusterer et al. obtained saturated map of the fertility restoration region *Rf1* [13]. Mapping data have confirmed the location of *Rf1* on LG13 near marker ORS1030. According to Yue et al. *Rf1* is in the interval between markers ORS511 and ORS799 of linkage group 13 [20]. Based on this, the most likely location for the candidate *Rf1* genes appears to be chromosome 13.

Within chromosome 13, based on the results of the GWAS analysis, a 7.72 Mb long section (coordinates170494693–178217103), can be distinguished, in which eight significant SNPs are located with p-values ranging from 5.69×10^{-9} to 1.53×10^{-18} (Table 1, Figure S2).

Table 1. List of Single nucleotide polymorphisms at linkage group 13, significantly associated with the ability to suppress CMS phenotype after Bonferroni correction.

Marker	Position	p-Value
S13_170494693	170494693	1.01×10^{-15}
S13_171053833	171053833	1.53×10^{-18}
S13_173268042	173268042	3.46×10^{-18}
S13_173832391	173832391	5.69×10^{-9}
S13_174474103	174474103	1.22×10^{-14}
S13_174474122	174474122	1.22×10^{-14}
S13_174809087	174809087	1.10×10^{-13}
S13_178217103	178217103	2.03×10^{-14}

To compare localization of 7.72 Mb region identified in this study with previously reported data we blasted PCR primer sequences of the ORS511, ORS799 and ORS1030 markers against the reference genome. ORS511 and ORS1030 were mapped in close proximity to each other on LG13 in according to Tang et al. [14]. Complete sequences of ORS1030 forward and reverse primers were mapped with 100% identity twice in the genome. Forward primer mapped to the positions 169535691–169535666 and 169655088–169655063 and reverse primer to the positions 169535262–169535287 and 169654659–169654684 of LG 13. For ORS511 complete sequences of forward primer have no hits on the 100% identity threshold. Reverse primer of ORS511 was mapped at 169733686–169733704 of LG13. For the ORS799 marker complete sequences of forward and reverse primers were uniquely mapped to the genome in position 186516272–186516291 and 186516418–186516399 of LG13 respectively.

These data suggest that identified 7.72 Mb region (coordinates170494693–178217103) is located within segment of chromosome 13 flanked by SSR markers ORS799 and ORS1030 (coordinates 169535262–186516418).

3.2. Identification of Rf1 Candidate Genes

Within identified 7.72 Mb region in the HanXRQr1.0 reference genome sequence [36], 11 *PPR* genes are located, which are the most likely candidate genes for the fertility restorer gene *Rf1*. Almost all *Rf* genes in various plant species that have been identified so far belong to this family [46–49].

PPR genes are thought to be present in all eukaryotes, but they are most common in the genomes of terrestrial plants, where they form one of the largest gene families [50]. For example, in the genome of *Arabidopsis thaliana* L. there are about 450 genes of this family [51,52], about 500 in the maize genome [53], and more than 600 in the genome of *Oryza sativa* L. [25]. Proteins of this family are characterized by the presence of multiple helix-turn-helix domains, forming a supercoil with a central groove [50,52]. This allows the protein to bind to RNA and participate in the RNA-protein interactions. Pentatricopeptide repeat (PPR) proteins play a significant role in regulating gene expression in the organelles at the RNA level [50,54,55].

The total number of annotated *PPR* genes in the sunflower genome HanXRQr1.0 is 333. Therefore, the identified region of 7.72 Mb (comprising 0.214% of the genome length) contains 3.3% of all annotated *PPR* genes and is rich in *PPR* genes. In addition, within this region, 10 genes of the *TPR* family are annotated. It is known that the sequences of PPR proteins are similar to the sequences of the TPR-family proteins and it is assumed that the tetratricopeptide repeat (*TPR*)- family genes gave rise to *PPR* genes at the early stages of the evolution of eukaryotes [56].

Therefore, it was decided to include the gene sequence of both the *PPR* and *TPR* families in the further analysis. It should be noted that genome sections 7.72 Mb in length, flanking the region of the chromosome 13 mentioned above, did not contain any annotated sequence of the *PPR* family, and only a single sequence belonging to the *TPR* family (HanXRQChr13g0421851).

The analysis of the translated amino acid sequences of 22 genes of the *PPR* and *TPR* families located in the identified region and its flanking regions was conducted using ScanProsite tool of ExPASy SIB Bioinformatics Resource Portal (SIB Swiss Institute of Bioinformatics, Lausanne, Switzerland). As a result, in all 11 amino acid sequences of the PPR family and in 10 of the 11 sequences of the TPR family, Pentatricopeptide (PPR) repeats were identified. Therefore, within the 7.72 Mb region and the flanking regions, 21 genes were detected, their protein products demonstrating the primary structure characteristic of the sequences of the *PPR* family. Meanwhile, in addition to PPR repeats, the amino acid sequence of the protein product of one of the genes revealed a region of homology with UDP-glycosyltransferases, and therefore this gene was excluded from the list of possible candidate genes for *Rf1*.

In addition to *PPR* genes, a gene annotated as Probable aldehyde dehydrogenase 5F1 was detected in the 7.72 Mb region of chromosome 13. It was previously shown that *Rf2* gene of maize is the gene encoding aldehyde dehydrogenase [57]. Therefore, this gene is also a possible candidate *Rf* gene. The list of identified candidate genes is shown in Table 2, and their arrangement within the 7.72 Mb region is shown in Figure 3.

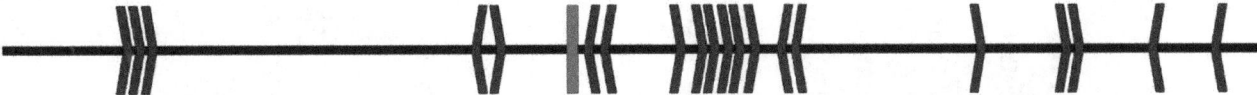

Figure 3. Schematic localization of the candidate *Rf1* genes within the 7.72 Mb region. Green arrows indicate the gene sequences of the *PPR* family. The direction of the arrow reflects the orientation of the sequence in the genome. Red box indicates the location of the Probable aldehyde dehydrogenase 5F1gene.

The number of PPR repeats in the sequence and the length of the protein products of the candidate PPR family genes varied from 2 to 15 and from 110 to 756 amino acids, respectively.

Genomic regions with increased LD could be recognized as signatures of strong selection pressure on the traits encoded within these regions. The results of the analysis showed the presence of an extended section of elevated LD in 13 LG (Figure 4), of which the identified 7.72 Mb region forms part. This fact is an indirect proof of the localization of candidate genes in this region of the genome.

Table 2. The list of candidate *Rf1* genes identified within the 7.72 Mb region.

Gene	Start	End	Strand	Product	Gene Bank Accession Number of Translated Protein	Hits for All PROSITE (Release 2018_11) Motifs
HanXRQChr13g0418841	170850155	170852002	+	Putative pentatricopeptide repeat	OTG02960	PS51375 Pentatricopeptide (PPR) repeat
HanXRQChr13g0418861	170908019	170909110	+	Putative pentatricopeptide repeat	OTG02962	PS51375 Pentatricopeptide (PPR) repeat
HanXRQChr13g0419621	173473487	173475525	−	Probable pentatricopeptide repeat-containing protein At2g41080	OTG03034	PS51375 Pentatricopeptide (PPR) repeat
HanXRQChr13g0419631	173484455	173500401	+	Putative pentatricopeptide repeat	OTG03035	PS51375 Pentatricopeptide (PPR) repeat
HanXRQChr13g0419931	174209661	174217234	−	Putative pentatricopeptide repeat	OTG03064	PS51375 Pentatricopeptide (PPR) repeat
HanXRQChr13g0420121	174799667	174801481	+	Probable pentatricopeptide repeat (PPR) superfamily protein	OTG03081	PS51375 Pentatricopeptide (PPR) repeat
HanXRQChr13g0420241	174944047	174945506	+	Putative pentatricopeptide repeat	OTG03093	PS51375 Pentatricopeptide (PPR) repeat
HanXRQChr13g0420261	174962084	174962512	+	Putative pentatricopeptide repeat	OTG03095	PS51375 Pentatricopeptide (PPR) repeat
HanXRQChr13g0420351	175219425	175219886	−	Putative pentatricopeptide repeat	OTG03099	PS51375 Pentatricopeptide (PPR) repeat
HanXRQChr13g0420811	176970038	176972308	+	Probable pentatricopeptide repeat (PPR) superfamily protein	OTG03141	PS51375 Pentatricopeptide (PPR) repeat
HanXRQChr13g0421081	178216563	178219635	−	Probable putative pentatricopeptide repeat-containing protein At4g17915	OTG03166	PS51375 Pentatricopeptide (PPR) repeat
HanXRQChr13g0418851	170877322	170879307	+	Putative tetratricopeptide-like helical domain	OTG02961	PS51375 Pentatricopeptide (PPR) repeat
HanXRQChr13g0419881	174159006	174160682	−	Putative tetratricopeptide-like helical domain	OTG03060	PS51375 Pentatricopeptide (PPR) repeat
HanXRQChr13g0420277	175002640	175003793	+	Putative tetratricopeptide-like helical domain	OTG03096	PS51375 Pentatricopeptide (PPR) repeat
HanXRQChr13g0420281	175016437	175018065	+	Putative tetratricopeptide-like helical domain	OTG03097	PS51375 Pentatricopeptide (PPR) repeat
HanXRQChr13g0420301	175055952	175057826	+	Putative tetratricopeptide-like helical domain	OTG03098	PS51375 Pentatricopeptide (PPR) repeat
HanXRQChr13g0420371	175253986	175294219	−	Putative tetratricopeptide-like helical domain	OTG03101	PS51375 Pentatricopeptide (PPR) repeat
HanXRQChr13g0420861	177597409	177599211	+	Putative tetratricopeptide-like helical domain	OTG03145	PS51375 Pentatricopeptide (PPR) repeat
HanXRQChr13g0420881	177609240	177611054	+	Putative tetratricopeptide-like helical domain	OTG03147	PS51375 Pentatricopeptide (PPR) repeat
HanXRQChr13g0421271	178655189	178657150	−	Probable tetratricopeptide repeat (TPR)-like superfamily protein	OTG03183	PS51375 Pentatricopeptide (PPR) repeat
HanXRQChr13g0419821	174082899	174091500	−	Probable aldehyde dehydrogenase 5F1	OTG03054	NA

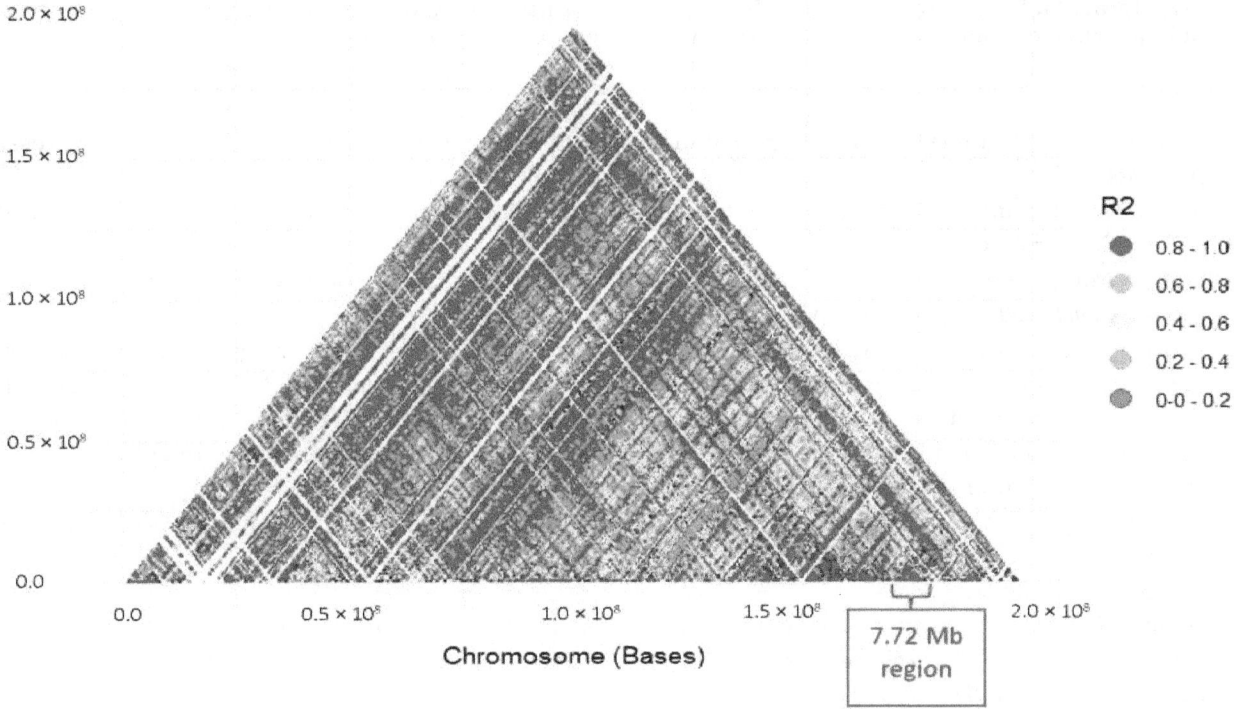

Figure 4. Pairwise Linkage Disequilibrium (LD) Plot of the LG13. Individual data points reflect squared allele frequency correlations (R2) for all possible pairs of polymorphic SNP markers of LG13. The x- and y-axes correspond to the coordinates within 13 LG. Location of 7.72 Mb indicated by curly bracket.

It should be noted that the reference genome used in the analysis was obtained by sequencing the XRQ line, which is a cytoplasmic male sterility maintainer (PET1 type) [36]. At the same time, it is known that the *Rf* locus may undergo complex evolutionary events [46] and the structure of the identified site may differ in the genome of the fertility restorer lines. Therefore, to identify the *Rf1* gene, to determine the sequence of the dominant alleles of the *Rf1* gene, and to understand the evolution of the sunflower *Rf1* locus, the additional analysis of the structure of the 7.72 Mb region in the genome of fertility restorer lines is required.

4. Conclusions

In this work, high-throughput genotyping of sunflower lines, differing by the ability to suppress CMS phenotype was carried out and a genome-wide association study was performed. The GWAS results made it possible to isolate a segment 7.72 Mb in length on chromosome 13, in which 21 candidate *Rf1* fertility restorer genes were identified, including 20 *PPR*-family genes and one Probable aldehyde dehydrogenase gene. The results will serve as a basis for further study of the genetic nature and molecular mechanisms for pollen fertility restoration in sunflower, as well as for the search of selection markers.

Supplementary Materials: Table S1: List of the sunflower lines selected for the study, Table S2: List of the sequenced sunflower samples, corresponding flowcell lanes and barcodes, Table S3: List of SNPs, significantly associated with the ability to suppress CMS phenotype after FDR correction, Figure S1: Quantile-quantile plot of associations with the ability to suppress CMS phenotype, Figure S2: Distribution of the ability to suppress CMS phenotype across sunflower samples with different allelic states for 8 statistically significant markers: (A) S13_170494693, (B) S13_171053833, (C) S13_173268042, (D) S13_173832391, (E) S13_174474103, (F) S13_174474122, (G) S13_174809087, (H) S13_178217103.

Author Contributions: Conceptualization, S.V.G. (Svetlana V. Goryunova), I.N.A.; methodology, N.I.B., Y.N.D., P.E.K.; formal analysis, D.V.G., S.V.G. (Svetlana V. Goryunova), E.A.S.; investigation, A.I.C., E.U.M., S.V.B., A.F.A., E.A.G.; writing—original draft preparation, D.V.G., S.V.G. (Svetlana V. Goryunova), I.N.A., V.A.G.; writing—review and editing, A.A.S., P.V.M., R.F.G., S.V.G. (Sergey V. Garkusha); supervision, P.E.K.; project administration, D.A.P.; funding acquisition, D.V.G., S.V.G. (Svetlana V. Goryunova), P.E.K., Z.M.M.

Acknowledgments: We are grateful to S.V. Nuzhdin, S.A. Spirin and A.V. Alexeyevsky for valuable discussion. We would also like to thank I.N. Benko and A. Yu. Donchenko for technical support.

References

1. Rhoades, M.M. Cytoplasmic inheritance of male sterility in *Zea mays*. *Science* **1931**, *73*, 340–341. [CrossRef] [PubMed]

2. Schnable, P. The molecular basis of cytoplasmic male sterility and fertility restoration. *Trends Plant Sci.* **1998**, *3*, 175–180. [CrossRef]

3. Ivanov, M.K.; Dymshits, G.M. Cytoplasmic male sterility and restoration of pollen fertility in higher plants. *Russ. J. Genet.* **2007**, *43*, 354–368. [CrossRef]

4. Anisimova, I.N.; Gavrilova, V.A.; Rozhkova, V.T.; Timofeeva, G.I.; Tikhonova, M.A. Molecular markers in identification of sunflower pollen fertility restorer genes. *Russ. Agric. Sci.* **2009**, *35*, 367–370. [CrossRef]

5. Leclercq, P. Une sterilite male cytoplasmique chez le tournesol. *Ann. Amel. Plantes* **1969**, *19*, 99–106.

6. Horn, R.; Köhler, R.H.; Zetsche, K. A mitochondrial 16 kDa protein is associated with cytoplasmic male sterility in sunflower. *Plant Mol. Biol.* **1991**, *17*, 29–36. [CrossRef] [PubMed]

7. Hans Köhler, R.; Horn, R.; Lössl, A.; Zetsche, K. Cytoplasmic male sterility in sunflower is correlated with the co-transcription of a new open reading frame with the *atpA* gene. *MGG Mol. Gen. Genet.* **1991**, *227*, 369–376. [CrossRef]

8. Miller, J.F.; Fick, G.N. Genetics of sunflower. In *Sunflower Technology and Production*; American Society of Agronomy, Crop Science Society of America, Soil Science Society of America: Madison, WI, USA, 1997; pp. 441–495. ISBN 978-0-89118-135-4.

9. Anaschenko, A.V.; Duca, M.V. Studies of sunflower (*Helianthus annuus* L.) CMS–*Rf* genetic system: II. Male fertility restoration in hybrids based on CMSP. *Russ. J. Genet.* **1985**, *21*, 1999–2004.

10. Serieys, H. Identification, study and utilization in breeding programs of new CMS sources. *Helia* **1996**, *19*, 144–158.

11. Gentzbittel, L.; Vear, F.; Zhang, Y.-X.; Bervillé, A.; Nicolas, P. Development of a consensus linkage RFLP map of cultivated sunflower (*Helianthus annuus* L.). *Theor. Appl. Genet.* **1995**, *90*, 1079–1086. [CrossRef]

12. Jan, C.C.; Vick, B.A.; Miller, J.F.; Kahler, A.L.; Butler, E.T. Construction of an RFLP linkage map for cultivated sunflower. *Theor. Appl. Genet.* **1998**, *96*, 15–22. [CrossRef]

13. Kusterer, B.; Horn, R.; Friedt, W. Molecular mapping of the fertility restoration locus *Rf1* in sunflower and development of diagnostic markers for the restorer gene. *Euphytica* **2005**, *143*, 35–42. [CrossRef]

14. Tang, S.; Yu, J.-K.; Slabaugh, B.; Shintani, K.; Knapp, J. Simple sequence repeat map of the sunflower genome. *Theor. Angew. Genet.* **2002**, *105*, 1124–1136. [CrossRef]

15. Horn, R.; Kusterer, B.; Lazarescu, E.; Prüfe, M.; Friedt, W. Molecular mapping of the *Rf1* gene restoring pollen fertility in PET1-based F1 hybrids in sunflower (*Helianthus annuus* L.). *Theor. Appl. Genet.* **2003**, *106*, 599–606. [CrossRef] [PubMed]

16. Markin, N.; Usatov, A.; Makarenko, M.; Azarin, K.; Gorbachenko, O.; Kolokolova, N.; Usatenko, T.; Markina, O.; Gavrilova, V. Study of informative DNA markers of the *Rf1* gene in sunflower for breeding practice. *Czech J. Genet. Plant Breed.* **2017**, *53*, 69–75. [CrossRef]

17. Anisimova, I.N.; Gavrilova, V.A.; Rozhkova, V.T.; Port, A.I.; Timofeeva, G.I.; Duka, M.V. Genetic diversity of sources of sunflower pollen fertility restorer genes. *Russ. Agric. Sci.* **2011**, *37*, 192–196. [CrossRef]

18. Sykes, T.; Yates, S.; Nagy, I.; Asp, T.; Small, I.; Studer, B. In-silico identification of candidate genes for fertility restoration in cytoplasmic male sterile perennial ryegrass (*Lolium perenne* L.). *Genome Biol. Evol.* **2016**, *9*, 351–362.

19. Horn, R.; Hamrit, S. Gene cloning and characterization. In *Genetics, Genomics and Breeding of Sunflower*; CRC Press: Boca Raton, FL, USA, 2010; pp. 173–219. ISBN 978-1-138-11513-2.

20. Yue, B.; Vick, B.A.; Cai, X.; Hu, J. Genetic mapping for the *Rf1* (fertility restoration) gene in sunflower (*Helianthus annuus* L.) by SSR and TRAP markers. *Plant Breed.* **2010**, *129*, 24–28. [CrossRef]

21. Anisimova, I.N.; Alpatieva, N.V.; Rozhkova, V.T.; Kuznetsova, E.B.; Pinaev, A.G.; Gavrilova, V.A. Polymorphism among RFL-PPR homologs in sunflower (*Helianthus annuus* L.) lines with varying ability for the suppression of the cytoplasmic male sterility phenotype. *Russ. J. Genet.* **2014**, *50*, 712–721. [CrossRef]

22. Cui, X.; Wise, R.P.; Schnable, P.S. The *rf2* nuclear restorer gene of male-sterile T-cytoplasm maize. *Science* **1996**, *272*, 1334–1336. [CrossRef]

23. Itabashi, E.; Iwata, N.; Fujii, S.; Kazama, T.; Toriyama, K. The fertility restorer gene, *Rf2*, for Lead Rice-type cytoplasmic male sterility of rice encodes a mitochondrial glycine-rich protein: *Rf2* for CMS rice encodes a glycine-rich protein. *Plant J.* **2011**, *65*, 359–367. [CrossRef] [PubMed]

24. Melonek, J.; Stone, J.D.; Small, I. Evolutionary plasticity of restorer-of-fertility-like proteins in rice. *Sci. Rep.* **2016**, *6*, 35152. [CrossRef] [PubMed]

25. Lurin, C. Genome-wide analysis of *Arabidopsis* pentatricopeptide repeat proteins reveals their essential role in organelle biogenesis. *Plant Cell* **2004**, *16*, 2089–2103. [CrossRef] [PubMed]

26. Cheng, S.; Gutmann, B.; Zhong, X.; Ye, Y.; Fisher, M.F.; Bai, F.; Castleden, I.; Song, Y.; Song, B.; Huang, J.; et al. Redefining the structural motifs that determine RNA binding and RNA editing by pentatricopeptide repeat proteins in land plants. *Plant J. Cell Mol. Biol.* **2016**, *85*, 532–547. [CrossRef] [PubMed]

27. Fujii, S.; Bond, C.S.; Small, I.D. Selection patterns on restorer-like genes reveal a conflict between nuclear and mitochondrial genomes throughout angiosperm evolution. *Proc. Natl. Acad. Sci. USA* **2011**, *108*, 1723–1728. [CrossRef] [PubMed]

28. Kaur, P.; Verma, M. Insights into *PPR* Gene Family in *Cajanus cajan* and other legume species. *J. Data Min. Genom. Proteom.* **2016**, *7*. [CrossRef]

29. Dahan, J.; Mireau, H. The *Rf* and *Rf*-like PPR in higher plants, a fast-evolving subclass of *PPR* genes. *RNA Biol.* **2013**, *10*, 1469–1476. [CrossRef]

30. Gaborieau, L.; Brown, G.G.; Mireau, H. The propensity of pentatricopeptide repeat genes to evolve into restorers of cytoplasmic male sterility. *Front. Plant Sci.* **2016**, *7*, 1816. [CrossRef]

31. Rieseberg, L.H.; Blackman, B.K. Speciation genes in plants. *Ann. Bot.* **2010**, *106*, 439–455. [CrossRef]

32. Zhigunov, A.V.; Ulianich, P.S.; Lebedeva, M.V.; Chang, P.L.; Nuzhdin, S.V.; Potokina, E.K. Development of F1 hybrid population and the high-density linkage map for European aspen (*Populus tremula* L.) using RADseq technology. *BMC Plant Biol.* **2017**, *17*, 180. [CrossRef]

33. Bolger, A.M.; Lohse, M.; Usadel, B. Trimmomatic: A flexible trimmer for Illumina sequence data. *Bioinformatics* **2014**, *30*, 2114–2120. [CrossRef] [PubMed]

34. Bradbury, P.J.; Zhang, Z.; Kroon, D.E.; Casstevens, T.M.; Ramdoss, Y.; Buckler, E.S. TASSEL: Software for association mapping of complex traits in diverse samples. *Bioinform. Oxf. Engl.* **2007**, *23*, 2633–2635. [CrossRef] [PubMed]

35. Langmead, B.; Trapnell, C.; Pop, M.; Salzberg, S.L. Ultrafast and memory-efficient alignment of short DNA sequences to the human genome. *Genome Biol.* **2009**, *10*, R25. [CrossRef] [PubMed]

36. Badouin, H.; Gouzy, J.; Grassa, C.J.; Murat, F.; Staton, S.E.; Cottret, L.; Lelandais-Brière, C.; Owens, G.L.; Carrère, S.; Mayjonade, B.; et al. The sunflower genome provides insights into oil metabolism, flowering and Asterid evolution. *Nature* **2017**, *546*, 148–152. [CrossRef]

37. Wickham, H. *ggplot2*; Springer: New York, NY, USA, 2009; ISBN 978-0-387-98140-6.

38. Zhang, Z.; Ersoz, E.; Lai, C.-Q.; Todhunter, R.J.; Tiwari, H.K.; Gore, M.A.; Bradbury, P.J.; Yu, J.; Arnett, D.K.; Ordovas, J.M.; et al. Mixed linear model approach adapted for genome-wide association studies. *Nat. Genet.* **2010**, *42*, 355–360. [CrossRef] [PubMed]

39. Chang, C.C.; Chow, C.C.; Tellier, L.C.; Vattikuti, S.; Purcell, S.M.; Lee, J.J. Second-generation PLINK: Rising to the challenge of larger and richer datasets. *GigaScience* **2015**, *4*, 7. [CrossRef] [PubMed]

40. Purcell, S.; Neale, B.; Todd-Brown, K.; Thomas, L.; Ferreira, M.A.R.; Bender, D.; Maller, J.; Sklar, P.; de Bakker, P.I.W.; Daly, M.J.; et al. PLINK: A tool set for whole-genome association and population-based linkage analyses. *Am. J. Hum. Genet.* **2007**, *81*, 559–575. [CrossRef]

41. Turner, S.D. QQman: An R package for visualizing GWAS results using QQ and manhattan plots. *BioRxiv* **2014**. [CrossRef]

42. Tang, S.; Leon, A.; Bridges, W.C.; Knapp, S.J. Quantitative trait loci for genetically correlated seed traits are tightly linked to branching and pericarp pigment loci in sunflower. *Crop Sci.* **2006**, *46*, 721. [CrossRef]

43. Nambeesan, S.U.; Mandel, J.R.; Bowers, J.E.; Marek, L.F.; Ebert, D.; Corbi, J.; Rieseberg, L.H.; Knapp, S.J.; Burke, J.M. Association mapping in sunflower (*Helianthus annuus* L.) reveals independent control of apical vs. basal branching. *BMC Plant Biol.* **2015**, *15*, 84. [CrossRef]

44. Yu, J.-K.; Tang, S.; Slabaugh, M.B.; Heesacker, A.; Cole, G.; Herring, M.; Soper, J.; Han, F.; Chu, W.-C.; Webb, D.M.; et al. Towards a saturated molecular genetic linkage map for cultivated sunflower. *Crop Sci.* **2003**, *43*, 367. [CrossRef]

45. Kusterer, B.; Rozynek, B.; Brahm, L.; Prüfe, M.; Tzigos, S.; Horn, R.; Friedt, W. Construction of a genetic map and localization of major traits in sunflower (*Helianthus annuus* L.). *Helia* **2004**, *27*, 15–23. [CrossRef]

46. Hernandez Mora, J.R.; Rivals, E.; Mireau, H.; Budar, F. Sequence analysis of two alleles reveals that intra-and intergenic recombination played a role in the evolution of the radish fertility restorer (*Rfo*). *BMC Plant Biol.* **2010**, *10*, 35. [CrossRef] [PubMed]

47. Kazama, T.; Toriyama, K. A fertility restorer gene, *Rf4*, widely used for hybrid rice breeding encodes a pentatricopeptide repeat protein. *Rice* **2014**, *7*, 28. [CrossRef] [PubMed]

48. Madugula, P.; Uttam, A.G.; Tonapi, V.A.; Ragimasalawada, M. Fine mapping of *Rf2*, a major locus controlling pollen fertility restoration in sorghum A1 cytoplasm, encodes a PPR gene and its validation through expression analysis. *Plant Breed.* **2018**, *137*, 148–161. [CrossRef]

49. Bentolila, S.; Alfonso, A.A.; Hanson, M.R. A pentatricopeptide repeat-containing gene restores fertility to cytoplasmic male-sterile plants. *Proc. Natl. Acad. Sci. USA* **2002**, *99*, 10887–10892. [CrossRef]

50. Schmitz-Linneweber, C.; Small, I. Pentatricopeptide repeat proteins: A socket set for organelle gene expression. *Trends Plant Sci.* **2008**, *13*, 663–670. [CrossRef]

51. Aubourg, S.; Boudet, N.; Kreis, M.; Lecharny, A. In *Arabidopsis thaliana*, 1% of the genome codes for a novel protein family unique to plants. *Plant Mol. Biol.* **2000**, *42*, 603–613. [CrossRef]

52. Small, I.D.; Peeters, N. The PPR motif—A TPR-related motif prevalent in plant organellar proteins. *Trends Biochem. Sci.* **2000**, *25*, 46–47. [CrossRef]

53. Wei, K.; Han, P. Pentatricopeptide repeat proteins in maize. *Mol. Breed.* **2016**, *36*. [CrossRef]

54. Manna, S. An overview of pentatricopeptide repeat proteins and their applications. *Biochimie* **2015**, *113*, 93–99. [CrossRef] [PubMed]

55. Small, I.D.; Rackham, O.; Filipovska, A. Organelle transcriptomes: Products of a deconstructed genome. *Curr. Opin. Microbiol.* **2013**, *16*, 652–658. [CrossRef] [PubMed]

56. Barkan, A.; Small, I. Pentatricopeptide repeat proteins in plants. *Annu. Rev. Plant Biol.* **2014**, *65*, 415–442. [CrossRef] [PubMed]

57. Liu, Z.; Wang, D.; Feng, J.; Seiler, G.J.; Cai, X.; Jan, C.-C. Diversifying sunflower germplasm by integration and mapping of a novel male fertility restoration gene. *Genetics* **2013**, *193*, 727–737. [CrossRef] [PubMed]

13

Overexpression of Soybean Transcription Factors *GmDof4* and *GmDof11* Significantly Increase the Oleic Acid Content in Seed of *Brassica napus* L.

Qinfu Sun, Jueyi Xue, Li Lin, Dongxiao Liu, Jian Wu, Jinjin Jiang and Youping Wang *

Jiangsu Provincial Key Laboratory of Crop Genetics and Physiology, Yangzhou University, Yangzhou 225009, China; D150105@yzu.edu.cn (Q.S.); jueyi.xue@student.unsw.edu.au (J.X.); M160780@yzu.edu.cn (L.L.); DX120170119@yzu.edu.cn (D.L.); wu_jian@yzu.edu.cn (J.W.); jjjiang@yzu.edu.cn (J.J.)
* Correspondence: wangyp@yzu.edu.cn

Abstract: Rapeseed (*Brassica napus* L.) with substantial lipid and oleic acid content is of great interest to rapeseed breeders. Overexpression of *Glycine max* transcription factors *Dof4* and *Dof11* increased lipid accumulation in *Arabidopsis* and microalgae, in addition to modifying the quantity of certain fatty acid components. Here, we report the involvement of *GmDof4* and *GmDof11* in regulating fatty acid composition in rapeseeds. Overexpression of *GmDof4* and *GmDof11* in rapeseed increased oleic acid content and reduced linoleic acid and linolenic acid. Both qPCR and the yeast one-hybrid assay indicated that GmDof4 activated the expression of *FAB2* by directly binding to the *cis*-DNA element on its promoters, while *GmDof11* directly inhibited the expression of *FAD2*. Thus, *GmDof4* and *GmDof11* might modify the oleic acid content in rapeseed by directly regulating the genes that are associated with fatty acid biosynthesis.

Keywords: *Brassica napus*; *GmDof4*; *GmDof11*; oleic acid; fatty acid composition

1. Introduction

Rapeseed (*Brassica napus* L.) is among the most important oil crops worldwide, providing high-quality edible oils and industrial raw materials [1–3]. The production and yield of rapeseed has rapidly increased in China in recent years [4]. Rapeseed oil is principally a mixture of seven main fatty acids [5], namely palmitic acid ($C_{16:0}$), stearic acid ($C_{18:0}$), oleic acid ($C_{18:1}$), linoleic acid ($C_{18:2}$), linolenic acid ($C_{18:3}$), eicosenoic acid ($C_{20:1}$), and erucic acid ($C_{22:1}$), of which oleic acid is the most abundant and has the highest nutritional value [6]. Therefore, creating new rapeseed varieties with a high seed oil content that is rich in oleic acid content is a primary goal for rapeseed breeders [7]. Remarkable progress in increasing the content of seed oil and proportion of oleic acid has been reported by traditional breeding and putative candidate genes have been dissected using quantitative trait loci mapping and molecular markers [8–11].

Genetic engineering is a potentially efficient method of modifying the expression of single or multiple genes that are involved in lipid metabolism [7,12]. In *B. napus*, the overexpression of genes encoding glycerol-3-phosphate dehydrogenase [13], acyl-CoA: lysophosphatidic acid acyltransferase [14], mitochondrial pyruvate dehydrogenase kinase [15], and diacylglycerol acyltransferases [16–19] significantly increased seed oil content. Liu et al. [20] overexpressed triacylglyceride (TAG) synthesis pathway genes in *B. napus*, including *BnGPDH*, *BnGPAT*, *BnDGAT*, and *ScLPAAT*, and found that the overexpression of a single gene could increase the content of seed oil, but the simultaneous overexpression of multiple genes may result in more substantial changes in oil composition.

Besides lipid synthase, a number of genes encoding seed-specific transcription factors (TFs) have been shown to play important roles in the regulation of lipid biosynthesis [7,21]. Previous reports have suggested that altering the expression levels of the plant-specific B3 domain family members LEAFY COTYLEDON 2, FUSCA3 and ABSCISIC ACID INSENSITIVE 3 [22–24]; NF-YB type TF LEAFY COTYLEDON 1 [7,25]; AP2/EREB domain TF WRI1 [26,27], *Arabidopsis* 6b-interacting protein 1-like 1 [28]; BnGRF2 (GRF2-like gene from *B. napus*) [21]; and, SHOOTMERISTEMLESS [29] resulted in a change in the proportions of seed storage materials. These genes could be used to genetically improve the oil content and composition of rapeseed.

Fatty acid dehydrogenase (FAD) and fatty acid elongase (FAE) are the key enzymes that determine fatty acid composition in seed oil. FAD catalyzes the biosynthesis of polyunsaturated fatty acids, such as linoleic and linolenic acid [30,31], while FAE catalyzes the chain elongation reaction, resulting in the formation of long-chain fatty acids, including eicosenoic and erucic acid [30,32]. Previous studies have suggested that inhibition of *FAE1* expression increases oleic acid and reduces erucic acid content in rapeseed seed oil [33,34], as did inhibition the expression of the *BnFAD2* gene in transgenic seeds [34]. Jung et al. [35] found that expression of the *B. rapa BrFAD2* gene in an antisense orientation increased the synthesis of oleic acid in *B. napus*. FAD3 desaturase is responsible for the synthesis of linolenic acid [36], in *BnFAD3* mutants of *B. napus*, the concentration of linolenic acid was significantly reduced [37].

Dof (DNA binding with one finger) is an important family of TFs in plants, with its members being widely involved in seed development, plant growth, morphogenesis, nutrient metabolism, and other processes [36,38–40]. As far as we know, comprehensive analysis of Dof family factors in *B. napus* has not been previously performed, with few reports of the function of *Dof* genes in *B. napus* [41], even though genome-wide analysis has been performed in other *Brassica* plants [42]. In soybean, 28 Dof members have been identified [43], and eight of them, including *GmDof4* and *GmDof11*, are strongly expressed in the flowers and pods of soybean. Wang et al. [44] found that fatty acid and seed oil content, and seed weight were significantly increased in *GmDof4* and *GmDof11* overexpressing lines of *A. thaliana*. Further studies showed that *GmDof4* and *GmDof11* directly downregulated the expression of the seed storage protein gene *CRA1*. Moreover, *GmDof4* and *GmDof11* have been shown to induce the expression of the β-subunit of the ACCase encoding gene *acetyl CoA carboxylase* (*accD*) and *long-chain-CoA synthetase gene 5* (*LACS5*), respectively [44]. These results indicate that *GmDof4* and *GmDof11* can simultaneously increase seed oil content by upregulating genes that are involved in fatty acid synthesis and downregulating genes associated with the accumulation of seed protein in Arabidopsis. In addition, increased lipid accumulation was demonstrated after heterologous expression of *GmDof4* in *Chlorella ellipsoidea* [45], indicating that *GmDof4* regulates seed oil content and composition both in higher and lower plants.

In the current study, using the rapeseed cultivar 'Yangyou 6' as receptor, we created *GmDof4* and *GmDof11* overexpression *B. napus* lines via *Agrobacterium*-mediated transformation. 'Yangyou 6' is a double low variety, which is widely planted in the Jiangsu province of China. Our results demonstrated that, when compared with non-transgenic lines, the content of oleic acid in the transgenic lines increased significantly, whereas the content of linoleic acid and linolenic acid were reduced. We found that *GmDof4* and *GmDof11* could activate or inhibit genes that are involved in fatty acid synthesis by directly binding to promoter regions. These findings indicate that *GmDof4* and *GmDof11* have the potential to improve the quality of rapeseed oil.

2. Materials and Methods

2.1. Plant Growth and Transformation

B. napus cv. 'Yangyou 6' plants were grown at 24 °C using a 16-h photoperiod in a growth chamber. *GmDof4* (Accession No: DQ857254) and *GmDof11* (Accession No: DQ857261) DNA sequences were cloned using the primer pairs: GmDof4-F: GACGCACTCACTGACATCAACACTAG, GmDof4-R:

GGTGAGATAAGATTTAGAAGAGGCGTG, and GmDof11-F: GAGACTTCGCAATTTGCATGACTC, GmDof11-R: CTAGCTACTGCTAGAGTGAAGTCATTG, respectively, as designed by Wang, et al. (2007) [44]. The Soybean cultivar 8904 was used for cloning *GmDof*. The overexpression vectors pBIN438-*GmDof4* and pBIN438-*GmDof11* were kindly gifted by Professor Shouyi Chen (Institute of Genetics and Developmental Biology, Chinese Academy of Sciences). The vectors contained a *neomycin phosphate transferase II* (*NPT II*) gene as a selection marker. The *GmDof4* and *GmDof11* genes were driven using a CaMV 35S promoter. The vectors were introduced into the *Agrobacterium tumefaciens* strain GV3101 for genetic transformation into *B. napus*.

GmDof4 and *GmDof11* overexpressing lines of *B. napus* were generated as described by De Block et al. [46] with some modifications. Certified, uniform, and healthy seeds were surface-sterilized with sodium hypochlorite solution and then rinsed in sterile distilled water. The seeds were germinated in the dark on 1/2 MS basal medium containing 2% (*w*/*v*) sucrose. Seven-day-old hypocotyl explants (~15 mm) were prepared and cultured on co-cultivation medium [MS medium supplemented with 2% (*w*/*v*) sucrose, 1 mg/L 2,4-D (2,4-dichlorophenoxyacetic acid), and 1 mg/L benzyladenine (BA); pH 5.8] for three days. Explants were then transferred to selection medium [MS medium supplemented with 2% (*w*/*v*) sucrose, 1 mg/L 2,4-D, 1 mg/L BA, 300 mg/L cephalosporin (Cef) and 30 mg/L G418; pH 5.8] and incubated at 25 °C. The explants with shoot initials were transferred to shoot outgrowth medium [MS medium supplemented with 2% (*w*/*v*) sucrose, 0.3 mg/L NAA, and 300 mg/L Cef; pH 5.8]. Finally, green shoots were transferred to root initiation medium [MS medium supplemented with 2% (*w*/*v*) sucrose, 0.3 mg/L NAA, and 300 mg/L Cef; pH 5.8]. All of the regenerated plantlets were transferred into pot containing nutritious soil after becoming fully developed.

2.2. PCR, Semi-Quantitative, and Quantitative Real-Time PCR Analyses

Total DNA was extracted from the young leaves of each transgenic plant using the CTAB method, as described by Porebski et al. [47]. PCR was performed to identify positive transformants using specific primers.

Total RNA was extracted from non-transgenic and *GmDof*-overexpression seedlings, and the young seeds of *B. napus* using an RNA isolator (Vazyme, Nanjing, China) in accordance with the manufacturer's instructions. First-strand cDNA synthesis was performed using the first-strand cDNA synthesis kit HiScript Q RT SuperMix with oligo(dT)$_{23}$ (50 µM) and Random hexamers (50 ng/µL) as primers for semi-quantitative PCR analysis (Vazyme, Nanjing, China). T_2 generation seeds were collected at 30 days after flowering (DAF), from which 30 seeds were randomly selected and used for RNA extraction.

Semi-quantitative PCR was performed, as follows: 95 °C for 3 min then 32 cycles of 95 °C for 30 s, annealing (56 or 58 °C; detailed information shown in Supplemental Table S1) for 30 s, polymerization at 72 °C for 30 s, followed by 72 °C for 5 min. Real-time PCR was performed in an Mx3500p (Agilent, Santa Clara, CA, USA) using FastStart Universal SYBR Green Master (Rox) (Roche Applied Science, Penzberg, Germany). *BnActin* transcripts were used as the internal reference [18,22]. Relative gene expression was calculated using the $2^{-\Delta\Delta Ct}$ method. qPCR was performed with three biological replicates and three technical replicates for every sample.

To identify whether the expression of genes related to fatty acid metabolism in transgenic seeds, namely, *FAB2* (*Fatty acid biosynthesis 2*), *FAD2*, *FAD3*, *FAD6*, *FAD7*, *FAD8*, *FAE1*, and *FAE7*, were regulated, the expression pattern of genes involved in lipid and fatty acid synthesis were analyzed in seeds at 30 DAF. Amplification primers of these genes were designed to amplify all homologous of specific gene. The primers used for qPCR and the gene ID are listed in Supplementary Table S1.

2.3. Determination of Seed Oil and Fatty Acid Composition

Total seed oil content of the transgenic and non-transgenic plants was determined using near-infrared reflectance (NIR) spectroscopy [48]. Fatty acid concentration was measured using the method that was described by Taylor et al. [49]. The seeds of transgenic (T_2 generation) and non-transgenic plants were ground to a fine powder in a mortar. Five mL of *iso*-propanol were mixed with 0.5 g seed powder and incubated at 100 °C for 5 min. The solution was immediately cooled on ice and 2.5 mL of dichloromethane added. The samples were shaken at 200 rpm for 30 min at room temperature after which, 4 mL of dichloromethane and 4 mL of 1 mol/L KCl in 0.2 mol/L H_3PO_4 were sequentially added into each tube to separate the organic and aqueous phases. The samples were vortexed and centrifuged at 2000 rpm for 20 min. The supernatant was washed twice with 4 mL of dichloromethane, and the original organic phase combined with the washes and dried under nitrogen to yield triacylglycerol. The triacylglycerol was hydrolyzed and the fatty acid esterified, as described by Fatima et al. [50].

Fatty acid composition was analyzed using a gas chromatography-mass spectrometer (Trace GC DSQII, Thermo, Waltham, MA, USA) with a DB-WAX capillary column (30 m × 0.25 mm ID × 0.25 μm df) [49]. The peaks were identified by reference to the identified retention times of internal standard FAMEs (Sigma, Lot No.: 18919-1AMP, St. Louis, MO, USA). GC was performed using a gas carrier (helium) flow rate of 30 mL·min^{-1} and a column and injector temperature of 250 °C. Running temperatures were as follows: 50 °C for 2 min, increasing to 220 °C at a rate of 4 °C/min, and held at 220 °C for 7 min. Each experimental material was biologically replicated three times.

2.4. Detection of DNA Binding Specificity of GmDof4 and GmDof11 by Yeast One-Hybrid Assay

The yeast strain, Y1HGold (*MATα, ura3-52, his3-200, ade2-101, trp1-901, leu2-3, 112, gal4Δ, gal80Δ, met-,* and *MEL1*) containing the AbA^r reporter gene was used as the assay system. *GmDof4* and *GmDof11* were amplified and fused to the GAL4 DNA binding domain on the pGADT7 plasmid. Two or three copies of the *cis*-DNA elements of interest at the promoter of potential targeted genes were synthesized, annealed, and cloned into the "prey" plasmid pAbAi. Then, the recombinant "prey" plasmid was then digested using *Bst*BI for 1 h and transfected into yeast Y1HGold cells according to the manufacturer's protocol (Clontech, Mountain View, CA, USA). The PCR-identified recombinant Y1HGold strains were then used to introduce pGADT7-*GmDof4* or pGADT7-*GmDof11* plasmids. The transfected yeast cells were then cultured in SD/-Leu/-Ura plates. Finally, cultures were placed on SD/-Leu/-Ura + AbA (0.2 mg/L) plates. Strains growing in colonies indicated positive *GmDof* binding on the corresponding *cis*-DNA element.

2.5. Statistical Analysis

All experimental data, including seed oil and fatty acid content analysis, were compared statistically using one-way analysis of variance (ANOVA) followed by Student's *t* test to determine significant differences among the means of different groups using Statistical Product and Service Solutions (SPSS) v16.0 software.

3. Results

3.1. Dof Family Numbers in B. napus

Based on the *Arabidopsis* annotated *Dof* genes, 134 homologous genes of *AtDof* were identified using BlastP (E-value ≤ 1 × 10^{-5}, identity ≥50% and coverage ≥50%) in the *B. napus* reference genome of Darmor-bzh [51] (Supplementary Table S3). BlastP results showed no homology of either *GmDof4* or *GmDof11* in the *B. napus* reference genome using the DNA sequences of *GmDof4* and *GmDof11* as query terms (results not shown).

3.2. Generation and Identification of B. napus Transgenic Plants

To investigate whether *GmDof4* and *GmDof11* could regulate lipid biosynthesis in rapeseed, they were transfected into rapeseed plants, under the control of the CaMV 35S promoter, while using the *Agrobacterium*-mediated method. *B. napus* L. cultivar "Yangyou 6" was used as the receptor and hypocotyl explants were prepared from seven-day-old seedlings. Each experiment was performed using approximately 650 explants. Twenty and 12 rooted plantlets were obtained for *GmDof4* and *GmDof11* transformation, respectively. The existents of *GmDof4* and *GmDof11* in plantlets were identified using PCR (Figures S1 and S2), with the putative transformants then transferred into nutritious soil and placed in a green house. The RNA of the plantlets was extracted and first-strand cDNA synthesized, with the expression of *GmDof4* and *GmDof11* genes in individual transgenic plants detected by semi-quantitative PCR (Figure 1). Finally, seven and five overexpression lines of *GmDof4* and *GmDof11* were obtained, respectively. Two *GmDof4* transformants (*DOF4-2* and *DOF4-20*) demonstrated no *GmDof4* expression, and the expression of *GmDof11* in the *DOF11-12* transformant was very low. Based on the expression levels of *GmDof4* and *GmDof11*, the transgenic plants *DOF4-9*, *DOF4-13*, *DOF11-1*, and *DOF11-6* were further analyzed. When compared with the non-transgenic plants, no significant difference was observed in their growth and development. The presence of the *GmDof* transgene in T_1 generation transgenic plants was confirmed by PCR.

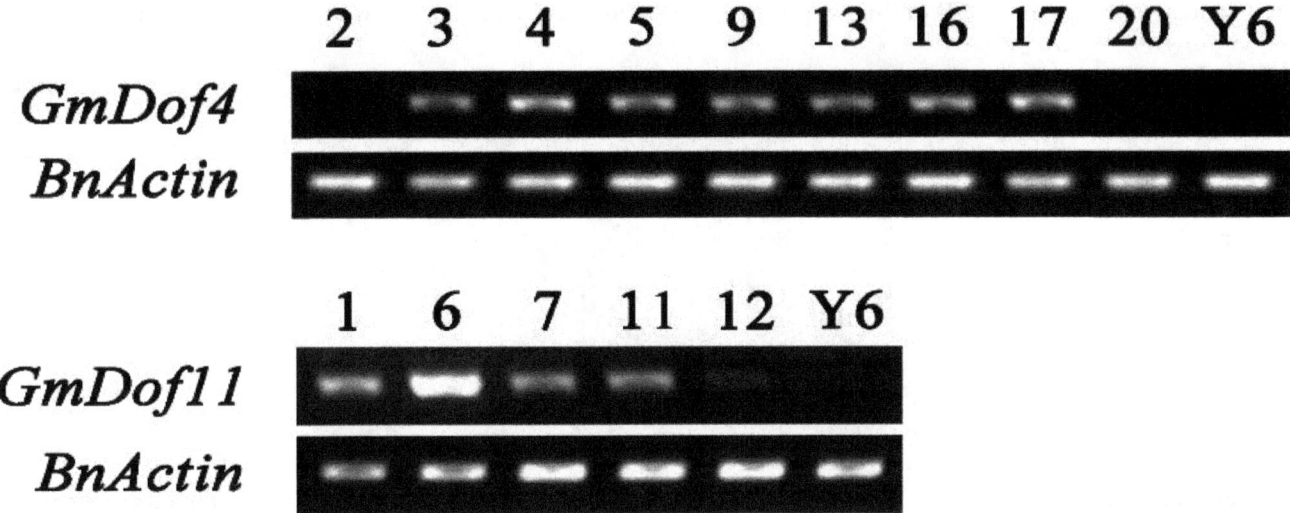

Figure 1. Semi-quantitative RT-PCR analysis of *GmDof4* and *GmDof11* transgenic plants. *BnActin* expression is displayed as the internal control. The lengths of the amplification products of *BnActin*, *GmDof4* and *GmDof11* were 144 bp, 135 bp and 186 bp respectively.

3.3. Changes in the Fatty Acid Composition of GmDof4 and GmDof11 Overexpressing Lines of B. napus

The relative content of the principal fatty acids in the seeds of *GmDof* transgenic and non-transgenic lines was analyzed using gas chromatography (GC). The T_1 and T_2 progenies of *DOF4-9*, *DOF4-13*, *DOF11-1*, and *DOF11-6* lines were produced by self-pollination. The seeds of homozygous T_1 lines that had no gene segregation were used for the fatty acid content determination. The results showed that the quantities of major unsaturated fatty acids, such as oleic, linoleic, and linolenic acid, underwent significant alteration in four transgenic lines compared with the non-transgenic plants. However, the content of the two principal saturated fatty acids, namely, palmitic acid and stearic acid, were consistent with those of the non-transgenic lines (Figure 2, Supplementary Table S2). Among the fatty acids, the content of oleic acid in four overexpression lines was significantly increased, from 62.8% in the non-transgenic rapeseed to 67.11–71.32%. Conversely, the content of linoleic and linolenic acid in the four overexpression lines was significantly lower than in the non-transgenic lines. In addition, total lipid content was measured in the seeds of the overexpression and non-transgenic plants by NIR. Seed oil content of the *GmDof4* and *GmDof11* overexpression lines was ~39%, being not significantly different than the non-transgenic lines (Figure 3). These results indicate that the expression of *GmDof4* and *GmDof11* stimulated the accumulation of oleic acid and regulated the fatty acid composition of rapeseed, but, neither gene could increase total seed oil content.

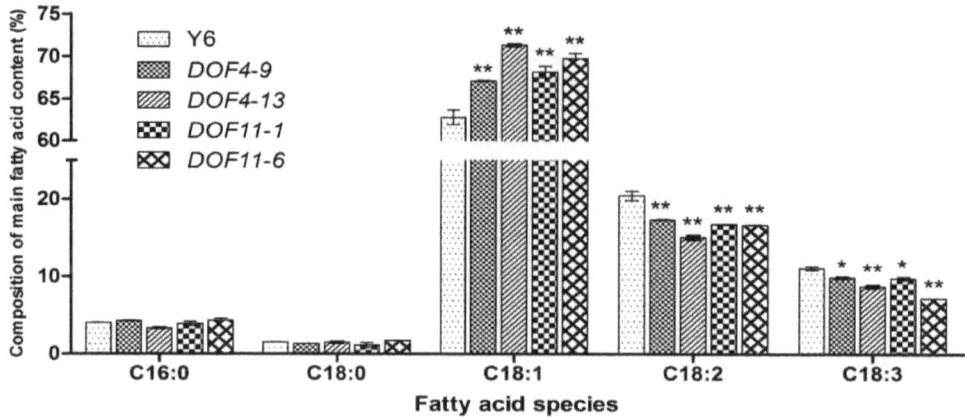

Figure 2. Composition of the major fatty acids in *GmDof4* and *GmDof11* transgenic seeds. The data represent the means ± SD of three replicate experiments and were analyzed by Student's *t*-test ($n = 3$). * $p < 0.05$; ** $p < 0.01$.

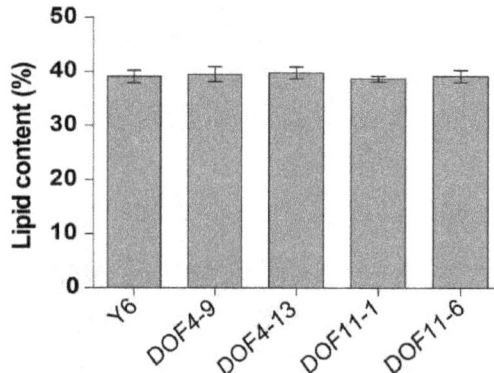

Figure 3. Total lipid contents in the seeds of *GmDof4* and *GmDof11* transgenic plants. The data represent the means ± SD of three replicate experiments and were analyzed by Student's *t*-test ($n = 3$).

3.4. Changes in the Expression of Fatty Acid Metabolism-Related Genes in GmDof4 and GmDof11 Transgenic Plants

Referring to previous reports, the expression of target genes that were directly controlled by *GmDof4* and *GmDof11* were additionally analyzed, including the 12S storage protein subunit encoding gene *CRA1*, the ACCase β subunit encoding gene *accD*, and *LCAS5*. The expression of *accD* was significantly upregulated in two *GmDof4* transgenic seeds compared with its expression in the non-transgenic plants. *FAB2*, which is responsible for oleic acid synthesis, *FAD3* and *FAD8*, which is responsible for the synthesis of linolenic acid from linoleic acid, were significantly upregulated by more than threefold in both lines (Figure 4). However, no significant difference was found in the expression of the other genes, except for *LACS5*, which was slightly upregulated. The expression of *accD* was also upregulated by approximately threefold in two *GmDof11* transgenic seeds. However, the expression of *FAD2* and *FAD6*, which are the coding genes responsible for the synthesis of linoleic acid, was inhibited (Figure 4). These results indicate that GmDof4 and GmDof11 do upregulate the expression of *accD*, and both genes jointly and specifically upregulate or downregulate the genes that are involved in the synthesis of fatty acids.

In addition, qPCR results demonstrated that the gene expression levels of *FAE1* and *FAE7*, which are responsible for the synthesis of eicosanoic acid, were lower than the detection limit of the qPCR technique (Ct > 40). Moreover, there was no expression of the *CRA1* gene in both the

GmDof4 transgenic and non-transgenic seeds, but it was detected in the two *GmDof11* transgenic seeds with slight expression (expression level relative to *BnActin* $\approx 10^{-4}$), indicating that *CRA1* might be upregulated slightly in the *GmDof11* transgenic plants.

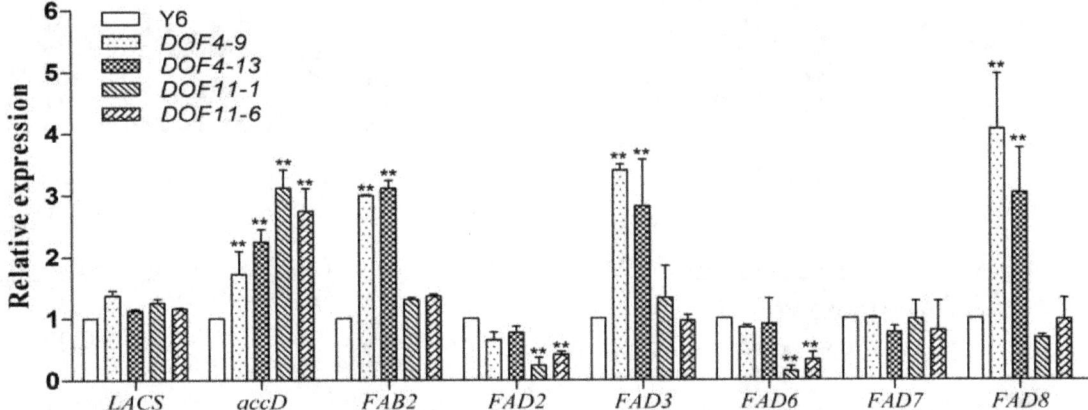

Figure 4. Gene expression detected by qRT-PCR in transgenic and non-transgenic seeds. RT-qPCR was used to determine the relative expression of genes related to lipid and fatty acid metabolism. Bars indicate SD ($n = 3$), Significant differences between transgenic and non-transgenic seeds are labelled with asterisks: ** $p < 0.01$ (Student's t test).

3.5. Yeast One-Hybrid Assay to Detect Target Genes of GmDof4 and GmDof11

Based on the transcriptome data in various tissues and organs of *B. napus* obtained earlier by our group, more than one copy of the *CRA1*, *FAB2*, and *FAD2* genes was found in *B. napus* (Table 1). Analysis of the promoters demonstrates numerous Dof-binding *cis*-DNA elements in the promoter regions of these genes (Table 1). To test whether GmDof4 and GmDof11 regulated the expression of the aforementioned genes by directly binding to their promoter regions, we investigated part of the putative Dof binding elements in the 1.5 kb promoter region of the *CRA1*, *FAB2*, and *FAD2* genes according to the binding features of GmDof4 and GmDof11. The results showed that GmDof4 protein could bind strongly to the *FAB2-1* and *FAB2-2 cis*-DNA elements (Figure 5a). These results suggest that GmDof4 protein may regulate *FAB2* by binding directly to their promoters. Analysis of GmDof11 protein binding activity demonstrated that GmDof11 binds strongly to *FAD2-1* but weakly to *CRA1-1* and *FAD2-2* (Figure 5a). These results indicate that the GmDof11 protein can directly regulate *CRA1* and *FAD2*.

Table 1. Number of transcripts of *GmDof* regulated genes in seeds at 34 days after flowering (DAF) and the *cis*-DNA elements of these genes.

		Cis-DNA Element				Sum
		AAAAG	TAAAG	CTTTT	CTTTA	
FAB2	BnaA03g20420D	1	4	3	3	11
	BnaA05g03490D	6	5	3	5	19
	BnaC03g24420D	5	3	2	2	12
	BnaC04g03030D	3	0	3	2	8
FAD2	BnaA05g26900D	7	4	3	5	19
	BnaAnng09250D	4	3	5	2	14
	BnaC05g40970D	9	6	11	4	30

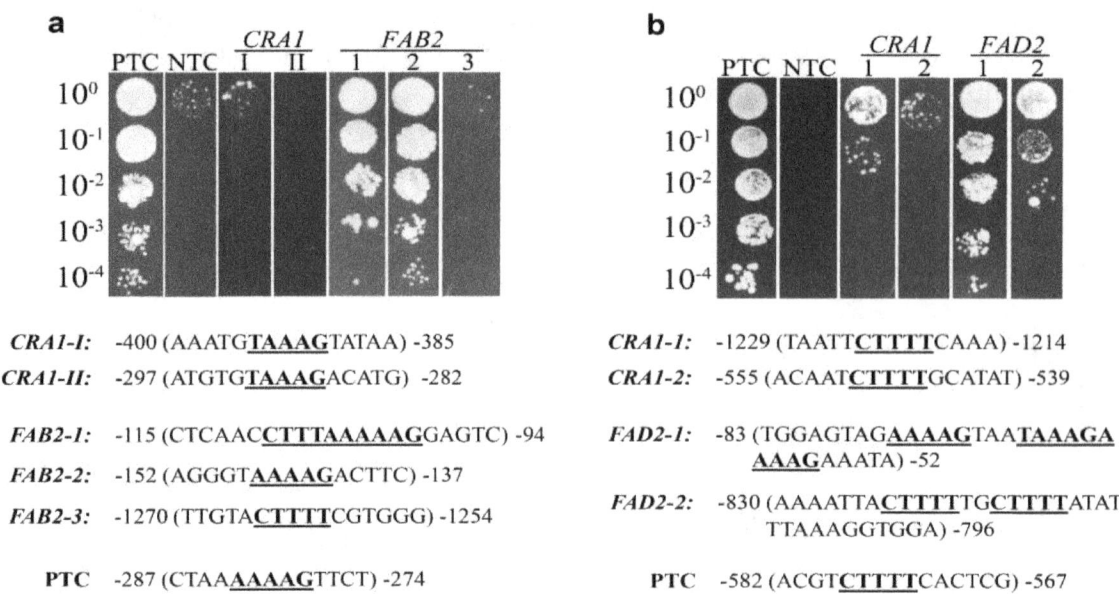

CRA1-I: -400 (AAATG<u>**TAAAG**</u>TATAA) -385

CRA1-II: -297 (ATGTG<u>**TAAAG**</u>ACATG) -282

FAB2-1:   -115 (CTCAAC<u>**CTTTAAAAAG**</u>GAGTC) -94

FAB2-2:   -152 (AGGGT<u>**AAAAG**</u>ACTTC) -137

FAB2-3:   -1270 (TTGTA<u>**CTTTT**</u>CGTGGG) -1254

PTC   -287 (CTAA<u>**AAAAG**</u>TTCT) -274

CRA1-1:   -1229 (TAATT<u>**CTTTT**</u>CAAA) -1214

CRA1-2:   -555 (ACAAT<u>**CTTTT**</u>GCATAT) -539

FAD2-1:   -83 (TGGAGTAG<u>**AAAAG**</u>TAA<u>**TAAAGA**</u>
        <u>**AAAG**</u>AAATA) -52

FAD2-2:   -830 (AAAATTA<u>**CTTTT**</u>TG<u>**CTTTT**</u>ATAT
        TTAAAGGTGGA) -796

PTC   -582 (ACGT<u>**CTTTT**</u>CACTCG) -567

Figure 5. GmDof4 and GmDof11 interact with the *cis*-DNA elements in the promoter regions of downstream genes in the transgenic plants (**a**) Interaction between GmDof4 and the *cis*-DNA elements in the promoter regions of *CRA1* and *FAB2*. The bolded and underlined sequences indicate the core sequence of the Dof-binding elements. The putative Dof-binding elements were cloned into pAbAi, and these plasmids were transfected into yeast Y1HGold cells with pGADT7-*GmDof4*. Growth of the transfected yeast cells on the SD/-Leu/-Ura + AbA (0.2 mg/L) plates indicates that GmDof4 protein can bind to its corresponding *cis*-DNA element. PTC is a strain that contains pGADT7-*GmDof4* and a pAbAi plasmid with an element in the promotor region of *AtaccD*, which was confirmed to interact with the GmDof4 protein. NTC is a strain that contains pGADT7-*GmDof4* and an empty pAbAi plasmid. (**b**) Interaction between GmDof11 and the *cis*-DNA elements in the promoter regions of *CRA1* and *FAD2*. The bolded and underlined sequences indicate the core sequence of the Dof-binding elements. The putative Dof-binding elements were cloned into pAbAi, and these plasmids were transfected into yeast Y1HGold cells with pGADT7-*GmDof11*. Growth of the transfected yeast cells on SD/-Leu/-Ura + AbA (0.2 mg/L) plates indicates that the GmDof11 protein can bind to its corresponding *cis*-DNA element. PTC is a strain that contains pGADT7-*GmDof11* and a pAbAi plasmid with an element in the promotor region of *AtCRA1*, which was confirmed to be interacting with GmDof11 protein. NTC is a strain that contains pGADT7-*GmDof11* and an empty pAbAi plasmid.

4. Discussion

4.1. Overexpression of GmDof4 and GmDof11 Augmented the Oleic Acid in B. napus Seed Oil

GmDof4 and GmDof11 are TFs involved in the regulation of seed oil synthesis in soybean. Overexpression of *GmDof4* or *GmDof11* augments oil synthesis in transgenic *Arabidopsis* and the single-cell microalga *C. ellipsoidea* [44,45]. *GmDof4* and *GmDof11* overexpressed rapeseed plants were produced using *Agrobacterium*-mediated genetic transformation. However, no significant change was found in the seed oil content in the four transgenic lines. Comparison of the fatty acids in *GmDof4* and *GmDof11* transgenic and non-transgenic plants showed a significant change in fatty acid composition. The relative level of the monounsaturated fatty acid oleic acid increased, while the relative levels of the polyunsaturated fatty acids linoleic and linolenic acid decreased significantly in the *GmDof* transgenic plants compared with the non-transgenic plants. These results suggest that GmDof4 and GmDof11 may play a role in the late stage of fatty acid synthesis in *B. napus*, by regulating the synthesis of a few specific fatty acids rather than the carbon metabolic flux that would alter the relative levels of the major fatty acids. This phenomenon is different from those in *Arabidopsis* and *C. ellipsoidea*. The total lipid content increased significantly in transgenic *Arabidopsis* seeds and *C. ellipsoidea* cells, but the relative levels of each fatty acid did not change, except for linoleic acid in *Arabidopsis* overexpressing

GmDof4 [44,52]. The difference may be due to the large genome of *B. napus* and the complex network regulation of the synthesis and accumulation of oil during seed development. Therefore, increasing the oil content of rapeseed may require precise and more targeted genetic engineering.

4.2. GmDof4 and GmDof11 Regulated the Genes of Fatty Acid Synthesis by Binding to the Cis-DNA Elements in the Promoter Region of These Genes

In oil crops, lipid and fatty acid synthesis involves a number of enzymes [5,7,12]. *FAB2*, which encodes a stearoyl-ACP desaturase, catalyzes the synthesis of oleic acid. The expression level of *FAB2* affects the content of oleic acid [53,54]. Kachroo et al. found that the stearic acid content in the *FAB2* gene mutant (*ssi2*) was approximately 18 times higher than that in WT plants, and the content of oleic, linoleic, and linolenic acid was significantly reduced. Meanwhile, *FAB2* is involved in the activation of NPR1-dependent and -independent defense responses [55–57]. FAD2 and FAD3 are two important enzymes in the synthesis of unsaturated fatty acids in seed oils. They are integrated into the endoplasmic reticulum and they are responsible for the catalysis of the conversion of oleic acid to linoleic acid and then linolenic acid. Furthermore, in plastids FAD6 is an isoenzyme of FAD2, while FAD7 and FAD8 are isoenzymes of FAD3 [31,58]. FAD6, FAD7, and FAD8 are closely related to the synthesis of unsaturated fatty acids on chloroplast membranes [59]. The expression of these genes affect leaf lipids in *Arabidopsis* [31,56]. Therefore, we examined the expression of these desaturases in transgenic lines.

accD, which encodes the β-subunit of ACCase, was upregulated in seeds of the *GmDof4* transgenic plants at 30 DAF. This result is consistent with studies in *Arabidopsis*. In addition, *FAB2*, *FAD3*, and *FAD8* genes were significantly upregulated. This finding suggests that *GmDof4* likely increased oleic acid synthesis by increasing the expression of genes that are related to oleic acid synthesis, in spite of the upregulated *FAD3* not increasing the content of $C_{18:3}$, which is possibly due to the complexity of the regulation of fatty acid accumulation in seeds. The expression of *FAD2* and *FAD6* were downregulated in *GmDof11* transgenic seeds. The downregulation of *FAD2* may have caused a decrease in the content of linoleic acid and linolenic acid, thereby increasing the proportion of oleic acid. It is not clear whether the change in the expression of *FAD6* and *FAD8* in the overexpression plants changed the response to stress. Expression of *FAE1* and *FAE7* genes, which are responsible for erucic acid ($C_{22:1}$) synthesis, were not detected in all plants. This result may be due to the fact that the rapeseed variety that is used here has low erucic acid characteristics, and erucic acid synthesis genes are severely inhibited. In addition, although the expression of *accD* was upregulated in the seeds of the four transgenic lines, seeds oil content of did not increase. This result indicates that other regulation mechanisms in the seeds of *B. napus* related to the accumulation of oil.

In *Arabidopsis*, GmDof4 binds directly to the Dof-binding *cis*-DNA element in the promoter regions of the *accD* and *CRA1* genes, and GmDof11 directly regulates the expression of *LCAS* and *CRI1* genes [44]. In this study, we found that GmDof4 bound to the *cis*-DNA element in the promoter region of *FAB2*, whereas GmDof11 bound to the *cis*-DNA element in the promoter region of *CRA1* and *FAD2*. These results indicate that *GmDof4* and *GmDof11* regulated components of fatty acid synthesis in seed oil by regulating the expression of specific genes. Whether the slight upregulation of the *CRA1* gene in *GmDof11* transgenic seeds was caused by the direct interaction of GmDof11 and the *cis*-DNA element of *CRA1* should be further examined using a dual luciferase reporter system. Evaluation of the number of Dof binding elements (A/T)TTTG or CAAA(A/T) at the promoter regions of the potential target genes revealed that they contained a large number of Dof binding elements (Table 1). While considering that there are 134 putative *Dof* genes in *B. napus*, the existence of those elements indicates that the specific spatial and temporal expression of these genes may be regulated by various *Dof* TFs. This makes it possible to regulate the expression of the genes that are involved in lipid and fatty acid synthesis in *B. napus* by GmDof4 and GmDof11. Interestingly, except for *BnaC.accD.c* (BnaC09g27690D), which showed incomplete genome sequencing at the promoter region, the 1.5 kb promoter region of three *accD* duplicates in *B. napus* were identical (Figure S3). The homology of the *accD* gene to *Arabidopsis* was 96.74%. In addition, it has been suggested that the 5′-untranslated region (UTR) of plant *FAD2* genes is evolutionarily conserved [35,60]. These results strongly suggest that the regulation of expression of the genes involved in fatty acid synthesis might also be highly conserved. GmDof4 and GmDof11 proteins increased oleic acid content in seed oil by activating or inhibiting genes that are associated with fatty acid synthesis in *B. napus*. Both proteins may be used as a genetic resource to improve the quality of rapeseed oil.

Supplementary Materials: Figure S1: PCR analysis of putative transformants of *GmDof4*, Figure S2: PCR analysis of putative transformants of *GmDof11*, Figure S3: Homology of the *accD* gene promoters in *B. napus* and *A. thaliana*, Table S1: Amplification primers used for Semi-quantitative and quantitative RT-PCR, Table S2: Composition of the major fatty acids in *GmDof4* and *GmDof11* transgenic seeds, Table S3: the *Dof* genes in *B. napus*

Author Contributions: Y.W. designed the experiment, Q.S. and J.X. performed experiments, L.L. and D.L. created the materials, J.W. and J.J. analyzed data and revised the manuscript. All the authors approved the final manuscript.

References

1. Lin, L.; Allemekinders, H.; Dansby, A.; Campbell, L.; Durance-Tod, S.; Berger, A.; Jones, P.J.H. Evidence of health benefits of canola oil. *Nutr. Rev.* **2013**, *71*, 370–385. [CrossRef] [PubMed]

2. Aldhaidhawi, M.; Chiriac, R.; Badescu, V. Ignition delay, combustion and emission characteristics of Diesel engine fueled with rapeseed biodiesel—A literature review. *Renew. Sustain. Energy Rev.* **2017**, *73*, 178–186. [CrossRef]

3. Saka, S.; Kusdiana, D. Biodiesel fuel from rapeseed oil as prepared in supercritical methanol. *Fuel* **2001**, *80*, 225–231. [CrossRef]

4. Hu, Q.; Hua, W.; Yin, Y.; Zhang, X.K.; Liu, L.J.; Shi, J.Q.; Zhao, Y.G.; Qin, L.; Chen, C.; Wang, H.Z. Rapeseed research and production in China. *Crop J.* **2016**, *5*, 127–135. [CrossRef]

5. Wang, X.D.; Long, Y.; Yin, Y.T.; Zhang, C.Y.; Gan, L.; Liu, L.Z.; Yu, L.J.; Meng, J.L.; Li, M.T. New insights into the genetic networks affecting seed fatty acid concentrations in *Brassica napus*. *BMC Plant Biol.* **2015**, *15*, 91. [CrossRef] [PubMed]

6. Gillingham, L.G.; Harris-Janz, S.; Jones, P.J.H. Dietary monounsaturated fatty acids are protective against metabolic syndrome and cardiovascular disease risk factors. *Lipids* **2011**, *46*, 209–228. [CrossRef] [PubMed]

7. Weselake, R.J.; Taylor, D.C.; Rahman, M.H.; Shah, S.; Laroche, A.; McVetty, P.B.E.; Harwood, J.L. Increasing the flow of carbon into seed oil. *Biotechnol. Adv.* **2009**, *27*, 866–878. [CrossRef] [PubMed]

8. Zhao, J.Y.; Dimov, Z.; Becker, H.C.; Ecke, W.; Möllers, C. Mapping QTL controlling fatty acid composition in a doubled haploid rapeseed population segregating for oil content. *Mol. Breed.* **2008**, *21*, 115–125. [CrossRef]

9. Delourme, R.; Falentin, C.; Huteau, V.; Clouet, V.; Horvais, R.; Gandon, B.; Specel, S.; Hanneton, L.; Dheu, J.E.; Deschamps, M.; et al. Genetic control of oil content in oilseed rape (*Brassica napus* L.). *Theor. Appl. Genet.* **2006**, *113*, 1331–1345. [CrossRef] [PubMed]

10. Javed, N.; Geng, J.F.; Tahir, M.; McVetty, P.B.E.; Li, G.; Duncan, R.W. Identification of QTL influencing seed oil content, fatty acid profile and days to flowering in *Brassica napus* L. *Euphytica* **2016**, *207*, 191–211. [CrossRef]

11. Teh, L.; Möllers, C. Genetic variation and inheritance of phytosterol and oil content in a doubled haploid population derived from the winter oilseed rape Sansibar × Oase cross. *Theor. Appl. Genet.* **2016**, *129*, 181–199. [CrossRef] [PubMed]

12. Katavic, V.; Shi, L.; Yu, Y.Y.; Zhao, L.F.; Haughn, G.W.; Kunst, L. Investigation of the contribution of oil biosynthetic enzymes to seed oil content in *Brassica napus* and *Arabidopsis thaliana*. *Can. J. Plant Sci.* **2014**, *94*, 1109–1112. [CrossRef]

13. Vigeolas, H.; Waldeck, P.; Zank, T.; Geigenberger, P. Increasing seed oil content in oil-seed rape (*Brassica napus* L.) by over-expression of a yeast glycerol-3-phosphate dehydrogenase under the control of a seed-specific promoter. *Plant Biotechnol. J.* **2007**, *5*, 431–441. [CrossRef] [PubMed]

14. Zou, J.; Katavic, V.; Giblin, E.M.; Barton, D.L.; MacKenzie, S.L.; Keller, W.A.; Hu, X.; Taylor, D.C. Modification of seed oil content and acyl composition in the brassicaceae by expression of a yeast sn-2 acyltransferase gene. *Plant Cell* **1997**, *9*, 909–923. [CrossRef] [PubMed]

15. Zou, J.; Qi, Q.; Katavic, V.; Marillia, E.-F.; Taylor, D.C. Effects of antisense repression of an Arabidopsis thaliana pyruvate dehydrogenase kinase cDNA on plant development. *Plant Mol. Biol.* **1999**, *41*, 837–849. [CrossRef] [PubMed]

16. Taylor, D.C.; Zhang, Y.; Kumar, A.; Francis, T.; Giblin, E.M.; Barton, D.L.; Ferrie, J.R.; Laroche, A.; Shah, S.; Zhu, W.M.; et al. Molecular modification of triacylglycerol accumulation by over-expression of *DGAT1* to produce canola with increased seed oil content under field conditions. *Botany* **2009**, *87*, 533–543. [CrossRef]

17. Peng, D.; Zhang, L.; Tan, X.F.; Yuan, D.Y.; Liu, X.M.; Zhou, B. Increasing seed oil content and altering oil quality of *Brassica napus* L. by over-expression of diacylglycerol acyltransferase 1 (*SsDGAT1*) from *Sapium sebiferum* (L.) Roxb. *Mol. Breed.* **2016**, *36*, 136. [CrossRef]

18. Zhao, C.Z.; Li, H.; Zhang, W.X.; Wang, H.L.; Xu, A.X.; Tian, J.H.; Zou, J.T.; Taylor, D.C.; Zhang, M. BnDGAT1s function similarly in oil deposition and are expressed with uniform patterns in tissues of *Brassica napus*. *Front. Plant Sci.* **2017**, *8*, 2205. [CrossRef] [PubMed]

19. Aznar-Moreno, J.; Denolf, P.; Van Audenhove, K.; De Bodt, S.; Engelen, S.; Fahy, D.; Wallis, J.G.; Browse, J. Type 1 diacylglycerol acyltransferases of *Brassica napus* preferentially incorporate oleic acid into triacylglycerol. *J. Exp. Bot.* **2015**, *66*, 6497–6506. [CrossRef] [PubMed]

20. Liu, F.; Xia, Y.P.; Wu, L.; Fu, D.H.; Hayward, A.; Luo, J.L.; Yan, X.H.; Xiong, X.J.; Fu, P.; Wu, G.; et al. Enhanced seed oil content by overexpressing genes related to triacylglyceride synthesis. *Gene* **2015**, *557*, 163–171. [CrossRef] [PubMed]

21. Liu, J.; Hua, W.; Yang, H.L.; Zhan, G.M.; Li, R.J.; Deng, L.B.; Wang, X.F.; Liu, G.H.; Wang, H.Z. The *BnGRF2* gene (*GRF2-like* gene from *Brassica napus*) enhances seed oil production through regulating cell number and plant photosynthesis. *J. Exp. Bot.* **2012**, *63*, 3727–3740. [CrossRef] [PubMed]

22. Tan, H.L.; Yang, X.H.; Zhang, F.X.; Zheng, X.; Qu, C.M.; Mu, J.Y.; Fu, F.Y.; Li, J.N.; Guan, R.Z.; Zhang, H.S.; et al. Enhanced seed oil production in canola by conditional expression of *Brassica napus LEAFY COTYLEDON1* and *LEC1-LIKE* in developing seeds. *Plant Physiol.* **2011**, *156*, 1577–1588. [CrossRef] [PubMed]

23. Elahi, N.; Duncan, R.W.; Stasolla, C. Decreased seed oil production in *FUSCA3 Brassica napus* mutant plants. *Plant Physiol. Biochem.* **2015**, *96*, 222–230. [CrossRef] [PubMed]

24. Baud, S.; Mendoza, M.S.; To, A.; Harscoët, E.; Lepiniec, L.; Dubreucq, B. WRINKLED1 specifies the regulatory action of LEAFY COTYLEDON2 towards fatty acid metabolism during seed maturation in Arabidopsis. *Plant J.* **2007**, *50*, 825–838. [CrossRef] [PubMed]

25. Elahi, N.; Duncan, R.W.; Stasolla, C. Modification of oil and glucosinolate content in canola seeds with altered expression of *Brassica napus LEAFY COTYLEDON1*. *Plant Physiol. Biochem.* **2016**, *100*, 52–63. [CrossRef] [PubMed]

26. Li, Q.; Shao, J.H.; Tang, S.H.; Shen, Q.W.; Wang, T.H.; Chen, W.L.; Hong, Y.Y. Wrinkled1 accelerates flowering and regulates lipid homeostasis between oil accumulation and membrane lipid anabolism in *Brassica napus*. *Front. Plant Sci.* **2015**, *6*, 1–15. [CrossRef] [PubMed]

27. Liu, J.; Hua, W.; Zhan, G.M.; Wei, F.; Wang, X.F.; Liu, G.L.; Wang, H.Z. Increasing seed mass and oil content in transgenic *Arabidopsis* by the overexpression of *wri1*-like gene from *Brassica napus*. *Plant Physiol. Biochem.* **2010**, *48*, 9–15. [CrossRef] [PubMed]

28. Gao, M.J.; Lydiate, D.J.; Li, X.; Lui, H.L.; Gjetvaj, B.; Hegedus, D.D.; Rozwadowski, K. Repression of seed maturation genes by a trihelix transcriptional repressor in *Arabidopsis* seedlings. *Plant Cell* **2009**, *21*, 54–71. [CrossRef] [PubMed]

29. Elhiti, M.; Yang, C.C.; Chan, A.; Durnin, D.C.; Belmonte, M.F.; Ayele, B.T.; Tahir, M.; Stasolla, C. Altered seed oil and glucosinolate levels in transgenic plants overexpressing the *Brassica napus SHOOTMERISTEMLESS* gene. *J. Exp. Bot.* **2012**, *63*, 4447–4461. [CrossRef] [PubMed]

30. Ohlrogge, J.; Browse, J. Lipid biosynthesis. *Plant Cell* **1995**, *7*, 957–970. [CrossRef] [PubMed]

31. Wallis, J.G.; Browse, J. Mutants of *Arabidopsis* reveal many roles for membrane lipids. *Prog. Lipid Res.* **2002**, *41*, 254–278. [CrossRef]

32. Beisson, F.; Koo, A.J.K.; Ruuska, S.; Schwender, J.; Pollard, M.; Thelen, J.J.; Paddock, T.; Salas, J.J.; Savage, L.; Milcamps, A.; et al. Arabidopsis Genes Involved in Acyl Lipid Metabolism. A 2003 Census of the Candidates, a Study of the Distribution of Expressed Sequence Tags in Organs, and a Web-Based Database. *Plant Physiol.* **2003**, *132*, 681–697. [CrossRef] [PubMed]

33. Shi, J.; Lang, C.; Wu, X.; Liu, R.; Zheng, T.; Zhang, D.; Chen, J.; Wu, G. RNAi knockdown of fatty acid elongase1 alters fatty acid composition in *Brassica napus*. *Biochem. Biophys. Res. Commun.* **2015**, *466*, 518–522. [CrossRef] [PubMed]

34. Shi, J.H.; Lang, C.X.; Wang, F.L.; Wu, X.L.; Liu, R.H.; Zheng, T.; Zhang, D.Q.; Chen, J.Q.; Wu, G.T. Depressed expression of *FAE1* and *FAD2* genes modifies fatty acid profiles and storage compounds accumulation in *Brassica napus* seeds. *Plant Sci.* **2017**, *263*, 177–182. [CrossRef] [PubMed]

35. Jung, J.H.; Kim, H.; Go, Y.S.; Lee, S.B.; Hur, C.-G.; Kim, H.U.; Suh, M.C. Identification of functional *BrFAD2-1* gene encoding microsomal delta-12 fatty acid desaturase from *Brassica rapa* and development of *Brassica napus* containing high oleic acid contents. *Plant Cell Rep.* **2011**, *30*, 1881–1892. [CrossRef] [PubMed]

36. Yanagisawa, S. The *Dof* family of plant transcription factors. *Trends Plant Sci.* **2002**, *7*, 555–560. [CrossRef]

37. Bocianowski, J.; Mikołajczyk, K.; Bartkowiak-Broda, I. Determination of fatty acid composition in seed oil of rapeseed (*Brassica napus* L.) by mutated alleles of the FAD3 desaturase genes. *J. Appl. Genet.* **2012**, *53*, 27–30. [CrossRef] [PubMed]

38. Noguero, M.; Atif, R.M.; Ochatt, S.; Thompson, R.D. The role of the DNA-binding One Zinc Finger (DOF) transcription factor family in plants. *Plant Sci.* **2013**, *209*, 32–45. [CrossRef] [PubMed]

39. Kurai, T.; Wakayama, M.; Abiko, T.; Yanagisawa, S.; Aoki, N.; Ohsugi, R. Introduction of the ZmDof1 gene into rice enhances carbon and nitrogen assimilation under low-nitrogen conditions. *Plant Biotechnol. J.* **2011**, *9*, 826–837. [CrossRef] [PubMed]

40. Diaz, I.; Martinez, M.; Isabel-Lamoneda, I.; Rubio-Somoza, I.; Carbonero, P. The DOF protein, SAD, interacts with GAMYB in plant nuclei and activates transcription of endosperm-specific genes during barley seed development. *Plant J.* **2005**, *42*, 652–662. [CrossRef] [PubMed]

41. Xu, J.Y.; Dai, H.B. *Brassica napus* Cycling Dof Factor1 (*BnCDF1*) is involved in flowering time and freezing tolerance. *Plant Growth Regul.* **2016**, *80*, 315–322. [CrossRef]

42. Ma, J.; Li, M.Y.; Wang, F.; Tang, J.; Xiong, A.S. Genome-wide analysis of Dof family transcription factors and their responses to abiotic stresses in Chinese cabbage. *BMC Genom.* **2015**, *16*, 33. [CrossRef] [PubMed]

43. Tian, A.G.; Wang, J.; Cui, P.; Han, Y.J.; Xu, H.; Cong, L.J.; Huang, X.G.; Wang, X.L.; Jiao, Y.Z.; Wang, B.J.; et al. Characterization of soybean genomic features by analysis of its expressed sequence tags. *Theor. Appl. Genet.* **2004**, *108*, 903–913. [CrossRef] [PubMed]

44. Wang, H.W.; Zhang, B.; Hao, Y.J.; Huang, J.; Tian, A.G.; Liao, Y.; Zhang, J.S.; Chen, S.Y. The soybean Dof-type transcription factor genes, *GmDof4* and *GmDof11*, enhance lipid content in the seeds of transgenic Arabidopsis plants. *Plant J.* **2007**, *52*, 716–729. [CrossRef] [PubMed]

45. Zhang, J.H.; Hao, Q.; Bai, L.L.; Xu, J.; Yin, W.B.; Song, L.Y.; Xu, L.; Guo, X.J.; Fan, C.M.; Chen, Y.H.; et al. Overexpression of the soybean transcription factor GmDof4 significantly enhances the lipid content of *Chlorella ellipsoidea*. *Biotechnol. Biofuels* **2014**, *7*, 128. [CrossRef] [PubMed]

46. De Block, M.; De Brouwer, D.; Tenning, P. Transformation of *Brassica napus* and *Brassica oleracea* Using *Agrobacterium tumefaciens* and the Expression of the bar and neo Genes in the Transgenic Plants. *Plant Physiol.* **1989**, *91*, 694–701. [CrossRef] [PubMed]

47. Porebski, S.; Bailey, L.G.; Baum, B.R. Modification of a CTAB DNA extraction protocol for plants containing high polysaccharide and polyphenol components. *Plant Mol. Biol. Rep.* **1997**, *15*, 8–15. [CrossRef]

48. Tkachuk, R. Oil and protein analysis of whole rapeseed kernels by near infrared reflectance spectroscopy. *J. Am. Oil Chem. Soc.* **1981**, *58*, 819–822. [CrossRef]

49. Taylor, D.C.; Barton, D.L.; Giblin, E.M.; MacKenzie, S.L.; van den Berg, C.; McVetty, P. Microsomal *Lyso*-phosphatidic acid acyltransferase from a *Brassica oleracea* cultivar incorporates erucic acid into the *sn*-2 position of seed triacylglycerols. *Plant Physiol.* **1995**, *109*, 409–420. [CrossRef] [PubMed]

50. Fatima, T.; Snyder, C.L.; Schroeder, W.R.; Cram, D.; Datla, R.; Wishart, D.; Weselake, R.J.; Krishna, P. Fatty acid composition of developing sea buckthorn (*Hippophae rhamnoides* L.) berry and the transcriptome of the mature seed. *PLoS ONE* **2012** *7*, e34099. [CrossRef] [PubMed]

51. Chalhoub, B.; Denoeud, F.; Liu, S.; Parkin, I.A.P.; Tang, H.; Wang, X.; Chiquet, J.; Belcram, H.; Tong, C.; Samans, B.; et al. Early allopolyploid evolution in the post-Neolithic *Brassica napus* oilseed genome. *Science* **2014**, *345*, 950–953. [CrossRef] [PubMed]

52. Zhang, D.; Hua, Y.; Wang, X.; Zhao, H.; Shi, L.; Xu, F. A High-Density Genetic Map Identifies a Novel Major QTL for Boron Efficiency in Oilseed Rape (*Brassica napus* L.). *PLoS ONE* **2014**, *9*, e112089. [CrossRef] [PubMed]

53. Kachroo, P.; Shanklin, J.; Shah, J.; Whittle, E.J.; Klessig, D.F. A fatty acid desaturase modulates the activation of defense signaling pathways in plants. *Proc. Natl. Acad. Sci. USA* **2001**, *98*, 9448–9453. [CrossRef] [PubMed]

54. Song, N.; Hu, Z.R.; Li, Y.H.; Li, C.; Peng, F.X.; Yao, Y.Y.; Peng, H.R.; Ni, Z.F.; Xie, C.J.; Sun, Q.X. Overexpression of a wheat stearoyl-ACP desaturase (SACPD) gene *TaSSI2* in *Arabidopsis ssi2* mutant compromise its resistance to powdery mildew. *Gene* **2013**, *524*, 220–227. [CrossRef] [PubMed]

55. Kachroo, A.; Lapchyk, L.; Fukushige, H.; Hildebrand, D.; Klessig, D.; Kachroo, P. Plastidial fatty acid signaling modulates salicylic acid- and jasmonic acid-mediated defense pathways in the *Arabidopsis ssi2* mutant. *Plant Cell* **2003**, *15*, 2952–2965. [CrossRef] [PubMed]

56. Kachroo, P.; Venugopal, S.C.; Navarre, D.A.; Lapchyk, L.; Kachroo, A. Role of salicylic acid and fatty acid desaturation pathways in *ssi2*-mediated signaling. *Plant Physiol.* **2005**, *139*, 1717–1735. [CrossRef] [PubMed]

57. Kachroo, A.; Shanklin, J.; Whittle, E.; Lapchyk, L.; Hildebrand, D.; Kachroo, P. The Arabidopsis stearoyl-acyl carrier protein-desaturase family and the contribution of leaf isoforms to oleic acid synthesis. *Plant Mol. Biol.* **2007**, *63*, 257–271. [CrossRef] [PubMed]

58. Nishiuchi, T.; Hamada, T.; Kodama, H.; Iba, K. Wounding changes the spatial expression pattern of the arabidopsis plastid omega-3 fatty acid desaturase gene (*FAD7*) through different signal transduction pathways. *Plant Cell* **1997**, *9*, 1701–1712. [CrossRef] [PubMed]

59. Gibson, S.; Arondel, V.; Iba, K.; Somerville, C. Cloning of a temperature-regulated gene encoding a chloroplast ω-3 desaturase from *Arabidopsis thaliana*. *Plant Physiol.* **1994**, *106*, 1615–1621. [CrossRef] [PubMed]

60. Kim, M.J.; Kim, H.; Shin, J.S.; Chung, C.-H.; Ohlrogge, J.B.; Suh, M.C. Seed-specific expression of sesame microsomal oleic acid desaturase is controlled by combinatorial properties between negative cis-regulatory elements in the SeFAD2 promoter and enhancers in the 5′-UTR intron. *Mol. Genet. Genom.* **2006**, *276*, 351–368. [CrossRef] [PubMed]

14

Genomic Selection in Cereal Breeding

Charlotte D. Robertsen [1,2,*], Rasmus L. Hjortshøj [1] and Luc L. Janss [2]

[1] Sejet Plant Breeding I/S, 8700 Horsens, Denmark; rlh@sejet.dk
[2] Center for Quantitative Genetics and Genomics, Aarhus University, 8830 Tjele, Denmark; luc.janss@mbg.au.dk
*   Correspondence: cdr@mbg.au.dk

Abstract: Genomic Selection (GS) is a method in plant breeding to predict the genetic value of untested lines based on genome-wide marker data. The method has been widely explored with simulated data and also in real plant breeding programs. However, the optimal strategy and stage for implementation of GS in a plant-breeding program is still uncertain. The accuracy of GS has proven to be affected by the data used in the GS model, including size of the training population, relationships between individuals, marker density, and use of pedigree information. GS is commonly used to predict the additive genetic value of a line, whereas non-additive genetics are often disregarded. In this review, we provide a background knowledge on genomic prediction models used for GS and a view on important considerations concerning data used in these models. We compare within- and across-breeding cycle strategies for implementation of GS in cereal breeding and possibilities for using GS to select untested lines as parents. We further discuss the difference of estimating additive and non-additive genetic values and its usefulness to either select new parents, or new candidate varieties.

Keywords: crops; quantitative genetics; estimated breeding value; genomic prediction; plant breeding; breeding scheme; pedigree; genetic value

1. Introduction

Agronomically important quantitative traits are often controlled by many small-effect genes, which have been difficult to take advantage of in practical breeding [1]. The small-effect genes are difficult to map, and, if mapping is successful, often multiple quantitative trait loci (QTL) are present, which are difficult to use simultaneously in breeding. As a consequence, marker-assisted-selection (MAS), when defined as the use of mapped genes in breeding, has had limited success in improving such traits [2]. A key example quantitative trait is yield, which has shown difficult to improve in nearly all plant crops [3,4]. Gene editing, like CRISPR (Clustered Regularly Interspaced Short Palindromic Repeats), will likely not offer a solution either, because, like MAS, they are conditional on first identifying mutations or modifications with large effect. Genomic (or genome-wide) selection (GS) is a method that has promised to overcome the limitations of MAS for quantitative traits [2]. The objective of GS is to determine the genetic potential of an individual instead of identifying the specific QTL. GS was originally developed in livestock breeding as a method to predict breeding values of individuals based on markers covering the whole genome using simulated data [5]. Initial studies on the application of GS in a dairy cattle breeding program showed promising improvements in the accuracy of selection [6]. Promisingly, GS have been indicated to outperform MAS using the same economic investment, even at low accuracies [7,8]. It may be noted that before GS was established, plant breeders already developed ideas with similar ingredients as Meuwissen et al.'s GS [5]. Notably, Bernardo [9] developed a multi-marker MAS version with random marker effects, but it was developed within the MAS paradigm using only markers flanking identified QTL, while the main break-through in Meuwissen et al.'s GS [5] was to avoid identifying QTL.

Decreasing costs of genotyping using high-density single nucleotide polymorphism (SNP)-arrays and development of statistical methods to accurately predict marker effects have led to the breakthrough of GS. Selection decisions based on GS results have been indicated to improve the accuracy of selection and speed of genetic improvement. GS is now used in dairy cattle breeding programs around the world and included in the marketing of bulls [10]. Plant breeders have often been relying on phenotypic selection (PS) to choose the best offspring to continue in the breeding program. One of the first studies on the prospects of GS in plant breeding was carried out in maize (*Zea mays* L.) by Bernardo et. al [8] using simulated data. Predictions have also been carried out in cereals as wheat (*Triticum aestivum* L.) [11], barley (*Hordeum vulgare* L.) [12], and oat (*Avena sativa* L.) [13]. The potential of GS has been explored in both hybrid breeding [14,15] and inbred or double haploid (DH) lines [16], and in most cases authors conclude that prediction accuracies are sufficient to make GS more efficient than PS.

To capture the total genetic variance, the effect of each marker in the whole marker-set is estimated in GS regardless of the significance threshold, assuming that markers are in linkage disequilibrium (LD) with the QTLs. Marker effects are estimated using individuals with both genotypic and phenotypic information. The estimated marker effects are combined with marker information of an individual to give the genomic-estimated-breeding-value (GEBV). The predictive ability of the model is calculated based on a cross-validation (CV) system using a training- and a test-population to optimize the model. Marker effects are calculated based on genotypes and phenotypes from the training-population. Subsequently, GEBVs are estimated for the test-population based on these calculations. The predictive ability of the model is then calculated as the correlation between GEBV and phenotypes of the test-population (Figure 1).

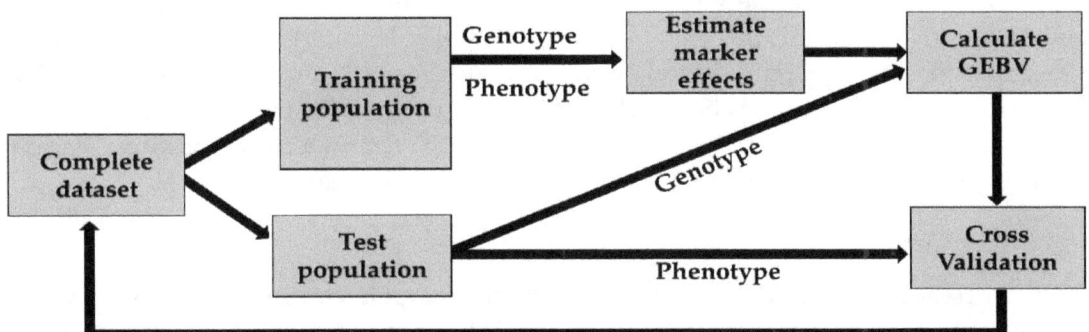

Figure 1. Overview of genomic selection with cross validation using a training population to estimate marker effects in order to get a genomic estimated breeding value (GEBV) of lines in the test-population.

Many papers have now established that GS is a promising approach in different plant species, and several reviews have considered the basic approaches of GS [2], and comparison of statistical methods for GS [17]. There is still limited attention for the ways that GS can fit in plant-breeding programs, how information would flow, the relevance of close and distant relationships in GS implementation, and where to improve accuracy or speed of the program. It is currently a timely moment to review in more detail how GS could be used to improve plant-breeding systems. Most research on GS in plants has ignored pedigree-information, unlike animal GS applications, and it appears useful to discuss the use of pedigree in plants as well.

This review will explore features of GS set-up in cereal breeding, including size of training population, the relationship between individuals in the training and the test data, and marker density. The paper also includes a comparison of implementation strategies for GS in a breeding program, making predictions within and across breeding cycles, as well as the potential to select parents purely based on GEBV. We use breeding of cereals, like wheat and barley based on DH lines, as our primary example to describe GS in plant breeding, but most of our discussion will equally apply to breeding of other inbred and self-fertilizing species.

2. The Set-Up of Genomic Selection

This section describes in more detail the set-up of GS as shown in Figure 1, and factors affecting the accuracy of genomic predictions.

2.1. Size of Training Data

Several studies have shown that prediction accuracies are influenced by the training population size. It is highly important for breeders to determine the number of lines to be genotyped and phenotyped to establish a suitable training data set, because set-up of the first training data is often a large investment. In a study of spring-barley, Nielsen et al. [18] observed a reduced accuracy of GS as a result of reducing the training data set, and moreover, accuracies appeared less stable with CV rounds showing larger variation in accuracy for a small training data set. Cericola et al. [19] found an increase in prediction accuracy with increasing training population size, reaching a plateau at ~700 lines consisting of full-sib, half-sib, and less related wheat lines from 3 consecutive breeding cycles. However, the optimal training set size was found to be higher in a study by Norman et al. [20] using training set sizes varying from 250 to 8300 lines with differing relationships. An increase in prediction accuracy as a results of increased training population size was also observed by Meuwissen [21] using distantly related individuals.

2.2. Relatedness between Training and Test Individuals

The accuracy of GS models has been shown to be affected by the relatedness of individuals between the training and test population [22]. Isidro et al. [23] found the highest prediction accuracies when training data represented the whole population and had a strong relationship to the test data. Relatedness between individuals has also been a subject of interest in MAS. Gowda et al. found that the relatedness between individuals severely impacts QTL-estimation using MAS in hybrid-wheat [24]. Decreasing prediction accuracies for GS with less related individuals in the training and test population was also observed by Lorenz et al. [25], and Nielsen et al. [18] found a decrease in prediction accuracy when using less related individuals in a leave-family-out CV strategy.

2.3. Cross-Validation Strategies

To get an initial assessment of genomic prediction accuracies, most studies run a CV within the collected training data, as introduced for GS studies by Meuwissen et al. [5]. CV makes predictions for individuals, excluding their own phenotypes from the prediction model. The two main CV strategies are leave-one-out (LOO) and k-fold CV, where k-fold CV can be further subdivided as using random folds or stratified folds. In LOO, one line is left out and predicted based on the remaining population, which is repeated for every line as performed by Nielsen et al. [18]. In a random k-fold CV, the population is divided into a number (k) of random groups and one group is left out and predicted based on the remaining ones. In a stratified k-fold CV, grouping can be based on, for instance, families, breeding cycles, locations, environments, etc., and again one group is left out and predicted based on the remaining ones. Random and stratified k-fold CV strategies, where the stratified version was based on breeding cycles, have been carried out by Cericola et al. [19].

Each of the CV approaches will be relevant for a particular use in breeding. In LOO and random k-fold CV, the line(s) predicted typically have closely related individuals in the training data, as well as many samples from the same year, location, and environment as the ones left-out, and are available in the training set. This situation is relevant to test predictions for lines that have not been phenotyped for some trait, based on phenotypes from closely related material in the same breeding cycle. It can, for instance, apply to quality traits that are only measured on a subset to supply predictions for the lines without the quality trait measurements. In a stratified k-fold CV, situations can be tested, such as forward prediction from an older generation as training to a newer generation as test, or predictions across locations or across environments. Results from stratified k-fold CV can be relevant to test GS strategies that shorten the breeding cycle, or that reduce testing in locations and environments.

The effect of relationships on prediction accuracy can also be seen in CV, where LOO and random k-fold CV often show higher prediction accuracy than stratified k-fold CV [26]. The different levels of relationships between training and test data is one major factor to explain these differences, because LOO and random k-fold CV tend to have higher relationships between training and test individuals than the stratified k-fold CV strategies. Additionally, genotype by environment interaction (GxE) may contribute to poorer prediction accuracy when the stratification is across years or environments.

2.4. Marker Density

Several studies have investigated the possibility to use a reduced marker set for GS without affecting prediction accuracies remarkably. Using a smaller marker set would reduce the genotyping costs for each line in the training population, making it feasible to genotype more individuals for the same expense. Meuwissen [21] found that prediction accuracies are increasing with an increase in marker density. It can be argued that at least one marker should be in LD with each QTL to capture all the genetic variation in a population. This is especially the case for unrelated lines, as LD between markers may vary between the training and the test population. Using genetically related barley lines, Nielsen et al. [18] observed a remarkable decrease in prediction accuracy when using less than 1000 markers, however this was dependant on the trait. Cericola et. al [19] also concluded that using 1000 randomly selected and spaced markers was enough to reach maximum prediction accuracy in wheat breeding lines. High marker density has been observed to be more critical when predicting more distant relatives [20].

2.5. Prediction of Genomic Estimated Breeding Values (GEBV)

Estimation of breeding values by Best Linear Unbiased Prediction (BLUP) in a mixed model using a pedigree-based relationship matrix was already introduced in animal breeding for selection based on phenotypes and pedigree [27]. This pedigree-based BLUP serves as the basis of one of the most-popular practical approaches to estimate GEBV by using "genomic BLUP" (GBLUP). In GBLUP, the pedigree-matrix is replaced by a G-matrix representing the genomic relationship between individuals, as described by VanRaden et al. [28]. GBLUP is a mixed model, which in the most basic form can be written as:

$$Y = X\beta + Z\alpha + \varepsilon \qquad (1)$$

where y is a vector of phenotypes, β is a vector of fixed-effects, α is a vector of genomic breeding values, X and Z are design matrices, and ε is a vector of residual effects. In the mixed model (1), genomic breeding values have the multivariate Normal distribution $\alpha \sim N\left(0, G\sigma_g^2\right)$. GBLUP can directly provide GEBV for an individual without phenotypes by simply adding it in the G-matrix. Kernel-methods, similarly to GBLUP, apply similarity (or distance) matrices and are more versatile than GBLUP in that they also capture non-additive effects (see Box 1 for details).

Genomic prediction can also be based on models that estimate marker-effects for all genome-wide markers simultaneously. A basic model for this approach can be described as in (1), but where Z contains genotypes and α are the marker effects. Regression on markers from the whole genome will often face the problem of the number of markers being much larger than the number of observations, causing a lack of degrees of freedom when estimating the marker effects simultaneously with the least square method. The problem is also known as *"large p − small n"* and is often solved in GS by using a mixed model treating marker effects random to obtain BLUP of marker effects [5], or by one of many Bayesian regression models, known as the "Bayesian alphabet" (reviewed in [29] and described in more detail in Box 1). The difference between the BLUP of marker effects and the Bayesian regression models lies in the assumption about the distribution of marker-effects. BLUP assumes that marker effects follow a Normal distribution with an equal variance for all loci. In the Bayesian methods, heavy-tailed prior distributions or mixture distributions are used as the distribution of maker-effects (see Box 1 for details), allowing for some markers to contribute more to genomic variance than others. Bayesian methods often rely on using Markov-Chain Monte Carlo (MCMC) to estimate the model parameters.

The models for genomic prediction have been extensively compared. Meuwissen et al. [5] originally compared four different statistical methods for GS, least-square estimation (LS), BLUP, and two Bayesian estimation methods, BayesA and BayesB. In their study, BLUP outperformed LS remarkably having a correlation between estimated and true breeding values of 0.732 and 0.318, respectively. Additional increases in accuracies compared to GBLUP were observed for BayesA (~9%) and BayesB (~16%). More extensive comparisons of prediction models can be found in Heslot et al. [30], Maltecca et al. [31], and De Los Campos et al. [17], typically concluding that when predicting close relatives and considering a trait affected by many genes of small effect, differences between the methods are small, and methods like (G)BLUP and ridge regression are effective and robust; when traits have some larger QTL or when considering prediction of distant relatives, improvements in prediction accuracy can be obtained from Bayesian and machine learning methods, where in particular BayesB and BayesC(pi) are popular. In plant breeding, the kernel-methods are also popular, in particular to predict non-additive effects and to handle complex multi-environment multi-trait models [32,33]. Also, the popular "deep learning" or deep belief networks have been applied and compared recently, but performed poorer than existing genomic prediction methods [34]. For situations where Bayesian or machine-learning methods prove useful to improve prediction accuracy, but computational time for these methods prohibits fast routine use, Su et al. [35] introduced a weighted GBLUP (WGBLUP) using SNP-weights based on results from a Bayesian model.

Box 1. Statistical models for genomic prediction.

Mixed models estimating marker effects as "random regressions" by BLUP [5].

Bayesian Lasso [36] and Bayesian regression models from the "Bayesian alphabet" [29], such as BayesA and BayesB [5], BayesC, BayesCpi [37], and BayesR [38]. These are all multiple-regression models, like the mixed model, but with Bayesian shrinkage approaches applied to treat the marker-effects. Bayesian Lasso and BayesA apply non-differential shrinkage by applying a long-tail distribution to marker effects, being LaPlace and student-t, respectively. It has been recognized that the shrinkage in Bayesian Lasso and BayesA is still quite uniform [39], like the mixed model, which has led to variations when applying more extreme long-tailed distributions, such as the normal exponential gamma [40], the power-exponential distribution (Power Lasso, [41]), and the modifications proposed by Fang et al. [39]. The BayesB, BayesC(pi), and BayesR methods apply a mixture of distributions as the prior distribution of marker effects, one of which can be a spike at zero, and where BayesB and BayesC(pi) use two distributions, and BayesR uses four. Other Bayesian models applying mixture distributions fall in this same category, such as SSVS [42] and the methods based on George and McCulloch 's [43] Bayesian Variable Selection applied in Kapell et al. [44] and Gao et al. [41], the latter also using a four-mixture distribution as in BayesR.

Statistical and machine learning methods for high-dimensional data, such as support vector machines [45], ridge regression [46,47], and Bayesian additive regression trees [48].

Methods that do not estimate marker effects but collapse marker data into relationship or similarity or distance matrices, such as the mixed model GBLUP [28] and kernel methods [49,50]. In these methods, the kernel-methods can be seen as modifications and extensions of GBLUP by implicitly considering multiple and different relationship measures than only the additive relationships considered in GBLUP. The kernel-methods have been shown to also capturing epistatic and other non-additive relationships [51]. It is possible to also extend GBLUP in similar ways, i.e., by including a second relationship matrix, which is the Hadamard product of G, a mixed model is obtained that also captures (two-way) epistatic interactions [51,52].

Non-parametric methods [53] and PLS [54] have also been considered for genomic prediction.

---

3. Strategies for Implementation of Genomic Selection

3.1. Basic Breeding Scheme in Cereals

We will first describe an example of a standard barley breeding-scheme using double DHs, shown in Figure 2. In this standard breeding scheme, two parents are crossed to make an F1 progeny, and from pollen culture of the F1 a large set of fully inbred DH progeny can be developed. Every DH will have a unique mosaic of the parental genomes, and the main task of the breeder is to sort among the DH progeny to identify the ones with the best combination of parental alleles. However, agronomic traits, such as yield, cannot be determined on the single DH plants, and s seed of each DH is multiplied to allow sowing yield trials for each DH genotype. After the first multiplication step, there is enough seed to sow one small plot called Preliminary Yield Trial (PYT); after the second multiplication step, there is enough seed for about 3 replicates at 2 locations called Advanced Yield Trials (AYT). Since barley is a self-pollinating crop, the seed from harvested yield trials can be used to perform trials in multiple locations called Elite Yield Trial (EYT) in the following crop cycle. At each multiplication step, breeders will select and reduce the number of DH lines retained, because limited space and resources for field trials will not allow progressing all DH progeny to the final EYT stage. The optimal way to implement GS in plant breeding programs is not straightforward and multiple different strategies have been discussed in the literature [55,56].

Figure 2. Standard breeding scheme showing one cross using double haploid lines, e.g., barley. Triangles indicate steps where material is selected and reduced using genomic selection. P1 = Parent one, P2 = Parent 2, F1 = offspring/hybrid, DH = Double Haploids, PYT = Preliminary Yield Trial, AYT = Advanced Yield Trial, YET = Elite Yield Trial. Photos of field trials on breeding station.

3.2. Across-Breeding Cycle Genomic Selection

GS was first introduced with prediction across generations in animals, where marker effects are calculated on the basis of one generation and used for selection of individuals with an unknown phenotype in the upcoming generation [5]. The equivalent of this strategy in plant breeding would be to predict across breeding cycles. Improvement of genetic gains in animal breeding have mainly been due to selection of traits where phenotypes cannot be directly measured, such as sex-limited traits or traits related to meat-quality [57]. The analogy in plant breeding would be to improve selection in the early stages of the breeding program for traits that are difficult to measure with a low amount of seeds, such as yield and some quality parameters. For instance, malting quality in barley has been considered a good target for GS [58] and baking quality in wheat [59]. Figure 3 shows how data from PYT can be used for GS2 of DH-lines in the next generation. Subsequent selection steps, for instance from PYT to AYT, can similarly be based on data from the previous year. In these steps own data for each line also becomes available, reducing the need to rely on previous generations' data to compute GEBV. Only a few studies have used phenotypic data from single plot PYT in GS models [60], while most GS studies have evaluated use of concluded 2-year EYT data as training data. As shown in Figure 3, there would be a 4-year lag when using concluded EYT data to make predictions on DH, and when using concluded EYT data in later selection steps, the lag would be more than 4 years.

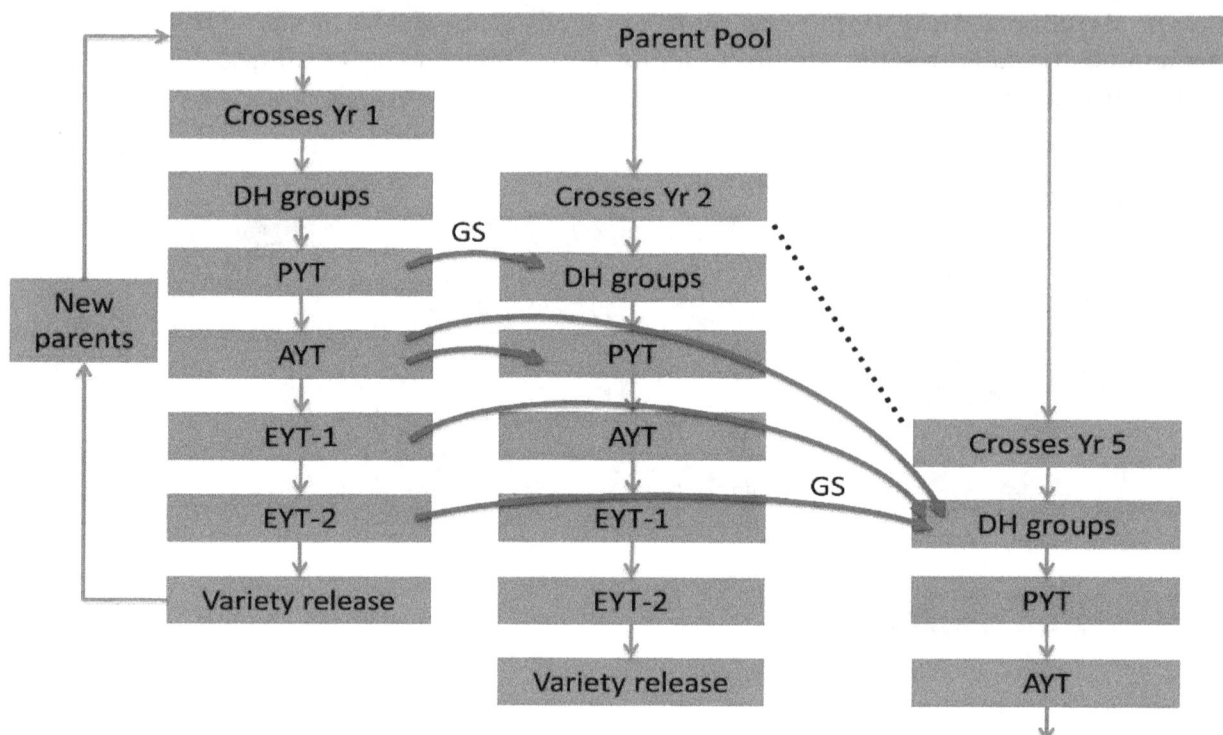

Figure 3. Use of genomic selection across generations based on a standard cereal breeding scheme. Red curved arrows show how information for GS could be used across generations. DH=Double Haploids, PYT = Preliminary Yield Trial, AYT = Advanced Yield Trial, YET = Elite Yield Trial, Yr = Year.

As described before, most studies on GS in plant breeding have found that good prediction accuracies are only obtained when the training and test data are well related. Hence, an important requirement for the across-breeding cycle GS to work well, is that the relationships between subsequent years are high, i.e., there must be many of the same parents used in the crossings of subsequent years, or the crossings must be based on progeny of previous years that are re-used as parents. Often, varieties released by other breeders enter the breeding program to supply new genetic material, which could make it a challenge to keep the breeding material sufficiently related for across-breeding cycle GS to work well [61], and without modifications in the breeding program, there is a time-lag of 6 years for the use of own progeny as new parents. Using data from a real breeding program, Cericola et al. [19] observed low prediction accuracy between breeding cycles, which could indeed be attributed to low overlap of parents and low relationship between breeding cycles.

3.3. Within-Breeding Cycle Genomic Selection

Another way to implement GS is within the generations of one breeding cycle, as shown on Figure 2. Here, lines from the same breeding cycle are used as a training population for GS, for instance to predict sister-lines with missing phenotypes. This type of prediction of sister-lines can be optimized by purposely reducing phenotyping, or by omitting environments, or to measure expensive (quality) traits on only a part of the progeny and predict the rest. Additionally, GEBV combined with phenotypes will improve accuracy of line selection to continue in the breeding program, and with subsequent generation, new phenotypic information will become available for the individuals making GEBVs more accurate, thereby further assisting in the selection process. Predictions within a generation will often have a high relationship between lines as multiple lines from each family are tested. The accuracy of selection is increased with GS within generation, especially for early years where each line has a limited number of phenotypic repeats and information can be borrowed from full-sib and half-sibs.

3.4. Genomic Selection Using Untested Parents for Breeding

A drastic way to use GS is to completely skip phenotypic testing, at least for some part of the breeding program, and select new parents purely based on GEBV. We call this the use of "untested parents", because the lines will not have been tested in the field when they start being used as parents. If breeding cycles are long due to extensive phenotypic testing, use of untested parents can often significantly shorten the breeding cycle and realize faster genetic progress per year. The use of untested parents was suggested by Schaeffer et al. [62] for selection of bulls in dairy cattle breeding and has now been widely adopted and revolutionized dairy cattle breeding. Many plant breeding programs use extensive phenotypic testing and use of untested parents could similarly revolutionize plant breeding programs. Longin et al. [63] found an increased genetic gain when selecting parents based entirely on GEBV, however, this was only the case for highly heritable traits. In our cereal breeding example scheme, this type of fast-cycle breeding could be implemented at the DH stage, selecting DH with good GEBV directly as new parents. Combined with special reproductive techniques to reduce generation time [64], this could reduce each breeding cycle to less than 2 years.

Breeding using untested parents would completely rely on GS using previous years' data, as shown in Figure 3, and as such would be even more sensitive to concerns about sufficient levels of relationships between the subsequent years. However, the fast breeding cycles can compensate for poorer prediction accuracy, as is also the case in dairy cattle breeding [62].

4. Pedigree Information

Using pedigree in selection models has been widely adopted in animal breeding as an important factor in genetic selection programs [65]. Selection based on pedigree alone has not gained the same interest in plant breeding, which quickly moved the focus from phenotypic selection to GS using markers. A comprehensive understanding of the gains from phenotypic selection to pedigree selection in plants is therefore not available. A few studies in plant breeding have investigated the effect of pedigree selection compared to selection using markers. In theory, using markers can give the realized genetic relationship taking account of Mendelian sampling, as opposed to the expected genetic relationship from a pedigree derived relationship matrix. Juliana et al. [66] found similar accuracies for two GS models using pedigree and markers, respectively. However, Cericola et al. [19] found slightly higher prediction accuracies when using markers compared to pedigree. Some gains in prediction accuracy have been seen when pedigree and genomic information is used collectively for GS compared to GS using only markers [67,68]. Single-step methods have been introduced in pig breeding, which combines pedigree and genotype information in a single matrix, making it possible to include non-genotyped lines with a known pedigree in the GS model [69]. Similar accuracies have been found for selection with pedigree, marker, and single-step models for prediction in wheat [70]. However, using pedigree gives the possibility to make predictions on non-genotyped lines with a known pedigree. The additional use of pedigree and single-step methods will be a straightforward improvement in the within-breeding cycle GS; however, for across-breeding cycle GS, it is often seen that complete pedigrees are not available in plant breeding and the use of pedigree information will be more problematic.

5. Use of Additive and Non-Additive Genetic Effects

In species where the same genotype can be replicated by cloning, or by selfing of an inbred individual, it is relevant to distinguish the additive genetic value and the total genetic value (TGV), where the latter also includes all non-additive effects (see Box 2). Most GS focuses on predicting the additive genetic value, and this value is relevant to determine the value of an individual "as parent". However, the value of a variety in the market is determined by its TGV. Ideally, a breeding program, therefore, should focus on obtaining both accurate additive genetic values, as well as accurate TGV of individuals in their breeding trials. Individuals with good additive genetic value are candidates to become new parents within the breeding program, while individuals with good TGV are candidates for marketing.

Obtaining additive genetic (breeding) values is relatively straightforward. The described standard methods for (genomic) breeding value estimation, such as BLUP using pedigree or BLUP using genomic relationships (GBLUP), produce estimates of additive genetic values. These methods flexibly combine all information from relatives into individual breeding values, whether the individual has own data or not, and whether the relatives are parents, progeny, sibs, or other relatives.

Obtaining the TGV is less straightforward. It is not modelled or predicted in a standard BLUP or GBLUP model and must be based on either own data of the individual, or on specialized prediction models that also capture non-additive effects [49,71,72]. When using own data to estimate TGV, estimates of TGV become available as soon as individuals start accumulating plot data, with multiple replicates from generation 4 of breeding scheme 1. For DH lines selected without field-testing, TGV would not be available in this way.

It will be very interesting if genomic information can supply accurate estimates of TGV, as this could predict early-stage breeding material with good market value. Such breeding material can then be put on track for market-development. Models are available that estimate epistatic interactions by using the Hadamart product (GxG) of the genomic relationship matrix [52]), or by using kernel methods, such as Reproducing Kernel Hilbert Space (RKHS) [49]. The use of the Hadamard product of G relies on assumptions that all interactions contribute equally to the TGV, and GxG implies capturing pair-wise interactions only, ignoring all higher-order interactions, while the kernel-methods, such as RKHS, are more flexible and versatile. Perez-Rodriguez et al. [73] compared the prediction accuracy of different linear and non-linear models, including RKHS, GBLUP, and Bayesian models, using a random cross-validation scheme in wheat. Their study [73] found higher prediction accuracy of non-linear models, such as RKHS, which might be attributed to better capturing of higher-order genes by gene interactions. The use of such approaches to predict TGV is thus promising.

Box 2. Breeding values and genetic values in plant breeding.

Additive genetic value (AGV), breeding value or General Combining Ability (GCA): the genetic value based on only the additive effects, or (average) allele substitution effects at loci. In practice, the AGV can be retrieved as the mean of a large progeny group from matings with many different parents, and this is also the basic definition of "breeding value" (in animal genetics) or General Combining Ability (in plant genetics). The AGV is also the genetic value estimated using BLUP methods with pedigree or genomic data (GBLUP).

Total genetic value (TGV): the genetic value based on additive effects at loci, and all interactions within and between loci (for inbreds, only the interactions between loci, epistasis, is relevant). In practice it can be retrieved as the mean performance of a genotype over a large number of plots, replicating the same genotype by cloning or selfing of an inbred individual. In species where varieties are marketed by cloning or seed-multiplication by selfing, the TGV is the value of the variety in the market.

Special Combining Ability (SCA): the progeny mean of a particular combination of two parents, deviated from the mean AGV of the parents. SCA can be expressed as the average TGV of progeny of two parents, and this can differ from the mean AGV of the two parents due to interaction effects. For one parent, SCA effects a large set of other parents average to zero, because the mean progeny performance averaged over matings with many other parents is the AGV of that one parent.

6. Discussion

We have reviewed the main factors that determine prediction accuracy in genomic selection (GS), with a focus on results from plant breeding studies. Overall, most studies find good prediction accuracies, indicating GS is a useful approach in plant breeding. Several publications indicate that prediction across (breeding) cycles is more difficult in plants than in animals [19]. This may be attributed to two main factors: (1) the relatedness of breeding material across breeding cycles may generally be lower in plants than in animals, because every year plant breeders use new parents with unknown background from competitors, while animal breeders work in closed populations; (2) genotype-by-environment interaction (GxE) is stronger in plants than in animals and will make it more difficult to consistently predict a next year's performance. Multiple studies have reported higher prediction accuracies of GxE models compared to models that do not include the interaction term [74,75]. The interaction term has also been explored in unbalanced datasets, giving higher prediction accuracies of lines that had been tested in some environments and not in others [76]. Sukumaran et al. [77] also found higher prediction accuracies when using a GxE model to predict lines across environments. Lopez-Cruz et al. [78] reported the highest prediction accuracies with GxE models when the environments were positively correlated. The superiority of GxE models have proven to be especially pronounced for complex traits as yield compared to less complex traits, such as thousand-grain weight [79]. Further development of GS in plant breeding may therefore need to focus more on how to incorporate unknown parents in a breeding program, and to find and implement efficient GS prediction accounting for GxE.

We have also described three main ways GS could be used in plant breeding programs—the within breeding-cycle GS, across breeding-cycle GS, and the extreme case of using (phenotypically) untested parents, based purely on their GEBV. Across breeding-cycle GS allows for a direct selection on traits which are not measurable in early generations. However, GS studies in plants have mostly tested the within breeding-cycle GS by evaluating accuracy of prediction with a *k*-fold or LOO CV method. Most results are therefore not suitable to indicate feasibility of across breeding-cycle GS, or GS using untested parents. Only a few studies have investigated GS using a CV system that is more suited to determine prediction performance across breeding cycles. Song et al. [16] observed a remarkable decrease in prediction accuracy when predicting yield across cycles compared to within-cycle. The study [16] was performed on DH wheat lines from a biparental cross giving individuals a high relatedness. Michel et al. [61] found a strong upwards bias when predicting within breeding cycles compared to predictions across breeding cycles. Comparison of LOO, leave-family-out (LFO), and leave-subset (cycle) -out (LSO) CV strategies have shown differences in predictive abilities, with the highest predictive ability obtained with LOO and the lowest predictive ability obtained with LSO [26]. Leaving out an entire family or a subset (breeding cycle) from the training population would create a lower relationship between training and test population. The poorer prediction results for predicting across cycles or sets may make it challenging to take advantage of GS in across-breeding-cycle GS.

Within-breeding-cycle GS currently appears to be a feasible approach. In within-breeding-cycle generation GS, a close relationship between individuals will usually give good to high prediction accuracies. The main benefit from using within-breeding-cycle GS should come from a more accurate selection of individuals to continue in the breeding program, so that better lines are retained, and the breeding program may use a smaller field-testing capacity compared to phenotypic selection strategies. However, within-breeding-cycle GS will not shorten the breeding cycle, which will limit the potential impact of using GS compared to across-breeding-cycle GS. The breeding stage for genotyping and using BVs for selection is essential for application of GS. Using GS in earlier generations, before PYT, have proven to give better long-term results in a simulation study [56]. The size of training population and marker-density should be considered according to the GS system used, as the optimum tends to be affected by the relationship between training- and test set.

Use of untested parents, which have a genomic breeding value but no phenotypes, is the GS system with potentially the largest advantage, mainly by reducing the number of years from crossing to marketing. However, predicting ahead of generations and maintaining high accuracies seems challenging. Bayesian models have proven to be superior to BLUP, when the training-population is separated from the test-population by a number of generations [80]. Meuwissen [21] also obtained more accurate SNP effect estimates over generations with a decreasing relationship when using Bayesian models compared to GBLUP. As plant breeding studies currently still indicate poor predictive abilities across breeding cycles, more research is needed to evaluate Bayesian models and their performance to predict distant related material.

The prediction accuracy of different GS models has been compared [30]. Today, most plant breeding programs seems to be using the GBLUP model. A great advantage of GBLUP is that routine-evaluation of breeding values can be done without iterations (using fixed variance components), making it less computationally intensive than Bayesian models. GBLUP is often argued to be best suited for traits controlled by many genes due to the assumption of normally distributed marker-effects. Even quantitative traits are often influenced by a minor fraction of markers, which is not in accordance with the GBLUP model. Similar accuracies have been reported for BLUP and Bayesian models for prediction of close relatives. However, Bayesian models have proven to be superior to GBLUP in the case of distantly related training and test populations [81]. Lower prediction accuracies have been observed for BLUP models compared to Bayesian models in across-population prediction [82]. A few plant breeding studies have also included pedigree information, and compared use of pedigree, genomic, or pedigree and genomic information for prediction. These few studies indicate that pedigree alone can predict quite well, with sometimes only a small or no advantage from adding genomic information. Since pedigree information is very cheap compared to genomic information, plant breeders should more often also consider pedigree information, and evaluate carefully if, where, and how additional genomic information is useful. Additionally, compared to the GS models used in animal breeding, plant breeding may also benefit from extending GS models with non-additive effects.

7. Conclusions

Genomic selection (GS) using markers covering the whole genome to predict genomic-estimated breeding values of individuals is a powerful tool for plant breeders. However, the optimal implementation of GS is an on-going debate. High selection accuracies can be utilized from predictions within-breeding-cycle in the breeding program, whereas selections across-breeding-cycle can suffer from a low relationship between the training and test population making prediction less accurate. More studies on prediction of distantly related individuals are needed. Lower accuracies can also be expected for GS combined with use of untested parents due to the lack of accuracy of prediction ahead of multiple breeding cycles. The optimal solution for application of GS in plant breeding programs might rely on a combination of different strategies. GS could benefit from inclusion of pedigree information for higher prediction accuracies and obtaining breeding values of non-genotyped lines. The size of training population and marker set is affected by the trait and relationship of individuals and should thus be considered independently before implementation of GS in a breeding program. GS is generally used to predict the additive genetic value of individuals and to disregard non-additive genetic variance, which indicates how a line performs as parent. Upcoming studies are investigating the estimation of TGV, which is more suitable for the marketing of a variety. We conclude that within-generation GS is currently a promising and feasible option, where investments in genotyping could be recovered by making better selection decisions and by reducing phenotyping and reducing the candidates that are kept in the breeding program. Across-breeding-cycle GS, and in particular use of untested parents, needs to be investigated in more detail, because prediction accuracies in such systems may be low. We also conclude that plant breeders could benefit more from using pedigree data, and combined pedigree-genomic data, than they currently do.

Author Contributions: Conceptualization by C.D.R. and L.L.J., writing, editing and literature collection for draft manuscript by C.D.R. with contributions from LLJ (methods) and RLH (breeding programs), review editing by C.D.R. and L.L.J., supervision by L.L.J., funding acquisition by L.L.J. and R.L.H., project administration by R.L.H.

Acknowledgments: The authors wish to thank Sejet Breeding I/S, 8700 Horsens, Denmark, for contributing with photos of breeding trials.

References

1. Lande, R.; Thompson, R. Efficiency of marker-assisted selection in the improvement of quantitative traits. *Genetics* **1990**, *124*, 743–756. [PubMed]
2. Heffner, E.L.; Sorrells, M.E.; Jannink, J.L. Genomic selection for crop improvement. *Crop Sci.* **2009**, *49*, 1–12. [CrossRef]
3. Laidig, F.; Piepho, H.P.; Rentel, D.; Drobek, T.; Meyer, U.; Huesken, A. Breeding progress, environmental variation and correlation of winter wheat yield and quality traits in German official variety trials and on-farm during 1983–2014. *Theor. Appl. Genet.* **2017**, *130*, 223–245. [CrossRef] [PubMed]
4. Sharma, R.C.; Crossa, J.; Velu, G.; Huerta-Espino, J.; Vargas, M.; Payne, T.S.; Singh, R.P. Genetic gains for grain yield in CIMMYT spring bread wheat across international environments. *Crop Sci.* **2012**, *52*, 1522–1533. [CrossRef]
5. Meuwissen, T.H.E.; Hayes, B.J.; Goddard, M.E. Prediction of total genetic value using genome-wide dense marker maps. *Genetics* **2001**, *157*, 1819–1829. [PubMed]
6. Su, G.; Guldbrandtsen, B.; Gregersen, V.R.; Lund, M.S. Preliminary investigation on reliability of genomic estimated breeding values in the Danish Holstein population. *J. Dairy Sci.* **2010**, *93*, 1175–1183. [CrossRef] [PubMed]
7. Heffner, E.L.; Lorenz, A.J.; Jannink, J.L.; Sorrells, M.E. Plant breeding with genomic selection: Gain per unit time and cost. *Crop Sci.* **2010**, *50*, 1681–1690. [CrossRef]
8. Bernardo, R.; Yu, J.M. Prospects for genomewide selection for quantitative traits in maize. *Crop Sci.* **2007**, *47*, 1082–1090. [CrossRef]
9. Bernardo, R. A model for marker-assisted selection among single crosses with multiple genetic markers. *Theor. Appl. Genet.* **1998**, *97*, 473–478. [CrossRef]
10. Hayes, B.J.; Bowman, P.J.; Chamberlain, A.J.; Goddard, M.E. Invited review: Genomic selection in dairy cattle: Progress and challenges. *J. Dairy Sci.* **2009**, *92*, 433–443. [CrossRef] [PubMed]
11. Crossa, J.; Perez, P.; Hickey, J.; Burgueno, J.; Ornella, L.; Ceron-Rojas, J.; Zhang, X.; Dreisigacker, S.; Babu, R.; Li, Y.; et al. Genomic prediction in CIMMYT maize and wheat breeding programs. *Heredity* **2014**, *112*, 48–60. [CrossRef] [PubMed]
12. Lorenz, A.J.; Smith, K.P.; Jannink, J.L. Potential and optimization of genomic selection for fusarium head blight resistance in six-row barley. *Crop Sci.* **2012**, *52*, 1609–1621. [CrossRef]
13. Asoro, F.G.; Newell, M.A.; Beavis, W.D.; Scott, M.P.; Tinker, N.A.; Jannink, J.L. Genomic, marker-assisted, and pedigree-BLUP selection methods for beta-glucan concentration in elite oat. *Crop Sci.* **2013**, *53*, 1894–1906. [CrossRef]
14. Lariepe, A.; Moreau, L.; Laborde, J.; Bauland, C.; Mezmouk, S.; Decousset, L.; Mary-Huard, T.; Fievet, J.B.; Gallais, A.; Dubreuil, P.; et al. General and specific combining abilities in a maize (*Zea mays* L.) test-cross hybrid panel: Relative importance of population structure and genetic divergence between parents. *Theor. Appl. Genet.* **2017**, *130*, 403–417. [CrossRef] [PubMed]
15. Riedelsheimer, C.; Czedik-Eysenberg, A.; Grieder, C.; Lisec, J.; Technow, F.; Sulpice, R.; Altmann, T.; Stitt, M.; Willmitzer, L.; Melchinger, A.E. Genomic and metabolic prediction of complex heterotic traits in hybrid maize. *Nat. Genet.* **2012**, *44*, 217–220. [CrossRef] [PubMed]
16. Song, J.Y.; Carver, B.F.; Powers, C.; Yan, L.L.; Klapste, J.; El-Kassaby, Y.A.; Chen, C. Practical application of genomic selection in a doubled-haploid winter wheat breeding program. *Mol. Breed.* **2017**, *37*, 117. [CrossRef] [PubMed]
17. De los Campos, G.; Hickey, J.M.; Pong-Wong, R.; Daetwyler, H.D.; Calus, M.P.L. Whole-genome regression and prediction methods applied to plant and animal breeding. *Genetics* **2013**, *193*, 327–345. [CrossRef]
18. Nielsen, N.H.; Jahoor, A.; Jensen, D.; Orabi, J.; Cericola, F.; Edriss, V.; Jensen, J. Genomic prediction of seed quality traits using advanced barley breeding lines. *PLoS ONE* **2016**, *11*, e0164494. [CrossRef]
19. Cericola, F.; Jahoor, A.; Orabi, J.; Andersen, J.R.; Janss, L.L.; Jensen, J. Optimizing training population size and genotyping strategy for genomic prediction using association study results and pedigree information. A case of study in advanced wheat breeding lines. *PLoS ONE* **2017**, *12*, e0169606. [CrossRef]
20. Norman, A.; Taylor, J.; Edwards, J.; Kuchel, H. Optimising genomic selection in wheat: Effect of marker density, population size and population structure on prediction accuracy. *G3-Genes Genomes Genet.* **2018**, *8*, 2889–2899. [CrossRef]

21. Meuwissen, T.H.E. Accuracy of breeding values of 'unrelated' individuals predicted by dense SNP genotyping. *Genet. Sel. Evol.* **2009**, *41*, 35. [CrossRef] [PubMed]

22. Habier, D.; Tetens, J.; Seefried, F.R.; Lichtner, P.; Thaller, G. The impact of genetic relationship information on genomic breeding values in German Holstein cattle. *Genet. Sel. Evol.* **2010**, *42*, 5. [CrossRef] [PubMed]

23. Isidro, J.; Jannink, J.L.; Akdemir, D.; Poland, J.; Heslot, N.; Sorrells, M.E. Training set optimization under population structure in genomic selection. *Theor. Appl. Genet.* **2015**, *128*, 145–158. [CrossRef] [PubMed]

24. Gowda, M.; Zhao, Y.; Wuerschum, T.; Longin, C.F.H.; Miedaner, T.; Ebmeyer, E.; Schachschneider, R.; Kazman, E.; Schacht, J.; Martinant, J.P.; et al. Relatedness severely impacts accuracy of marker-assisted selection for disease resistance in hybrid wheat. *Heredity* **2014**, *112*, 552–561. [CrossRef] [PubMed]

25. Lorenz, A.J.; Smith, K.P. Adding genetically distant individuals to training populations reduces genomic prediction accuracy in barley. *Crop Sci.* **2015**, *55*, 2657–2667. [CrossRef]

26. Kristensen, P.S.; Jahoor, A.; Andersen, J.R.; Cericola, F.; Orabi, J.; Janss, L.L.; Jensen, J. Genome-wide association studies and comparison of models and cross-validation strategies for genomic prediction of quality traits in advanced winter wheat breeding lines. *Front. Plant Sci.* **2018**, *9*, 69. [CrossRef] [PubMed]

27. Henderson, C.R. Best linear unbiased estimation and prediction under a selection model. *Biometrics* **1975**, *31*, 423–447. [CrossRef] [PubMed]

28. VanRaden, P.M. Efficient methods to compute genomic predictions. *J. Dairy Sci.* **2008**, *91*, 4414–4423. [CrossRef] [PubMed]

29. Gianola, D. Priors in whole-genome regression: The Bayesian alphabet returns. *Genetics* **2013**, *194*, 573–596. [CrossRef]

30. Heslot, N.; Yang, H.P.; Sorrells, M.E.; Jannink, J.L. Genomic selection in plant breeding: A comparison of models. *Crop Sci.* **2012**, *52*, 146–160. [CrossRef]

31. Maltecca, C.; Parker, K.L.; Cassady, J.P. Application of multiple shrinkage methods to genomic predictions. *J. Anim. Sci.* **2012**, *90*, 1777–1787. [CrossRef] [PubMed]

32. Sousa, M.B.E.; Cuevas, J.; Couto, E.G.D.; Perez-Rodriguez, P.; Jarquin, D.; Fritsche-Neto, R.; Burgueno, J.; Crossa, J. Genomic-enabled prediction in maize using kernel models with genotype x environment interaction. *G3-Genes Genomes Genet.* **2017**, *7*, 1995–2014. [CrossRef]

33. Cuevas, J.; Granato, I.; Fritsche-Neto, R.; Montesinos-Lopez, O.A.; Burgueno, J.; Bandeira, M.B.E.; Crossa, J. Genomic-enabled prediction kernel models with random intercepts for multi-environment trials. *G3-Genes Genomes Genet.* **2018**, *8*, 1347–1365. [CrossRef] [PubMed]

34. Bellot, P.; de los Campos, G.; Perez-Enciso, M. Can deep learning improve genomic prediction of complex human traits? *Genetics* **2018**, *210*, 809–819. [CrossRef] [PubMed]

35. Su, G.; Christensen, O.F.; Janss, L.; Lund, M.S. Comparison of genomic predictions using genomic relationship matrices built with different weighting factors to account for locus-specific variances. *J. Dairy Sci.* **2014**, *97*, 6547–6559. [CrossRef]

36. Park, T.; Casella, G. The Bayesian LASSO. *J. Am. Stat. Assoc.* **2008**, *103*, 681–686. [CrossRef]

37. Habier, D.; Fernando, R.L.; Kizilkaya, K.; Garrick, D.J. Extension of the Bayesian alphabet for genomic selection. *BMC Bioinform.* **2011**, *12*, 186. [CrossRef]

38. Erbe, M.; Hayes, B.J.; Matukumalli, L.K.; Goswami, S.; Bowman, P.J.; Reich, C.M.; Mason, B.A.; Goddard, M.E. Improving accuracy of genomic predictions within and between dairy cattle breeds with imputed high-density single nucleotide polymorphism panels. *J. Dairy Sci.* **2012**, *95*, 4114–4129. [CrossRef]

39. Fang, M.; Jiang, D.; Li, D.D.; Yang, R.Q.; Fu, W.X.; Pu, L.J.; Gao, H.J.; Wang, G.H.; Yu, L.Y. Improved LASSO priors for shrinkage quantitative trait loci mapping. *Theor. Appl. Genet.* **2012**, *124*, 1315–1324. [CrossRef]

40. Hoggart, C.J.; Whittaker, J.C.; De Iorio, M.; Balding, D.J. Simultaneous analysis of all SNPs in genome-wide and re-Sequencing association studies. *PLoS Genet.* **2008**, *4*, e1000130. [CrossRef]

41. Gao, H.; Su, G.; Janss, L.; Zhang, Y.; Lund, M.S. Model comparison on genomic predictions using high-density markers for different groups of bulls in the Nordic Holstein population. *J. Dairy Sci.* **2013**, *96*, 4678–4687. [CrossRef] [PubMed]

42. Verbyla, K.L.; Hayes, B.J.; Bowman, P.J.; Goddard, M.E. Accuracy of genomic selection using stochastic search variable selection in Australian Holstein Friesian dairy cattle. *Genet. Res.* **2009**, *91*, 307–311. [CrossRef] [PubMed]

43. George, E.I.; McCulloch, R.E. Variable selection via Gibbs sampling. *J. Am. Stat. Assoc.* **1993**, *88*, 881–889. [CrossRef]

44. Kapell, D.; Sorensen, D.; Su, G.S.; Janss, L.L.G.; Ashworth, C.J.; Roehe, R. Efficiency of genomic selection using Bayesian multi-marker models for traits selected to reflect a wide range of heritabilities and frequencies of detected quantitative traits loci in mice. *BMC Genet.* **2012**, *13*, 42. [CrossRef] [PubMed]

45. Abraham, G.; Tye-Din, J.A.; Bhalala, O.G.; Kowalczyk, A.; Zobel, J.; Inouye, M. Accurate and robust genomic prediction of celiac disease using statistical learning. *PLoS Genet.* **2014**, *10*, e1004137. [CrossRef]

46. Piepho, H.P. Ridge regression and extensions for genomewide selection in maize. *Crop Sci.* **2009**, *49*, 1165–1176. [CrossRef]

47. Endelman, J.B. Ridge regression and other kernels for genomic selection with R package rrBLUP. *Plant Genome* **2011**, *4*, 250–255. [CrossRef]

48. Waldmann, P. Genome-wide prediction using Bayesian additive regression trees. *Genet. Sel. Evol.* **2016**, *48*, 42. [CrossRef]

49. Gianola, D.; van Kaam, J. Reproducing kernel Hilbert spaces regression methods for genomic assisted prediction of quantitative traits. *Genetics* **2008**, *178*, 2289–2303. [CrossRef]

50. Morota, G.; Koyama, M.; Rosa, G.J.M.; Weigel, K.A.; Gianola, D. Predicting complex traits using a diffusion kernel on genetic markers with an application to dairy cattle and wheat data. *Genet. Sel. Evol.* **2013**, *45*, 17. [CrossRef]

51. Jiang, Y.; Reif, J.C. Modeling epistasis in genomic selection. *Genetics* **2015**, *201*, 759–768. [CrossRef]

52. Martini, J.W.R.; Wimmer, V.; Erbe, M.; Simianer, H. Epistasis and covariance: How gene interaction translates into genomic relationship. *Theor. Appl. Genet.* **2016**, *129*, 963–976. [CrossRef] [PubMed]

53. Howard, R.; Carriquiry, A.L.; Beavis, W.D. Parametric and nonparametric statistical methods for genomic selection of traits with additive and epistatic genetic architectures. *G3-Genes Genomes Genet.* **2014**, *4*, 1027–1046. [CrossRef]

54. Du, C.; Wei, J.L.; Wang, S.B.; Jia, Z.Y. Genomic selection using principal component regression. *Heredity* **2018**, *121*, 12–23. [CrossRef] [PubMed]

55. Bassi, F.M.; Bentley, A.R.; Charmet, G.; Ortiz, R.; Crossa, J. Breeding schemes for the implementation of genomic selection in wheat (*Triticum* spp.). *Plant Sci.* **2016**, *242*, 23–36. [CrossRef]

56. Gaynor, R.C.; Gorjanc, G.; Bentley, A.R.; Ober, E.S.; Howell, P.; Jackson, R.; Mackay, I.J.; Hickey, J.M. A two-part strategy for using genomic selection to develop inbred lines. *Crop Sci.* **2017**, *57*, 2372–2386. [CrossRef]

57. Meuwissen, T.; Hayes, B.; Goddard, M. Accelerating improvement of livestock with genomic selection. *Annu. Rev. Anim. Biosci.* **2013**, *1*, 221–237. [CrossRef] [PubMed]

58. Schmidt, M.; Kollers, S.; Maasberg-Prelle, A.; Grosser, J.; Schinkel, B.; Tomerius, A.; Graner, A.; Korzun, V. Prediction of malting quality traits in barley based on genome-wide marker data to assess the potential of genomic selection. *Theor. Appl. Genet.* **2016**, *129*, 203–213. [CrossRef]

59. Michel, S.; Kummer, C.; Gallee, M.; Hellinger, J.; Ametz, C.; Akgol, B.; Epure, D.; Gungor, H.; Loschenberger, F.; Buerstmayr, H. Improving the baking quality of bread wheat by genomic selection in early generations. *Theor. Appl. Genet.* **2018**, *131*, 477–493. [CrossRef] [PubMed]

60. Michel, S.; Ametz, C.; Gungor, H.; Akgol, B.; Epure, D.; Grausgruber, H.; Loschenberger, F.; Buerstmayr, H. Genomic assisted selection for enhancing line breeding: Merging genomic and phenotypic selection in winter wheat breeding programs with preliminary yield trials. *Theor. Appl. Genet.* **2017**, *130*, 363–376. [CrossRef]

61. Michel, S.; Ametz, C.; Gungor, H.; Epure, D.; Grausgruber, H.; Loschenberger, F.; Buerstmayr, H. Genomic selection across multiple breeding cycles in applied bread wheat breeding. *Theor. Appl. Genet.* **2016**, *129*, 1179–1189. [CrossRef]

62. Schaeffer, L.R. Strategy for applying genome-wide selection in dairy cattle. *J. Anim. Breed. Genet.* **2006**, *123*, 218–223. [CrossRef]

63. Longin, C.F.H.; Mi, X.F.; Wurschum, T. Genomic selection in wheat: Optimum allocation of test resources and comparison of breeding strategies for line and hybrid breeding. *Theor. Appl. Genet.* **2015**, *128*, 1297–1306. [CrossRef] [PubMed]

64. Watson, A.; Ghosh, S.; Williams, M.J.; Cuddy, W.S.; Simmonds, J.; Rey, M.D.; Hatta, M.A.M.; Hinchliffe, A.; Steed, A.; Reynolds, D.; et al. Speed breeding is a powerful tool to accelerate crop research and breeding. *Nat. Plants* **2018**, *4*, 23–29. [CrossRef] [PubMed]

65. Weigel, K.A.; VanRaden, P.M.; Norman, H.D.; Grosu, H. A 100-Year Review: Methods and impact of genetic selection in dairy cattle—From daughter–dam comparisons to deep learning algorithms. *J. Dairy Sci.* **2017**, *100*, 10234–10250. [CrossRef] [PubMed]

66. Juliana, P.; Singh, R.P.; Singh, P.K.; Crossa, J.; Huerta-Espino, J.; Lan, C.X.; Bhavani, S.; Rutkoski, J.E.; Poland, J.A.; Bergstrom, G.C.; et al. Genomic and pedigree-based prediction for leaf, stem, and stripe rust resistance in wheat. *Theor. Appl. Genet.* **2017**, *130*, 1415–1430. [CrossRef]

67. Burgueno, J.; de los Campos, G.; Weigel, K.; Crossa, J. Genomic prediction of breeding values when modeling genotype x environment interaction using pedigree and dense molecular markers. *Crop Sci.* **2012**, *52*, 707–719. [CrossRef]

68. Legarra, A.; Aguilar, I.; Misztal, I. a relationship matrix including full pedigree and genomic information. *J. Dairy Sci.* **20009**, *92*, 4656–4663. [CrossRef]

69. Christensen, O.F.; Lund, M.S. Genomic prediction when some animals are not genotyped. *Genet. Sel. Evol.* **2010**, *42*, 2. [CrossRef]

70. Perez-Rodriguez, P.; Crossa, J.; Rutkoski, J.; Poland, J.; Singh, R.; Legarra, A.; Autrique, E.; de los Campos, G.; Burgueno, J.; Dreisigacker, S. Single-step genomic and pedigree genotype x environment interaction models for predicting wheat lines in international environments. *Plant Genome* **2017**, *10*. [CrossRef]

71. Bouvet, J.M.; Makouanzi, G.; Cros, D.; Vigneron, P. Modeling additive and non-additive effects in a hybrid population using genome-wide genotyping: Prediction accuracy implications. *Heredity* **2016**, *116*, 146–157. [CrossRef] [PubMed]

72. El-Dien, O.G.; Ratcliffe, B.; Klapste, J.; Porth, I.; Chen, C.; El-Kassaby, Y.A. Implementation of the realized genomic relationship matrix to open-pollinated white spruce family testing for disentangling additive from nonadditive genetic effects. *G3-Genes Genomes Genet.* **2016**, *6*, 743–753. [CrossRef]

73. Perez-Rodriguez, P.; Gianola, D.; Gonzalez-Camacho, J.M.; Crossa, J.; Manes, Y.; Dreisigacker, S. Comparison between linear and non-parametric regression models for genome-enabled prediction in wheat. *G3-Genes Genomes Genet.* **2012**, *2*, 1595–1605. [CrossRef] [PubMed]

74. Jarquin, D.; Crossa, J.; Lacaze, X.; Du Cheyron, P.; Daucourt, J.; Lorgeou, J.; Piraux, F.; Guerreiro, L.; Perez, P.; Calus, M.; et al. A reaction norm model for genomic selection using high-dimensional genomic and environmental data. *Theor. Appl. Genet.* **2014**, *127*, 595–607. [CrossRef] [PubMed]

75. Cuevas, J.; Crossa, J.; Montesinos-Lopez, O.A.; Burgueno, J.; Perez-Rodriguez, P.; de los Campos, G. Bayesian genomic prediction with genotype x environment interaction kernel models. *G3-Genes Genomes Genet.* **2017**, *7*, 41–53. [CrossRef]

76. Jarquin, D.; da Silva, C.L.; Gaynor, R.C.; Poland, J.; Fritz, A.; Howard, R.; Battenfield, S.; Crossa, J. Increasing genomic-enabled predictionaccuracy by modeling genotype x environment interactions in Kansas wheat. *Plant Genome* **2017**, *10*. [CrossRef] [PubMed]

77. Sukumaran, S.; Jarquin, D.; Crossa, J.; Reynolds, M. Genomic-enabled prediction accuracies increased by modeling genotype x environment interaction in durum wheat. *Plant Genome* **2018**, *11*. [CrossRef]

78. Lopez-Cruz, M.; Crossa, J.; Bonnett, D.; Dreisigacker, S.; Poland, J.; Jannink, J.L.; Singh, R.P.; Autrique, E.; de los Campos, G. Increased prediction accuracy in wheat breeding trials using a marker x environment interaction genomic selection model. *G3-Genes Genomes Genet.* **2015**, *5*, 569–582. [CrossRef]

79. Sukumaran, S.; Crossa, J.; Jarquin, D.; Reynolds, M. Pedigree-based prediction models with genotype x environment interaction in multienvironment trials of CIMMYT wheat. *Crop Sci.* **2017**, *57*, 1865–1880. [CrossRef]

80. Zhong, S.Q.; Dekkers, J.C.M.; Fernando, R.L.; Jannink, J.L. Factors affecting accuracy from genomic selection in populations derived from multiple inbred lines: A barley case study. *Genetics* **2009**, *182*, 355–364. [CrossRef]

81. Wu, X.; Lund, M.S.; Sun, D.; Zhang, Q.; Su, G. Impact of relationships between test and training animals and among training animals on reliability of genomic prediction. *J. Anim. Breed. Genet.* **2015**, *132*, 366–375. [CrossRef] [PubMed]

82. Thavamanikumar, S.; Dolferus, R.; Thumma, B.R. Comparison of genomic selection models to predict flowering time and spike grain number in two hexaploid wheat doubled haploid populations. *G3-Genes Genomes Genet.* **2015**, *5*, 1991–1998. [CrossRef] [PubMed]

Permissions

All chapters in this book were first published by MDPI; hereby published with permission under the Creative Commons Attribution License or equivalent. Every chapter published in this book has been scrutinized by our experts. Their significance has been extensively debated. The topics covered herein carry significant findings which will fuel the growth of the discipline. They may even be implemented as practical applications or may be referred to as a beginning point for another development.

The contributors of this book come from diverse backgrounds, making this book a truly international effort. This book will bring forth new frontiers with its revolutionizing research information and detailed analysis of the nascent developments around the world.

We would like to thank all the contributing authors for lending their expertise to make the book truly unique. They have played a crucial role in the development of this book. Without their invaluable contributions this book wouldn't have been possible. They have made vital efforts to compile up to date information on the varied aspects of this subject to make this book a valuable addition to the collection of many professionals and students.

This book was conceptualized with the vision of imparting up-to-date information and advanced data in this field. To ensure the same, a matchless editorial board was set up. Every individual on the board went through rigorous rounds of assessment to prove their worth. After which they invested a large part of their time researching and compiling the most relevant data for our readers.

The editorial board has been involved in producing this book since its inception. They have spent rigorous hours researching and exploring the diverse topics which have resulted in the successful publishing of this book. They have passed on their knowledge of decades through this book. To expedite this challenging task, the publisher supported the team at every step. A small team of assistant editors was also appointed to further simplify the editing procedure and attain best results for the readers.

Apart from the editorial board, the designing team has also invested a significant amount of their time in understanding the subject and creating the most relevant covers. They scrutinized every image to scout for the most suitable representation of the subject and create an appropriate cover for the book.

The publishing team has been an ardent support to the editorial, designing and production team. Their endless efforts to recruit the best for this project, has resulted in the accomplishment of this book. They are a veteran in the field of academics and their pool of knowledge is as vast as their experience in printing. Their expertise and guidance has proved useful at every step. Their uncompromising quality standards have made this book an exceptional effort. Their encouragement from time to time has been an inspiration for everyone.

The publisher and the editorial board hope that this book will prove to be a valuable piece of knowledge for researchers, students, practitioners and scholars across the globe.

List of Contributors

Sunny Ahmar and Muhammad Uzair Qasim
National Key Laboratory of Crop Genetic Improvement, College of Plant Science and Technology, Huazhong Agricultural University, Wuhan 430070, Hubei, China

Rafaqat Ali Gill
Oil Crops Research Institute, Chinese Academy of Agriculture Sciences, Wuhan 430070, China

Ki-Hong Jung
Graduate School of Biotechnology & Crop Biotech Institute, Kyung Hee University, Yongin 17104, Korea

Aroosha Faheem
State Key Laboratory of Agricultural Microbiology and State Key Laboratory of Microbial Biosensor, College of Life Sciences Huazhong Agriculture University, Wuhan 430070, China

Mustansar Mubeen
State Key Laboratory of Agricultural Microbiology and Provincial Key Laboratory of Plant Pathology of Hubei Province, College of Plant Science and Technology, Huazhong Agricultural University, Wuhan 430070, China

Weijun Zhou
Institute of Crop Science and Zhejiang Key Laboratory of Crop Germplasm, Zhejiang University, Hangzhou 310058, China

Wanwei Hou, Qingbiao Yan, Ping Li, Weichao Sha, Yingying Tian and Yujiao Liu
Qinghai Academy of Agricultural and Forestry Sciences, State Key Laboratory of Plateau Ecology and Agriculture and Qinghai Research Station of Crop Gene Resource & Germplasm Enhancement, Qinghai University, Xining 810016, Qinghai, China

Xiaojuan Zhang
State Key Laboratory of Plateau Ecology and Agriculture and College of Eco-Environmental Engineering, Qinghai University, Xining 810016, Qinghai, China

Yue Han, Dengjie Luo, Babar Usman, Gul Nawaz, Neng Zhao and Fang Liu
College of Agriculture, State Key Laboratory for Conservation and Utilization of Subtropical Agro-Bioresources, Guangxi University, Nanning 530004, China

Rongbai Li
College of Agriculture, State Key Laboratory for Conservation and Utilization of Subtropical Agro-Bioresources, Guangxi University, Nanning 530004, China
Guangxi Academy of Agricultural Sciences, Guangxi, Nanning 530007, China

Jieqiong Wang and Yuanfei Zhang
Luoyang Academy of Agriculture and Forestry Science, Luoyang Key Laboratory of Crop Molecular Biology and Germplasm Enhancement, Luoyang 471000, Henan, China

Shuzuo Lv and Shaofeng Peng
Luoyang Academy of Agriculture and Forestry Science, Luoyang Key Laboratory of Crop Molecular Biology and Germplasm Enhancement, Luoyang 471000, Henan, China
BGI Luoyang Agricultural Innovation Center, Luoyang 471023, Henan, China

Kewei Feng, Jianxin Bian and Xiaojun Nie
State Key Laboratory of Crop Stress Biology in Arid Areas, College of Agronomy and Yangling Branch of China Wheat Improvement Center, Northwest A&F University, Yangling 712100, Shaanxi, China

Ahmad Ali, Jiajia Cao, Hao Jiang and Liaqat Shah
College of Agronomy, Anhui Agricultural University, Hefei 230036, China
Key Laboratory of Wheat Biology and Genetic Improvement on South Yellow and Huai River Valley, Ministry of Agriculture, Hefei 230036, China
National Engineering Laboratory for Crop Stress Resistance Breeding, Hefei 230036, China

Cheng Chang, Hai-Ping Zhang and Chuanxi Ma
College of Agronomy, Anhui Agricultural University, Hefei 230036, China
Key Laboratory of Wheat Biology and Genetic Improvement on South Yellow and Huai River Valley, Ministry of Agriculture, Hefei 230036, China
National Engineering Laboratory for Crop Stress Resistance Breeding, Hefei 230036, China
Anhui Key Laboratory of Crop Biology, Hefei 230036, China

SalmaWaheed Sheikh
School of Life Sciences, Anhui Agricultural University,
Hefei 230036, China

Cecilie S. L. Christensen and Søren K. Rasmussen
Department of Plant and Environmental Sciences,
University of Copenhagen, DK-1871 Frederiksberg C,
Denmark

**Natalia Cristina Aguirre, Carla Valeria Filippi, Juan
Gabriel Rivas, Cintia Vanesa Acuña, Pamela Victoria
Villalba, Martín Nahuel García, Sergio González,
Máximo Rivarola, María Carolina Martínez, Andrea
Fabiana Puebla, Norma Beatriz Paniego and Susana
Noemí Marcucci Poltri**
Instituto de Agrobiotecnología y Biología Molecular—
IABiMo—INTA-CONICET, Instituto de Biotecnología,
Centro de Investigaciones en Ciencias Agronómicas
y Veterinarias, Instituto Nacional de Tecnología
Agropecuaria, Dr. Nicolás Repetto y de los Reseros S/N,
Hurlingham B1686IGC, Argentina

Giusi Zaina and Michele Morgante
Department of Agricultural, Food, Environmental and
Animal Sciences, University of Udine, 33100 Udine,
Italy

Horacio Esteban Hopp
Instituto de Agrobiotecnología y Biología Molecular—
IABiMo—INTA-CONICET, Instituto de Biotecnología,
Centro de Investigaciones en Ciencias Agronómicas
y Veterinarias, Instituto Nacional de Tecnología
Agropecuaria, Dr. Nicolás Repetto y de los Reseros S/N,
Hurlingham B1686IGC, Argentina
Laboratorio de Agrobiotecnología, FBMC, Facultad de
Ciencias Exactas y Naturales, Universidad de Buenos
Aires, Ciudad Universitaria, Buenos Aires C1428EHA,
Argentina

Francesca Taranto and Pasquale De Vita
CREA Research Centre for Cereal and Industrial
Crops, 71121 Foggia, Italy

Alessandro Nicolia and Nunzio D'Agostino
CREA Research Centre for Vegetable and Ornamental
Crops, 84098 Pontecagnano Faiano, Italy

Stefano Pavan
Department of Soil, Plant and Food Sciences, University
of Bari Aldo Moro, 70126 Bari, Italy
Institute of Biomedical Technologies, National Research
Council (CNR), 70126 Bari, Italy

**Jun Zhang, Hao Zheng, Xiaoqin Zeng, Hui Zhuang,
Honglei Wang, Jun Tang, Huan Chen, Yinghua Ling
and Yunfeng Li**
Rice Research Institute, Key Laboratory of Application
and Safety Control of Genetically Modified Crops,
Academy of Agricultural Sciences, Southwest University,
Chongqing 400715, China

**S. Marisol L. Basile, Jorge A. Cardozo, Horacio R.
Dalla Valle and W. John Rogers**
Laboratorio de Biología Funcional y Biotecnología
(CICPBA-BIOLAB AZUL) CIISAS, Facultad de
Agronomía, Universidad Nacional del Centro de la
Provincia de Buenos Aires (UNCPBA),CONICET-
INBIOTEC, Av. Rep. Italia 780, CC 47, 7300 Azul,
Province of Buenos Aires, Argentina

**Mike M. Burrell, Heather J. Walker, Chloe Steels and
Felix Kallenberg**
Biomics Facility, Department of Animal and Plant
Sciences, Alfred Denny Building, University of Sheffield,
Western Bank, Sheffield S10 2TN, UK

Jorge A. Tognetti
Laboratorio de Fisiología Vegetal, Facultad de Ciencias
Agrarias, Universidad Nacional de Mar del Plata, Ruta
226 Km 73,5, Balcarce, Provincia de Buenos Aires, y
Comisión de Investigaciones Científicas de la Provincia
de Buenos Aires, 7620 Balcarce, Province of Buenos
Aires, Argentina

**Cintia V. Acuña, Juan G. Rivas, Silvina M. Brambilla,
Martín N. García, Pamela V. Villalba, Natalia C.
Aguirre, Julia V. Sabio y García, María C. Martínez
and Susana N. Marcucci Poltri**
Instituto de Agrobiotecnología y Biología Molecular—
IABiMo—INTA-CONICET, Instituto de Biotecnología,
Centro de Investigaciones en Ciencias Agronómicas
y Veterinarias, Instituto Nacional de Tecnología
Agropecuaria, Dr. Nicolás Repetto y de los Reseros S/N,
Hurlingham, Buenos Aires B1686IGC, Argentina

Teresa Cerrillo
EEA Delta del Paraná, Río Paraná y Canal Laurentino
Comas, Campana (2804), Buenos Aires B1686IGC,
Argentina

Enrique A. Frusso
Instituto de Recursos Biológicos, Centro de Investigación
de Recursos Naturales, Instituto Nacional de Tecnología
Agropecuaria, Dr. Nicolás Repetto y de los Reseros S/N,
Hurlingham, Buenos Aires B1686IGC, Argentina

Esteban H. Hopp
Instituto de Agrobiotecnología y Biología Molecular—IABiMo—INTA-CONICET, Instituto de Biotecnología, Centro de Investigaciones en Ciencias Agronómicas y Veterinarias, Instituto Nacional de Tecnología Agropecuaria, Dr. Nicolás Repetto y de los Reseros S/N, Hurlingham, Buenos Aires B1686IGC, Argentina
Lab. Agrobiotecnología, FBMC, Facultad de Ciencias Exactas y Naturales, Universidad de Buenos Aires, Intendente Güiraldes 2160, Ciudad Universitaria, CABA C1428EGA, Argentina

Alina I. Chernova, Elena U. Martynova, Asiya F. Ayupova, Rim F. Gubaev, Pavel V. Mazin, Elena A. Gurchenko, Artemy A. Shumskiy, Daria A. Petrova and Philipp E. Khaitovich
Skolkovo Institute of Science and Technology, Moscow 121205, Russia

Denis V. Goryunov
Skolkovo Institute of Science and Technology, Moscow 121205, Russia
Belozersky Institute of Physico-Chemical Biology, Lomonosov Moscow State University, Moscow 119992, Russia

Irina N. Anisimova and Vera A. Gavrilova
N.I. Vavilov All-Russian Research Institute of Plant Genetic Resources, Saint Petersburg (ex Leningrad) 190000, Russia

Evgeniia A. Sotnikova
Department of Computer Science and Control Systems, Bauman Moscow State Technical University, Moscow 105005, Russia

Stepan V. Boldyrev and Svetlana V. Goryunova
Skolkovo Institute of Science and Technology, Moscow 121205, Russia
Institute of General Genetics, Russian Academy of Science, Moscow 119333, Russia

Sergey V. Garkusha and Zhanna M. Mukhina
All-Russia Rice Research Institute, Krasnodar 350921, Russia

Nikolai I. Benko
Breeding and Seed Production Company "Agroplazma", Krasnodar 350012, Russia

Yakov N. Demurin
Pustovoit All-Russia Research Institute of Oil Crops, Krasnodar 350038, Russia

Qinfu Sun, Jueyi Xue, Li Lin, Dongxiao Liu, Jian Wu, Jinjin Jiang and Youping Wang
Jiangsu Provincial Key Laboratory of Crop Genetics and Physiology, Yangzhou University, Yangzhou 225009, China

Rasmus L. Hjortshøj
Sejet Plant Breeding I/S, 8700 Horsens, Denmark

Luc L. Janss
Center for Quantitative Genetics and Genomics, Aarhus University, 8830 Tjele, Denmark

Charlotte D. Robertsen
Sejet Plant Breeding I/S, 8700 Horsens, Denmark
Center for Quantitative Genetics and Genomics, Aarhus University, 8830 Tjele, Denmark

Index

Printed in the USA
CPSIA information can be obtained
at www.ICGtesting.com
JSHW052312231023
50683JS00006BA/85

9 781641 167673